Molecular Biology

For Postgraduate and Undergraduate Students

Dr. P.S. Verma
M.Sc., Ph.D., F.E.S.I., F.A.Z.
Former Reader
Department of Zoology
Meerut College, Meerut

Dr. V. K. Agarwal
M.Sc., Ph.D.
Former Head
Department of Zoology
Meerut College, Meerut

S Chand And Company Limited
(ISO 9001 Certified Company)

S Chand And Company Limited

(ISO 9001 Certified Company)

Head Office: D-92, Sector–2, Noida – 201301, U.P. (India), Ph. 91-120-4682700

Registered Office: A-27, 2nd Floor, Mohan Co-operative Industrial Estate, New Delhi – 110 044, Phone: 011-49731800

www.**schandpublishing.com;** e-mail: **info@schandpublishing.com**

Marketing Offices:

Chennai : Ph: 23632120; chennai@schandpublishing.com
Guwahati : Ph: 2738811, 2735640; guwahati@schandpublishing.com
Hyderabad : Ph: 40186018; hyderabad@schandpublishing.com
Jalandhar : Ph: 4645630; jalandhar@schandpublishing.com
Kolkata : Ph: 23357458, 23353914; kolkata@schandpublishing.com
Lucknow : Ph: 4003633; lucknow@schandpublishing.com
Mumbai : Ph: 25000297; mumbai@schandpublishing.com
Patna : Ph: 2260011; patna@schandpublishing.com

First Edition 2009
Reprints 2020, 2021, 2022 (Twice), 2023 (Twice)

Reprint 2025

ISBN: 978-81-219-3191-5 **Product Code:** H6MBY68BIOL10ENAA09O

PRINTED IN INDIA

By Vikas Publishing House Private Limited, Plot 20/4, Site-IV, Industrial Area Sahibabad, Ghaziabad – 201 010 and Published by S Chand And Company Limited, A-27, 2nd Floor, Mohan Co-operative Industrial Estate, New Delhi – 110 044.

PREFACE

Molecular Biology deals with biomolecules, especially macromolecules such as DNA, RNAs and proteins. It revolves around the central dogma which holds that DNA, the genetic information containing and self-replicating molecule of a cell transcribes three main types of RNA molecules (*i.e.*, rRNA, tRNA and mRNA) which translate genetic information into protein molecules. Proteins, the polymers of amino acids may have a structural or functional role and organise the living entity.

This textbook of **Molecular Biology** contains 16 chapters. It is hoped that this book will open new vistas for the readers and will serve them immensely. With the aid of simple text, numerous figures, tables, boxes and photographs, authors have tried to craft a balanced account of Molecular Biology. Problems at the end of some chapters have tried to show our readers the way to solve Molecular Biological Problems. Answers to selected problems and Multiple Choice Questions may help our readers.

This textbook of Molecular Biology has been developed for B.Sc. students of all Indian Universities. It will also serve the purpose of the students of B.Sc. (Hons.), M.Sc. Zoology, Botany, Microbiology and Biotechnology. Students preparing for Medical entrance exams, I.F.S., P.C.S. and I.A.S., etc. will also benefit by this book.

This book is the outcome of sincere and combined tireless efforts of the authors and the editorial team of S. Chand and Company Pvt. Ltd.

Constructive suggestions for further improvement of the book will always be welcomed and entertained.

Dr. P. S. VERMA
"PREETI"
121, New Mohanpuri Colony
Meerut- 250 001

Dr. V.K. AGARWAL
"Neelkanth"
566, Brahmpuri
(Sharda Road)
Meerut- 250 002

Contents

Prologue

Viruses are noncellular biological entities. Only two basic types of cells are found in the cellular organisms : **Prokaryotic cells** (in Eubacteria and Archae bacteria) and **Eukaryotic cells** (in organisms of four kingdoms – animals, plants, fungi, and protists – the unicellular algae and protozoa).

Eubacteria (*e.g.*, *Escherichia coli*) are structurally defined as prokaryotes. These cells have a plasma membrane (a fluid-mosaic membrane of molecules of lipids and proteins), usually enclosed in a rigid cell wall, but no intracellular compartments. They have a single major circular chromosome. They may be unicellular or multicellular.

Archaea (*e.g.*, *Methanococcus jannaschii*) are structurally defined as prokaryotes but probably branched off from the eukaryotes after their common ancestor diverged from the eubacteria (**Cooper** and **Hausman**, 2007). They tend to inhabit extreme environments. They are biochemically closer to eubacteria in some ways but to eukaryotes in others. For example, the 1740 kb genome of the archaeon *M. jannaschii* encodes a maximum of 1738 proteins. Comparisons revealed that those (proteins) involved in energy production and metabolism are most like those of eubacteria, while those involved in replication, transcription and translation are more similar to those of eukaryotes.

Cells of Eukaryota possess well-defined subcellular compartment bounded by lipid-protein membrane (e.g., nuclei, mitochondria, chloroplasts, endoplasmic reticulum, Golgi apparatus, etc.). These organelles are sites of distinct biochemical processes and define the eukaryotes.

In most multicellular eukaryotes, groups of cells **differentiate** during development of the organism to provide specialized functions (*e.g.*, as in liver, brain, kidney). In most cases, they contain the same DNA but transcribe different genes. Like all other cellular processes, differentiation is controlled by genes. Coordination of the activities of different cell types requires communication between them. This involves **signaling molecules** such as neurotransmitters, hormones, and growth factors which are secreted by one tissue and act upon another through cell-specific receptors.

Molecular Biology seeks to explain the relationships between the structure and function of biological molecules and how these relationships contribute to the operation and control of biochemical processes. Of principal interest are the processes of replication, transcription and translation. The new experimental technologies involved in manipulating these molecules are central to modern molecular biology. The **central dogma** is the original proposal that 'DNA makes RNA and RNA makes protein', which happen via the processes of transcription and translation respectively. This is broadly correct, although a number of examples are known which contradict part of it. Retroviruses reverse transcribe RNA into DNA, a number of viruses are able to replicate RNA directly into an RNA copy, and a number of organisms can edit a messenger

RNA sequence so that the protein coding sequence is not directly specified by DNA sequence.

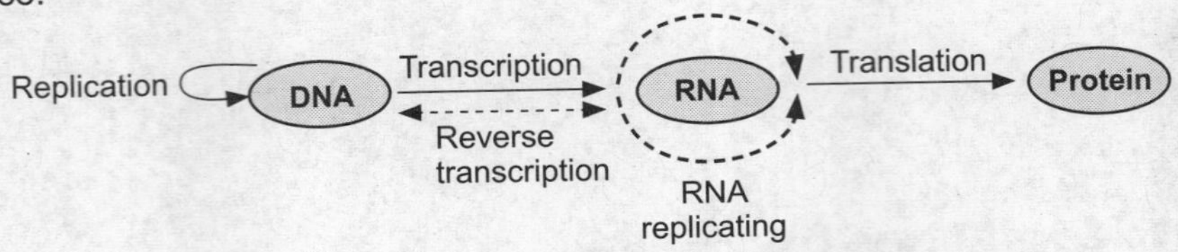

Fig. Flow of genetic information.

In case of **prokaryotic gene expression** transcription of a single gene of an operon starts at the promoter, ends at the terminator and produces monocistronic or polycistronic messenger RNA. The coding regions of the message are translated by the ribosomes from the start codon (close to the ribosome binding site) to the stop codon. Transfer RNAs deliver the proper amino acid, according to the genetic code, to the growing protein chain.

In case of **eukaroyotic gene expression**, mostly monocistronic messenger RNA's are transcribed from a gene, initiated at a promoter. The resulting pre-messenger RNA is capped at the 5'-end and has a poly (A) tail added to the 3'end. Introns are removed from exons by splicing before the mature mRNA (formed by joining of exons) is exported from the nucleus to be translated by ribosomes in the cytoplasm.

Mutations are heritable permanent changes in the base sequence of DNA. Point mutations may be transitions (*e.g.* G·C $\rightarrow$ A·T) or transversions (*e.g.*, G·C $\rightarrow$ T·A). Deletions and insertions involve the loss or addition of bases and can cause frame-shifts in reading the genetic code. Silent mutations have no phenotypic effects, while missense and nonsense mutations change the amino acid sequence of the encoded protein.

DNA repair involves photoreactivation and excision repair (Chapter 16). In **photoreactivation**, cleavage of the cyclobutane ring of pyrimidine dimers by DNA photolyases restores the original DNA structure. Photolyases (the photoreactivating enzymes) have chromophores which absorb blue light to provide energy for the reaction Photoreactivation is an example of direct reversal of a lesion of DNA and is error-free (**Turner** *et al.*, 2000). In **nucleotide excision repair**, an endonuclease makes nicks on either side of the lesion in DNA, which is then removed to leave a gap. This gap is filled by a DNA polymerase, and DNA ligase makes the final phosphodiester bond. In **base excision repair** the lesion is removed by a specific DNA glycosylase. The resulting **AP site** (AP = a purinic or a pyrimidinic) is cleaved and expanded by a gap by an AP endonuclease plus exonuclease. Thereafter the process is like nucleotide excision repair.

Cancer results from mutation that disrupts the controls regulating normal cell growth. The growth of normal cells is subjected to many different types of control which may be lost independently of each other, during the multistep progression to malignancy.

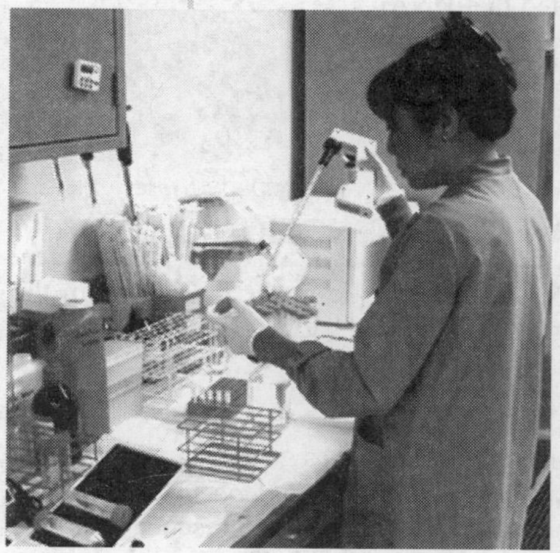

CHAPTER 1

Identification of the Genetic Materials

Chapter Contents

When it became evident that the chromosomes were the organs of heredity, because (*i*) they formed the only link between two generations, (*ii*) they carried linearly arranged genes, (*iii*) they occurred in every organism in specific number and had specific morphology for a particular species, and (*iv*) any variation in their number of morphology affected the phenotype of the species, then, various attempts were made by early molecular geniticists to identify the physical and chemical nature of genes. But the genes were found so minute structures that their physical identity remained almost impossible. However, the extensive chemical analysis of chromosomes of different organisms have revealed that chromosomes contain proteins and nucleic acids (DNA and RNA), and it was thought that genes might have either proteins or nucleic acids as their component molecules. Early molecular geneticists have assigned the informational roles of genes to the chromosomal proteins because, they found nucleic acids too simple to carry genetic informations. (This is called **tetranucleotide hypothesis).**

The controversy about the assignment of genetic role either to chromosomal proteins or to chromosomal DNA, existed up to 1949 when **A. Mirsky** and **H. Ris** had found that all cells of an organism appeared to contain the same amount of DNA, whereas different cell types contained quite different amounts and kinds of protein. Its constancy, therefore, favoured DNA as the genetic material. Around 1953, it was universally accepted that DNA is the genetic substance (*i.e.*, chemical of which genes are composed) of most microorganisms and higher organisms. Later on, RNA was found to be the genetic material of some viruses. The concept that DNA or RNA is the genetic material of most organisms has been developed and supported by the following evidences :

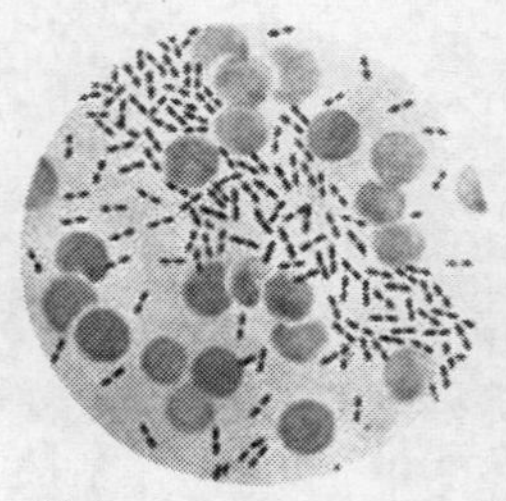

Diplococcus pneumoniae.

1.1. DIRECT EVIDENCES FOR DNA AS THE GENETIC MATERIAL

The most conclusive evidences in support of DNA as the genetic material came from the following three avenues of approach on microorganism : transformation of bacteria, mode of infection of bacteriophages and conjugation of bacteria.

1. The Transformation Experiments

In 1928, **Frederick Griffith** encountered a phenomenon now known as genetic transformation. Colonies of virulent strain (pathogenic) of pneumonia causing bacterium, *Diplococcus pneumoniae* grown on nutrient agar, have a **smooth (S)** glistening appearance owing to the presence of a type specific, polysaccharide (a polymer of glucose and glucuronic acid) capsule. The avirulent (non-pathogenic) strains, on the other hand, lack this capsule and they produce dull, **rough (R)** colonies. Smooth (S) and rough (R) characters are directly related to the presence or absence of the capsule and this trait is known to be genetically determined. Both S and R forms occur in several types and are designated as S-I, S-II, S-III, etc., and R-I, R-II, R-III, etc., respectively. All these subtypes of S and R bacteria differ with each other in the type of antigens, they produce. The kind of antigen produced is likewise genetically determined. Smooth (S) forms sometimes mutate to rough (R) forms, but this change has not been found reversible.

In the course of his work, **Griffith** injected laboratory mice with live R-II pneumococci; the mice suffered no illness because R-II pneumococci was avirulent. But, when the mice were injected with virulent S-III pneumococci, the mice suffered from pneumonia and died. However, when he injected the heat killed S-III bacteria in mice, they did not suffer from pneumonia. But, when the mice were injected with the mixture of living avirulent R-II and heat killed S-III virulent, the unexpected symptoms of pneumonia appeared and high mortality resulting in them. By postmorteming the dead mice, it was found that their heart blood had both R-II and S-III pneumococci. From these results, Griffith concluded that the presence of the heat-killed S-III bacteria must have caused a **transformation** of the living R-II bacteria, so as to restore to them the capacity for capsule formation they had earlier lost by gene mutation. This was called "**Griffith effect**" or more popularly "**bacterial transformation**".

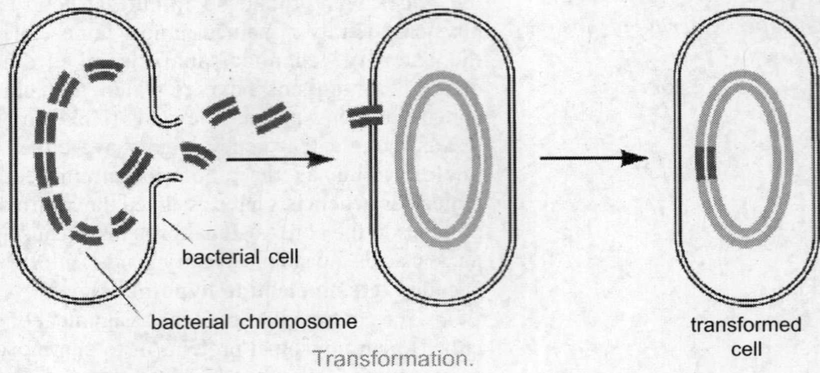

Transformation.

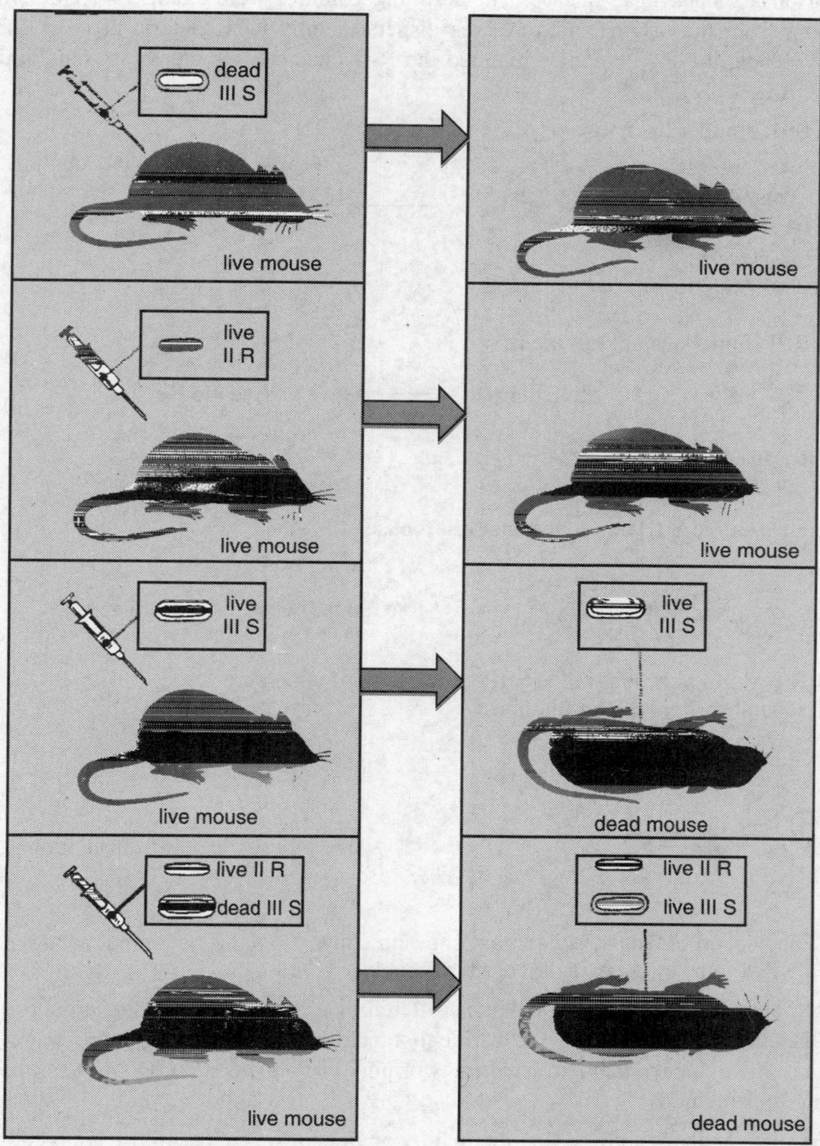

Fig. 1.1. Griffith experiment that demonstrated the principle of transformation.

Identification of the "transforming principle or substance". Griffith could not understand the cause of bacterial transformation and that is first of all identified by **Oswald Avery, Colin MacLeod** and **Maclyn McCarty** (1944). They partially purified the **transforming principle** from the cell extract (*i.e.,* cell free extract of S-III bacteria) and demonstrated that it was DNA. These workers modified the known schemes for isolating DNA and prepared samples of DNA from S-III bacteria. They added this DNA to a live R bacterial culture; after a period of time they placed a sample of S-III containing R-II bacterial culture on an agar surface and allowed it to grow to form colonies. Some of the colonies (about 1 in 10^4) that grew were S-III type (Fig. 1.3). To show that this was a permanent genetic change, they dispersed many of the newly formed S-III colonies and placed them on a second agar surface. The resulting colonies were again S-III type. If an R-II colony arising from the original mixture was dispersed, only R-II bacteria grew in subsequent generations. Hence, the R-II colonies retained the R-II character, whereas the transformed S-III colonies bred true as S-III.

A S III strain kills mouse

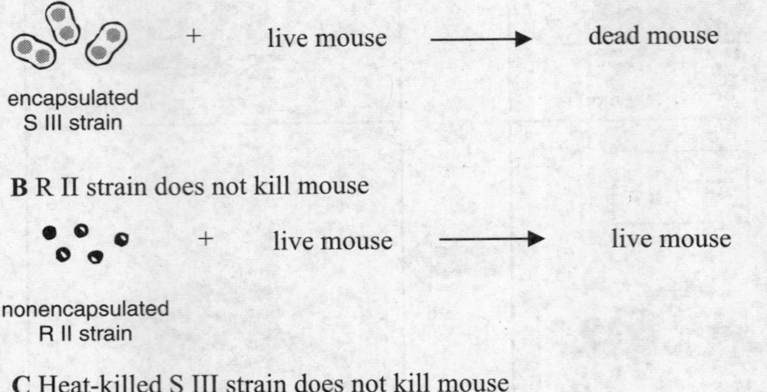

C Heat-killed S III strain does not kill mouse

D R II strain plus heat-killed S III strain, both of which are separately nonlethal, kill mouse

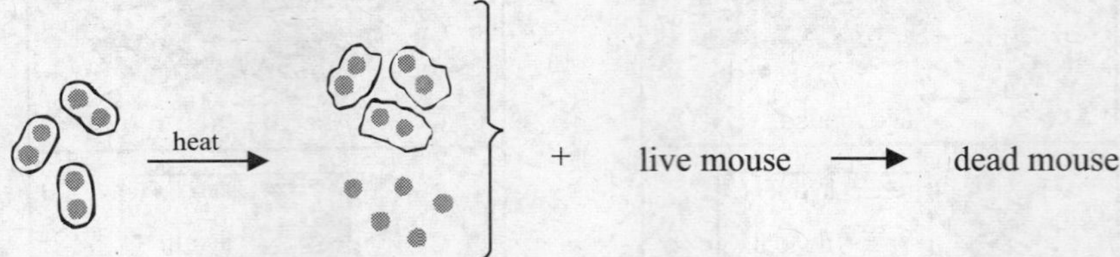

Fig. 1.2. Explanation of Griffith experiment showing conversion of a non-lethal bacterial strain to a lethal form by a cell extract (after Freifelder, 1985).

Further, because S-III and R-II colonies differed by a polysaccharide coat around each S-III bacterium, the ability of purified polysaccharide to transform was also tested, but no transformation was observed. The evidence that extracted transforming principle contain DNA, was provided by the following four methods :

1. Chemical analysis showed that the major component was deoxyribose containing nucleic acid.

2. Physical measurement showed that the sample contained a highly viscous substance having the properties of DNA.

3. Experiments demonstrated that transforming activity is not lost by reaction with either (*a*) purified proteolytic (protein - hydrolyzing) enzymes trypsin, chymotrypsin, or a mixture of both or (*b*) ribonuclease (an enzyme that depolymerizes RNA). The lipase too has no effect.

4. It was demonstrated that the treatment with materials, known to contain DNA depolymerizing activity (DNAase enzyme), inactivated the transforming principle. Later on, such a transforming substance, the DNA, was found in a variety of bacteria (*e.g., Hemophilus influenzae, Bacillus subtilus, Shigella paradysenteriae,* etc.) and several other organisms.

preparation of transforming principle from S strain

lysis, precipitation

cell-free extract

transforming principle from S strain

ecapsulated S strain

addition of transforming principle to R strain

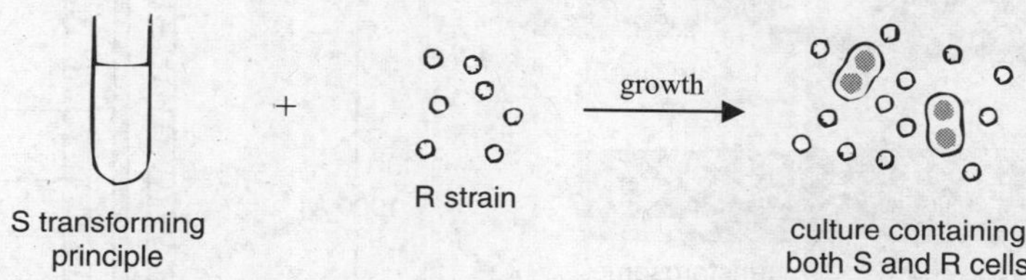

Fig. 1.3. Transformation experiment of **Avery, MacLeod** and **McCarthy** (after Freifelder, 1985).

2. The Blender Experiment

An elegant confirmation of the genetic nature of DNA came from an experiment with *E. coli* phage T_2. This experiment known as the **blender experiment** because a kitchen waring blender was used as a major piece of apparatus, was performed by **Alfred Hershey** and **Martha Chase** in 1952. They demonstrated that the DNA injected by a phage particle into a bacterium contains all the information required to synthesize progeny phage particles.

A single particle of phage T_2 consists of DNA (now known to be a single molecule) encased in a protein shell (Fig. 1.4). The DNA is the only phosphorus containing substance in the phage particle; the proteins of the shell, which contain the amino acids methionine and cysteine, have the only sulphur atoms. In these experiments, phage DNA was made radioactive by growing infected bacteria on a medium containing radioactive phosphate ($^{32}PO_4$). Since phage proteins do not contain phosphorus, only DNA would be labelled. Similarly, phage proteins were labelled with the help of $^{35}SO_4$. Since DNA does not contain sulphur, only protein would be labelled with ^{35}S. Such differential labelling would enable one to distinguish between DNA and proteins of the phage without performing any chemical tests. **Hershey** and **Chase** then allowed both kinds of labelled phage particles to infect *E. coli* bacteria (Fig. 1.5). The infected bacteria were immediately agitated in a waring blender. After shaking, only radioactive ^{32}P was found associated with bacterial cells and ^{35}S was found only in surrounding medium and not in bacterial cells. When phage progeny was studied for radioactivity in this experiment, it was found that the phage progeny carried labelled only ^{32}P. The progeny was not labelled with ^{35}S. This clearly indicates that only DNA and not protein is injected into bacterial cells. The empty protein coat (ghost) left outside. Thus, **Hershey** and **Chase** experiment proved that DNA entering the host cell carries all the genetic information for synthesis of new phage particles, hence, is the sole genetic material in DNA bacteriophages (*e.g.*, T_2).

3. Bacterial Conjugation

Another conclusive evidence for DNA as the genetic material came from the phenomenon of conjugation of bacteria. **Laderberg** and **Tatum** (1946) found that when an F^+ ('male') *E. coli* cell conjugated with an F^- ('female') *E. coli* cell, unidirectional transfer of F^+ factor of 'male' cell to F^-

or 'female' cell took place, so that the latter was found to be a fragment of DNA molecule which occurred in the cytoplasm of bacterial cell.

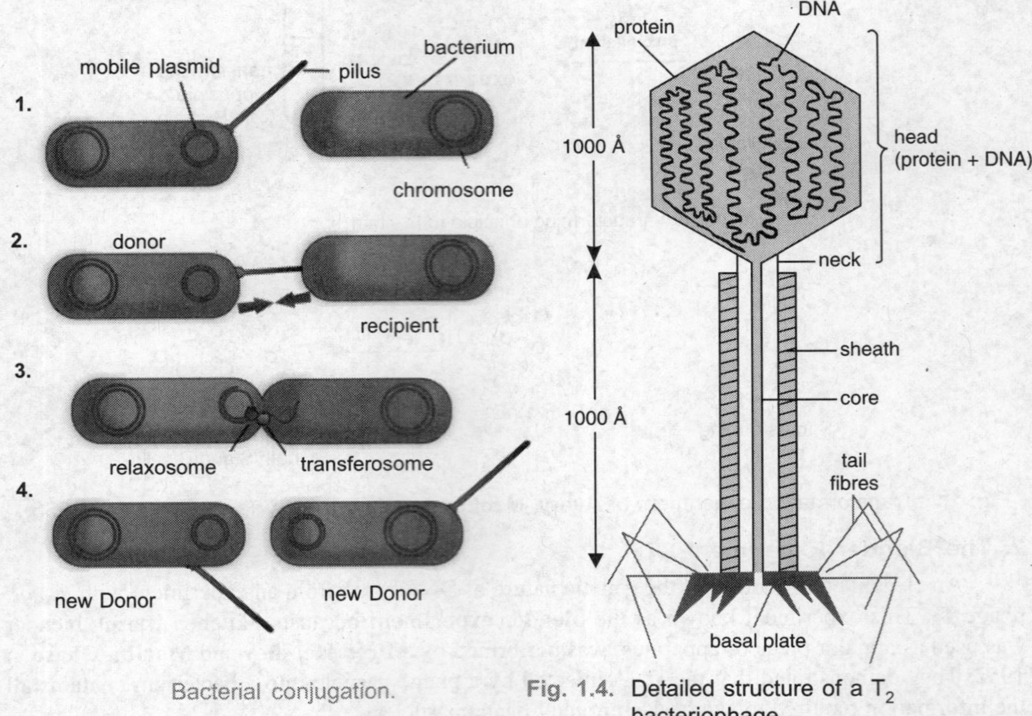

Bacterial conjugation.

Fig. 1.4. Detailed structure of a T_2 bacteriophage.

1.2. INDIRECT EVIDENCES FOR DNA AS THE GENETIC MATERIAL

The indirect evidences in favour of DNA as the genetic meterial came from higher organisms which are not as easy to manipulate as bacteria and viruses. The fact that DNA is the genetic material of higher organisms has been supported by the following facts :

1. The Feulgen techniques have shown that DNA entirely remains restricted to the chromosomes and it forms one of the major components of chromosomes.

Table 1.1. DNA content and ploidy (**Pollister** *et. al.*, 1951)

Cells	Mean DNA Feulgen content (Picograms)	Presumed chromosome set (Ploidy)
Spermatids	1.68	Haploid (2n)
Liver	3.16	Diploid (2n)
Liver	6.30	Tetraploid (4n)
Liver	12.80	Octaploid (8n)

2. Various quantitative measurements of amount of DNA in different cells have shown that there is a correlation between the amount of DNA and the number of chromosomes sets (ploidy).

Diploid cells contain twice as much DNA as do haploid cells of the same species as shown in the Table 1.1.

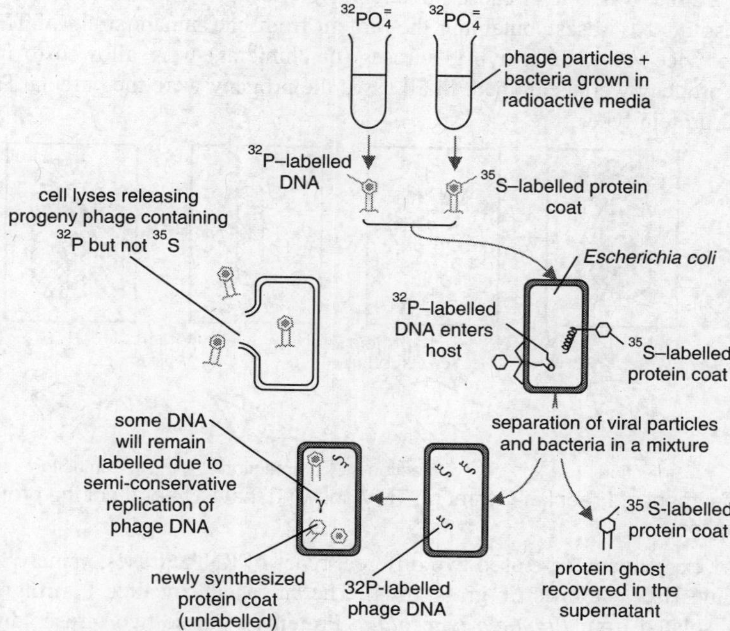

Fig. 1.5. A classical experiment of Hershey and Chase, where DNA of T_2 phage is labelled with ^{32}P and the protein is labelled with ^{35}S.

Table 1.2. DNA contents in picograms (gram × 10^{-12}) of various cells.

S.No.	Organism	Kidney	Liver	Erythrocytes	Sperm
1.	Chicken	2.4	2.5	2.5	1.3
2.	Bovine	6.4	6.4	—	3.3
3.	Carp	—	3.0	3.3.	1.6
4.	Human	5.6	5.6	—	2.5

3. The diploid amount of DNA is constant within the species but varies from one species to another, as shown in the Table 1.2.

4. In 1950 **Swift** found the amount of DNA per resting nucleus (interphase nucleus) in mitotically active tissues of animals ranged from twice (2n) the value found in the gametes (1n) to four times (4n) this value. He concluded that DNA synthesis occurred during the interphase.

This parallelism of behaviour in DNA and chromosomes clearly indicates that DNA is the genetic material of higher animals.

1.3. EVIDENCES FOR RNA AS THE GENETIC MATERIAL OF SOME VIRUSES

The demonstration that RNA is the genetic material in RNA - containing viruses came in 1956, when **A. Gierer** and **G. Schramm** showed that tobacco plants could be inoculated with purified RNA from the tobacco mosaic virus (TMV), and TMV-like lesions could later be identified on the tobacco leaves. A different approach was taken by **H. Fraenkel-Conrat** and **B. Singer** in experiments published in 1957. They first developed the techniques for separating TMV particles into RNA and proteins. Later by using RNA and proteins separately in tests for infectivity, it was shown that RNA alone was able to cause infection (Fig. 1.6). They then developed techniques for forming "reconstituted" viruses containing the protein from one mutant strain of TMV and RNA from another, or vice-versa. Such hybrid viruses (or chimeras) were allowed to infect tobacco leaves, and the progeny were examined. In all cases the progeny were the parental RNA type and not the parental protein type.

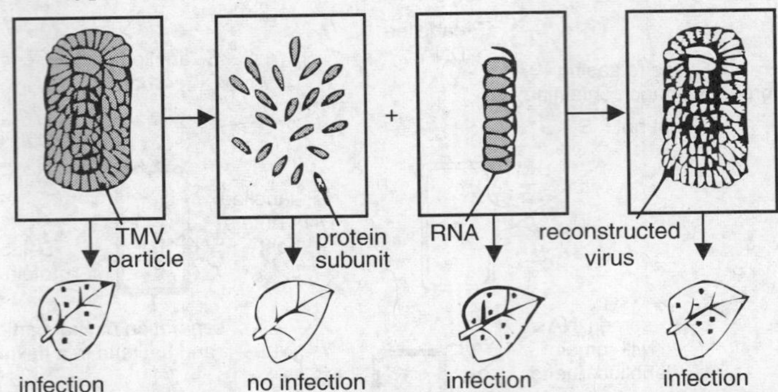

Fig. 1.6. Experiment of Fraenkel-Conrat on TMV showing that RNA and not the protein can cause infection.

In another experiment were used two different types of RNA viruses, namely TMV (tobacco mosaic virus) and HRV (Holmes rib-grass virus). The latter is also called **Plantago strain,** since this virus was isolated from *Plantago lanceolata.* Proteins of these two viruses differ in having different frequencies and sequences of amino acids. Also these viruses give different symptoms on leaves of a particular variety of tobacco, while TMV produces mottling of leaves, HRV produces distinct ring patterns. Reciprocal chimeras using RNA of one strain and protein of other strain were obtained. It was found that when these chimeras were used for infection, the progeny had proteins which correspond to the virus from which RNA of the infecting virus particle was derived (Fig.1.7.). This proved that specificity of virus proteins was determined by RNA alone and that proteins carried no genetic information.

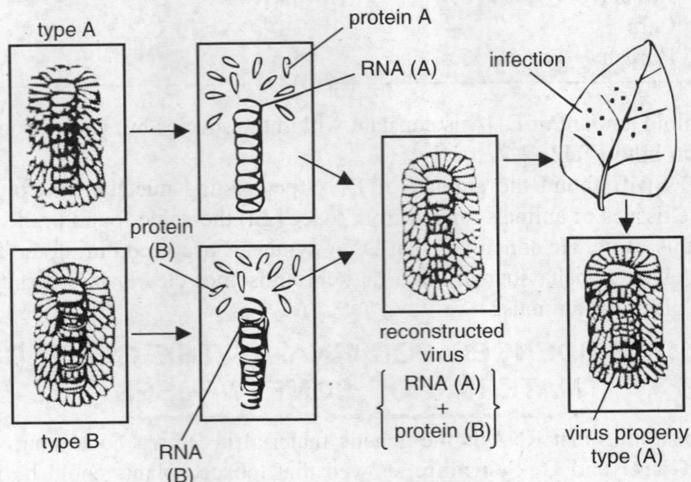

Fig. 1.7. Reconstruction of virus, showing that reconstructed virus gives progeny which resembles the virus from which RNA was derived.

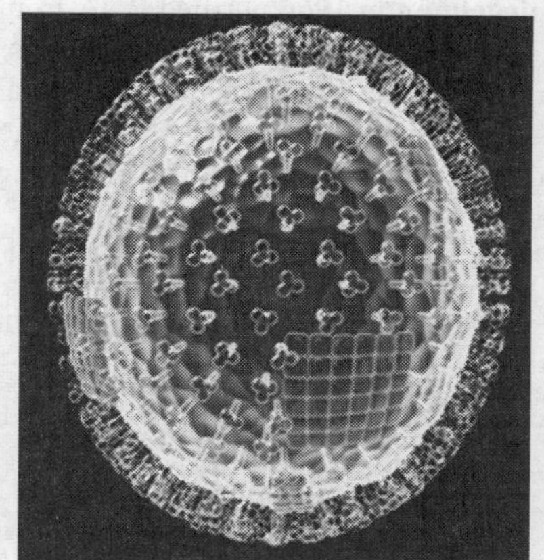

Influenza virus.

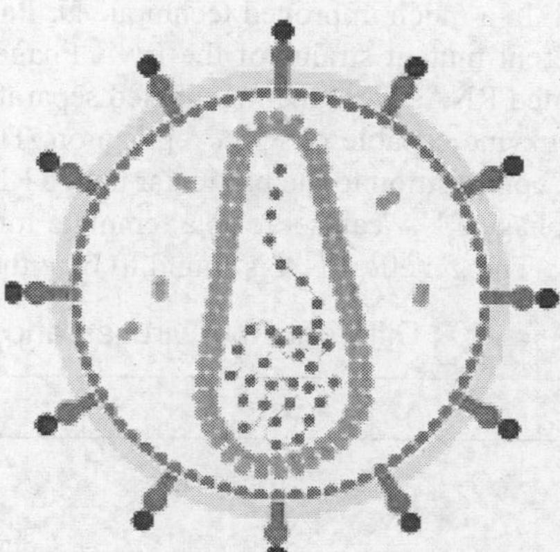

Rous Sarcoma Virus.

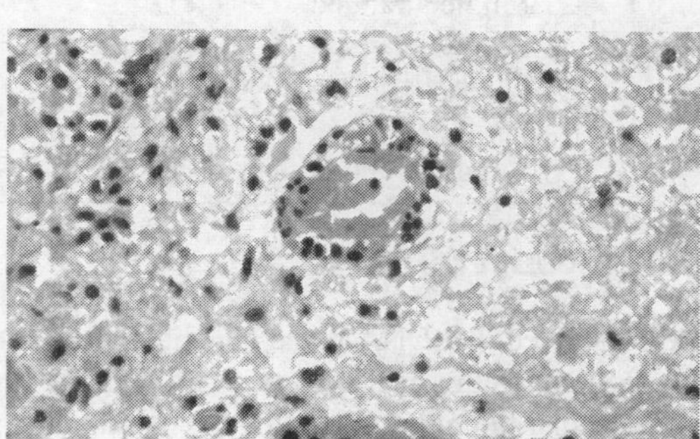

Poliomyelitis virus.

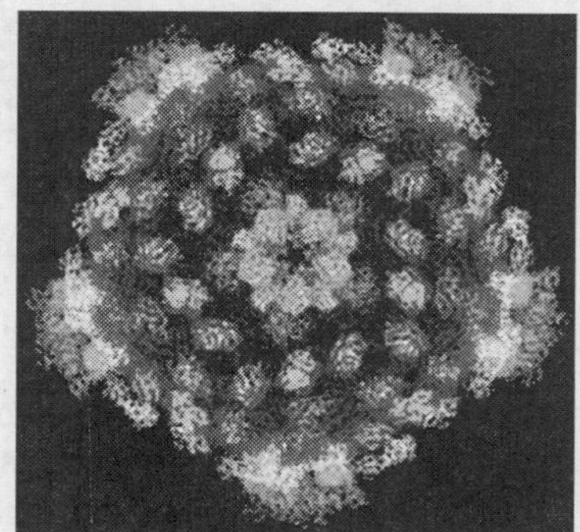

Reo virus.

ANIMAL VIRUSES

Tobacco mosaic virus.

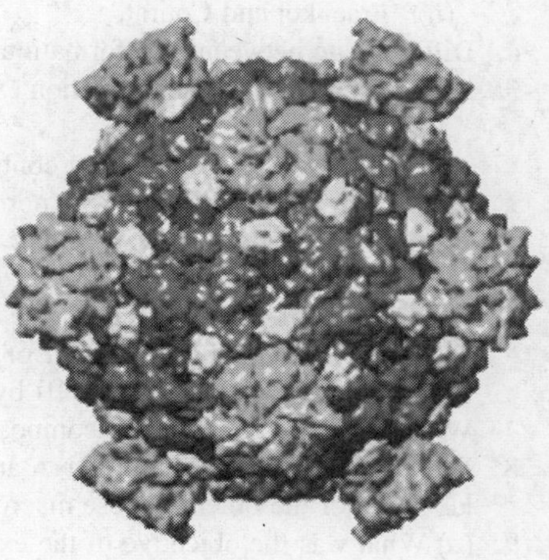

Wound tumour virus.

PLANT VIRUSES

In a much improved technique **N. Pace** and **S. Spiegelman** in 1966 purified RNA from two different mutant strains of the RNA Phage QB which had quite distinct base compositions. The isolated RNAs were then incubated separately in the presence of an *E. coli* cell extract containing an enzyme capable of RNA replication. The new RNA synthesized was in each case identical in base composition to the particular phage RNA presented to the *in vitro* system, thus, indicating that the phage RNA can serve as a template for its self-replication.

The genetic RNA is found to be either single-stranded or double-stranded (see Table 1.3).

Table 1.3. Different RNA viruses and the nature of genetic RNA associated with them.

	Virus	Types of RNA
1.	**Plant viruses**	
	TMV	Single stranded
	Wound tumour	Double stranded
2.	**Animal viruses**	
	Influenza virus	Single stranded
	Rous sarcoma	Single stranded
	Poliomyelitis	Single stranded
	Reovirus	Double stranded
3.	**Bacteriophages**	
	MS2, F2, r17	Single stranded

Questions And Problems

1. Describe two classical experiments which demonstrated that DNA is the genetic material.
2. Describe an experiment which established the genetic nature of RNA in certain viruses.
3. Write about the contribution of the following molecular biologists regarding the identification of the genetic material :
   (*i*) F. Griffith;
   (*ii*) O.T. Avery, C.M. Macleod and M. McCarthy;
   (*iii*) A.D. Hershey and M.J. Chase;
   (*iv*) Fraenkel and Conrat.
4. Differentiate between transformation and transduction.
5. (*a*) How did the transformation experiments of Griffith differ from those of Avery and his associates ?
   (*b*) What was the significant contribution of each ?
   (*c*) Why was Griffith's work not evidence for DNA as the genetic material whereas the experiments of Avery *et. al.* provided direct proof that DNA carried the genetic information ?
6. How could it be demonstrated that the mixing of heat-killed Type III pneumococcus with live Type II resulted in a transfer of genetic material from Type III to Type II rather than a restoration of viability to Type III by Type II ?
7. What is the macromolecular composition of a bacteriophage such as phage T2 ?
8. What chemical properties do DNA and proteins possess that allow researchers to specifically label one or the other of these macromolecules with a radioactive isotope?
9. (*a*) What was the objective of the experiment carried out by Hershey and Chase ?
   (*b*) How was the objective accomplished ?
   (*c*) What is the significance of this experiment ?

Answers to Problems

4. Transformation involves naked DNA from one cell becoming incorporated into another's DNA, whereas in transduction, a virus serves as the vector transferring DNA derived from one cell into another.
5. (*a*) Griffith's *in vivo* experiments demonstrated the occurrence of transformation in *Pneumococcus*. They provided no indication as to the molecular basis of the transformation phenomenon. Avery and colleagues carried out *in vitro* experiments, employing biochemical analyses to demonstrate that transformation was mediated by DNA.
   (*b*) Griffith showed that a transforming substance existed; Avery *et. al.* defined it as DNA.
   (*c*) Griffith's experiments detected the substance responsible for transformation. Avery *et. al.* isolated DNA in "pure" form and demonstrated that it could mediate transformation.
6. Purified DNA from Type III cells was shown to be sufficient to transform Type II cells. This occurred in the absence of any dead Type III cells.
7. About 1/2 protein, 1/2 DNA. A single long molecule of DNA is enclosed within a complex "coat" that is composed of many proteins.
8. DNA contains phosphorus (normally ^{31}P), but no sulphur, it can be labelled with radioactive phosphate ^{32}P. Proteins contain sulphur (normally ^{32}S), but usually no phosphorus; they can be labelled with radioactive sulphur ^{35}S.
9. (*a*) The objective was to determine whether the genetic material was DNA or protein.
   (*b*) By labelling phosphorus, a constituent of DNA, and sulphur, a constituent of protein, in a virus, it was possible to demonstrate that only the labelled phosphorus was introduced into the host cell during the viral reproductive cycle. This was enough to reproduce new phages.
   (*c*) Therefore DNA, not protein, is the genetic material.

Multiple Choice Questions

Choose the correct answer from the four alternatives given.

1. Who performed transformation experiments in bacteria?
   (*a*) Louis Pasteur
   (*b*) Griffith
   (*c*) Macleod
   (*d*) Meselson and Stahl
2. Bacterial transformation was discovered by
   (*a*) Avery *et. al.*
   (*b*) Beadle and Tatum
   (*c*) Robert Brown
   (*d*) Har Govind Khorana
3. The dye involved in Feulgen reaction to stain DNA is
   (*a*) Janus Green B
   (*b*) Neutral Red
   (*c*) Basic Fuchsin
   (*d*) Haematoxylin
4. *Pneumococcus* experiment proved that
   (*a*) bacteria undergo binary fission
   (*b*) bacteria do not reproduce asexually
   (*c*) RNA may sometimes control production of DNA and protein
   (*d*) DNA is the genetic material
5. The process in which DNA of a bacterial cell is transferred into another bacterial cell through a virus is
   (*a*) transduction
   (*b*) transformation
   (*c*) conjugation
   (*d*) reproduction
6. Transformation experiment was performed on which bacteria
   (*a*) *E. coli*
   (*b*) *Diplococcus pneumoniae*

(*c*) *Salmonella*
(*d*) *Pasteurella pestis*

7. Who proved that DNA is the basic genetic material?
(*a*) Griffith
(*b*) Boveri and Sutton
(*c*) Hershey and Chase
(*d*) Watson

8. Who established that RNA is the genetic material?
(*a*) Nirenberg and Holley
(*b*) Lederberg
(*c*) Fraenkel Conrat
(*d*) Griffith

9. A bacteriophage with radioactive DNA and protein when infects a bacterium the radioactivity inside the bacterium will be located
(*a*) in DNA
(*b*) in protein
(*c*) both in DNA and protein
(*d*) in all parts of bacterial cell

Answers

1. (*b*)	2. (*a*)	3. (*c*)	4. (*d*)	5. (*a*)	6. (*b*)
7. (*c*)	8. (*c*)	9. (*a*).			

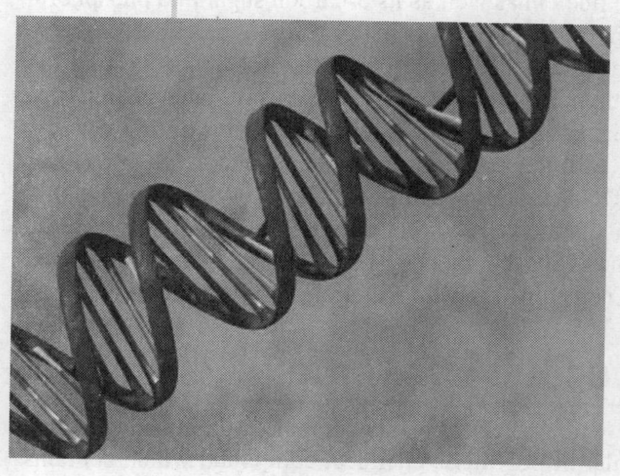

2

Chemical Nature of Genetic Materials (*i.e.*, DNA and RNA)

Chapter Contents

DNA (deoxyribonucleic acid) and RNA (ribonucleic acid), the principal genetic materials of living organisms, are chemically called **nucleic acids** and are complex molecules larger than most proteins and contain carbon, oxygen, hydrogen, nitrogen and phosphorus.

2.1. HISTORICAL

In 1869, **Friedrich Miescher**, 22-year old Swiss physician (working in the laboratory of Felix Hoppe-Seyler in Tubingen) had isolated from pus cells obtained from discarded bandages in the Franko-Prussian War, and from salmon sperm, a previously identified macromolecular substance, to which he gave the name **nuclein**. Although he was unaware of the structure and function of nuclein, he submitted his findings for publication. The editor who received the paper was doubtful about some aspects of the Miescher's report and delayed publication for two years while he tried repeating some of the more questionable aspects of 'Miescher's work. Finally, in 1871, Miescher's report was published, but it made little immediate impact. He continued his careful work up to his death in 1895, recognizing (with the help of his student **Richard Altmann** in 1889) that nuclein was of higher molecular weight and was associated in some way with a basic protein, to which he gave the name **protamine**.

In 1880, **Emil Fischer** identified pyrimidines and purines. The biochemist, **Albrecht Kossel** identified the constituent nitrogenous bases of nuclein as well as its 5-carbon sugar and phosphoric acid. It was **Altmann** who first suggested, in 1899, the use of the term **nucleic acid** to describe phosphorus-containing nuclein. **Kossel** was awarded the 1910 Nobel Prize for demonstrating the presence of two pyrimidines (cytosine and thymine) and two purines (adenine and guanine) in nucleic acids. Kossel's work and later investigations of **Ascoli**, **Levine** and **Jones** during the first quarter of the 1900s disclosed the two kinds of nucleic acids, deoxyribonucleic acid (DNA) and ribonucleic acid (RNA). Development of DNA-specific staining techniques by **Feulgen** and **Rossenbeck** in 1924 enabled Feulgen to demonstrate in 1937 that most of the DNA content of the cell is located in the nucleus. It was not until the 1950s that the inter-nucleotide bond was established by **A.R. Todd (Judson,**1979). Because DNA is the only genetic material of most living organisms, so deserves priority in discussion.

2.2. DEOXYRIBONUCLEIC ACID OR DNA

Highly purified DNA, extracted from a wide variety of plants, animals, bacteria and viruses, has been found to be complex macro-molecular or polymeric chemical compound which contains four kinds of smaller building blocks (monomers) called **deoxyribotids** or **deoxyribonucleotides**. Each deoxyribonucleotide is made up of three moieties: a **phosphoric acid molecule** (biologically called phosphate); a pentose sugar called **2 deoxyribose**; and **pyrimidine** and **purine** nitrogenous bases. Four major kinds of nitrogenous bases have been found in four kinds of deoxyribonucleotides of DNA : two are heterocyclic and two-ringed purines, **adenine (A)** and **guanine (G)**, and two are one ringed pyrimidines, **cytosine (C)** and **thymine (T)**. The pyrimidine ring can be numbered in two different ways (Fig 2.6). In this book, we will follow the old numbering system.

Fig. 2.1. Chemical formula of deoxyribose.

Fig. 2.2. Chemical formula of adenine (6-amino-purine mw = 151.13 daltons).

Fig. 2.3. Chemical formula of guanine (2-amino-6-hydroxipurine, purine, mw = 135.13 daltons).

Fig. 2.4. Chemical formula of cytosine (2-hydroxy-6-amino pyrimidine, mw = 111.10 daltons).

Fig. 2.5. Chemical formula of thymine (2,6-dihydroxy-5- methylpyrimidine, mw = 126.12 daltons).

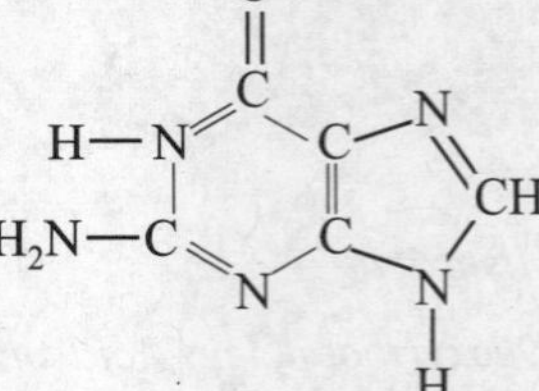

Fig. 2.6. Numbering of carbon atoms in a pyrimidine ring.

That part of each nucleotide which contains a nitrogenous base and deoxyribose is called **deoxyribonucleoside**. The four kinds of deoxyribonucleosides and deoxyribonucleotides can be tabulated as follows :

Table 2.1. Four nitrogen bases, nucleosides and nucleotides of DNA molecule.

Nitrogen base	Base + deoxyribose = deoxyribonucleoside	Deoxyribonucleoside + phosphoric acid = deoxyribonucleotide	Abbreviation for nucleotide
1. Adenine (A)	Deoxyadenosine	Deoxyadenylic acid (Deoxyadenosine monophosphate)	dAMP
2. Guanine (G)	Deoxyguanosine	Deoxyguanylic acid (Deoxyguanosine monophosphate)	dGMP
3. Cytosine (C)	Deoxycytidine	Deoxycytidylic acid (Deoxycytidine monophosphate)	dCMP
4. Thymine (T)	Thymidine	Thymidylic acid (Thymidine monophosphate)	TM

The four deoxyribonucleotides besides occurring in DNA molecule, occur also in nucleoplasm and cytoplasm, but in their triphosphate forms such as deoxyadenosine triphosphate (dATP), deoxyguanosine triphosphate (dGTP), deoxycytidine triphosphate (dCTP) and thymidine triphosphate (TTP). The significance of occurrence of deoxyribonucleotides in triphosphate forms lies in the fact that during DNA replication, the DNA polymerase enzyme can act only on triphosphate of deoxyribonucleotides.

NH_2 C N C N CH HC C N N H O HO — P — O — H_2C O OH H H H H OH H

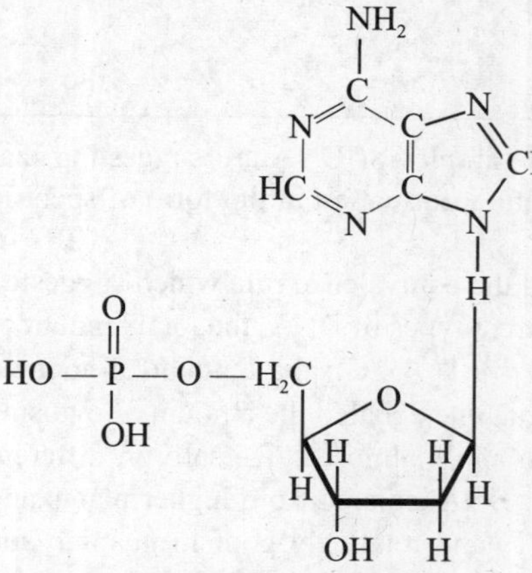

Fig. 2.7. Chemical formula of deoxycytidylic acid.

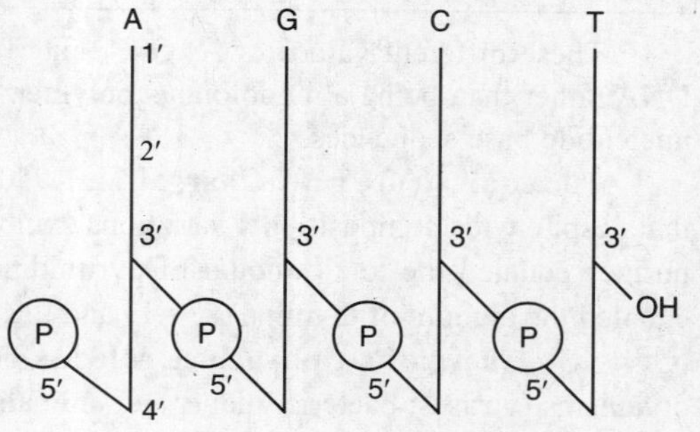

A G C T 1′ 2′ 3′ 3′ 3′ 3′ OH P P P P 5′ 5′ 5′ 5′ 4′

Fig. 2.8. A short hand system of representation of a polynucleotide with four nucleotides.

Molar Ratios of Nitrogen Bases in DNA Molecule

When many samples of DNA were isolated, purified and analysed by various techniques, such as paper chromatography, etc., it was found by **Hotchkiss** and **Chargaff** in 1948 that contrary to Levene's (1920's) tetranucleotide theory which considered DNA as a monotonous polymer having four DNA bases in approximately equal molar proportions, the four nucleotide bases are not necessarily present in DNA in exactly equal proportions. Thus, **Chargaff** reported that the DNA extracted from calf thymus nuclei contains the four bases in the following molar proportions: 28% adenine, 24% guanine, 20% cytosine and 28% thymine. When he analyzed the DNA samples of varied animals, he found that the exact base composition of DNA differs according to its biological source. The relative amounts of purines and pyrimidines in samples of DNA of different living organisms are tabulated as follows :

Table 2.2. Relative amounts of nitrogen bases in different samples of DNA (**Chargaff** and **Davidson**,1955).

S. N.	Source	Adenine	Guanine	Cytosine	Thymine	$\frac{A+T}{G+C}$
1.	Beef sperm	28.7	22.2	22.0	27.2	1.26
2.	Human thymus	30.9	19.9	19.8	29.4	1.52
3.	Human liver	30.3	19.5	19.9	30.3	1.53
4.	Human sperm	30.7	19.3	18.8	31.2	1.62
5.	Hen red cells	28.8	20.5	21.5	29.2	1.38
6.	Rat bone marrow	28.6	21.4	21.5	28.4	1.33
7.	Herring sperm	27.8	22.2	22.6	27.5	1.23
8.	*Paracentrotus lividus* (sea urchin) sperm	32.8	17.7	18.4	32.1	1.85
9.	Salmon	29.7	20.8	20.4	29.1	1.43
10.	Wheat germ	26.5	23.5	23.0	27.0	1.19
11.	Yeast	31.3	18.7	17.1	32.9.	1.79
12.	*Diplococcus pneumoniae*	29.8	20.5	18.0	31.6	1.59
13.	*K-12 Escherichia coli*	26.0	24.9	25.2	23.9	1.00
14.	*Mycobacterium tuberculosis*	15.1	34.9	35.4	14.6	0.42
15.	Bacteriophage T_2	32.5	18.2	16.7	32.6	1.86

These different values of A, G, C, and T in different samples of DNA are suggesting that DNA rather than being a monotonous polymer, carries genetic information in the form of specific nucleotide base sequences.

The equivalence rule. Chargaff, in 1950, formulated the equivalence rule which suggested that despite wide compositional variations exhibited by different types of DNA, the total amount of purines equaled the total amount of pyrimidines (A + G = T + C); the amount of adenine equaled the amount of thymine (A = T) and the amount of guanine equaled the amount of cytosine (G = C). **Chargaff's equivalence rule** has been found to apply almost universally in different organisms (viruses, bacteria, plants and animals). However, DNA isolated from higher plants and animals was in general rich in adenine and thymine (A : T) and relatively poor in guanine and cytosine (G : C) (*e.g.*, AT/GC ratio of DNA of man was 1.40 : 1); whereas DNA isolated from microorganisms (viruses, bacteria and lower plants and animals) was in general rich in guanine and cytosine and relatively poor in adenine and thymine (e.g., AT/GC ratio of DNA of *Mycobacterium tuberculosis* was 0.60 : 1). These differences in AT/GC ratios of microorganism and higher organisms undoubtedly reflect the difference in genetic information carried by these hereditary molecules and also phylogenetic, evolutionary and taxonomical significance of them.

Physical, Molecular or Geometrical Organization of DNA

The first person to give any thought to the three dimensional structure of DNA was **W.T. Astbury** who by his X-ray crystallographic studies of DNA molecule concluded in 1938 that because DNA has high density, so, its polynucleotide was a stack of flat nucleotides, each of which was oriented perpendicularly to the long axis of the molecule and was situated every 3.4 Å along the stack. The X-ray crystallographic studies of Astbury were continued by **Wilkins** and his associates who managed to prepare highly oriented DNA fibres that allowed them to obtain an X-ray diffraction photograph. One of his female associates **Rosalind Franklin** obtained a superior X-ray diffraction photograph of DNA which confirmed Astbury's earlier inference of 3.4 internucleotide distance. Such studies demonstrated that DNA was a helical structure with a diameter of 20 Å and a pitch (one round) of about 34 Å. **Watson** and **Crick** who were already engaged in constructing some suitable model for DNA structure, when observed Franklin's picture of DNA molecule, they immediately utilized that information in constructing a molecular model for DNA. In April 1953, **Watson** and **Crick** published their conclusions about the structure of the DNA in the same issue of **'Nature'**, in which **Wilkins** and his colleagues presented the X-ray evidence for that structure.

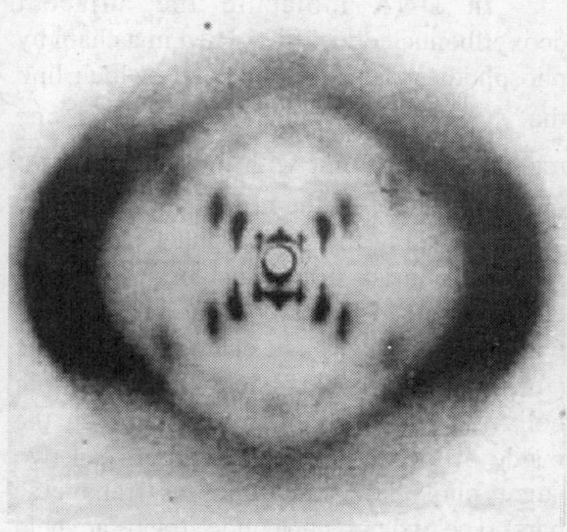

Franklin's original X-ray diffraction photograph.

Considerations of Watson and Crick in the construction of double helical structure of DNA molecule. Watson and **Crick** concluded directly from the X-ray diffraction photograph of DNA taken by **Franklin** that (1) the DNA polynucleotide chain has the form of a regular helix, (2) the helix has a diameter of about 20A° and, (3) the helix makes one complete turn every 34A° along its length, and hence, since the internucleotide distance is 3.4Å, consists a stack of ten nucleotides per turn. Considering the known density of the DNA molecule, **Watson** and **Crick** next concluded that the helix must contain two polynucleotide chains, or two stacks of ten nucleotides each per turn, since the density of a cylinder 20Å in diameter and 34A° long would be too low if it contained but a single stack of ten, and too high if it contained three or more stacks of ten nucleotides each. Before trying to arrange these two polynucleotide chains into a regular helix of the required dimensions, however, **Watson** and **Crick** placed a further restriction on their model– a restriction that derived from their knowledge that DNA is, after all, the genetic material. If DNA is to contain hereditary information as they reasoned, and if that information is inscribed as a specific sequence of the four bases along the polynucleotide chain, then the molecular structure of DNA must be able to accommodate any arbitrary sequence of bases along its polynucleotide chains. Otherwise, the capacity of DNA as an information carrier would be too severely limited. Hence, they felt the need of construction of such a regular helix that though is composed of two polynucleotide chains containing an arbitrary sequence of nucleotide bases every 3.4Å along their length, it would, nevertheless, have a constant diameter of 20A°. Since the dimension of the purine ring is greater than that of the pyrimidine ring, **Watson** and **Crick** hit upon the idea that the two-chain helix could have a constant diameter if there existed a complementary relation between the two nucleotide stacks, so that at every level one stack harbours a purine base and the other a pyrimidine base. Finally, to endow the helix with thermodynamics stability, the structure would have ample opportunities for the formation of hydrogen bonds between amino or hydroxyl-hydrogens and keto-oxygens or immino-nitrogens of the purines and pyrimidine bases. These considerations led them to construct a double helix model for the molecular structure of DNA molecule.

Watson and Crick's Structural Model of DNA

In DNA molecule the adjacent deoxyribonucleotides are joined in a chain by phosphodiester bridges or bonds which link the 5′ carbon of the deoxyribose of one mononucleotide unit with the 3′ carbon of the deoxyribose of the next mononucleotide unit. According to **Watson** and **Crick** DNA molecule consists of two such polynucleotide chains wrapped helically around each other, with the sugar-phosphate chain on the outside (forming ribbon-like backbone of double helix) and purines and pyrimidines on the inside of the helix (projecting between two sugar phosphate backbones as transverse bars). The two polynucleotide strands are held together by hydrogen bonds between specific pairs of purines and pyrimidines.

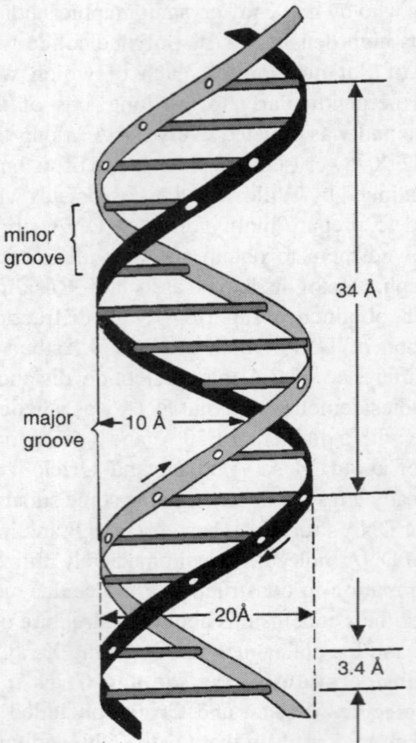

Fig. 2.9. Watson-Crick double helical DNA molecule.

The hydrogen bonds between purines and pyrimidines are such that adenine can bond only to thymine by two hydrogen bonds, and guanine can bond only to cytosine by three hydrogen bonds and no other alternative is possible between them. The specificity of the kind of hydrogen bonds that can be formed assures that for every adenine in one chain there will be thymine in the other. For every guanine in first chain there will be a cytosine in the other and so on. Thus, the two chains are complementary to each other; that is, the sequence of nucleotides in one chain dictates the sequence of nucleotides in the other. The two strands run anti-parallely–that is, have opposite directions. One strand has phosphodiester linkage in $3' \rightarrow 5'$ direction, while other strand has phosphodiester linkage in just reverse or $5' \rightarrow 3'$ direction. Further, both polynucleotides strands remain separated by 20A° distance. The coiling of double helix is right handed and a complete turn occurs every 34 A°. Since each nucleotide occupies 3.4 A° distance along the length of a polynucleotide strand, ten mononucleotides occur per complete turn. (The base pairs are rotated 36° with respect to each adjacent pair). The helix has two external grooves, a deep wide one, called **major**

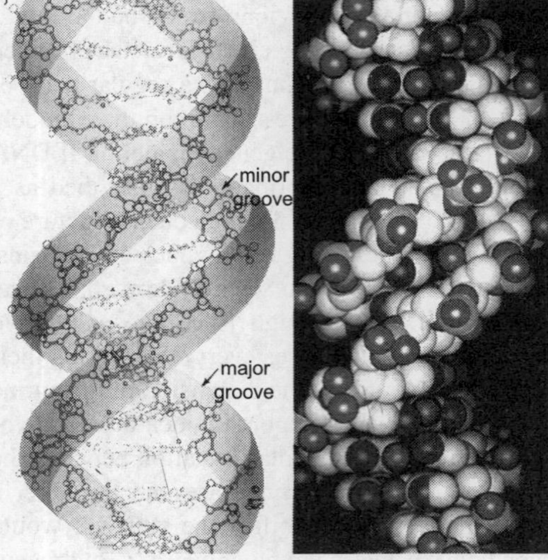

Double helical DNA molecule.

groove and a shallow narrow one, called **minor groove** : both of these grooves are large enough to allow protein molecules to come in contact with the bases.

Polymorphism of DNA Helix (Or Alternative Forms of DNA Double Helix)

For about 20 years after the discovery of the DNA double helix in 1953, DNA was thought to have the same monotonous structure, with exactly 36° of helical twist between its adjacent base pairs (10 nucleotide pairs per helical turn) and a uniform helix geometry. Subsequent experiments have shown that DNA is much more **polymorphic** than expected. Thus, DNA can be of the following types :

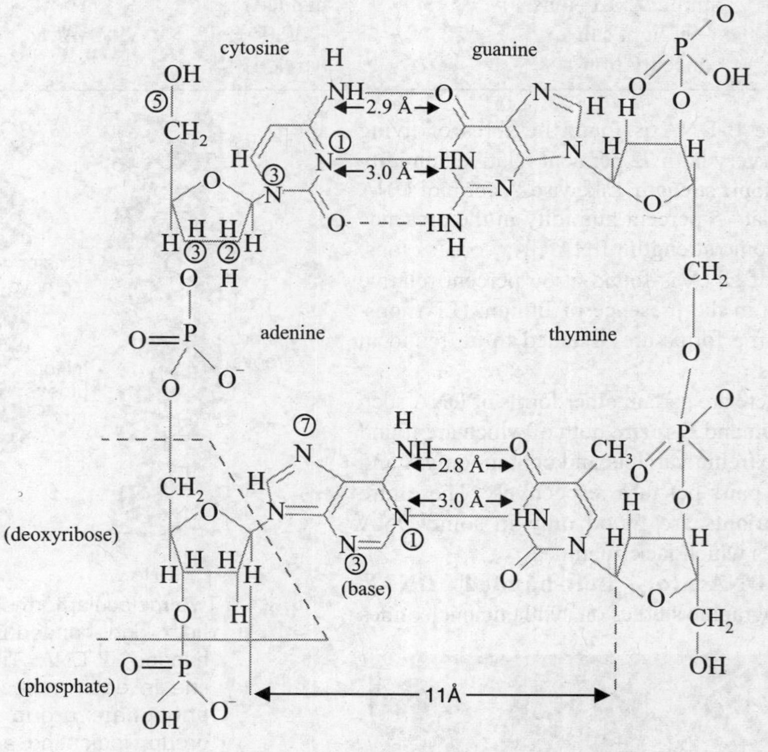

cytosine guanine adenine thymine (deoxyribose) (phosphate) (base) 2.9 Å 3.0 Å 2.8 Å 3.0 Å 11Å

Fig. 2.10. Segments of a DNA molecule.

The above described DNA molecule contains the right handed helical coiling and has been called **B-form** or **B-DNA**. It is a biologically important form of DNA that is commonly and naturally found in most living systems. This double helical structure of DNA is found to exist in other alternative forms (such as A-form and C-form) which differ in features such as (i) number of residues (monomers) per turn ("n") or (ii) the spacing of residues along the helical axis ("h") (Table 2.3). For example, **A - form DNA** (A-DNA) is right handed but less hydrated than the B-form DNA. A-DNA is more compact with 11 base pairs per turn of the helix and it is 23 angstroms in diameter. The bases are tilted more in relation to the axis of the helix than in the B-DNA. The A-DNA may occur under experimental conditions.

Table 2.3. Important features of different forms of DNA double helical structures.

Helix type	Conditions	Base pair per turn	Rotation per bp	Vertical rise per bp	Helical diameter
A	75 per cent relative humidity; Na^+,K^+, Cs^+, ions	11	+ 32.7° (right handed)	2.56 A°	23A°
B	92 per cent relative humidity, low ionic strength	10	+ 36.0° (right handed)	3.38A°	19A°
C	66 per cent relative humidity; Li^+ ions	9.33	+ 38.6° (right handed)	3.32A°	19A°
Z	Very light salt concentration	12	– 30.0° (left handed)	3.71 A°	18A°

The B-DNA is found in fibres of living cells at a very high 92 per cent relative humidity and low ionic strength. Likewise, A-form of DNA is found at 75 percent humidity in the presence of high ionic strength of NA^+, Ka^+ or Cs^+ ions. C-form of DNA is found at 66 percent relative humidity in the presence of lithium (Li^+) ions. These three forms are assumed to be found in all DNAs.

There are certain other forms of DNA such as **D-form** and **E-form**, both of which are found as rare extreme variants and contain only 8 and 7½ base pairs per turn respectively. These rare DNA variants are found only in some DNA molecules which lack guanine.

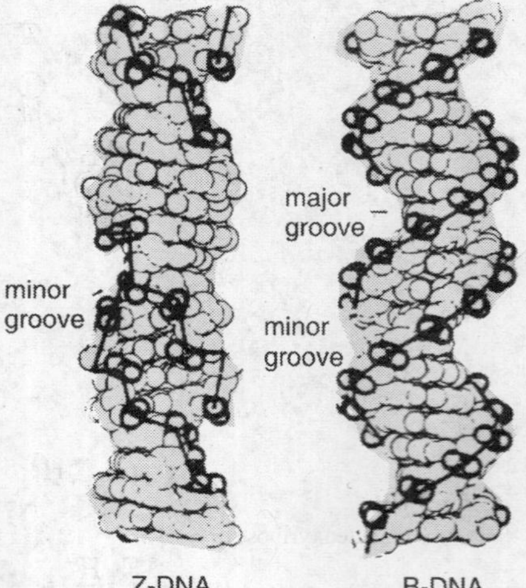

Fig. 2.11. Two molecular forms of DNA double helix, right-handed B-DNA and left-handed Z-DNA. The heavy black line in each molecule goes from phosphate group to phosphate group, indicating a smooth right-handed helix in B-DNA and irregular or zigzag left-handed helix in Z-DNA (after De Robertis and De Robertis Jr., 1987).

Z-DNA (or Left-handed DNA). Crystallographic studies on synthetic nucleotides consisting of alternating purines and pyrimidines such as GCGCGCGC have shown that left-handed DNA can also exist. This DNA is called **Z-DNA** because of its zigzag structure. Z-DNA can also be found in solutions of high-ionic strength, such as 2M NaCI. In Z-form of DNA, the molecule still consists of two anti-parallel chains, but, otherwise, it is quite different from

A DNA model.

the A or B form (Fig. 2.11). The helix of Z-DNA is 18 angstrom in diameter, containing 12 base pairs per turn. The differences and similarities of Z-DNA and B-DNA have been summarized in Table 2.3; some other features are the following :

A. Similarities between Z-DNA and B-DNA

(1) Both are double helical.

(2) In both DNAs, two polynucleotide strands of double helix are antiparallel.

(3) Both forms exhibit G ≡ C pairing.

B. Differences between Z-DNA and B-DNA

(i) Z-DNA has left-handed helical sense, while B-DNA has right-handed helical sense (Fig. 2.12).

(ii) The phosphate backbone of Z-DNA follows a zigzag course, while in B-DNA this backbone is regular.

(iii) In Z-DNA, the adjacent sugar residues have opposite orientation, while in B-DNA they have same orientation. Due to this, repeating unit is a dinucleotide in Z-DNA as against a mononucleotide unit in B-DNA (Fig. 2.13).

(iv) In Z-DNA, one complete helix (*i.e.*, a twist through 360°) has twelve base pairs or six repeating dinucleotide units, while in B-DNA one complete helix has only ten base pairs or 10 repeating units.

(v) The angle of twist (rotation) per repeating unit (dinucleotide) in Z-DNA is 60° than the 36° of mononucleotide in B-DNA.

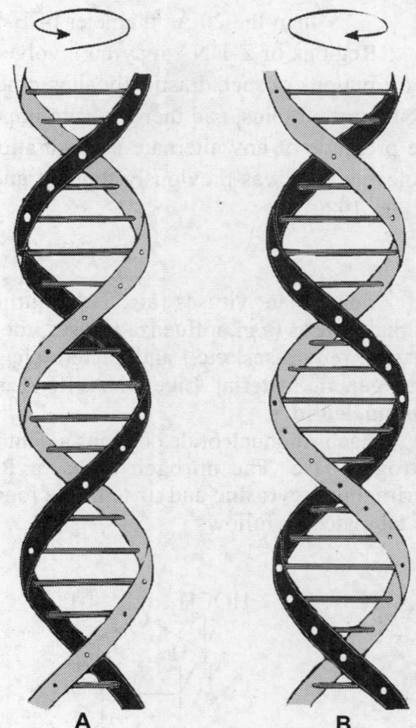

Fig. 2.12. The normal right handed helical sense of B-DNA (A) and rare left handed helical sense of Z-DNA(B).

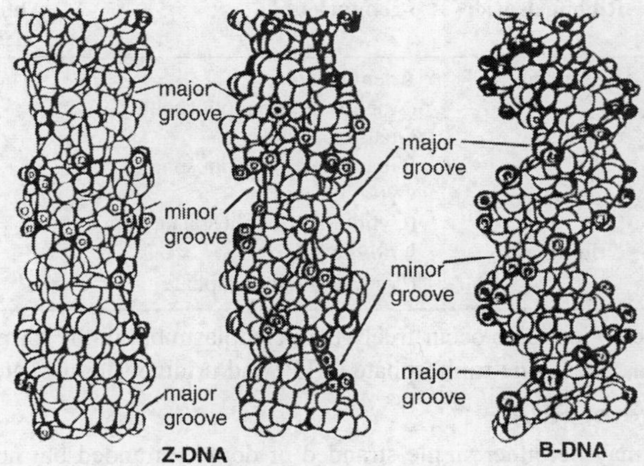

Fig. 2.13. Mode of orientation of adjacent sugar residues in Z-DNA and B-DNA. Z-DNA has opposite orientation resulting in dinucleotide units, while B-DNA has same orientation resulting in mononucleotide units.

(vi) In Z-DNA, one complete turn of helix is 45Å long, while in B-DNA it is 34Å long.

(vii) Since bases get more length spread out in Z-DNA and since the angle of tilt is 60°, they are more closer to the axis. Due to this fact the diameter of Z-DNA molecule is 18 A° than the 20 A° diameter of B-DNA.

Regions of Z-DNA may be involved in gene regulation in the cells of "higher" organisms. Short regions of such drastically altered helical geometry could be specifically recognized by gene regulatory proteins, and thereby have important biological roles (see **Alberts** *et al.*, 1989). Lastly, the presence of any alternate configuration (*e.g.*, Z-DNA) suggests that DNA is a more flexible molecule than was previously thought and that it can adopt in the genome of a variety of forms (**Rich**, 1980).

2.3. RIBONUCLEIC ACID (RNA)

Some plant viruses (*eg.*, TMV, turnip yellow mosaic viruses, wound tumour viruses, etc.,), animal viruses (*e.g.*, influenza viruses, foot and mouth viruses; rous sarcoma viruses, poliomyelitis viruses, reoviruses, etc.) and bacteriophages (*e.g.*, MS_2, etc.) contain ribonucleic acid (RNA) as their genetic material. Like DNA, RNA is polymeric nucleic acid of four monomeric **ribotids** or **ribonucleotides**.

Each ribonucleotide contains a pentose sugar (**D-ribose**); a molecule of phosphate group and nitrogen base. The nitrogen bases of RNA are two purines, **adenine** and **guanine** and two pyrimidines, **cytosine** and **uracil**. The four bases, ribonucleosides and ribonucleotides of RNA can be tabulated as follows :

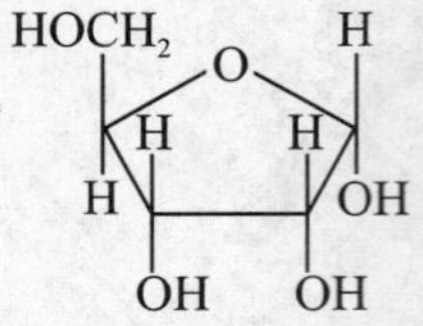

Fig. 2.14. Chemical formula of D-ribose.

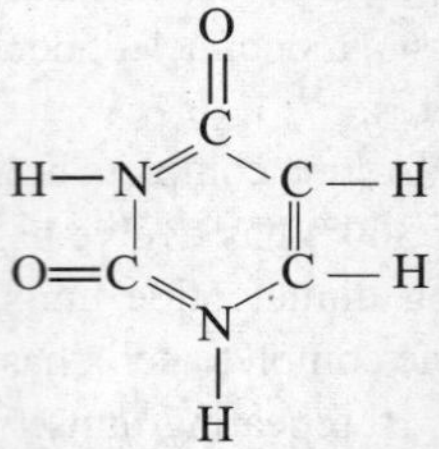

Fig. 2.15. Chemical formula of uracil.

Table 2.4. Four components of RNA.

Base	Ribonucleoside	Ribonucleotide	Abbreviation for ribonucleotide
1. Adenine (A)	Adenosine	Adenylic acid (Adenosine monophosphate)	AMP
2. Guanine (G)	Guanosine	Guanylic acid (Guanosine monophosphate)	GMP
3. Cytosine (C)	Cytidine	Cytidylic acid (Cytidine monophosphate)	CMP
4. Uracil (U)	Uridine	Uridylic acid (Uridine monophosphate)	UMP

The four ribonucleotides also occur freely in nucleoplasm but in the form of triphosphates of ribonucleosides such as adenosine triphosphate (ATP) and uridine triphosphate (UTP).

Molecular Structure of RNA

RNA molecule may be either single stranded or double stranded but not helical like DNA molecule. Single stranded RNA occurs as genetic material in plant viruses (e.g., TMV, TYM), animal viruses (e.g., influenza viruses, foot and mouth viruses, rous sarcoma viruses, poliomyelitis

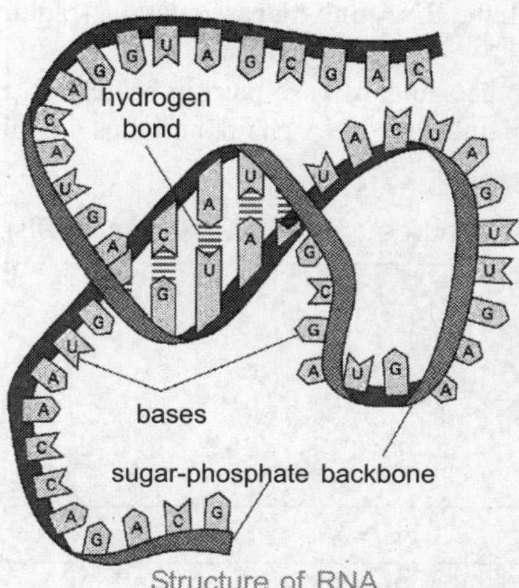

Structure of RNA.

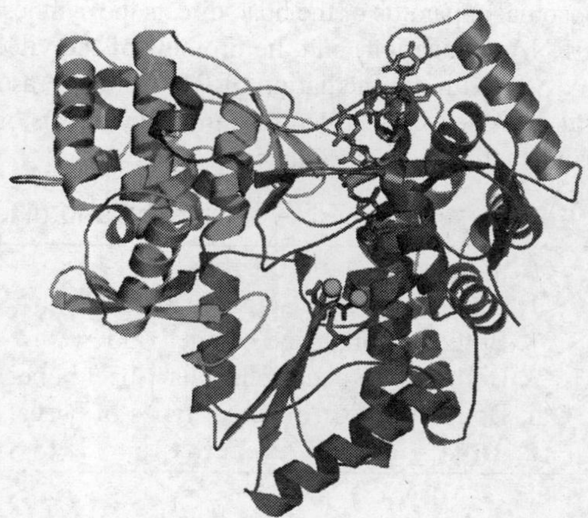

Hepatitis C virus polymerase with single-stranded RNA.

viruses) and bacteriophages (*e.g.*, MS_2). The non-genetic RNAs except tRNA of prokaryotes and eukaryotes, also have single stranded RNA molecules. The double stranded but non-helical RNA occurs as the genetic material in some plant viruses (*e.g.*, reoviruses). The transfer or soluble RNA (tRNA or sRNA) which is non-genetic RNA of prokaryotes and eukaryotes, is double stranded but non-helical structure.

Each strand of RNA is polynucleotidic, that is, made up of many ribonucleotides. In the polynucleotide strand of RNA, the ribose and phosphoric acids of nucleotides remain linked by phosphodiester bonds. The organisms which have only RNA, is called **genetic RNA**. While the organisms which have DNA along with RNA, they use the RNA in carrying the orders of DNA and in them because RNA has no genetic role, so called **non-genetic RNA**. The non-genetic RNA is heterogeneous and includes the following three genera: ribosomal RNA (rRNA), transfer RNA (tRNA) and messenger RNA (mRNA). Each genera of non-genetic RNA has **DNA-dependent replication** (transcription) of itself, that is, it is not self-replicating like DNA and is transcribed by DNA.

A model of double standed RNA.

Replication of Genetic RNA

The genetic RNA of viruses is self-replicating, that is, it can produce its own replica by itself. So, its mode of replication is called **RNA-dependent RNA synthesis**. The genetical research on genetic RNA has revealed the following facts about it.

(1) The viral RNA functions directly as a messenger RNA which, in association with the

ribosomal apparatus of the host, directs the synthesis of both the **RNA polymerase enzyme** (required for RNA replication) and the proteins of the viral coat.

(2) With the mediation of RNA polymerase and on the standard base-pairing principles, the viral RNA serves as a template in the synthesis of a complementary RNA chain, and thus a double stranded structure is produced.

Table 2.5. DNA - RNA similarities and differences. (Source : **Sylvia S.Mader,** 1998).

A. DNA - RNA similarities	
1. Both are nucleic acids.	
2. Both are composed of nucleotides.	
3. Both have a sugar-phosphate backbone.	
4. Both have four different types of bases.	
B. DNA - RNA differences	
DNA	**RNA**
1. Found in the nucleus	1. Found in nucleus and cytoplasm.
2. The genetic material	2. Helper to DNA.
3. Sugar is deoxyribose.	3. Sugar is ribose.
4. Bases are A,T,C,G..	4. Bases are A,U,C,G.
5. Double stranded.	5. Single stranded.
6. Is transcribed. (to give mRNA).	6. Is translated.(to give proteins).

Questions And Problems

1. Discuss, in brief, the structure of deoxyribonucleic acid and compare it with that of ribonucleic acid.
2. Give an account of Watson and Crick's double-stranded model of DNA molecule.
3. Do the two strands of DNA helix carry the same genetic information ? Explain.
4. What kind of evidence indicates (1) that DNA can reproduce itself and (2) that nucleotides occur in matched pairs in DNA molecule ?
5. Write short notes on the following :
   (*i*) Z-DNA, (*ii*) B-DNA,
   (*iii*) Tetranucleotide hypothesis, (*iv*) Left-handed DNA versus right-handed DNA.
6. Assume the following base sequence was found in a 20 base DNA strand :
   3′ ATT CGA CCT TAT TAC TGC AC 5′
   (*a*) What would be the first 5 bases in the 3′ end of the complementary strand ?
   (*b*) What would be the 10 bases of the 5′ end of the complementary strand ?
   (*c*) Assuming the presence of complementary strands, what is the per cent composition of the polymer with respect to AT base pairs and with respect to GC base pairs.
7. Analysis of four double-stranded DNA samples yielded the following information:
   1.15% cytosine ; 2.12% guanine ; 3.35% thymine ; 4.28% adenine.
   (*a*) What would be the percentage of the other bases in each sample ?
   (*b*) Could any of these samples have been obtained from the same organism ? If so, which ones ?
8. Four samples of nucleic acid were analyzed for the proportion of the different bases present, with the following results :
   (1) A = 30%, C = 30%, G = 20%, T = 20%.
   (2) A = 27.5%, C = 22.5%, G = 22.5%, T = 27.5%.
   (3) A = 18%, C = 32%, U = 32%, G = 18%.
   (4) A = 18%, C = 32%, U = 18%, G = 32%.

Which of these samples were DNA and which were RNA ? Which were double-stranded ?

9. Ratio of the bases present in different samples of nucleic acid yielded the following results :
 (1) $(A + C)/(T + G) = 1$, (2) $(A + C)/(U + G) = 0.8$,
 (3) $(A + G)/(T + C) = 1.5$.
 Which were RNA and which were DNA ? Which were single and double-stranded ?

10. If one DNA sample had a melting temperature of 85.5°C and another showed a melting temperature of 88°C. What might you conclude concerning the base composition of the two samples ?

11. Assume an average-sized gene consisting of a linear sequence of 1000 bases and there were 1000 genes in a bacterial chromosome.
 (*a*) How many bases would such a chromosome contain in each strand of the double helix ?
 (*b*) If 10 nucleotides = 34Å, how long would this chromosome would be in millimeters ?

12. (*a*) What background material did Watson and Crick have available for developing a model of DNA ?
 (*b*) What was their contribution to the building of the model ?

13. (*a*) Why was a double helix chosen for the basic pattern of the molecule ?
 (*b*) Why were hydrogen bonds placed in model to connect the base ?

14. What are differences between DNA and RNA ?

15. RNA was extracted from TMV (tobacco mosaic virus) particles and found to contain 20 per cent cytosine. Using this information, is it possible to predict what percentage of the bases in TMV are adenine ? If so, what percentage ? If not, why not ?

Answers to Problems

6. (*a*) Bases at the 3′ end of the opposite strand will be complementary to those at the end 5' end of the polymer given. Therefore, the first 5 bases will be 3′ GTGCA.
 (*b*) 5′ TAA GCT GGAA
 (*c*) A-T = 60 per cent, G-C = 40 per cent.

7. (*a*) 1.15% C, 15% G, 35% T, 35% A.
 2.12% G, 12% C, 38% T, 38% A.
 3.35% T, 35% A, 15% C, 15% G.
 4.28% A, 28% T, 22% G, 15% C.
 (*b*) Yes, samples 1 and 3.

9. (1) Double-stranded DNA, (2) Single-stranded RNA, (3) Single-stranded DNA.

10. The second sample has a higher G-C content.

11. (*a*) 1 million, (*b*) 3.4 mm.

12. (*a*) The ladderlike pattern was known from X-ray diffraction studies. Chemical analyses had shown that 1:1 relationship existed between the organic bases adenine and thymine and between cytosine and guanine. Physical data concerning the length of each spiral and the staking of bases were also available.
 (*b*) Watson and Crick developed the model of a double helix, with the rigid strands of sugar and phosphorus forming spirals around an axis, and hydrogen bonds connecting the complementary bases in base-pairs.

13. (*a*) A multistranded, spiral structure was suggested by the X-ray diffraction patterns. A double-stranded helix with specific base-pairing fits the 1:1 stoichiometry observed for A : T and G : C in DNA (*stoichiometry* is a branch of chemistry that treats of the proportions of elements or compounds involved in reactions, and the methods of calculating them).
 (*b*) Use of the known hydrogen-bonding potential of the bases provided a means of holding the two complementary strands in a stable configuration in such a double helix.

14. DNA has one atom less of oxygen than RNA in the sugar part of the molecule. In DNA, thymine replaces the uracil that is present in RNA. (In certain bacteriophages, DNA containing a uracil is present). DNA is most frequently double-stranded, but bacteriophages such as $\phi \times 174$ contain single-stranded DNA. RNA is most frequently single-stranded. Some viruses, such as reoviruses, however, contain double-stranded RNA chromosomes.
15. No. TMV RNA is single stranded. Thus, the base-pair stoichiometry of DNA does not apply.

MULTIPLE CHOICE QUESTIONS

Choose the correct answer from the four alternatives given.

1. In a sample, DNA is found to have the base composition (mole ratio) of adenine = 40, T = 22, G = 21 and cytosine = 17. It shows that
   (*a*) DNA is circular duplex
   (*b*) DNA is linear duplex
   (*c*) DNA is single stranded
   (*d*) DNA has melting point
2. In the double helix model of DNA, how far is each base pair from the next base pair ?
   (*a*) 3.4 nm (*b*) 0.34 nm
   (*c*) 2.0 nm (*d*) 34 nm
3. The scientist who developed cytochemical technique for identification DNA was
   (*a*) McClintok
   (*b*) Abbe
   (*c*) Fuelgen and Rossenbeck
   (*d*) Waldeyer
4. What is common point of similarity between DNA and RNA ?
   (*a*) both are double stranded
   (*b*) both have identical sugar molecules
   (*c*) both have identical pyrimidine bases
   (*d*) both are polymers of nucleotides
5. DNA strands are antiparallel because of
   (*a*) hydrogen bonds
   (*b*) phosphodiester bonds
   (*c*) disulphide bonds
   (*d*) glycosidic bonds
6. Chargaff's rule is applicable to
   (*a*) single stranded RNA
   (*b*) single stranded DNA and RNA
   (*c*) single stranded DNA
   (*d*) double stranded DNA
7. The diameter of Z-DNA molecule is
   (*a*) 18Å (*b*) 22Å
   (*c*) 45Å (*d*) 34Å
8. What is the type of coiling in DNA ?
   (*a*) right handed (*b*) left handed
   (*c*) opposite (*d*) zig-zag
9. Diameter of DNA helix is
   (*a*) 20Å (*b*) 10Å
   (*c*) 30Å (*d*) 40Å
10. Which ratio is constant for DNA ?
   (*a*) A + G / T + C
   (*b*) A + T / G + C
   (*c*) A + C / U + G
   (*d*) A + U / C + G

ANSWERS

1. (*c*)	2. (*b*)	3. (*c*)	4. (*d*)	5. (*b*)
6. (*d*)	7. (*a*)	8. (*a*)	9. (*a*)	10. (*a*).

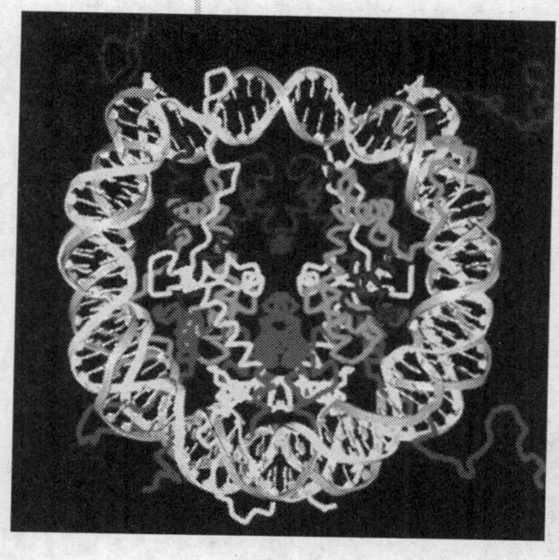

CHAPTER

3

Organization of Genetic Material

Chapter Contents

3.1. CHROMATIN

When chromatin is isolated from interphase nuclei, the individual chromosomes are not recognizeable. Instead one observes an irregular aggregate of nucleoprotein. Chemical analysis of isolated chromatin shows that it consists primarily of DNA and proteins and lesser amounts of RNA (see Table 3.1). These proteins are of two major classes:

(1) **Basic proteins** (positively charged at neutral pH) called **histones;** and (2) a heterogenous, largely acidic (negatively charged at neutral pH) group of proteins collectively referred to as **non-histone chromosomal proteins**.

Histones

Histones play a major structural role in chromatin. They are present in the chromatin of all higher eukaryotes in amounts equivalent to the amounts of DNA (weight to weight). Indeed histones are present in such enormous quantities in the cell (about 60 million molecules of each type per human cell) that their total mass in chromatin is about equal to that of DNA (see **Albert** *et al.,* 2002). The histones of all the higher plants and animals consists of five different major proteins. These five major histones, called **H1, H2a, H2b, H3** and **H4** are present in almost all cell types. The five histones are present in molar ratios of approximately 1H1:2H2a:2H2b:2H3:2H4. They are specifically complexed with DNA to produce

Table 3.1. The constituents of calf thymus chromatin (after **Tamarin,** 2002).

	Constituent	Relative Weight
1.	DNA	100
2.	Histone proteins	114
3.	Nonhistone proteins	33
4.	RNA	1

* Weight relative to 100 units of DNA.

the basic structural subunits of chromatin, small (approximately 110 A° in diameter by 60 A° high) ellipsoidal "beads" called **nucleosomes** (Fig. 3.1)by **P.Oudet** *et al.* (1975).

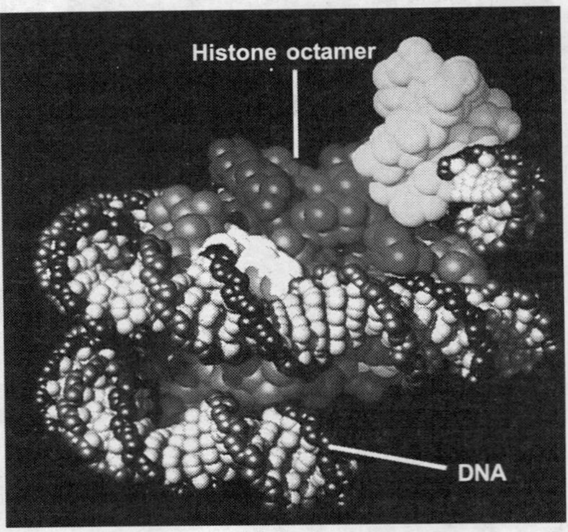

Histones.

The histones have been highly conserved during evolution, four of the five types histones being very similar in all higher eukaryotes. Indeed, the amino acid sequence of histone 4 from a pea and a cow differ at only at 2 of the 102 positions. This strong *evolutionary conservation* suggests that a change in any position is deleterious to the cell. This suggestion has been tested directly in yeast cells, in which it is possible to mutate a given histone gene *in vitro* and introduce it into the yeast genome in place of the normal gene. As might be expected, most changes in histone sequences are lethal; the few that are not lethal cause changes in the normal pattern of gene expression, as well as, other abnormalities.

The histones are basic because they contain 20 – 30 percent **arginine** and **lysine**, two positively charged amino acids (Table 3.2). The exposed – NH_3^+ group of arginine and lysine allow histones to act as **polycations**. This is important in their interaction with DNA, which is **polyanionic**, because of the negatively charged phosphate groups.

Table 3.2. Composition of histones (Source: **Tamarin,** 2002).

Fraction	Class	Number of amino acids	Percentage of basic amino acids
1. H1	Very lysine rich	213	30
2. H2A	Lysine, arginine rich	129	23
3. H2B	Moderately lysine rich	125	24
4. H3	Arginine rich	135	24
5. H4	Arginine, glycine rich	102	27

The remarkable constancy of histones H2A, H2B, H3 and H4 in all cell types of an organism and even between widely divergent species is consistent with the idea that they are important in chromatin structure ("DNA packaging" and are only *nonspecifically involved in the regulation of gene.* In contrast, the nonhistone protein fraction of chromatin consists of a large number of very heterogeneous proteins. The composition of the nonhistone chromosomal protein fraction varies widely among different cell types of the same organism. Thus, the nonhistone chromosomal proteins are likely candidate for roles in **the regulation of expression of specific genes or sets of genes.**

DNA Arrangement in Eukaryotic Chromosome: the Unineme Model

How is the 1 - 20 cm (10^4 to $2 \times 10^5 \mu m$) of DNA, which is present in an average eukaryotic chromosome, arranged in the highly condensed mitotic and meiotic structures that are seen with the light microscope? Are there many DNA molecules that run parallel through out the chromosome (the **"multineme"** or **"multistrand" model**), or is there just one DNA double helix extending from one end of the chromosome to the other end(the **"unineme"** or **"single – strand" Model**)? (*Note*: Here "strand" refers to the DNA double helix, not the single-strand of the DNA). Are there many DNA molecules joined end-to-end or arranged in some other fashion in the chromosome, or does one giant, continuous molecule of DNA extend from one end to the other in a highly coiled and folded form? The evidence supporting the unineme model of chromosome structure is now overwhelming. Thus, according to current view, each eukaryotic chromosome appears to contain a single, giant molecule of DNA that extends from one end through the centromere all the way to the other end of the chromosome.

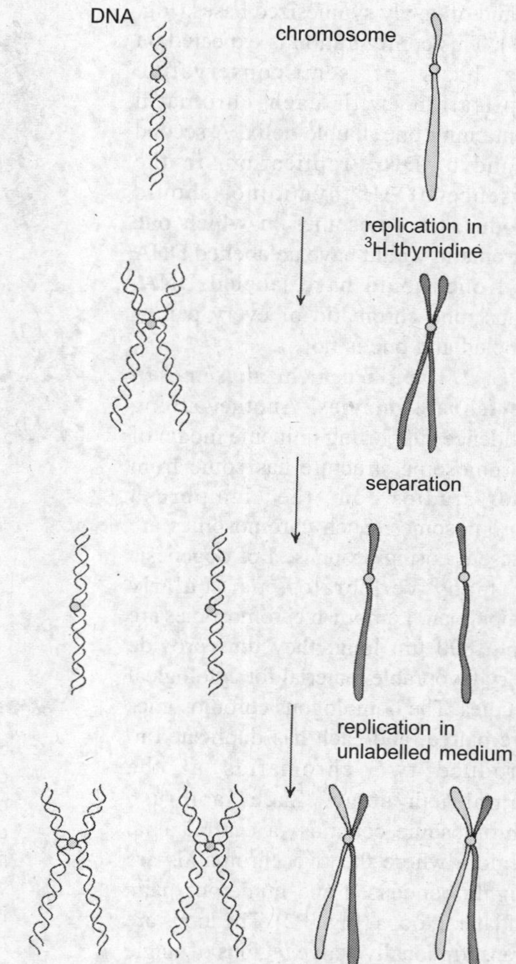

Fig. 3.1. Radioactive labeling of a uninemic eukaryotic chromosome following semiconservative replication. Replication occurs first in the presence of 3H – thymidine and then in its absence. After the second round of replication, one chromatid of each chromosome is labeled, whereas the other is not, confirming that there is only one DNA molecule per chromatid and that the chromosome is thus uninemic (after Tamarin, 2002).

Evidence in Support of Unineme Model

Evidence that the eukaryotic chromosome is **uninemic** comes from several sources as follows:

1. Radioactive experiments. The best data are provided by radioactive labeling studies, first done by **J.Taylor** and his colleagues in 1957. If a eukaryote is allowed to undergo one DNA replication in the presence of **tritiated (3H- 1) thymidine**, each of the daughter chromatids would be expected to contain a double helix with one unlabeled DNA template strand and one labeled strand of newly synthesized bases (Fig. 3.1). This configuration is expected on the basis of semi-conservative replication, with each chromatid containing one double helix. A second round of DNA replication, in the absence of 3H-thymidine, should produce chromosomes in which one chromatid would have unlabelled DNA and one would have labeled DNA. Thus, one chromatid of every pair is labeled and one is not.

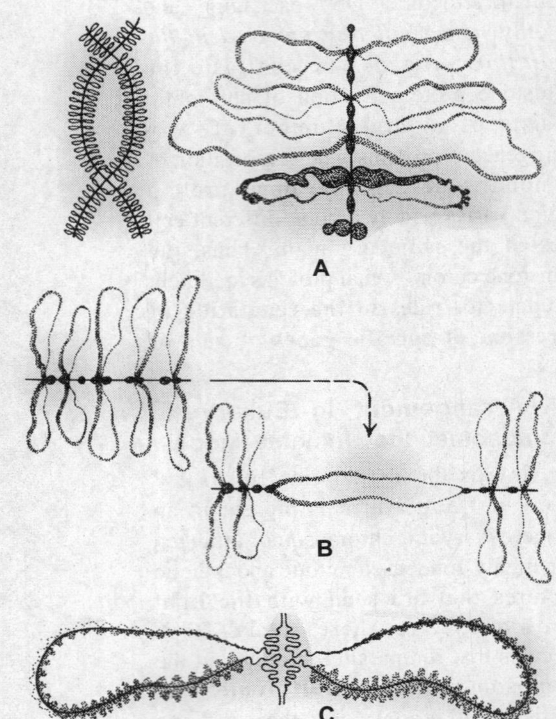

Fig. 3.2. Schematic representation of the structure of the giant lampbrush chromosomes of amphibian oocytes. A – semidiagrammatic view of the central chromosome axis with pairs of morphologically identical loops. Each chromosome has already duplicated in interphase. Each homologous chromosome thus contains two chromatids B – A portion of a lampbrush chromosome that has been stretched during preparation, revealing continuity of the loops and the central axis. C – A single pair of loops displaying the continuity of the single DNA molecule that extends through the loop and axial regions of each chromatid. The matrix is primarily nascent RNA that is synthesized on the extended DNA in the loop region (after Gardner *et al.*, 2001).

2. Lampbrush chromosomes in vertebrate oocytes. Another strong evidence supporting unineme model of chromosome structure has come from the studies on the lampbrush chromosomes. Such chromosomes are present during prophase I of oogenesis in many vertebrates, particularly amphibian. Lambrush chromosomes are upto 800 μm long; they thus provide very favourable material for cytological studies. The homologous chromosomes are paired, and each has duplicated to produce two chromatids at the lampbrush stage. Each lambrush chromosome contains a central axial region, where the two chromatids are highly condensed and numerous pairs of lateral loops (Fig. 3.2). The loops are transcriptionally active regions of single chromatids. The integrity of both the central axis and the lateral loops is dependent on DNA. Since treatment of both of these with **DNase enzyme** fragments them. In contrast, treatment with **RNase enzyme** or **protease enzyme** removes surrounding matrix material, but does not destroy the continuity of either the axis or the loops. Electron microscopy of RNase – and protease – treated lampbrush chromosomes reveals a central filament of just over 20 A° in diameter in the lateral loops. Since each loop is a segment of one chromatid, and since the diameter of a DNA double helix is 20A°, these lampbrush chromosomes must be unineme structures.

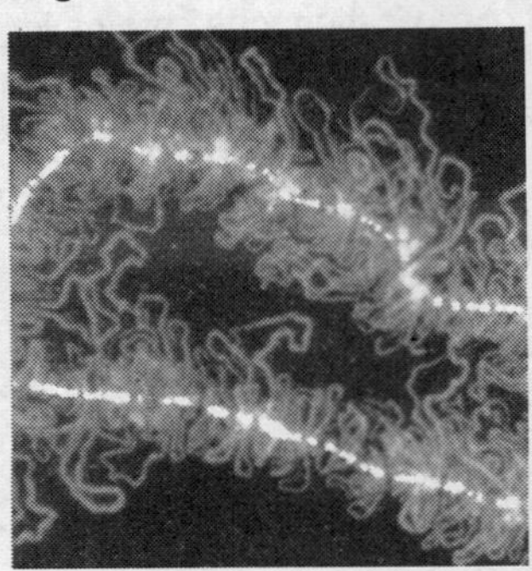

Lambrush chromosome.

Above conclusion is also supported by studies on the **kinetics of nuclear digestion** of lampbrush chromosomes. That is, the kinetics observed are those expected if a single double helix of DNA is the central filament in the loops. The axial region then contains two DNA molecules, one from each of the two tightly paired chromatids.

3. Viscoelsotometric evidence. In 1974, biophysicists such as **R.Kavenoff, L.Klotf** and **B. Zimm,** have used a technique called **viscoelastometry** (a procedure for analyzing viscosity parameters of molecules in solution) to estimate the sizes of DNA molecules from eukaryotic chromosomes. When DNA molecules in solution are exposed to a driving force (*e.g.,* by rotating a cylinder in the solution) they are stretched into an extended conformation. When the driving force is removed (*e.g.,* by stopping the rotation of the cylinder), the molecules will return to their lowest energy or relaxed state.

In the solution being analysed, the *recoil time* is a function of the size of DNA molecule. The recoil decays exponentially over long time intervals, and the time constant, called the **retardation time (T)**, is a sensitive function of the size of the *largest DNA molecules* in the solution. Thus, even if some of the DNA molecules are broken by shearing forces, the viscoelastometric procedure will permit one to estimate the size of the largest DNA molecules present. For minimizing the chance of shearing of the molecules when they are released from the cells, the cells can be lysed right in the chamber of the viscoelastometer.

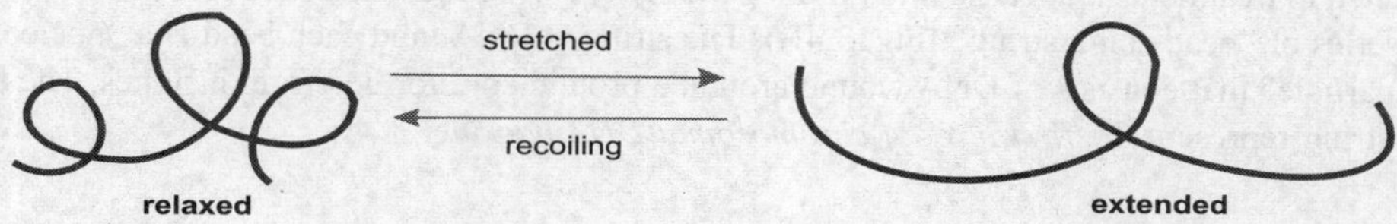

Fig. 3.3. Relaxed and extended state of a DNA molecule.

The largest chromosome in *Drosophila melanogaster* cells were estimated to have a mass of 4.1×10^{10} daltons by viscoelastometry. The largest chromosomes of *Drosophila* have been shown to contain about 4.3×10^{10} daltons of DNA by direct **microspectrophotometric** analysis of metaphase chromosome. Thus, the viscoelastometric estimate of the size of the largest DNA molecules in *D. melanogaster* nuclei correlated almost exactly with the total amount of DNA present in the largest chromosome.

4. Autoradiographic evidence. Researchers have also used the technique of **autoradiography** to attempt to detect chromosome size DNA molecules from eukaryotic cells in the same way that **Cairns** was able to visualise the intact *E.coli* chromosome. Similar to viscoelastometry, autoradiography permits one to lyse cells very gently so as to minimize shearing forces. DNA molecules from the lysed cells are permitted to diffuse onto a membrane that is used to pick up the DNA molecules and expose them to the β-particle – sentitive emulsion. Thus, breakage of molecules should not be a problem. The difficulty with autoradiographic examination of DNA molecules that are very large is that it is almost impossible to get all segments of the molecule sufficiently spread out on the membrane with no tangles or overlaps so that the entire graphic analysis of DNA molecules from *Drosophila* have been successful, and the results of these studies also support the concept of chromosome size DNA molecules (**Kavenoff** *et al.,* 1973).

Packaging the Gaint DNA Molecules into Chromosomes

1. DNA compaction ratio. All eukaryotic organisms have elaborate ways of packaging DNA into chromosomes. For example, human chromosome 22 contains about 48 million nucleotide pairs. Stretched out end to end, its DNA extend about 1.5 cm. yet, when it exists as a mitotic chromosome, chromosome 22 measures only about 2 μm in length, giving end-to-end **compaction ratio** of nearly 10,000-fold. This remarkable feat of compression is performed by proteins that successively coil and fold the DNA into higher and higher levels of organization. Although less condensed than mitotic chromosomes, the DNA of interphase chromosomes, called **chromatin,** is still tightly packed, with an overall compaction ratio of approximately 1000 – fold.

2. Nucleosomes. Chromosome structure is **dynamic**: not only do chromosomes globally condense in accord with the cell cycle, but different regions of the *interphase chromosomes condense and decondense* as the cells gain access to specific DNA sequences for gene expression, DNA repair and DNA replication. The packaging of chromosomes must therefore be accomplished in a way that allows rapid localized, on demand access to the DNA.

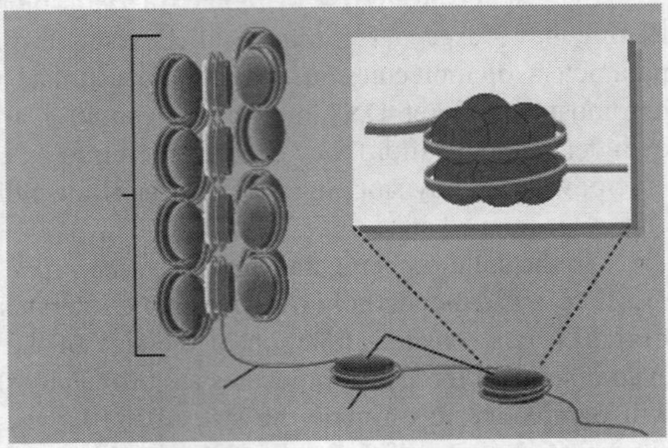

Nucleosomes.

When interphase nuclei are broken open very gently and their contents examined under the electron microscope, most of the chromatin is in the form of a fiber with a diameter of about 30nm (Fig. 3.4A). If this chromatin is subjected to treatments that cause it to unfold partially, it can be seen under the electron microscope as a series of "beads on a string" (Fig. 3.4B). The string is DNA, and each bead is a "nucleosome core particle" that consists of DNA wound around a protein core formed from histones. The beads on a string represent the *first level of chromosomal DNA packing.*

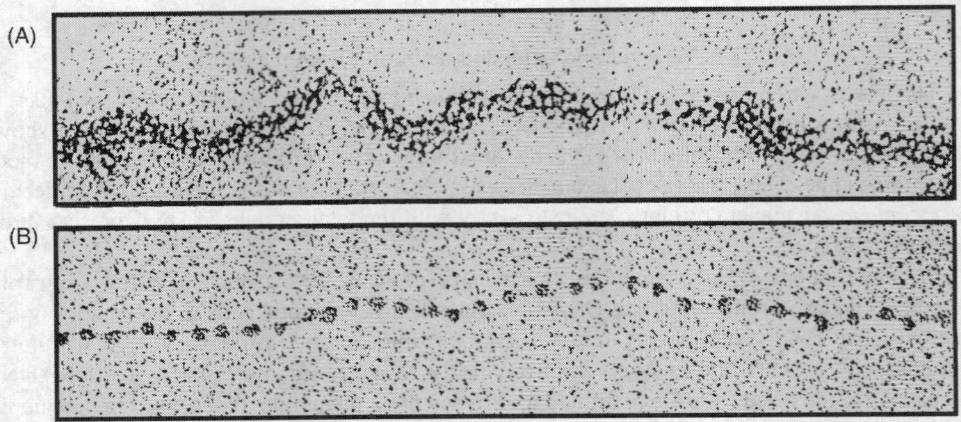

Fig. 3.4. Nucleosomes as seen in the electron microscope. A – Chromatin isolated directly from an interphase nucleus appears in the electron microscope as a thread 30 nm thick. B – This electron micrograph shows a length of chromatin that has been experimentally unpacked or decondensed, after isolation to show the nucleosomes.

The structural organization of nucleosomes was determined after first isolating them from unfolded chromatin by digestion with particular enzymes (called **micrococcal nuclease**) that break down DNA by cutting between the nucleosome core particles, the **linker DNA**, is degraded. Each individual nucleosome core particle consists of a octamer or complex of eight histone proteins – two molecules each of histones H2A, H2B, H3 and H4 and double-stranded DNA (called **core DNA**) that is 146 nucleotide pair long. The histone octamer forms a protein core (**spool** according to **Tamarin,** 2002) around which the double-stranded DNA is bound (Fig. 3.5).

Each nucleosome core particle is separated next by a region of linker DNA, which can vary in length from a few nucleotide pairs upto about 80 (the term **nucleosome** technically refers to a nucleosome core particle plus one of its adjacent DNA linker, but it is often used synonymously with nucleosome core particle; see **Alberts** *et al.*, 2002).

On an average, therefore, nucleosome repeats at intervals of about 200 nucleotide pairs. For example, a diploid human cell with 6.4×10^9 nucleotide pairs contains approximately 30 million nucleosomes. The formation of nucleosomes converts a DNA molecule into a **chromatin thread** about one-third of its initial length, and this provides the first level of DNA packing.

(*i*) **Ultrastructure of nucleosome core particle.** The high – resolution structure of a nucleosome core particle, solved in 1997 by **K.Luger** *et al.*, revealed a disc-shaped histone core around which the DNA was tightly wrapped 1.65 turns in a left-handed coil (Fig. 3.6). All four of the histones that make up the core of the nucleosome are relatively small proteins (102 – 135 aminoacids; see Table 3.2) and they share a structural motif, called **histone fold**, formed from three α helices connected by two loops (Fig. 3.7). In assembling a nucleosome, histone folds first bind to each other to form H3 – H4 and H2A – H2B dimers, and the H3 – H4 dimers combine to form tetramers.

An H3 – H4 tetramer then further combines with two H2A – H2B dimers to form the compact octamer core, around which the DNA is wound (Fig. 3.8 and Fig. 3.9)

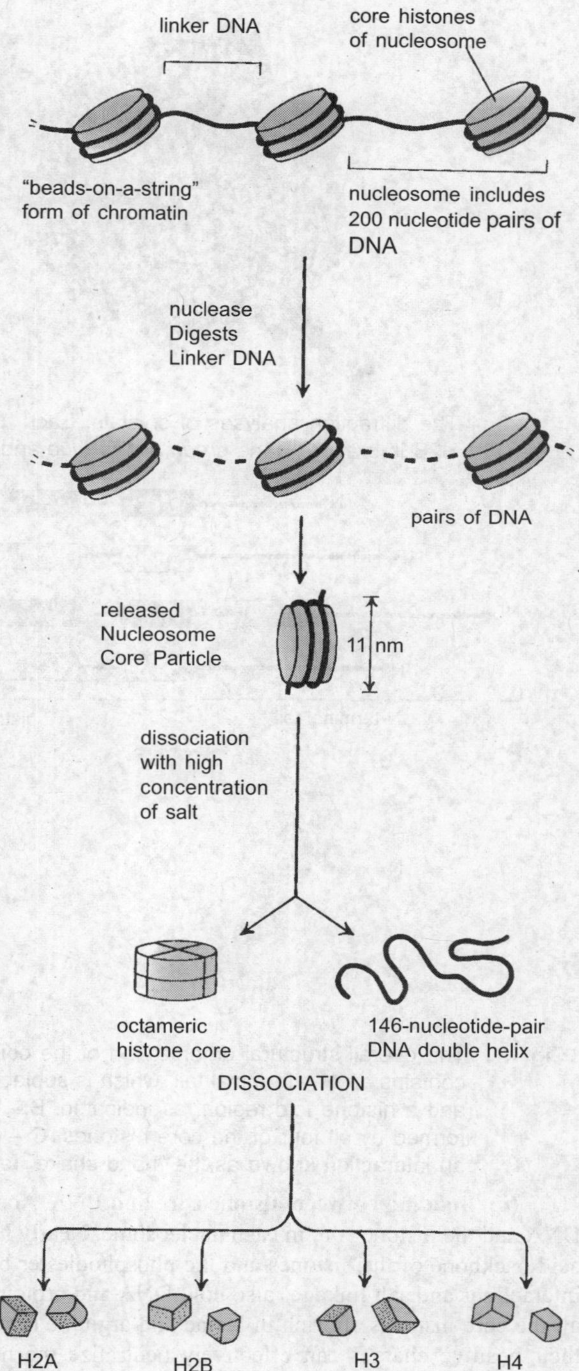

Fig. 3.5. Structural organization of the nucleosome (after Alberts *et al.*, 2002).

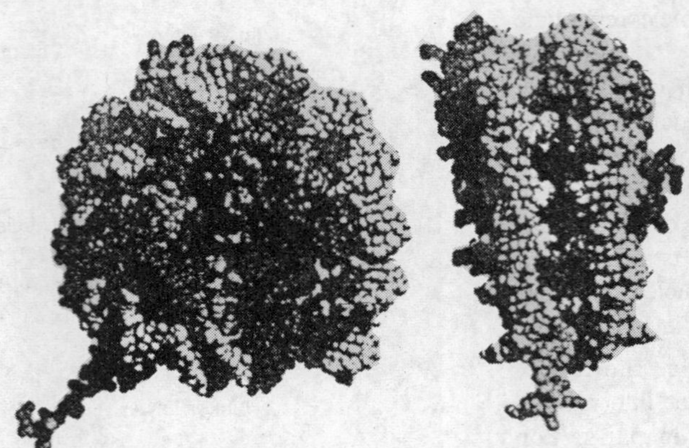

Fig. 3.6. The diffraction analyses of crystals. Each histone is coloured as follows: H2A is *red*, H2B is *orange,* H3 is *green,* H4 is *blue* and DNA is *light gray.*

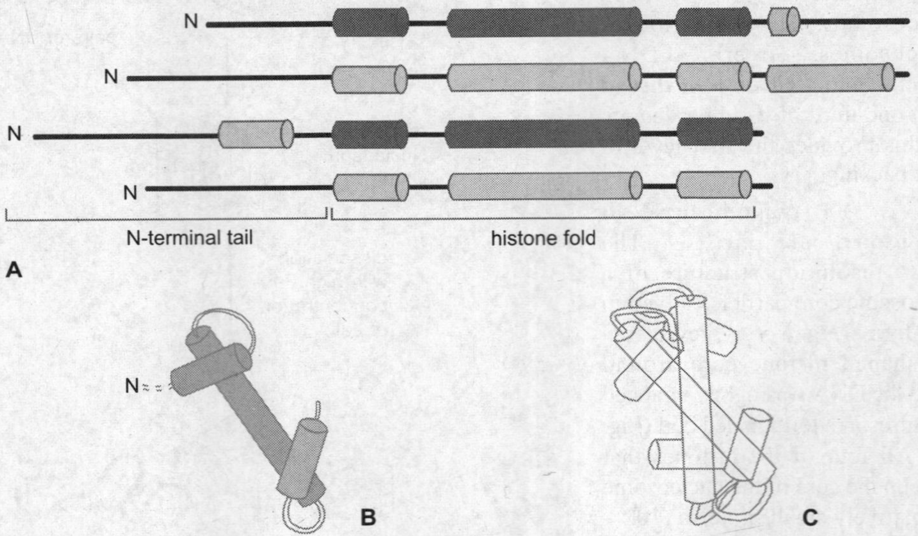

Fig.3.7. The overall structural organization of the core histones. A – Each of the core histones contains an N – terminal tail, which is subject to several forms of covalent modification, and a histone fold region, as indicated. B – The structure of the histone fold, which is formed by all four of the core histones. C – Histones 2A and 2B form a dimer through an interaction known as the "hand shake" (after Alberts *et al.,* 2002).

(ii) **Bonding between histone core and DNA.** About 142 hydrogen bonds are formed between DNA and the histone core in each nucleosome. Nearly half of these bonds form between the amino acid backbone of the histones and the phosphodiester backbone of DNA. Numerous hydrophobic interactions and salt linkages also hold DNA and protein together in the nucleosome. For example, all the core histones are rich in lysine and arginine (two amino acids with basic side chains), and their positive charges can effectively neutralize the negatively charged DNA backbone. These numerous interactions explain in part why DNA of virtually any sequence can be bound on a histone octamer core. The path of the DNA around the histone core is not smooth; rather, several kinks (or bends) are seen in the DNA; this is expected from the rough (or nonuniform) surface of histone core.

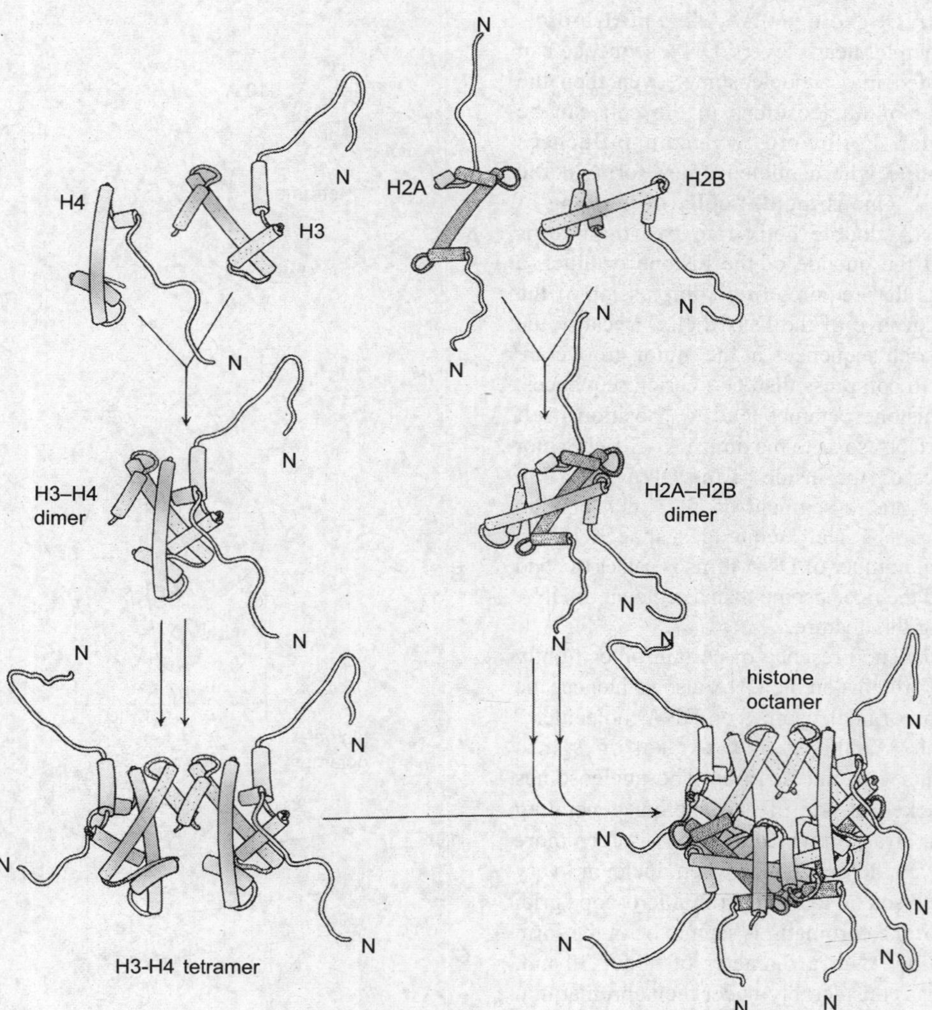

Fig. 3.8. The assembly of a histone octamer. The histone H3 – H4 dimer and the H2A – H2B dimer are formed from the hand shake interaction. An H3 – H4 tetramer forms the scaffold of the octamer onto which two H2A – H2B dimers are added, to complete the assembly. Note that all eight N – terminal tails of the histone protrude from the disc – shaped core structure (after Alberts *et al.,* 2002).

Further, each of the core histones has a long N-terminal amino acid "tail" which extends out from the DNA - histone core (Fig. 3.8). These histone tails are subject to several different types of covalent modifications, which control many aspects of chromatin structure.

(*iii*) Histone variants in sea urchin. Despite the high evolutionary conservation of the core histones, many eukaryotic organinsms also produce specialized variants of core histones that differ in amino acid sequences from the main ones. For example, the sea urchin has five histone **H2A variants**, each of which is expressed at a different time during development. It is believed that nucleosomes that have incorporated these variant histones differ in stability from regular nucleosomes, and they may be particularly well suited for the high rates of DNA transcription and DNA replication that occur during these early stages of development.

(*iv*) **Bending of DNA in a nucleosome.** In principle, nearly every DNA sequence can be folded into a nucleosome, even then the spacing of nucleosomes in the cell can be irregular. Following *two* main influences determine where nucleosomes form in the DNA: 1. One is the **difficulty of bending** of the DNA double helix into two tight turns around the outside of the histone octamer, a process that requires firm compression of the minor groove of the DNA helix. Because the A – T rich sequences in the minor groove are easier to compress than G – C rich sequences; each histone octamer tend to "position itself on the DNA so as to maximize A – T rich minor grooves on the inside of the DNA coil (Fig. 3.10). Thus, a segment of DNA that contain short A – T rich sequences spaced by an integral number of DNA turns is easier to bend around the nucleosome than a segment of DNA lacking this feature.

2. The presence of certain other tightly bound proteins on the DNA also influences the position of nucleosomes on DNA molecule.

3. Packaging of nucleosomes into compact chromatin fiber. The nucleosomes are packed on top of one another generating regular arrays in which the DNA is even more highly condensed. Thus, when nuclei are very gently lysed on to an electron microscope grid, most of the chromatin is seen to be in the form of a fiber with a diameter of about 30 nm, which is considerably wider then chromatin in the "beads on a string" form (Fig. 3.4).

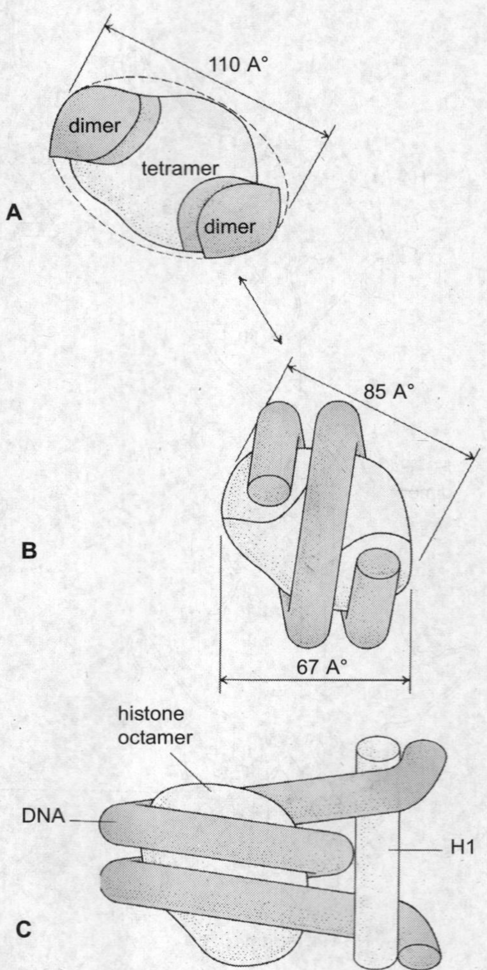

Fig. 3.9. Nucleosome structure. A – schematic comparison of the eight histones comprising the nucleosome in salt solution. A dimer consists of one H2A and one H2B histone molecule: a tetramer consists of two H3 and two H4 histones. B – DNA fits in surface grooves on the more compacted structure found in physiological conditions. C – The diagram shows the presumed position of the histone (after Tamarin, 2002).

The 30 nm chromatin fiber is found to have **zig – zag** structure (Fig. 3.11). The 30 nm structure found in chromosomes is probably a fluid mosaic of the different zig-zag variations. Differences, in linker lengths probably introduce further local perturbations into the zig-zag structure. Finally, the presence of other DNA-binding proteins and DNA sequences that are difficult to fold into nucleosome punctuate the 30 nm fiber with irregular features (Fig. 3.11).

Several mechanisms probably act together to form the 30 nm fiber from a linear string of nucleosomes. 1. A additional histone, called **histone H1**, is involved in this process. H1 is larger than the core histones and is considerably less well conserved. In fact, the cells of most eukaryotic organisms make several histone H1 proteins of related but quite distinct amino acid sequences. A single histone H1 molecule binds to each nucleosome, contacting both DNA and protein, and

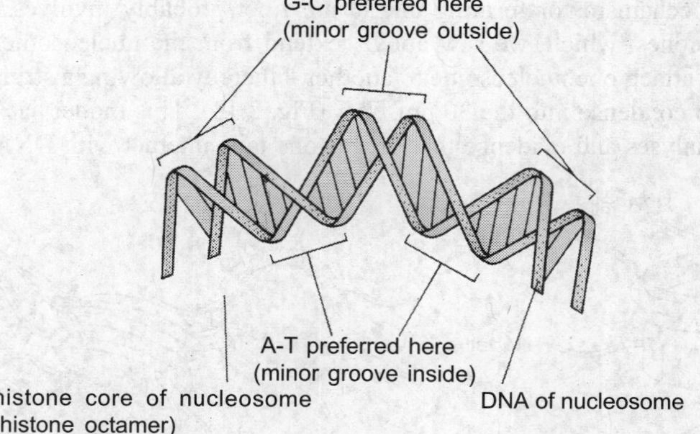

Fig. 3.10. Mode of bending of DNA in a nucleosome. The DNA helix makes 1.65 tight turns around the histone octamer. This diagram is showing how the minor groove is compressed on the inside of the turn. Owing to certain structural features of the DNA molecule, A – T base pairs are preferentially accommodated in such a narrow minor groove (after Alberts *et al.*, 2002).

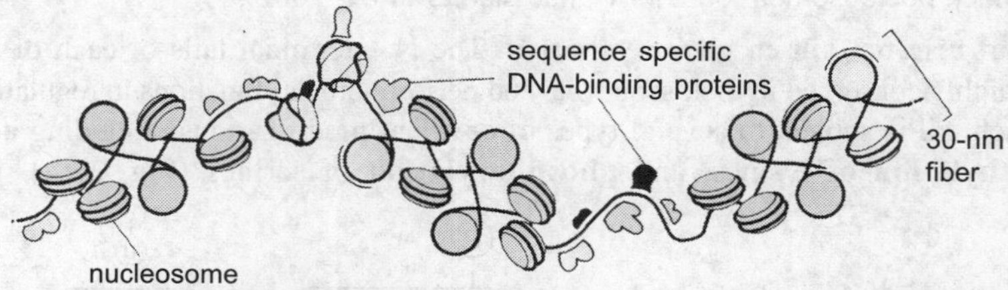

Fig. 3.11. Irregularities in the 30 nm fiber. This schematic view of the 30 nm fiber illustrates its interruption by sequence-specific DNA-binding proteins. The interruptions in the 30 nm fiber may be due to regions of DNA that lack nucleosomes altogether or more probably, to regions that contain altered or remodeled nucleosomes (after Alberts *et al.*, 2002).

changing the path of the DNA as it exits from the nucleosome. Histone H1 causes a change in the exit path in DNA which is crucial for compacting nucleosomal DNA and ultimately it interlocks to form the 30 nm fiber (Fig. 3.12).

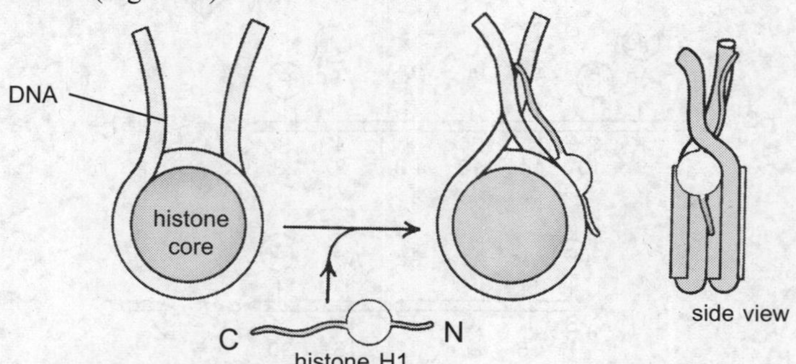

Fig. 3.12. A speculative model for how histone H1 change the path of DNA as it exits from the nucleosome. Histone H1 consists of a globular core and two extended tails. Part of the effect of H1 on the compaction of nucleosome organization may result from charge neutralization: like the core histone, H1 is positively charged (especially its C – terminal tail) and this helps to compact the negatively charged DNA. Unlike the core histones, H1 does not seem to be essential for cell viability; in one ciliated protozoan the nucleus expands nearly twofold in the absence of histone H1, but the cells otherwise appear normal (after Alberts *et al.*, 2002).

The second mechanism for forming the 30 nm fiber probably involves the highly flexible tails of the core histones, which, we saw above, extend from the nucleosome. It is thought that these tails may help attach one nucleosome to another – thereby allowing a string of them, with the aid of histone H1 to condense into the 30 nm fiber (Fig. 3.13). This model has been supported by X-ray diffraction analyses and evidence that the histone tails interact with DNA.

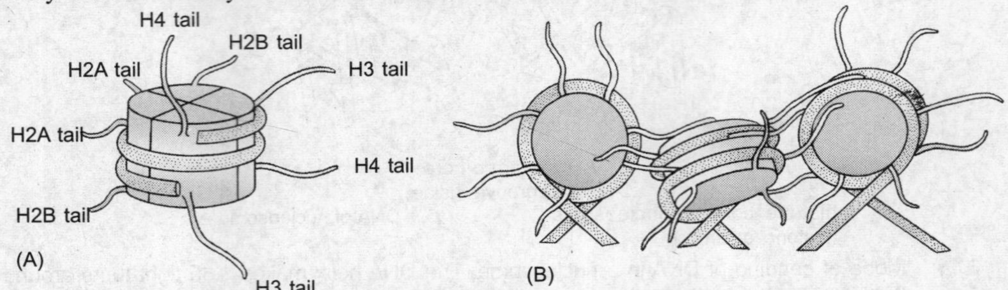

Fig. 3.13. A speculative model for histone tails in the formation of the 30 nm fiber. A – The approximate exit points of the eight histone tails(one from each histone subunit) from histone octamer. B – A speculative model showing how the histone tails may help to pack nucleosomes together (after Alberts *et al.,* 2002).

Roles of histone tails in gene regulation. The N – terminal tails of each of the four core histones are highly conserved in their sequence, and perform crucial functions in regulating chromatin structure. Each tail is subject to several types of covalent modifications, including **acetylation** of lysines, **methylation** of lysines, and **phosphorylation** of serines (Fig. 3.14). Histones are

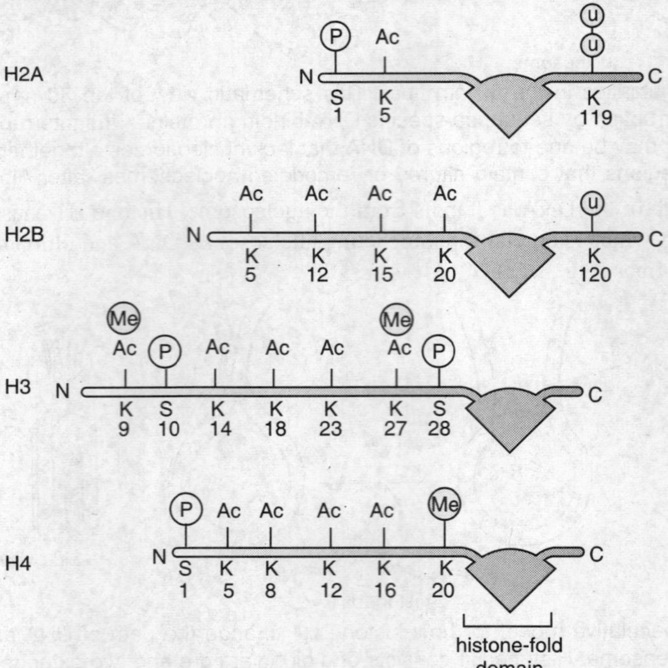

Fig. 3.14. Covalent modification of core histone tails. Known modifications of the four histone core proteins are indicated: K = lysine, S = serine, Me = methyl group, Ac= acetyl group, P = phosphate, u = ubiquitin. Note that some positions (*e.g.,* lysine of H3) can be modified in more than one way. Histone H2B can be modified by a single ubiquitin molecule; H2A can be modified by the addition of several ubiquitins (after Alberts *et al.,* 2002).

synthesized in the cytosol and then assembled into nucleosomes. Some of the modifications of histone tails occur just after their synthesis, but before their assembly. For us those modifications are significant however, which take place once the nucleosome has been assembled. These nucleosome modifications are added and removed by enzymes that reside in the nucleus; for example, acetyl groups are added to the histone tails by histone acetyl transferases (HATs) and taken off by histone deacetylase (HDACs).

The various modifications of the histone tails have many important outcomes. Although modifications of the tail have little direct effect on the stability of an individual nucleosome, they seem to affect the stability of the 30 nm chromatin fiber and of the higher order structures discussed below. For example, **histone acetylation** tends to destabilize the chromatin structure, perhaps in part because adding an acetyl group removes the positive charge from the lysine, thereby making it more difficult for histones to neutralize the charges on DNA as chromatin is compacted. However, the most acute effect of modified histone tails is their ability to attract specific proteins to a stretch of chromatin that has been appropriately modified. Depending on the precise tail modifications, these additional proteins can either cause further compaction of the chromatin or can facilitate access to the DNA.

Histone code hypothesis. This hypothesis was proposed by **C.D. Allis** and coworkers in 2000 and 2001. According this hypothesis, histone tails can be marked by different combinations of modifications. Each such marking conveys a specific meaning to the stretch of chromatin on which it occurs. Only a few of the meanings of the modifications are known. For example, a doublyacetylated H4 tail is "read" by a protein required for gene expression. H3 tail methylated at lysine 9 is recognized by a set of proteins that creates an especially compact form of chromatin, which silences gene expression (Fig. 3.15).

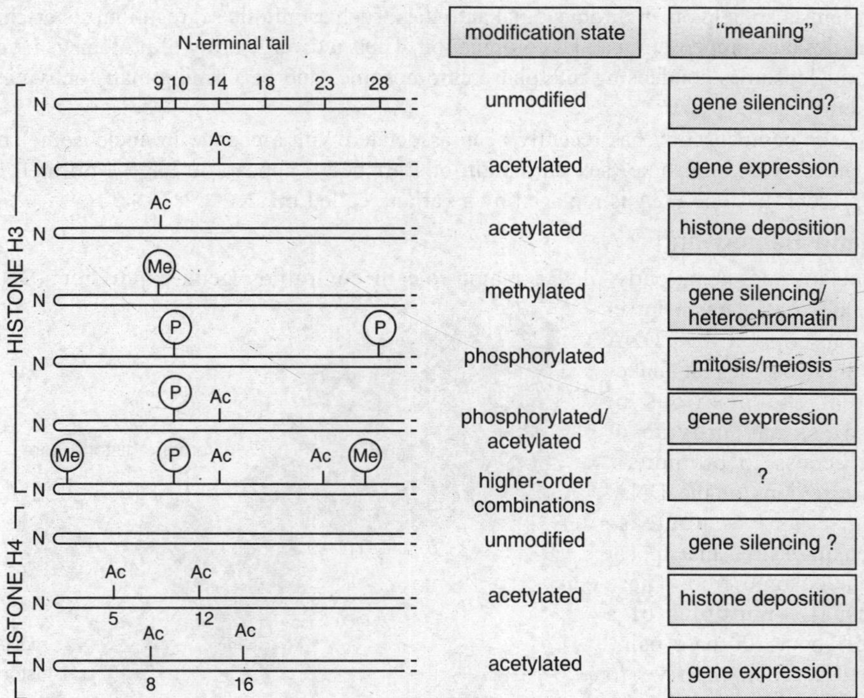

Fig. 3.15. Explanation of the histone code hypothesis. Ac-acetylation, Me-methylation, P-phosphorylation (after Alberts *et al.*, 2002).

Likewise, the acetylation of lysine 14 of histone H3 and lysine 8 and 16 of histone H4 – usually associated with gene expression – is performed by the *type A histone acetylases* (HATs) in the nucleus. In contrast, the acetylation of lysins 5 and 12 of histone H4 and a lysine of histone H3 takes place in the cytosol, after the histones have been synthesized but before they have been incorporated into nucleosomes; these modifications are catalyzed by the *type B histone acetylases*. These modified histones are deposited onto DNA after DNA replication, and their acetyl groups are taken off shortly afterwards by histone deacylases (HDACs). Thus, the acetylation at these positions signals newly replicated chromatin.

Further, modification of a particular position in a histone tail can take on different meanings depending on other features of the local chromatin structure. For example, the phosphorylation of position 10 of histone H3 is associated not only with the condensation of chromosomes that takes place in mitosis and meiosis but also with the expression of certain genes. Lastly, some histone tail modifications are interdependent. For example, methylation of H3 position 9 blocks the phosphorylation of H3 position 10 and vice versa.

Behavior of Nucleosomes During Chromosome Duplication

When DNA is replicated, twice as many nucleosomes are needed since one double helix becomes two. Recent studies indicate that a parental nucleosome is partly disassembled during DNA replication and reassembled on one or the other daughter strand, apparently randomly. The other DNA strand has a new nucleosome constructed of proteins called **chromatin assembly factors** which are of following *three* types: In fruit flies, a protein complex, called the **replication –coupling assembly factor** assembles new nucleosomes. In addition, a protein complex, called **condensin** is needed for the condensation of interphase chromosomes to mitotic chromosomes. This complex includes two **SMC proteins** (for structural maintenance of chromosomes) and two non-SMC proteins. SMC proteins also help other chromosomal activities, such as mitotic segregation, sister-chromatid adhesion, dosage compensation and recombination. Thus, a diverge array of proteins is involved in creating nucleosomes condensing interphase chromosomes and performing many activities of the chromosome.

Dosage compensation has recently been associated with a change in nucleosome structure. The activated X chromosome (sex chromatin or Barr body) appears to have a different type of histone present. Histone H2A is replaced by a variant, called **mH2A.**

Chromatin Remodeling

Nucleosomes apparently play a major role in controlling gene expression; DNA with nucleosomes has a much lower rate of transcription than DNA without nucleosomes. It makes sense that the positions of nucleosomes can provide or prevent access to promoters. There are regions of the DNA, known as **nuclease hypersensitive sites**, that appear to be nucleosome free. These sites, usually multiples of a region of about 200 base pairs, are particularly sensitive to digestion by different nuclease enzymes. When these regions are isolated they usually have

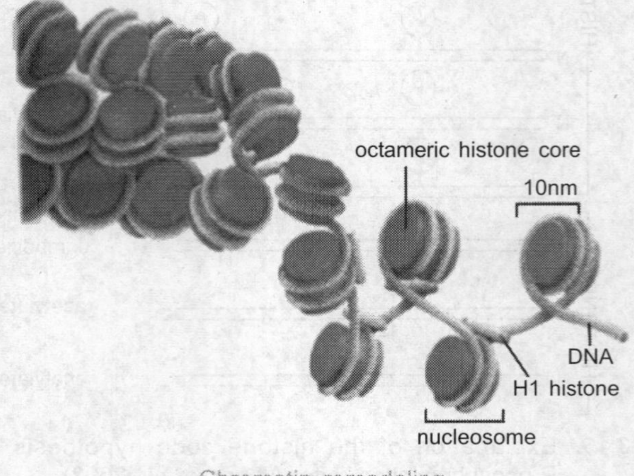

Chromatin remodeling.

sequences that control functions in *replication, transcription* and other activities of DNA. For example, numerous promoter regions in *Drosophila,* mouse and human DNA are in nuclease – hypersensitive sites. Hence, some specific DNA sequences are kept free of nucleosomes, and these sequences appear to be recognized by various enzymes such as RNA polymerase. In many other cases, however, nucleosomes do appear to cover promoters and repress transcription. For transcription to occur in these cases, some form of **chromatin remodeling** must take place.

Two general classes of proteins are involved in chromatin remodeling. *First* are proteins that **acetylate** the N – terminal tails of the histones, a process that may cause the nucleosomes to bind less tightly to the core and thus make it available for attachment of **transcription factors**. These enzymes are called **histone acetyl transferases (HATs)**. Deacetylaling enzymes have the reverse effect. They act to repress transcription. *Second,* a class of ATP dependent proteins such as **SWI/SNF complex** in yeast also affect chromatin remodeling. (**Note:** some workers called the proteins SWI because they are involved in mating type *switching,* and called them SNF for sucrose nonfermenting). The SWI/SNF complex is a group of eleven proteins involved in transcription activation in many genes, presumably allowing transcription factors to access the promoters by remodeling chromatin. These proteins are able to reposition a nucleosome on DNA by sliding the nucleosome down the DNA.

From above findings we can draw the conclusion that although nucleosomes serve as a general first order packing mechanism in eukaryotic DNA, they can be positioned precisely and can diminish transcription. However, once transcription begins, RNA polymerase enzyme apparently moves along nucleosomed DNA by translocation of the histone by 75 to 80 base pairs without the disrupting the nucleosome itself (**Studitsky** *et al.,* 1994). This seems to be accomplished by the RNA polymerase moving the DNA and then reforming the nucleosome in its wake (Fig. 3.16).

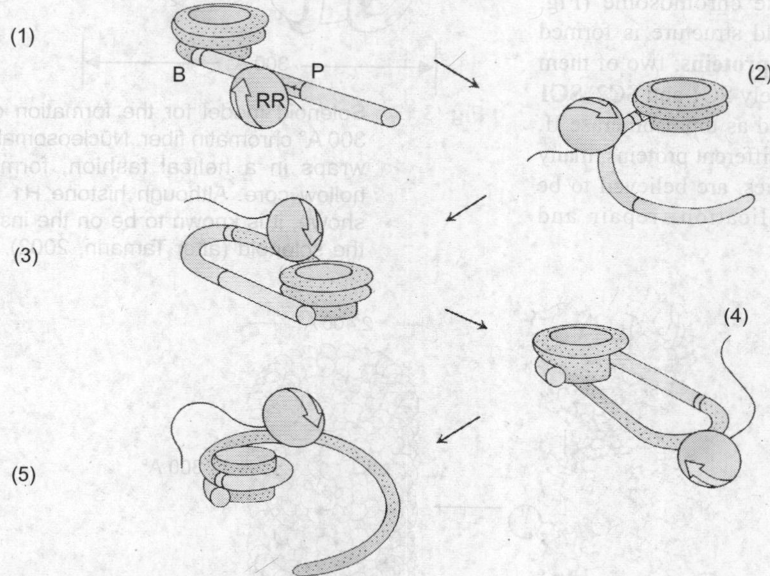

Fig. 3.16. RNA polymerase enzyme steps around a nucleosome without disrupting it. A – The RNA polymerase begins at a promoter (P) and heads for the nucleosomed DNA (*shaded*), whose border is noted with a line and letter B. As the polymerase encounters the nucleosome, it begins to unwrap the DNA from the histones. (2). The displaced DNA then reencounters the histones (3), about 75 to 80 base pairs upstream from the original point of nucleosome formation. The polymerase continues on its way (4 and 5) , and the nucleosome reforms in its displaced position without disrupting the histones or ever fully losing contact with the core DNA (after Tamarin, 2002).

Higher Order Structure of Chromatin

Since the nucleosome has a width of only 110A° and metaphase chromosomes appear to be constructed of a fibre having a diameter about 2400 A°, several additional levels of chromatin compaction lead to the metaphase chromosome. Various experiments, which change the ionic strength the chromatin is subject to, indicate that the 110 A° DNA spontaneously forms a 300 A°, solenoid-like fiber with increased ionic strength. It seems that this fiber results from the coiling of the nucleosomal DNA (Fig. 3.17). This 300 A° (30 nm) fiber is not, however, the final form of the DNA. We can account for the contraction of the 300 A° fiber to the 2,400 A° fiber found in metaphase chromosomes by the formation of a second solenoid-like structure from the winding of the 300 A° fiber (Fig. 3.18).

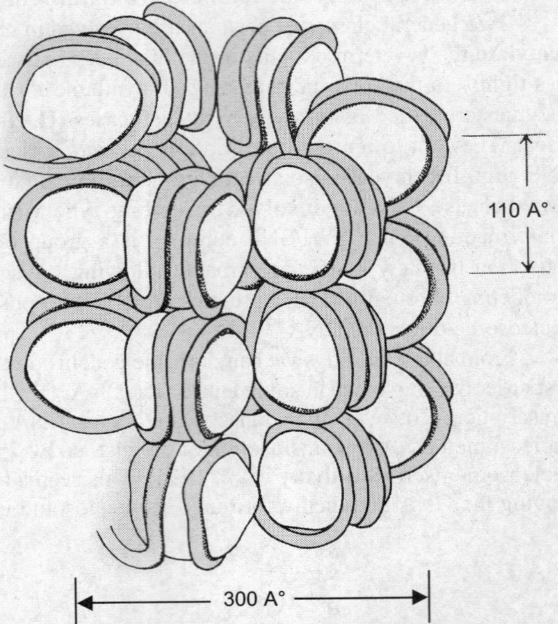

Fig. 3.17. Solenoid model for the formation of the 300 A° chromatin fiber. Nucleosomal DNA wraps in a helical fashion, forming a hollow core. Although histone H1 is not shown, it is known to be on the inside of the solenoid (after Tamarin, 2002).

When the histones are removed from a chromosome, the DNA billows out, leaving a proteinaceous structure termed a **scaffold** (Fig. 3.19). The scaffold is involved in condensing the 300 A° chromatin fiber into the tightly packed metaphase chromosome (Fig. 3.20). The scaffold structure is formed from **nonhistone proteins**; two of them predominate, namely SC1 and SC2. **SC1** has been identified as topoisomerase II. Several hundred different proteins, many in minute quantities, are believed to be involved in replication, repair and transcription.

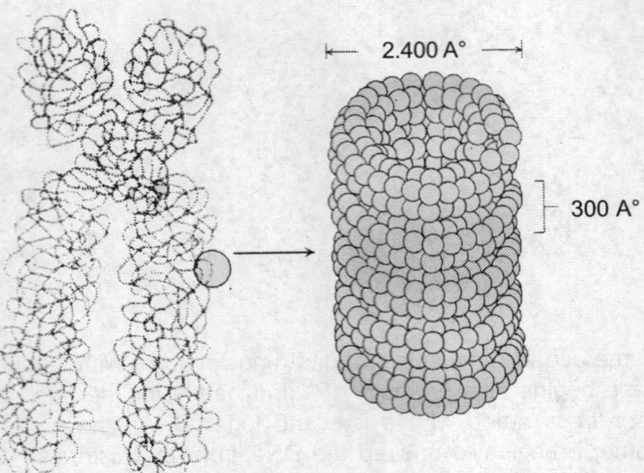

Fig. 3.18. The 2,400 A° fiber of the eukaryotic chromosome is a hollow, solenoid-like structure. It is formed by the coiling of the 300 A° fiber, which itself is solenoid (after Tamarin, 2002).

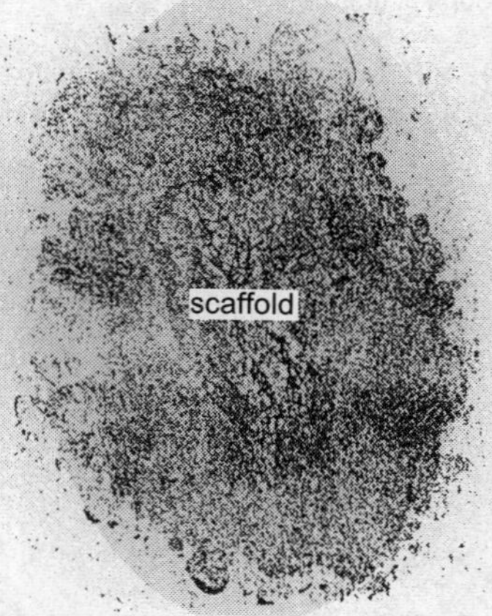

Fig. 3.19. Electron micrograph of a human metaphase chromosome to show the scaffold protein (**J. Paulson** and **U. Laemmli**, 1977). When the histones are removed from an eukaryotic chromosome, a fibrous scaffold remains. The DNA loops out from this scaffold.

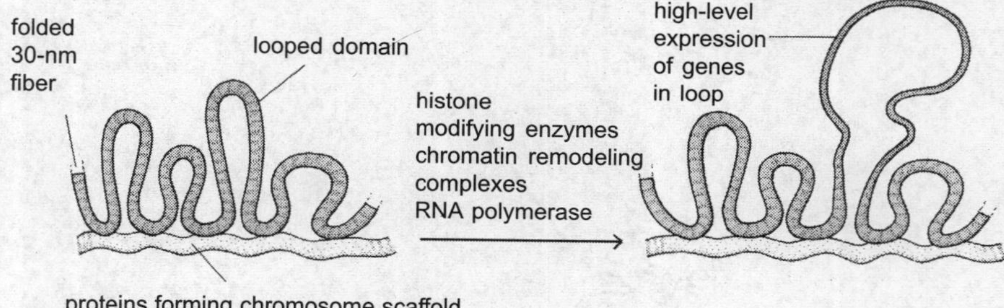

Fig. 3.20. A model for the structure of an interphase chromosome. A section of an interphase chromosome is shown folded into a series of looped domains, each containing 20,000 – 100,000 nucleotide pairs of double-helical DNA condensed into a 30 nm fiber, when the cell requires direct access to the DNA packaged in these loops. Such decondensation is brought about by enzymes that directly modify chromatin structure as well as proteins such as RNA polymerase, that act directly on the underlying DNA. It is not understood how the folded 30 nm fiber is anchored to the chromosome axis, but evidence suggests that the base of chromosomal loops is rich in DNA topoisomerases, which are enzymes that allow DNA to swivel when anchored (after Alberts *et al.,*, 2002).

3.2. CHROMOSOME BANDING

Various chromosomal staining techniques reveal consistent banding patterns. By means of these patterns, all of the human chromosomes can be distinguished. Of possibly greater importance is the fact that these staining techniques have provided some insight into the structure of the chromosome.

1. G – bands. These bands are obtained with **Giemsa stain**, a complex of stain specific for the phosphate groups of DNA. Treatment of fixed chromatin with trypsin or hot salts brings out the G –bands. Giemsa stain enhances banding that is already visible in mitotic chromosomes. The banding pattern is caused by the arrangement of chromosomes. Under careful observation, the major G – bands prove to consists of many smaller chromosomeres. G – banding appearance has led **Comings** (1978) to suggest the mechanism of chromosome folding (Fig. 3.21).

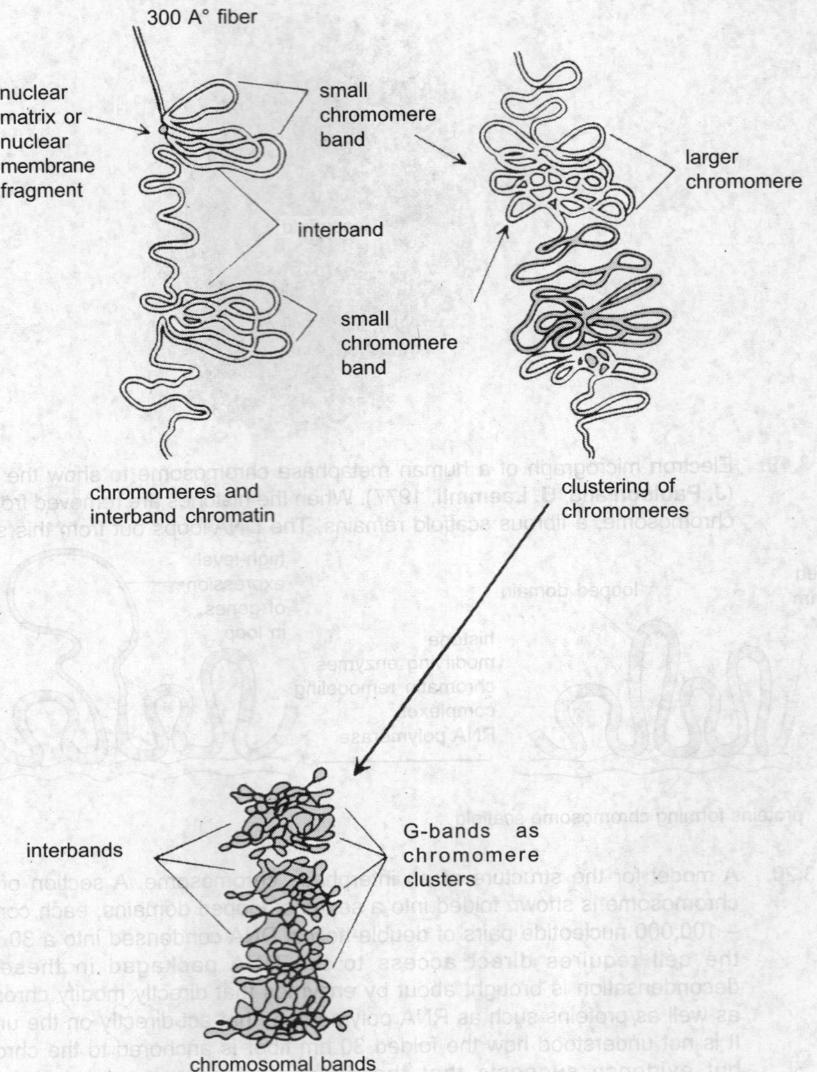

Fig. 3.21. Model of eukaryotic (mammalian) chromosomal banding. G – bands are chromomere clusters, which result from the contraction of smaller chromosomerers. These, in turn, result from looping of the 300 A° fiber (after Tamarin, 2002).

2. C – bands. These are giemsa – stained bands after the chromosomes are treated with NaOH. The C is for "centromere", because these bands represent constitutive heterochromatin surrounding the centromers. The DNA is also usually satellite rich.

Satellite DNA. This type of DNA differs in **buoyan density** from the major portion of cellular DNA. When eukaryotic DNA is isolated and centrifused in CsCl, forming a density gradient, the majority of the DNA forms one band in the gradient at a single buoyant density. The buoyancy is determined by the G – C content of the DNA. However, smaller secondary bands are also usually present, indicating regions of DNA having sequences different from the majority of the cell's DNA. DNA isolated by this way is referred to as **satellite DNA** because of the secondary, or satellite, bands formed in the density gradient. In a chromosome, the satellite DNA is found primarily around the centromeres and consists of numerous repetitions of a short sequence.

3. R – bands. These bands are visible with a technique that stains the regions between G – bands. The chromosomes are fixed, stained with Giemsa, and observed under the phase contrast microscope. Since the dark – light pattern is the opposite of the G – band pattern, these bands are called **reverse bands.**

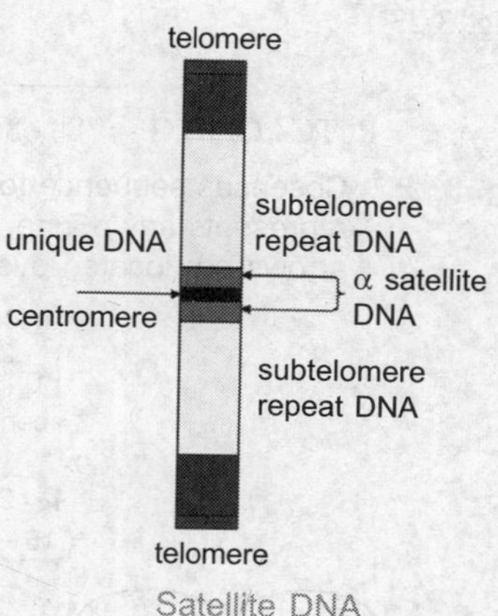

Satellite DNA

Based on the information obtained from these staining techniques, **Comings** (1978), distinguished between three basic chromatin types: euchromatin, constitutive heterochromatin and intercalary heterochromatin (Table 3.3). The only chromatin involved in transcription is **euchromatin. Constitutive heterochromatin surrounds** the centromere and is rich in satellite DNA. **Intercalary heterochromatin** is found throughout the chromosome. Thus, it becomes apparent that the eukaryotic chromosome is a relatively complex structure.

Table 3.3. The three major types of chromatin in eukaryotic chromosome (Source: **Tamarin,** 2002).

Feature	Euchromatin	Centromeric constitutive heterochromatin	Intercalary heterochromatin
1. Relation to bands	In R-bands	In C –bands	In G – bands
2. Location	Chromosome arms	Usually centromeric	Chromosome arms
3. Condition during interphase	Usually dispersed	Condensed	Condensed
4. Genetic activity	Usually active	Inactive	Probably inactive
5. Relation to chromomeres	Interchromomeric	Centromeric chromomere	Intercalary chromomeres

3.3. DNA OF CENTROMERE

The centromere is a specific region of eukryotic chromosome. It is involved in movement of the chromosome during mitosis and meiosis. As we know that, the terms *centromere* and *kinetochore*, while occasionally used interchangeably, are distinct. The **kinetochore** is the interface between the visible constriction in the chromosome (the centromere) and the microtubules of the spindle. The kinetochore of higher organisms (*e.g.,* mammals) contains protein and some RNA. Microscopically it is a trilaminar structure, attached to chromatin at the inner layer and to microtubules at the outer layer.

Most of the genetical and molecular biological information of centromere has been obtained by studies in the fungus, yeast (*Saccharomyces cervisiae*) and the plant, *Arabidopsis thaliana.* Cells did not maintain most artificially created yeast plasmids because they were lost during mitosis. However, some plasmids were isolated that did replicate normally during cell division. Presumably, they contained centromeres, allowing them to replicate and move in synchrony with the host's chromosomes. Genetic engineering also made it possible to isolate smaller and smaller regions that could serve as centromeres. For example, after sequencing the centromeres of fifteen or the sixteen yeast chromosomes, it was possible to conclude that the centromere from yeast is about 250 base pairs long with three consensus regions (**Clarke** and **Carbon,** 1985; Fig. 3.22). In the light of current molecular biological data, the centromere can be defined as a sequence of DNA, called the *C EN locus* or C E N region. This region may contain a single modified nucleosome associated with region II. The 250 base-pair length of the C E N regions of yeast chromosomes is about 200 A°, the, same as the diameter of a microtubule, indicating that only one microtubule attaches to each centromere during mitosis or meiosis in a yeast cell. This region is called a **point centromere** (Fig. 3.23).

I II III

⟵ ⟵ ⟶

PuTCACPuTG – 78 – 86 bp — TGTTTPyTGXTTTCCGAAAXXXXAAA

Fig. 3.22. Cosensus sequence for the three regions (I,II and III) of fifteen yeast cnetromeres. *Pu* represents any purine, *Py* represents any pyrimidine, and *X* represents any base. The arrows are located over the inverted repeat sequences (after Tamarin, 2002).

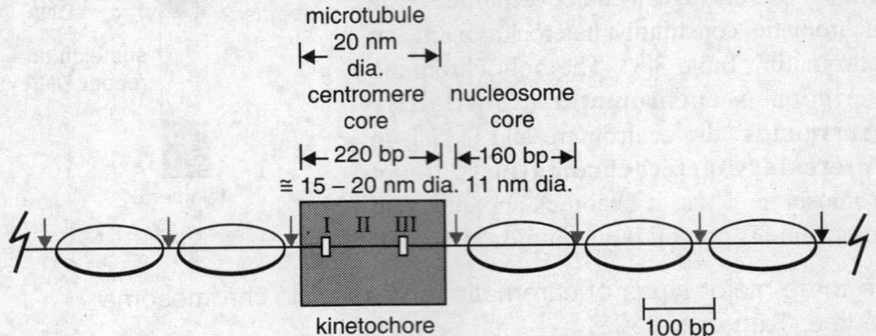

Fig. 3.23. Schematic view of a yeast centromeric region. The *arrows* indicate nuclease-hypersensitive sites. A microtubule is about the same width as the centromeric region (after Tamarin, 2002).

Higher eukaryotes have large centromeric regions each of which attaches more microtubules, *e.g.,* 4 to 7 in rat fetus and 70 to 150 in Plant *Haemanthus.* These regions are referred to as **regional centromeres.** Regional centromeres range from 19 to 100 kb, with unique and satellite (repeated sequence). DNA is heterochromatic and may include expressed genes.

Arabidopsis centromeres span 0.9 – 1.2 Mb of DNA (Box 3.1) and each one is made up largely of 180bp repeat sequences. In humans, the equivalent sequences are 171 bp and are called **alphoid DNA.** DNA of centromere of *Arabidopsis* contains multiple copies of genome-wide repeats, along with a few genes, the latter at a density of 7 – 9 per 100 kb compared with 25 genes per 100 bp for the noncentromeric regions of *Arabidopsis* chromosomes (**Copenhaver** *et al.,* 1999). The discovery that centromere DNA contains genes was a big surprise because it was thought that these regions were genetically inactive.

Box 3.1.

Unit of length for DNA molecules

Because DNA is a double –stranded, the lengths of molecules are described as so many base pairs (bp). A **kilobase pair (kb)** is 10^3 bp and a megabase pair (**Mb**) is 10^6 bp. A **giga base pair (GP)** is 10^9 MB. In summary:

1 kb = 1000 bp

1 Mb = 1000 kb = 1000 000 bp

1 Gb = 1000 Mb = 1000 000 000 bp

Length of RNA molecules cannot be expressed in bp because most RNAs are single – stranded: their lengths is described as so many nucleotides (see **Brown,** 2002).

The **special centromeric proteins** in humans include at least seven that are not found else where in the chromosome (**Warburton,** 2001). One of these proteins, **CENP-A** is very similar to histone H3 and is thought to replace this histone in the centromeric nucleosomes. Part of the human kinetochore is made up of alphoid DNA plus CENP_A and other proteins, but its structure has not been decribed in detail (**Vafa and Sullivan,** 1997).

3.4. DNA OF TELOMERES

It is already known that eukaryotic chromosomes are linear, each has two ends, referred to as **telomeres**. The telomeres not only mark the termination of the linear chromosome but also have several specific functions. For example, telomeres must prevent the chromosomal ends from acting in a "sticky" fashion the way the broken chromosomal ends act. In other words, chromosomal ends must also prevent the ends of chromosomes from being degraded by **exonucleases** and must allow chromosomal ends to be properly replicated.

Most telomeres isolated so far are repetitions of sequences of five to eight bases. In human beings, the telomeric sequence is T T A G G G, repeated 300 to 5000 times at the end of each chromosome. The human telomere was discovered by **Robert K. Moyzis** and his colleagues (1988) when they probed *the highly repetitive segment of human DNA.* When a probe for this sequence was applied to human chromosomes, the sequence was found at the tip of each chromosome in roughly the same quantity. This is a highly conserved sequence, found in all vertebrates studied as well as in unicellular trypanosomes. Similar sequences are found in various other eukaryotes (Table 3.4); the first sequence was isolated by **E.Blackburn** and **J.Gall** in 1978.

Table 3.4. Telomeric sequences in eukaryotes; the G-rich strand of the double helix is shown (Source: **Tamarin,** 2002).

Organism	Telomeric repeat
1. Human beings, other mammals, birds and reptiles	T T A G G G
2. Trypanosomes	T T A G G G
3. *Tetrahymena* (holotrichous ciliates)	G G G G T T
4. *Stylonychia* (hypotrichous ciliates)	G G G G T T T T
5. Yeast	G T, G G T and G G G T
6. Plants	T T T A G G G

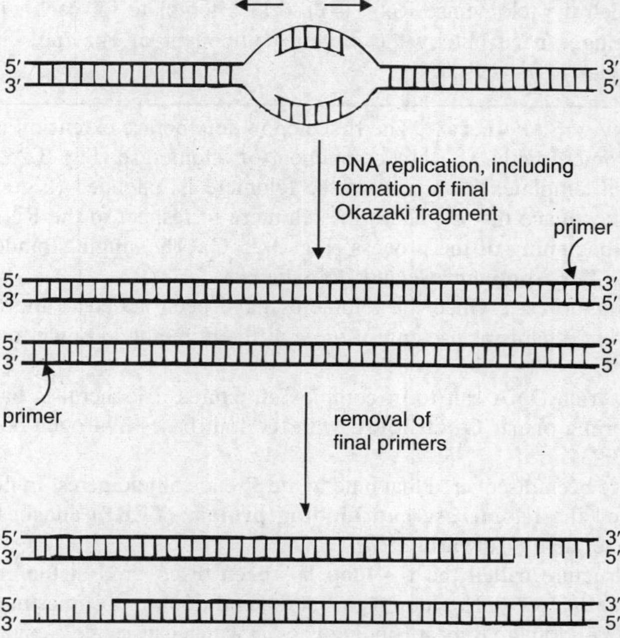

Fig. 3.24. Removal of final primers after the replication of linear DNA creates single stranded ends (after Tamarin, 2002).

When a linear DNA molecule is replicated, the 3' – 5' strand can be replicated up to the end. The 5' – 3' strand, however, is replicated with DNA primers that are then degraded, leaving a short gap on the progeny strand (Fig. 3.24). It is always the G-rich strand of telomeric DNA that ends up single stranded, forming a 3' overhang of twelve to sixteen nucleotides. Thus, *the normal replication process of a linear DNA molecule leaves an incomplete terminus.* Hence, molecular biologists suspected that there would be a unique mechanism for the replication of telomeres.

The Telomerase Enzyme

Telomeric sequences appear to be added *de novo* without DNA template assistance by an enzyme called **telomerase**. This enzyme was discovered by **Blackburn** and her colleagues (1990). This was seen when telomeres from another species were engineered into yeast cells. After a cell cycle, the yeast telomeric sequence had been added on at the ends of the foreign chromosome, the result presumably of the telomerase enzyme.

Further, when **Blackburn** and her colleagues (1992) isolated telomerase, they discovered that a segment of RNA, about 160 nucleotides in length is an integral part of the enzyme. Telomerase RNA has a region that is complementary of the G – rich repeat of the telomeric DNA sequence of the species. After careful experimentation, including the modifying the gene for the telomerase RNA, **Blackburn** and her colleagues conducted that telomerase enzyme uses its RNA as a template for adding telomeric repeats to the ends of chromosomes. Telomerase enzyme is thus a **reverse transcriptase**, using RNA nucleotides as a template to polymerize DNA nucleotides.

Box 3.2.

Telomerase is an unusual enzyme in that it consists of both protein and RNA. In the human enzyme the RNA component is 450 nucleotides in length and contains near its 5' end the sequence 5' – C U A A C C C U A A C – 3', whose central region is the reverse complement of the human telomere repeat sequence 5' – T T A G G G – 3' (**Feng** *et al.*, 1995). This enables telomerase to extend the telomeric DNA at the 3' end of a polynucleotide by the copying mechanism, in which the telomerase RNA is used as a template for each extension step, the DNA synthesis being carried out by the protein component of enzyme which is a reverse transcriptase (**Lingner** *et al.*, 1997).

Mode of working of telomerase. The first step in, telomerase extension is **hybridization** of the 3' end of the telomere with the RNA component of telomerase (Fig. 3.25A). Then, with the telomerase RNA as a template, the 3' end of the telomere is extended (Fig. 3.25 B). Finally, a translocation step takes place that displaces the telomere in respect to the RNA, returning to the configuration at the beginning of the process (Fig. 3.25 C). The single stranded C-rich strand is then synthesized with **DNA polymerase** and **DNA ligase**.

Protection of telomeres. Once the telomeres have been added to the ends of eukaryotic chromosomes, different organisms use any of *three* different method known to protect the ends of the chromosomes.

1. The guanine rich DNA can form complex structures. Biochemists have discovered that four guanines can form a planar **G-tetraplex,** with the four bases hydrogen bonded to each other (**Wang** and **Patel,** 1993).

2. Proteins have been discovered that bind to the 3' ends of telomeres. In the ciliate, *Oxytricha nova,* a protein called the **telomerase end binding protein (TEBP)** attaches to the 3' ends of telomeres and protects them (**Horvath,** 1998).

3. A novel structure called the **t – loop** has been discovered at the ends of mammalian telomeres (**Griffith** *et al.,*1999). The loop forms at the end of chromosomes under the direction of a protein called **TRF2** (Telomere Repeat Binding Factor) which causes the 3' end of the chromosome to loop around and interdigitate into the double helix, forming the loop (Fig. 3.26). A t-loop from mouse liver cell is about 10,000 bases around (**Griffith** *et al.*, 1999).

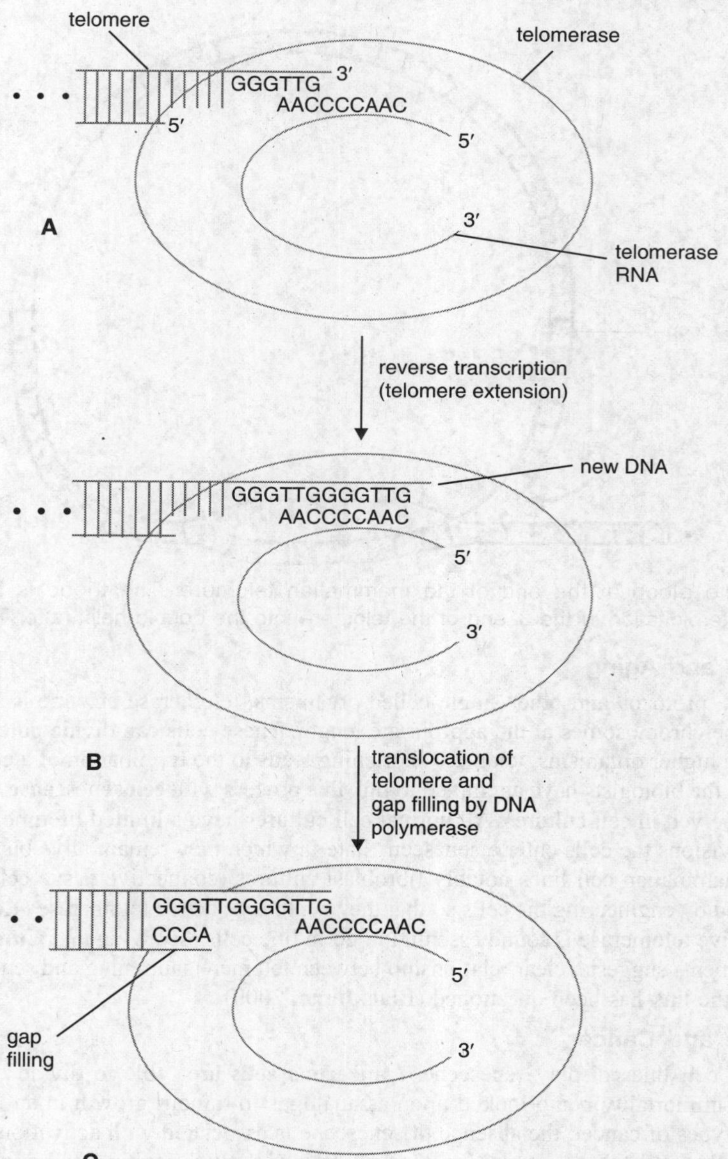

Fig. 3.25. Telomerase enzyme extends telomeres using telomerase RNA as a template. Gap filling by DNA polymerase and ligase complete the double helix (after Tamarin, 2002).

There are some telomeric proteins that are thought to form a linkage between the telomere and the periphery of the nucleus, the area in which the chromosome ends are localized (**Tham** and **Zakian,** 2000).

How do cells keep track of the number of their telomeric repeats? Proteins have isolated that bind to telomeres (Rap1 in yeast *Saccharomyces cerevisiae*, **Marcand** *et al.,* 1997; TRF1 in human beings. **Van Steensel** *et. al.*, 1997). By mutating these proteins or the telomeric sequences, molecular biologists have changed the equilibrium number of telomeric repeats. This led to the current model that the cell counts the number of these proteins bound to the telomeres, not the number of telomeres directly, to know whether telomeres should be added.

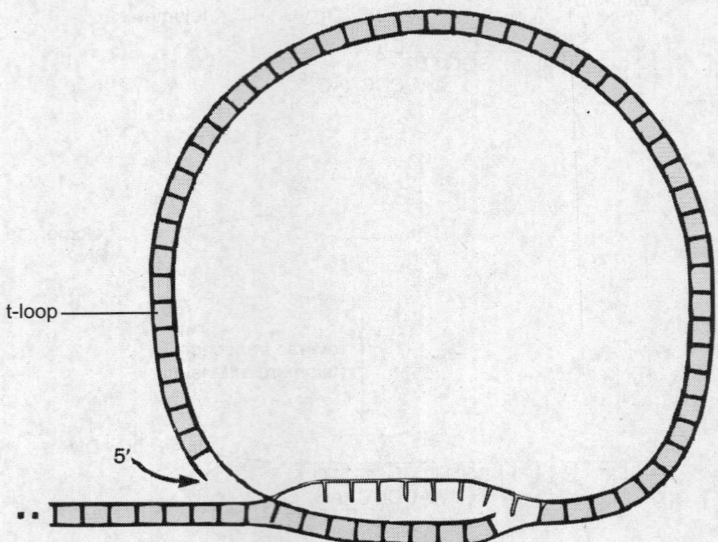

Fig. 3.26. The t-loop at the end of the mammalian telomere. The t-loop is formed by the interdigitation of the 3' end of the telomere into the double helix (after Tamarin, 2002).

Telomerase and Aging

In yeast, protozoa and other single-celled organisms, telomerase enzyme is active, keeping the ends of the chromosomes at the appropriate length. These cells can divide potentially forever. In contrast, in higher organisms, telomere shortening leads to the termination of a cell lineage. For several years the biologists have attempted to link this process with **cell senescense,** a phenomenon originally observed in cell cultures. All normal cell cultures have a limited lifetime: after a certain number of divisions the cells enter a senescent state in which they remain alive but cannot divide. With some mammalian cell lines notably fibroblast cultures (connective tissue cells), senescence can be delayed by engineering the cells so that they synthesize active telomerase (**Reddel** 1998). In mammals active telomerase is found essential in germ line cells (see **Klug** and **Cummings**, 2003). These experiments suggest a clear relationship between telomere shortening and senescence but the exactness of the link has been questioned (**Blackburn,** 2000).

Telomerase and Cancer

Not all cell lines display senescene. **Cancerous cells** are able to divide continuously in culture, their immortality being looked upon as analogus to tumour growth in an intact organism with several types of cancer, the absence of senescene is associated with activation of telomerase, sometimes to the extent that telomere length is maintained through multiple cell divisions, but often in such a way that the telomerase is overactive. In fact, human telomerase was isolated from an immortal cell line (HeLa) derived from cervical cancer cells.

This current information may be exploited for the clinical purpose : if, telomerase can be **deactivated** in tumour cells, the cells may stop dividing or die, thereby eliminating the cancer. One type of human cancer, **dyskeratosis cogenita**, appears to result from a mutation in the gene specifying the RNA component of human telomerase (**Marciniak** and **Curaente,** 2001).

3.5. THE C-VALUE PARADOX

Why do eukaryotes have so much DNA, and why is there huge variation in the DNA content between species of comparable complexity? These questions define the **C-value paradox**, in which C refers to the quantity of DNA in a cell. For an example of the paradox, although human beings have 3.3 billion base pairs in the haploid genome, an amoeba has more than 200 billion base pairs. Likewise, although an average bony fish has over 300 billion base pairs of DNA in its haploid genome, the Japanese puffer fish has less than half a billion base pairs. If the basic bony fish pattern can be created with less than half a billion base pairs, why does the average bony fish have over 600 times that much DNA? what is this excess DNA doing? To explain the C-value paradox, researches examined the **repetitiveness of DNA**, and more recently probed and sequenced DNA to understand its properties.

1. Unique DNA and Repetitive DNA

The **DNA-DNA hybridization** is the process of taking DNA from the same or different sources and heating and then cooling it, causing double helices to reform at homologous regions. This technique is useful for determining sequence similarities and degrees of repetitiveness among DNAs. **R.Britten** and **Kohne** (1968), using the technique of **DNA-DNA hybridization**, first systematically analyzed the repetitiveness of the DNA within eukaryotes. When DNA is heated, it cools, it **renatures.** The rate of renaturation depends on the DNA sequences. If the sample contains DNA with repeated sequences, it will hybridize faster than DNA that does not have repeated sequences. From these studies, **Britten** and **Kohne** (1968) found that eukaryotic chromosomes contain regions of unique, moderately repetitive and highly repetitive DNA. **Unique DNA** is, as its name implies, DNA with unrepeated sequences. **Repetitive DNA** is DNA whose sequences are repeated in the genome. During the last 39 years, hundreds of plants and animals have been analyzed for the relative proportion of repetitive DNA in their genome (Table 3.5).

Satellite DNA, found around centromeres is highly repetitive DNA with a unique repeat length of about two hundred base pairs. Unique DNA makes up most of the transcribed genes of an organism.

Table 3.5. Relative proportions (per cent of haploid genome) of repetitive sequences.

	Organism	Proportion of repetitive sequences (%)
I. Animals	(i) Sea urchin	38
	(ii) Snail	60
	(iii) Toad	45
	(iv) Mouse	40
	(v) Calf	45
	(vi) Human	30
II. Plants	(i) Bread wheat	80
	(ii) Oats	83
	(iii) Maize	78
	(iv) Garden pea	75
	(v) Barley	76
	(vi) Rye	92

The rest of the DNA is repetitive DNA in a few to several hundred thousand copies. This repetitive DNA comprises three categories. One is "junk" DNA, DNA that is not useful to the organism, made up of untranscribed and parasitic sequences (selfish DNA). Another category is transcribed genes in many copies that have diverged from each other, such as antibody, collagen, and globin genes. The term diver **gene family** has been given to the genes that have arisen by duplication, with or without divergence, from an ancestral gene. Finally, the transcribed genes in many copies that are virtually identical such as ribosomal RNA genes and histone genes, make up a third category of repetitive DNA.

2. Junk DNA

Transposons in prokaryotes are generally viewed as selfish or parasitic: they serve no purpose to the cell. The transposons replicate on their own, increasing in number. Eukaryotic transposons are mostly **retrotransposons**, *i.e.,* transposable elements that move by way of an RNA intermediate. In other words, the retrotransposon is transcribed into RNA and then, by reverse transcription, converted to a cDNA and then inserted into the genome. These elements can make up 50 per cent of the eukaryotic genome, existing in hundreds of thousands of copies. They generally fall into two categories: LINES and SINES.

Box 3.3.

The term **'junk DNA'** is falling out of favour of molecular biologists. This is partly because the number of surprises resulting from genome research over the last few years has meant that molecular biologists have become less confident in asserting that any part of the genome is unimportant simply because we not currently know what its function might be. One thing that is clear that the bulk of the intergenic DNA is made up of repeated sequences of one type or another (see **Brown,** 2002).

1. LINES or Long Interspersed Elements. They contain upto seven thousand pairs each and contain genes for **reverse transcription, RNA binding** and **endonuclease activity**. They thus have the ability to jump by way of an RNA intermediate. Human DNA is believed to be composed of about 15 per cent LINES.

2. SINES or Short Interspersed Elements. They are generally derivatives of transfer RNA (tRNA) genes and do not have the ability to retrotranspose on their own. That is, in the past, their transcripts were modified, converted to cDNA by reverse transcription and then reinserted into the host's genome. They depend on the reverse transcriptase enzyme provided by the genes of LINES or retroviruses. One group of SINES not derived from tRNA is derived from RNA of the signal recognition particle; members of this group occur in human beings in about five hundred thousand copies of a three hundred base pair sequence. Because these sequences are cleaved by the restriction endonuclease *alu I,* they are called the *Alu family*. The human genome is also pervaded by remnants of at least a dozen distinct families of ancient retroviruses scattered throughout our chromosomes.

At this juncture, some explanation can be forwarded for the C-value paradox. Much eukaryotic DNA is junk. Apparently doing no harm. In some cases, 97% of the host genome is composed of junk DNA. Recent work has explained that gross differences in DNA content between higher organisms may be due to the differing abilities of different species to rid themselves of this parasitic DNA. If it builds up without being removed, the DNA content of the species can rise. Thus, wide differences in DNA content among higher eukaryotes mentioned at the beginning of this section have little to do with the complexity of the organism, but rather with the ability of the organism to remove the junk DNA as it forms (see **Tamarin**, 2002).

3.6. SPLIT GENES

Split genes means that the (DNA) sequences containing actual information of the gene (called **exons**) are interrupted by other sequences (called **introns**) which are spliced out after transcription.

A. Discovery of Split Genes

With the discovery of DNA as the genetic material, a gene was regarded as a continuous segment of DNA. In 1964, it was also proved that there was co-linearity between sequence of nucleotides on DNA and sequence of amino acids in corresponding protein. Therefore, there was absolutely no doubt about the continuity of nucleotides in a gene represented by a DNA segment. This was proved to be true both for prokaryotes and eukaryotes. However, a big surprise came in 1977, when geneticists came to know that in some mammals, birds and amphibians, a gene may not be represented by a continuous sequence of nucleotides but may be interrupted by some intervening sequences which are not represented in mRNA transcribed from the gene and utilized for the synthesis of proteins. Such genes with intervening sequences were called **split genes** or **interrupted genes**

(Fig. 3.27). Split genes are also called **discontinuous** or **mosaic genes** (see **Brown**, 1998).

The discovery of split genes was made in 1977 by various groups of biologists in a variety of materials: (i) Two research groups separately headed by **Philip A. Sharp** and **Richard J. Roberts** studied genes of adenovirus 2. (ii) Research groups of **P.Chambon**, **P.Leader** and **R.A. Flavell** studied b globin genes, ovalbummin genes and tRNA genes. In all these cases the genes were found to be interrupted by intervening sequences. The credit for discovery of split genes, however, goes to **Philip Sharp** and **Richard Roberts**, who won in 1993 Nobel for Medicine or Physiology for their independent work on split genes. They analyzed the hybrids of *late mRNA* of adenovirus 2 with the adenovirus genomic DNA. When these mRNA-DNA hybrids were examined under electron microscope, the adjoining sequences of mRNA were found to be hybridized with discontinuous stretches of genomic DNA of adenovirus. The intervening DNA sequences were observed as loops and the phenomenon was later described as **R-looping**.

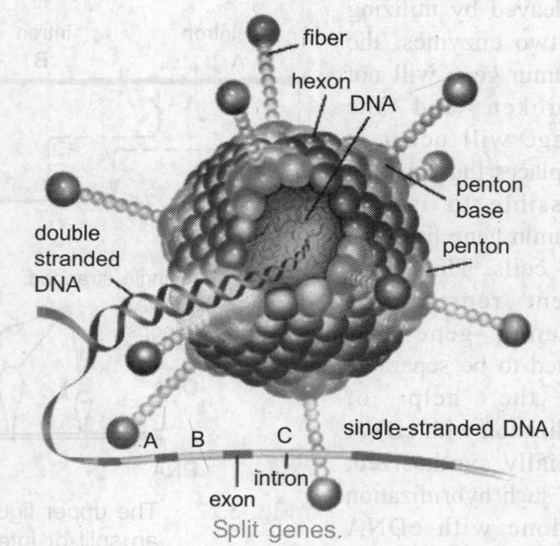

Split genes.

Chambon's group compiled sequences of the boundaries of introns from a large number of protein coding eukaryotic genes (not ribosomal RNA or tRNA genes), which revealed the presence of *consensus sequences* at the intron-exon junctions. Of these GT (guanine-thymine) was always found at the 5' side of the intron (**left splice junction**) and AG (adenine-guanine) at the 3' side (**right splice junction**). This became popularly known as **GT-AG rule** or **Chambon's rule** (*Note:* Rarely, AT-AC occur at intron-exon junctions). Some of the diseases (*e.g.,* thalessemia) are caused by mutations, which created or abolished these splice junctions.

Methods of Investigations of Split Genes

A detailed study was conducted on **ovalbumin gene** found in chicken. This ovalbumin gene is responsible for synthesis of a protein, called **ovalbumin** consisting of 386 amino acids and synthesized only by highly specialized tubular gland cells of the oviduct at the time when the hen is laying eggs. The expression of this ovalbumin gene is controlled by some female sex hormones. Chambon and his collegues synthesized artificial ovalbumin gene in order to study its regulation. Such an artificial gene of ovalbumin could be synthesized by using ovalbumin mRNA which could give rise to cDNA (complementary DNA) with the help of enzyme **reverse transcriptase**. This cDNA was inserted into a plasmid and cloned in *E.coli* for its multiplication. When this cDNA was compared with corresponding genomic DNA, it was discovered (through DNA hybridization) that the genomic DNA had additional intervening sequences.

Another ingenious technique used for the study of split gene was the use of **restriction enzymes** which have the property of cleaving DNA at unique sites. More than 100 such restriction endonuclease enzyme are now available. In view of this, restriction endonucleases could be used to find out the presence or absence of a certain sequence in a particular gene. For example, when *Eco*R1 and *Hind III* enzymes were used with cDNA for ovalbumin, it was found that no cleavage occurred suggesting that the sequences of six base pairs each recognized by these two restriction enzymes were absent. On the basis of this it was expected that if the DNA extracted from oviduct was cleaved by utilizing these two enzymes, the ovalbumin gene will not be broken and the cleavage will occur at other places thus making it possible to isolate ovalbumin gene from the living cells. This DNA segment representing ovalbumin gene was expected to be separated with the help of hybridization with cDNA artificially synthesized. When such hybridization was done with cDNA hybridized with different fragments of DNA rather than with a single fragment. Hybridization between the single stranded DNA having the gene for ovalbumin and its mRNA also showed the formation of distinct loops at specific sites as observed in electron microscope (Fig. 3.27). Such DNA which exists in the loops is obviously missing in mRNA. This kind of conclusion eventually led to the discovery of split genes in 1977.

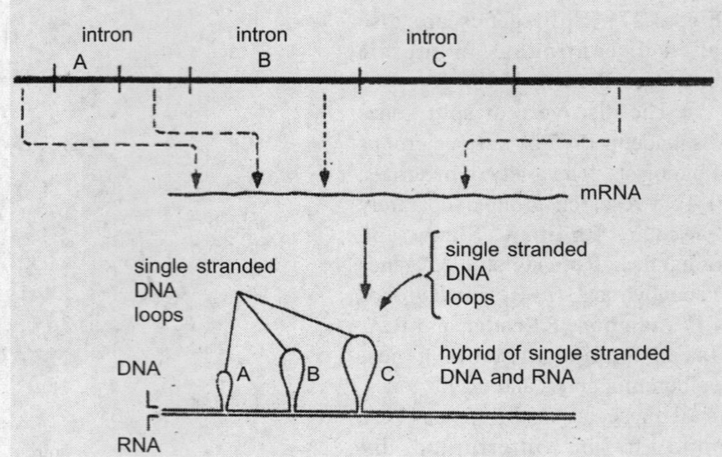

Fig. 3.27. The upper figure shows a DNA sequence representing an split or interrupted gene with their introns (A, B, C) and the synthesis of its mRNA; the bottom figure shows the result of hybridization of mRNA with single stranded DNA obtained after denaturation of native DNA (note the loops formed by three intron genes).

In subsequent years, it could be proved that split genes are present at least in two more cases, *i.e.,* gene for β-**globin** (a component of haemoglobin molecule) in rabbit and mouse and **immunoglobin gene** (antibody gene). Later on, split genes were reported to occur in higher organisms as a common phenomenon. For understanding the structure of a split gene, we can consider an example of a hypothetical split gene which has a DNA sequence including three pieces, called **exon1, exon 2** and **exon 3**(Fig. 3.28). These three exons are separated by two long intervening sequences, called **intron A** and **intron B**. The terms exon and intron were used by **Walter Gilbert** (1985) for the first time and are being followed ever since.

Figure 3.28 shows that in the transcription of DNA, the ultimate product, *i.e.,* mRNA had only those sequences which correspond to exons, the sequences representing introns being entirely absent. In subsequent years, it is discovered that both exons and introns are first transcribed and this primary transcript in then modified. The sequences

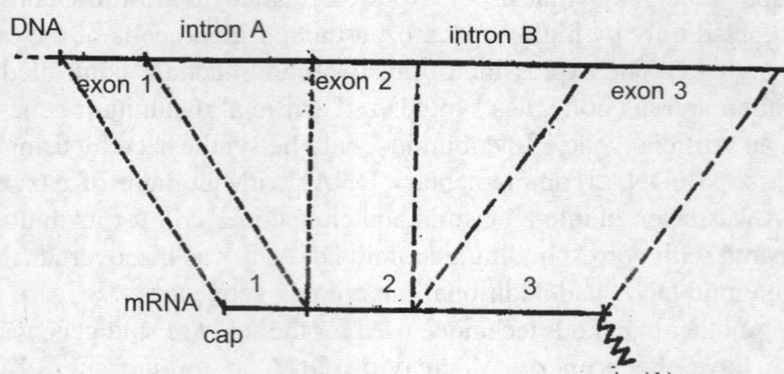

Fig. 3.28. Diagrammatic representation of a hypothetical split gene having three exons (exon1, exon2 and exon3) and two introns (intron A and intron B) and its relationship with mRNA. There are no sequences in mRNA corresponding to those in two introns.

corresponding to introns are removed from the transcript and the sequences corresponding to exons are joined together, in correct order to give rise to mRNA. A generalization has been made that the order of exons on DNA is the same as the order in which they are found in the processed mRNA.

Examples of split genes or interrupted genes occur in a variety of organisms: 1. nuclear genes for proteins; (ii) nuclear genes for rRNAs; (iii) nuclear genes for tRNAs; (iv) mitochondrial genes in yeast; (v) chloroplast genes in a wide variety of plants; (vi) genes in archaebaceria and (vii) genes in bacteriophages of *E.coli.* They were initially believed to be entirely absent in eubacterial genomes but in 1990s introns were discovered even in eubacteria.

B. Certain Examples of Split Genes

1. Structure of ovalbumin split gene of hen (chicken). In the split gene of ovalbumin there are eight exons and seven intron (Fig. 3.29 and fig. 3.30). The size of ovalbumin gene is 7,700 base pairs long although the mature mRNA for ovalbumin is only 1,872 nucleotides. The entire ovalbumin gene with its 7700 base pairs is first transcribed to a precursor RNA to which a cap consisting of 7 mG is added at the 5' end and poly A tail is added at the 3' end. After the addition of cap and tail, the introns are excised in the first step and the remaining two in second step (Fig. 3.31). The exons produced due to splicing are then joined with the help of enzyme ligase to produce the mature mRNA of ovalbumin gene.

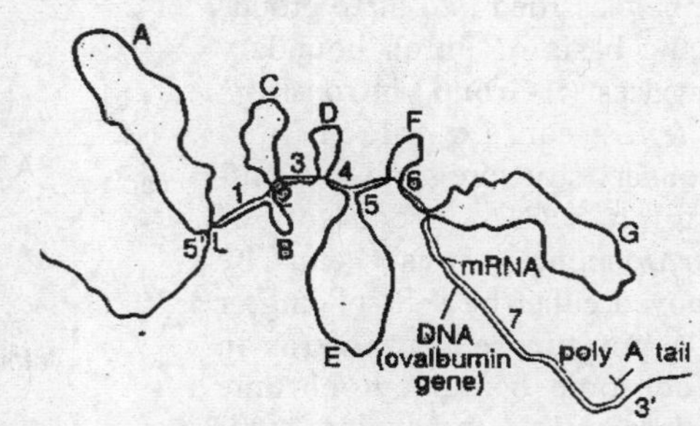

Fig. 3.29. The map of electron micrograph prepared after hybridizing mRNA (thin line) of the ovalbumin gene with its single stranded DNA(thick line). The loops A-G represent 7 different introns and segments L and 1-7 represent 8 exons.

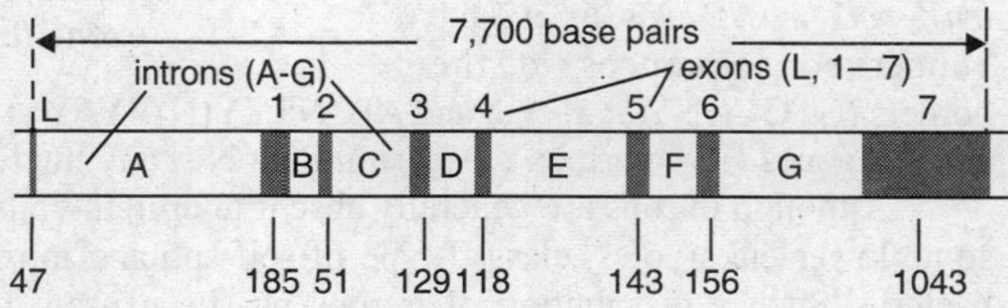

Fig. 3.30. Schematic representation of ovalbumin gene showing 7 introns (A-G) and 8 exons (L, 1 – 7); number of base pairs in each exon are also given.

2. Split genes in humans. In the human gene for the **cystic fibrosis** transmembrane regulator, the introns are much longer than the exons. This gene is 250 kb long and is split into 24 exons and 23 introns. The average length of the exons is 227 bp, so all the exons added together make up only 2.4% of the gene. These exons are scattered throughout the entrie length of gene, separated by introns that range in size from 2 to 35 kb (see **Brown,** 1998).

3. Split genes in fungal mitochondria. Split genes are also found in the mitochondria. Introns of these split genes in fungal mitochondria are of two type: (i) **Group I introns** which are found in majority of the fungal mitochondria split genes, do not carry any conserved sequence, called **internal guide sequence**. By internal pairing, this guide sequence brings the two intron-exon junctions together and help in splicing out the introns. These group I introns are also found in the nuclear gene coding for rRNA in *Tetrahymena* (a ciliate) and *Physarum*(a slime mold). Features similar to those of group I introns are also reported from introns of phage T4 genes. (ii) **Group II introns** resemble nuclear genes and have consensus sequences (GT and APy) and a branch sequence that resembles the TACTAAC box. These introns are excised as lariats.

4. Split Genes in Chloroplasts

Split genes for ribosomal RNA (rRNA), transfer RNAs (tRNA) and some proteins have also been reported in the chloroplast genomes of several plants including *Chlamydomonas* and *Nicotiana*. Introns found in chloroplast genes can be classified into three groups on the basis of intron boundary sequences. (i) **Group I introns**(*e.g.,* in *trnL*) can be folded in a secondary structure similar to self-splicing rRNA precursor of *Tetrahymena*. These can be removed either by self-splicing or by a **"maturase"** enzyme(as in cytochrome b and cytochrome oxidase mRNA precursors). (ii) **Group II introns** (*e.g.,* majority of genes including *trnA* and *trnI*) can be folded into a complex secondary structure (as in introns of mitochondrial genes for cytochrome oxidase in maize and yeast). (iii) **Group III introns** (*e.g., trnG, trnK, trnV, rpl2, rps12, rps16, etc.*)have conserved sequences at their borders (GTGCGNY at 5' end, and ATCNRYY(N)YYAY at 3' end), similar to those in the eukaryotic nuclear genes (R = purine; Y = pyrimidine; N = any nucleotide).

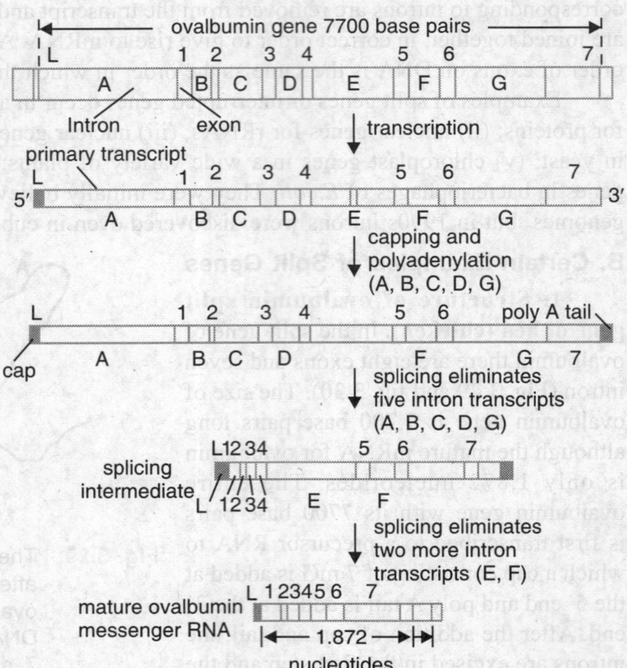

Fig. 3.31. Steps involved in the production of ovalbumin mRNA (1872 nucleotides long) from ovalbumin gene (7700 base pairs long); introns (A-G) and exons (L and 1-7).

Although introns are generally absent in protein-coding genes of bacteria, archaebacteria and some lower eukaryotes, class II type of self-splicing introns have recently been reported even in bacteria. Such a distribution of introns has been used for a study of the role of introns in the evolution of genes or alternatively for a study of the origin of genes or alternatively for a study of the origin of introns themselves.

C. Evolutionary Origin of Split Genes

Following three hypothesis/theories have been proposed for explaining origin of split genes: 1. Exon (domain) shuffling hypothesis; 2. Exon theory; and 3. Intron early hypothesis.

Exon shuffling hypothesis is an interesting idea which holds that each exon of split gene contains a different subcomponent of the biological information carried by the gene as a whole. Individual subcomponents are not sufficient on their own to code for a complete protein, but the information they contain is meaningful in that it specifies a recognizable portion of the proteins function, for example, enabling the protein to bind a particular substrate or attach to a specific site in the cell. The hypothesis is that during evolution, exons from different discontinuous genes can be "shuffled", creating new combinations of biological information (Fig. 3.32). **Exon shuffling** would be more likely to produce new functional proteins than would be entirely random rearrangement of existing genes.

Domain or exon shuffling is illustrated by **tissue plasminogen activator (TPA),** a protein in the blood of vertebrates and which is involved in the blood clotting response. The TPA gene has four exons, each coding for a different structural domain (Fig. 3.33). The upstream exon coding

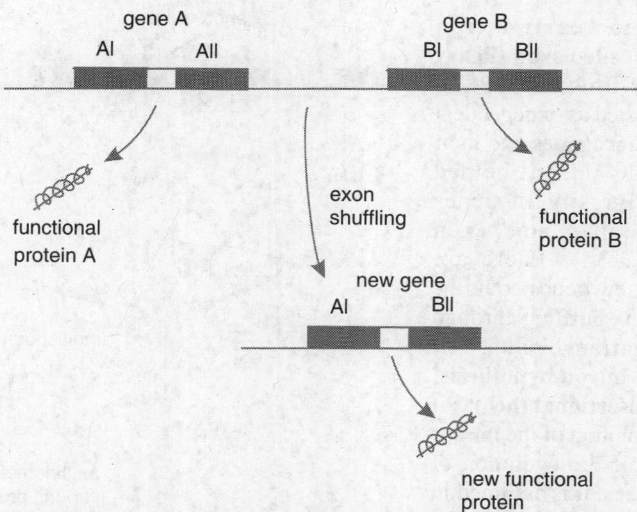

Fig. 3.32. The hypothesis of 'exon shuffling' suggests that new functional genes could be formed by rearranging exons of existing genes (after Brown, 1998).

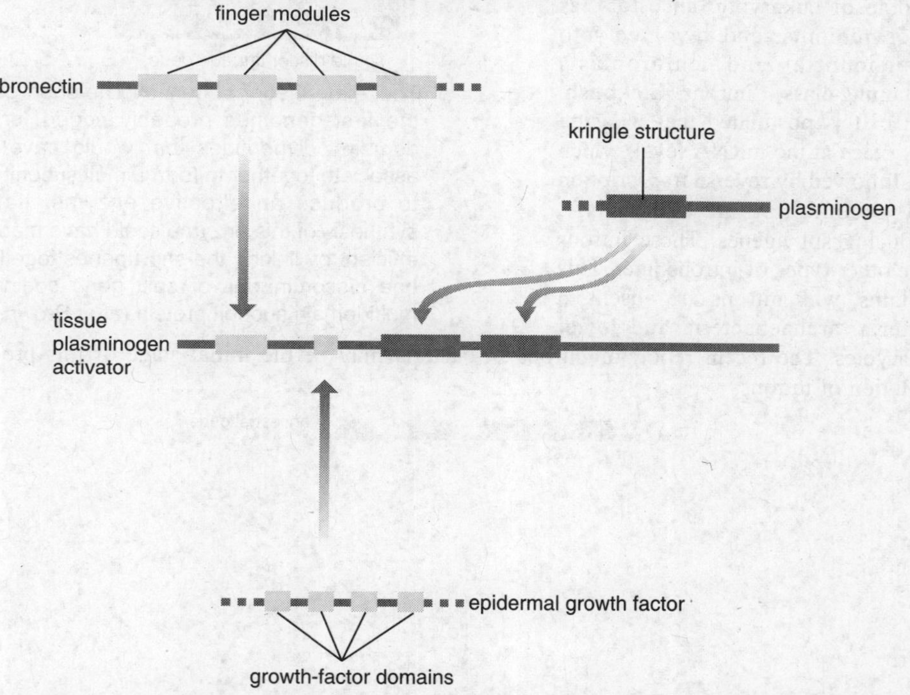

Fig. 3.33. The modular structure of the tissue plasminogen activator protein (after Brown, 2002).

for a 'finger' module that enables the TPA protein to bind to **fibrin**, a fibrous protein found in blood clots which activates TPA. This exon appears to be derived from a second fibrin-binding protein, **fibronectin,** and is absent from the gene for a related protein, **urokinase**, which is not activated by fibrin. The second TPA exon specifies a **growth-factor** domain which has apparently been obtained from the gene for **epidermal growth factor** and which may enable TPA to stimulate cell proliferation. The last two exons code for "Kringle" structures which TPA uses to bind to fibrin clots; these kringle exons come from the plasminogen gene (**Li** and **Graur,** 1991).

According to **"early-intron hypothesis"** (also called **exon theory** of genes by **W.Gilbert** in 1987), ancestral exons existed as independent genetic units of **microgenes** and their association with intron helped generation of diversity in genes through transplicing and exon shuffling (Fig. 3.34). There are examples, where new genes could be generated by exon shuffling through intron recombination events. (ii) According to **'late intron hypothesis'** (also called the **"insertional theory of intron orgin"**) evolution of the introns was independent of the evolution of protein-coding genes. It is assumed in this hypothesis that the self-splicing introns (class I introns) invaded the genome of eukaryotic ancestors, as endosymbionts, and revolved into mitochondria and chloroplasts containing class II introns (**Eickbush,** 2000). It is postulated that invasion took place at the mRNA level, which was followed by reverse transcription and recombination with DNA, producing split genes. These introns gave other types of introns later. This explains, why introns are absent in bacteria, archaebacteria and lower eukaryotes. The recent report of introns in bacteria may be the initial stage of this process of evolution of introns.

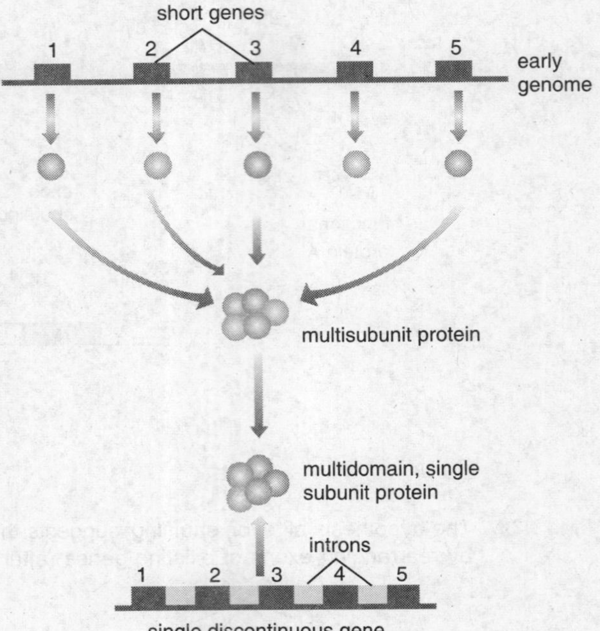

Fig. 3.34. The 'exon theory of genes'. The short genes of the first genomes probably coded for single domain polypeptides that would have had to associate together to form a multisubunit protein to produce an effective enzyme. Later the synthesis of this enzyme could have made more efficient by linking the short genes together into one discontinuous or split gene coding for a multidomain subunit protein (after Brown, 2002).

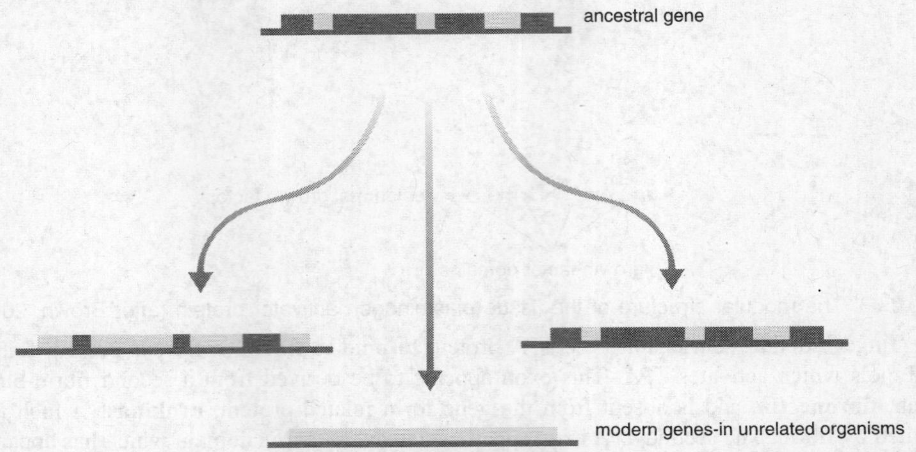

Fig. 3.35. One prediction of the "intron early" hypothesis is that the positions of introns in homologous genes should be similar in unrelated organisms, because all these genes are descended from an ancestral intron-containing gene (after Brown, 2002).

3.7. OVERLAPPING GENES

Bacteriophage ϕX174 contains a single stranded DNA approximately 5,400 nucleotides in length. Complete sequence of these 5,400 nucleotides was worked out by **Fredrick Sanger** and his colleagues. The genome of ϕX174 phage consists of nine cistrons. Amino acid sequences of major proteins coded by the genome of ϕX174 phage have also been worked out and even their molecular weights are known. From the information about protein coded, an estimate could be made of the number of nucleotides required. This estimate of number of nucleotides exceeds 6,000 which is much higher than the actual number of nucleotides present, *i.e.,* 5,400. Thus, it was difficult to explain how these proteins could be synthesized from a DNA segment which is not long enough to code for the required number of amino acids. When detailed study of system was done then it was discovered that sequences in the same segment could be utilized by two different cistrons coding for different proteins. Such **overlapping of genes** (cistrons) will be theoretically possible if the two cistron have two functions at different times and their nucleotide sequences are translated in two different reading frames.

In 1976 **Barrell** and his coworkers discovered that in ϕX174 phage which have nine cistrons (*A, B, C, D, E, J, F, G, H*), cistron *E* is present between *D* and *J* and that the cistron *E* overlap cistron *D*. This could be done by working out the nucleotide sequences of DNA segment representing gene *D, E* and *J*. It could be shown that amber (codon for chain termination) mutations in cistron *E* lie within cistron *D* and these amber mutations do not influence the translation of cistron *D* into its protein. Similarly some nonsense mutations for cistron *D* lie in cistron *E* suggesting that cistrons *D* and *E* overlap in the DNA sequences and that the cistrons *D* and *E* are translated in two different reading frames so that amber codon in mRNA of one cistron will be read as termination of the other cistron.

Among genes *D, E,* and *J* in ϕX174 phage, cistron *D* codes for a simple protein, cistron *E* codes for a protein for lysis and cistron *J* codes for a particle component. It was shown that *enough space is not available where these three genes are located,* thus, conforming that there was overlapping between *D* and *E* cistrons. Overlapping has also been suggested to be present in regions of genes *A* and *B*.

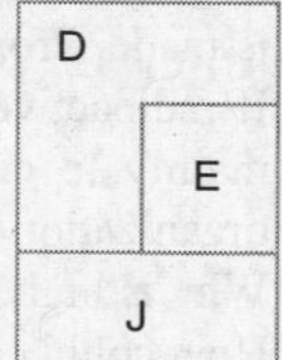

Fig. 3.36. Overlapping of *E* and *D* genes in phage ϕ X174.

Overlapping sequences (=genes) have also been discovered in tryptophan mRNA of *E.coli* and in many other cases. In view of wide occurence of overlapping genes, it seems that phenomenon of overlapping genes, is an economic device to make better use of genetic information is less DNA.

3.8. PSEUDOGENES

In multicellular organisms, a wide variety of DNA sequences are found, which are of no apparent use for the organism. Some of these sequences are defective copies of functional gene and are therefore called **pseudogenes**. Such pseudogenes are found to occur in human beings, mouse (*e.g.,* α globin pseudogenes), *Drophilla* (*e.g.,* histone pseudogenes), etc. For example, human α – **globin** and **â- globin pseudogenes** (ψ), found in each of the two globin gene clusters. Complete nucleotide sequence of alpha globin pseudogene is now known. It has been also reported that both of these genes (α-globin and â-globin) are non-translatable, since they may have mutations in initiation codons and also frame-shift mutations along their length.

3.9. CRYPTIC GENES

Cryptic genes are phenotypically, silent DNA sequences that are not expressed during the normal life cycle of an organism. They may be activated in a few individuals of a population, due to events such as mutation, recombination, transposition and other genetic mechanisms due to specific altered environmental conditions and this leads to their detection. For example, genes for α-glucoside utilization in *E.coli* and *Salmonella.* Four cryptic genetic systems (*bgl, cel, ase, arbT*) involved in the α –glucoside utilization have been characterized in *E.coli.*

Questions And Problems

1. Summerise the evidence that the eukaryotic chromosome is uninemic.
2. What are the major protein components of the eukaryotic chromosome? What are their functions?
3. What evidence is used to determine the length of DNA associated with a nucleosome? What is a nuclease-hypersensitive site? What functions are associated with these sites?
4. What is the protein composition of a nucleosome? What function does histone H1 have?
5. What is the relationship among the 110A°, 300A° and 2,400A° fibers of the eukaryotic chromosome?
6. What do we know about the sequence within a yeast centromere?
7. What is a telomere? What are its functions? What is its structure?
8. Describe three ways in which cells protect their telomere.
9. What functions exist in unique, repetitive and highly repetitive DNAs?
10. Can nucleosome contain the DNA for one gene? Explain.
11. What is the C-value paradox, and how is it explained?
12. What is satellite DNA? What does it signify?
13. If chromatin is digested with an endonuclease enzyme to produce two hundred base-pair fragments, and these fragments are then used for transcription experiments, very little RNA is made. Provide an explanation for this observation.
14. If radioactive probes are made from highly repetitive DNA, these probes hybridize *in situ* mainly to centromeric and telomeric regions. What does this result suggest about the organization of chromosomes?
15. Why is higher-order chromosomal structure expected in eukaryotes but not in prokaryotes?
16. How could comparative DNA studies aid us in understanding the roles of the different kinds of DNA present in the eukaryotic chromosomes?

Answers to Problems

3. The length of DNA associated with nucleosomes was determined by footprinting, in which free DNA was digested, leaving only those segments protected by nucleosomes. Nucleosome hypersensitive sites are sites not in a nucleosomal state; they seems to be sites involved in initiation, transcription and other DNA activities.
5. See Fig. 3.17 and Fig. 3.18.
7. Telomeres are repetitive DNA sequences at chromosomal end. They are repetition of five or eight base sequence. Most telomeres are G-rich. Telomeres protect the ends of chromosomes and probably provide signals on senescence of cells.
9. Highly repetitive DNA usually makes up the centromeric and telomeric regions of the chromosome. Unique DNA, Making up the bulk of structural genes, has a large component that is transcribed. Repetitive DNA is composed of dispersed DNA (*e.g.,* short and long interspersed elements – SINES and LINES), multiple copies of trascribed DNA (*e.g.,* ribosomal RNA, histones) and diverged copies of ancestral genes (*e.g.,* globin family genes).
11. The C-value paradox involves the issues of the excessive amounts of DNA in eukaryotic

species that seem to have similar complexity. It is explained by the large amount of structural DNA in chromosomes as well as the large amounts of short and long interpersed elements(SINES and LINES).

12. Satellite DNA differs in its base sequence from the main quantity of DNA and thus forms a satellite band during buoyant density analysis. It is usually centromeric heterochromatin, composed of a highly repetitive DNA.
13. Spaces between the nucleosomes must contain many promoter sequences. DNA to be digested, it must be unprotected. Since we see little transcription, the promoters must be missing and must have been destroyed
14. Highly repetitive DNA must be located in these chromosomal regions. Since most highly repetitive DNA is not transcribed, the results suggest that centromeric and telomeric regions are not transcribed.
15. The simplest explaination is the difference in amount of the genetic material in prokaryotes and eukaryotes. Since the average human chromosome has several centimeters of DNA, that DNA must be contracted to a size in which it can be moved during mitosis and meiosis, without tangling and breaking. Nucleosomes form the first order of coiling of the nucleosomed DNA bring it down to a manageable size for nuclear divisional processes.
16. Comparative DNA studies can be helpful in understanding the roles of the various types of DNA in the eukaryotic chromosomes if there are cases in which there are remarkably large differences in the amount of DNA in similar species. It can be then inferred that the basic developmental plan of an organism is contained in the one with the lower amount of DNA, and the extra DNA in the species with more DNA may be superfluous. We do have cases in which amphibian differ by as much as one hundred time the amount of DNA found in similar species. The puffer fish has only one-sixth amount of DNA as other higher eukaryotes.

Multiple Choice Questions

Choose the correct answer from the four alternatives given.

1. Chromatin consists of
   (a) RNA
   (b) DNA
   (c) RNA and histones
   (d) DNA and histones
2. DNA is associated with highly basic proteins called
   (a) histones
   (b) non-histones
   (c) albumins
   (d) all of these
3. A nucleosome is made of
   (a) DNA
   (b) histone
   (c) histone wrapped over octameric core of nucleic acid
   (d) DNA wrapped over octameric core of histone
4. Solenoid is a structure of
   (a) nucleosomal organisation with 10nm diameter.
   (b) condensed chromatin fiber with 30nm diameter
   (c) highly condensed form of chromatid with 300 nm diameter
   (d) well organized chromosome with 1400 nm thickness
5. The terminal end of a chromosome is called
   (a) metamere
   (b) telomere
   (c) centromere
   (d) basal granule
6. A gene that specifies the amino acid sequence of a polypeptide chain is termed as
   (a) structural gene
   (b) regulator gene
   (c) operator gene
   (d) split gene
7. The gene not expressing any protein is known as
   (a) epistatic gene
   (b) hypostatic gene

(c) pseudogene
(d) none of these

8. The intervening sequence of 'gene' are known as
(a) introns
(b) exons
(c) cistron
(d) codons

9. In split genes, the coding sequences are called
(a) introns
(b) operons
(c) exons
(d) cistrons

10. Exon part of mRNA have codes for
(a) protein
(b) lipid
(c) carbohydrate
(d) phospholipid

ANSWERS

1. (*d*)	2. (*a*)	3. (*d*)	4. (*b*)	5. (*b*)
6. (*a*)	7. (*c*)	8. (*a*)	9. (*c*)	10. (*a*).

CHAPTER 4

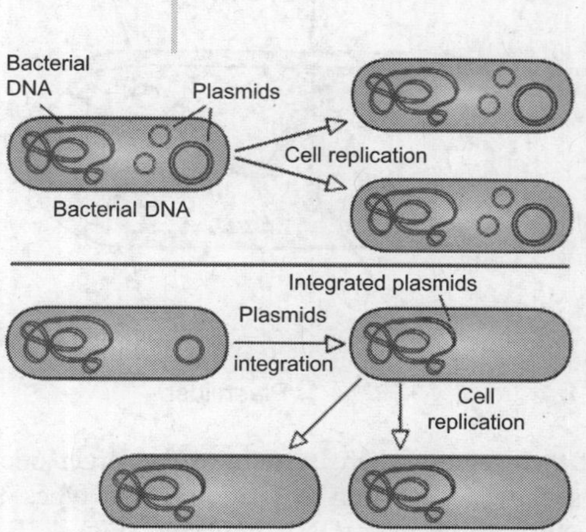

Plasmids, IS Elements, Transposons and Retroelements

Chapter Contents

4.1. PLASMIDS

Plasmids are autonomous self-replicating double-stranded circular DNA molecules that are maintained in the cell in a stable and characteristic number of copies. A plasmid is a replicon that is stably inherited in an extrachromosomal state. Most, but not all, plasmids are dispensable as they are not essential for the survival of bacterial cells, except under certain environmental conditions.

The word **plasmid** was originally used by **J. Lederberg** (1952) for any extrachromosomal hereditary determinat. This term is now restricted for those accessory DNA circles, which are found in bacteria (or in cell organelles such as mitochondria in eukaryotes), in addition to the main chromosome.

The nucleoid DNA molecule is not the only repository for genes in the bacterial cell. Bacteria often also harbour small independent DNA molecules, the plasmids, which carry genes not found on the main DNA molecule. Plasmids are circular molecules that lead an independent existence in the bacterial cell (Fig. 4.1).

Plasmids carry one or more genes which are often responsible for a useful characteristic. For example, the ability of bacteria to survive in normally toxic concentrations of antibiotics such as chloramphenicol or amplicillin is often due to the presence in bacterium of a plasmid that carries antibiotic – resistance genes. Genes carried by plasmids also provide to bacteria the ability to utilise certain complex compounds such as toulene as a carbon source.

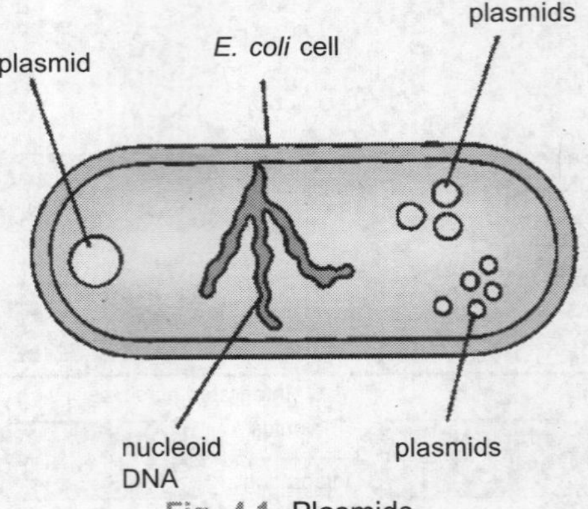

Fig. 4.1. Plasmids.

All plasmids possess at least one DNA sequence that can act as an origin of replication (replicon), so they are able to multiply in the cell independently of the main DNA molecule (Fig. 4.2A). The smaller plasmids make use of the cells own DNA replicative enzymes to make copies of themselves, whereas some of the larger ones carry genes that code for special enzymes that are specific for plasmid replication. A few types of plasmids are also able to replicate by inserting them into the main DNA molecule (Fig. 4.2B). These integrative plasmids are called **episomes** and they may be stably maintained in this form through numerous cell divisions, but always at some stage exist as independent elements.

Characteristics of Plasmids

Plasmids share the following general properties :

1. They are genetic elements made up of DNA; 2 they are smaller than and separate from the main chromosome, and 3. They are capable of replication. Some plasmids such as sex factors have unique properties such as 1. They control bacterial conjugation, 2. They are reversibly inserted in the bacterial chromosome and 3. They are transferred from one bacterium to another.

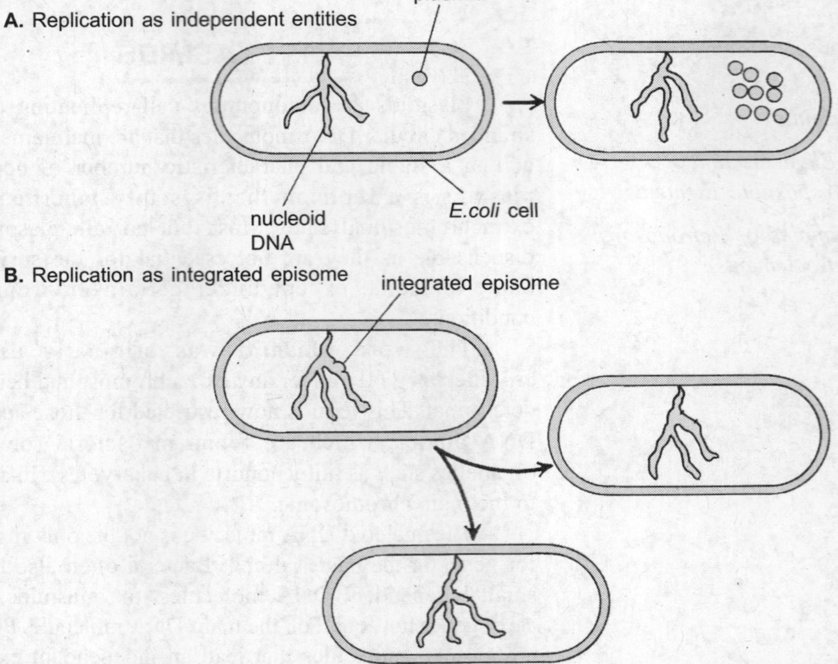

Fig. 4.2. Two methods of replication of the plasmids (after Brown, 1998).

Further, bacteriophages which are included in the category of episomes are sometimes excluded from the category of plasmids, since they show unrestricted replication, leading to lysis. But some phages such as λ (lambda) phage share most of the properties of plasmids, are grouped with them by some workers.

Classification of Plasmids

Almost all species of bacteria harbour plasmids and a large number of different types are known (Table 4.1). They are classified according to the genes they carry and the characteristics that these genes confer on the host bacterium. According to this classification there are following *five* main types of plasmids :

Table 4.1. Types of special feature of some bacterial plasmids.

	Type of plasmid	Special properties
1.	*F, R1, Col*	Fertility
2.	*Col E1*	Bacteriocin production
3.	*R6*	Heavy metal resistance
4.	*Ent*	Entrotoxin production
5.	*Com*	Metabolism of camphor
6.	*Ti* (in *Agrobacterium tumefaciens*)	Tumorigenicity in plants.

1. Fertility (F) plasmids. These plasmids direct conjugation between different bacteria. They are also called **conjugative** or **transmissible plasmids.** An example is the *F* plasmid of *E. coli.* During conjugation of *E. coli,* when F^+ strain (carrying *F* factor) comes in contact with F^- strain, two processes may follow :

One involving transfer of *F* factor, so that F^- becomes F^+ and second involving transfer of a segment of bacterial chromosome, which is facilitated by insertion of *F* factor within the bacterial chromosome at one or other sites, leading to production of *Hfr* (high frequency recombinant). Some F plasmids behave like the **episomes,** *i.e.,* they can exist in 'free' or 'integrated' state.

2. Resistance (R) plasmids. These plasmids carry genes imparting to the host bacterium the resistance to one or more antibacterial agents such as streptomycin, tetracycline, chloramphenicol, ampicillin, sulphonamide or mercury. R plasmids are very important in clinical microbiology because their spread through natural populations can have acute consequences for the treatment of bacterial infections. An example is RP4, commonly found in *Pseudomonas* but also occurring in other bacteria.

R plasmids was discovered in Japan in 1956, when a strain of bacteria called *Shigella* was shown to acquire resistance against several drugs in one step. This was shown to be due to transmission of resistance from one strain to another and was later attributed to R plasmid. R plasmid not only confers resistance, but also enables bacteria carrying them to conjugate (even in the absence of *F* factor) and transfer the R plasmid to other bacteria, related or quite unrelated. For example, R plasmids of *E. coli* are known to be transferred to *Proteus, Salmonella, Haemophilus, Pasteurella, Shigella,* etc. Due to further investigations, the genes for drug resistance were found to be a **transposon** (called Tn3) with terminal inverted repeat (IR) sequences.

Box 4.1.

Compound R Plasmids

IS sequences and transposable (*Tn*) elements are responsible for the transfer of genes controlling resistance to antibiotics and other drugs from one genetic element to the other. Such transpositions generate R plasmids having resistance genes for a variety of antibiotics. Such R plasmids are called **compound R plasmids** and many such compound R plasmids have been isolated and characterized.

Sometimes the R plasmids are found to contain two distinct parts : one being genes for the resistance (called R factor) and the other responsible for transfer of plasmid (called **resistance transfer factor** or **RTF**). But in some other cases such as penicillin resistance in *Staphylococcus,* the resistance may not be associated with ability to transfer (*RTF*). In such a case transfer of drug resistance may be brought about by **transduction.**

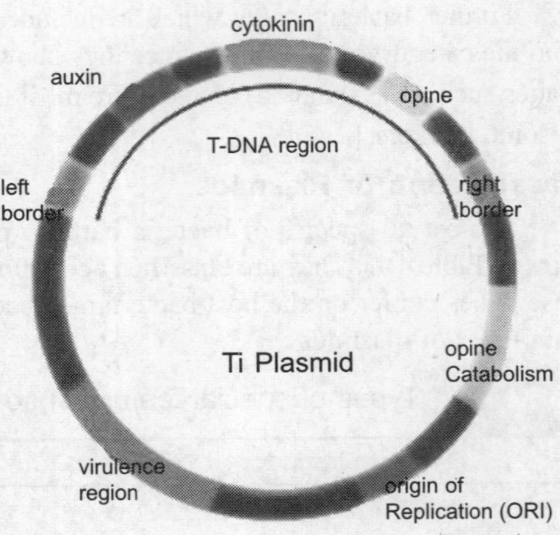

Ti Plasmid of *A. tumefaciens*

3. Col plasmids. These plasmids carry genes coding for colicins, the toxic proteins that kill other bacteria. An example is plasmid *Col E1* of *E. coli.* Colicin is produced by *Escherichia coli, Shigella* and *Salmonella* due to their *Col* plasmids. *Col* plasmids resemble *F* factor or *R* factor. Like transfer of *R* factor, transfer of *Col* factor does not depend on the *F* factor. Since transfer of *Col* factor can occur between two F^- strains of bacteria, one carrying *Col* and other lacking it. There are, however, some *col* plasmids (*e.g., Col* E1) which require the help of sex factor for its transfer. Due to this property, *R* plasmids and *Col* plasmids are also called **nonconjugative** or **nontransmissible plasmids.**

4. Degradative plasmids. These plasmids allow the host bacterium to metabolize unusual molecules such as toluene and solicylic acid. Several examples occur in the *Pseudomonas* genus of bacteria, for instance TOL of *Pseudomonas putida.*

5. Virulence plasmids. These plasmids confer pathogenicity on the host bacterium. The best known example is the *Ti* plasmid of *Agrobacterium tumefaciens,* which induces crown gall disease in dicotyledonous plants.

Replication, Transfer and Recombination in Plasmids

All plasmids discovered so far are small circular DNA molecules. The DNA circles may be single or multiple structures which are linked together in chains. The size of these circular DNA molecules varies from 30μm (*e.g.,* F^+) in plasmid of *E. coli,* making 20% of the main chromosome or can be less than 1μm as in P16C of *E. coli* (Table 4.2). Colicin plasmids such as *Col E1, E2* and *E3* are usually having a length of 2.3 μm. The number of plasmids in a bacterial cell varies from a few in case of *F* and *R* plasmids to as many as 20 – 30 in case of *Staphylococcus* plasmids (Box 4.2).

Table 4.2. Properties of various plasmids

Plasmid	Type	Size (kb)	Copy number	Bacterium
1. *F*	Fertility	95	1 – 2	*E. coli*
2. *Rbk*	Resistance	37	12	*E. coli*
3. *Col E1*	Colicin	6.36	10 – 15	*E. coli*
4. *TOL*	Degradative	117	1 – 2	*Pseudomonas putida*
5. *Ti*	Virulence	213	1 – 2	*Agrobacterium tumefaciens*

Box 4.2.

Copy Number of Plasmids

Each plasmid is maintained in the bacterial cell at a characteristic **copy number** mainly due to its replication control system. There are following methods for the regulation of plasmid's copy number in a cell :

1. Single Copy plasmids. They are maintained in one copy per bacterial cell. Their replication control is the same as that of the bacterial host chromosome so that they replicate and segregate with the bacterial chromosome. Such a replication control is called **stringent control.**

2. Multicopy plasmids. They exist in several, but a characteristic number of copies (typically 10 to 20 copies) per cell. Their replication control permits more than one replications to occur while the host cell chromosome replicates once. This type of replication control is called **relaxed control;** it allows a pool of origins to exist in each cell.

The DNA of plasmids replicates in **semi-conservative** manner. While initiation of replication is controlled by plasmid genes, other stages of replication such as elongation (polymerization) and termination are controlled by bacterial genes. Further, when a plasmid is inserted in the bacterial chromosome, bacterial control on plasmid replication is absolute, but when it is free, the control is less than absolute.

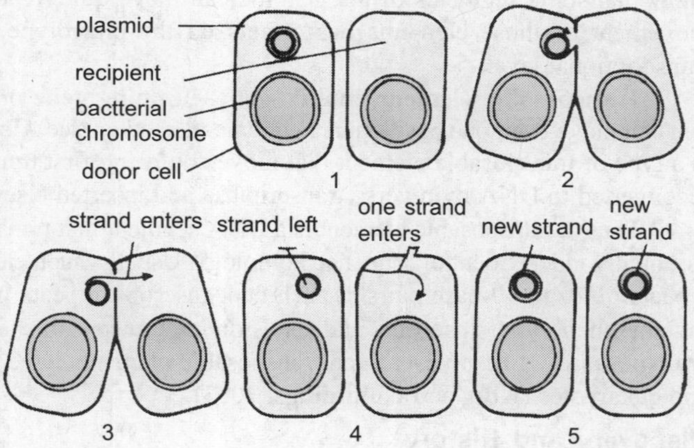

plasmid
recipient
bacterial chromosome
donor cell
1
2
strand enters
strand left
one strand enters
new strand
new strand
3
4
5

Fig. 4.3. Mechanism of transfer of plasmid (*e.g., F, R*) from one bacterium to another.

The property of transfer of plasmid from one bacterium to another is facilitated by the formation of a hair-like extracellular growth, called **sex pilus.** Sex pilus is formed of outer membrane of the donor bacterium. The *F* sex pilus is up to 20 µm in length and 8 nm wide. In *E. coli,* transfer is controlled by plasmid genes *tra-A, tra-B, tra-C,* etc., and mutations in these genes lead to failure of transfer. But, when two plasmids containing mutations for different *tra-* genes are introduced in the same cell, pilus may be formed and transfer may occur due to complementation between mutants.

In the process of transfer, one of the two strands of plasmid DNA is nicked (broken) at a site called **origin of transfer** (*Ori-T*) and the linear strand thus formed moves to the recipient bacteria. The single strands in the donor as well as in recipient cell then synthesize complementary strands, so that duplex DNA plasmid are reconstituted in both cells (Fig. 4.3).

Lastly, plasmid DNA may undergo **recombination** either with another plasmid DNA or with bacterial DNA. In the bacterium chromosome, there are **insertion sequences (IS),** which are homologous in recombination. There are atleast 17 such sites in *E. coli* genome. **Mapping** of genes in plasmid DNA has also been achieved. Methods employed in the plasmid mapping involved **recombination** and **complementation** between strains have overlapping deletions.

4.2. IS ELEMENTS AND TRANSPOSABLE ELEMENTS

A great part of classical genetic analysis has been devoted to the localization of genes on chromosomes. Accordingly, if we map a gene today, we except to see it in the same place tomorrow. However, our discovery of mobile genetic elements has modified that view to some extent. We now

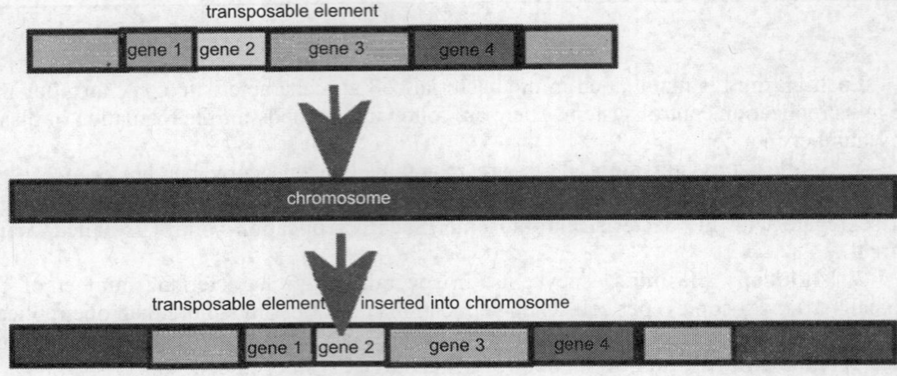

Transposons.

know that some segments of the genome can move quite readily from one place to another. The movement of those elements has effects on the phenotype of the organism, primarily at the transcriptional level.

Transposable element, mobile gene, jumping gene or **transposon** is a segment of DNA that can move from one position in a genome to another (see **Alberts** *et. al.,* 2004). **Retrotransposon** is a type of transposable element that moves by being first transcribed into RNA copy that is then reconverted to DNA by reverse transcriptase and inserted elsewhere in the chromosome.

Thus, a transposable element is a DNA segment that translocates to other sites in the genome, essentially independent of sequence homology. Usually, such elements are flanked by short, inverted repeats of 20 to 40 base pairs at each end. Insertion repeats into a structural gene can produce a mutant phenotype. Insertion and excision of transposable elements depend on two enzymes, **transposase** and **resolvase.** Such transposable elements have been identified in both prokaryotes and eukaryotes (**Klug** and **Cummings,** 2003).

Discovery and History

In the 1940s, in a study of the genetics of kernel mottling in maize, **Barbara McClintock** discovered an element that not only regulated the mottling but also caused breakage of the chromosome carrying the genes for colour and consistency of the kernel. The element was called **dissociation** (*Ds*).

Mapping data showed that the chromosome breakage always occurs at or very near the location of *Ds*. McClintock's critical observation was that *Ds* does not have a constant location but occasionally moves to a new position (*transposition*), causing chromosome breakage at the new site. Furthermore, *Ds* moves only if a second element, called *activator* (*Ac*) is also present. *Ac* itself moves within the genome and can cause in the expression of genes at or near its insertion site, alternations similar to the modifications that result from the presence of *Ds*. McClintock (1956) referred to *Ac* and *Ds* as **controlling elements.** Both of these form a two - factor system of transposable elements of maize. The term **transposable elements** was coined by another greater geneticist **Alexander Brink.**

In the mid - 1960s **Susum Mitsuhashi** first suggested that the gene responsible for resistance to several antibiotics were mobile and could move between bacterial plasmids and chromosomes. The first observation of jumping genes at the molecular level came in the early 1970s. A number of independent researchers, including **Peter Starlinger** and **James Shapiro,** visualised a unique class of mutations affecting different genes in various bacterial strains. The term **transposon** was coined by **R.W. Hedges** and **A.E. Jacob** in 1974 for a DNA segment which could move from one DNA molecule (or chromosome) to other and carried resistance for antibiotic ampicillin. In 1975, **David Hogness** and his colleagues **David Finnegan, Gerald Rubin** and **Michael Young** identified a class of mobile genes in *Drosophila melanogaster* that they designate as *Copia.*

Transposition and Site-specific Recombination

In general type of recombination, DNA rearrangements occur between DNA segments that are very similar in sequence. Although these rearrangements can result in the exchange of alleles between chromosomes, the order of the genes on the interacting chromosome typically remain the same. A second type of recombination, called **site - specific recombination,** moves specialised nucleotide sequences, the *mobile genetic elements,* between nonhomologous sites within a genome. Due to this reason, this type of recombination is also called **illegimate recombination** (**Brown,** 1998). Here, movement can occur between two different positions in a single chromosome, as well as between two different chromosomes.

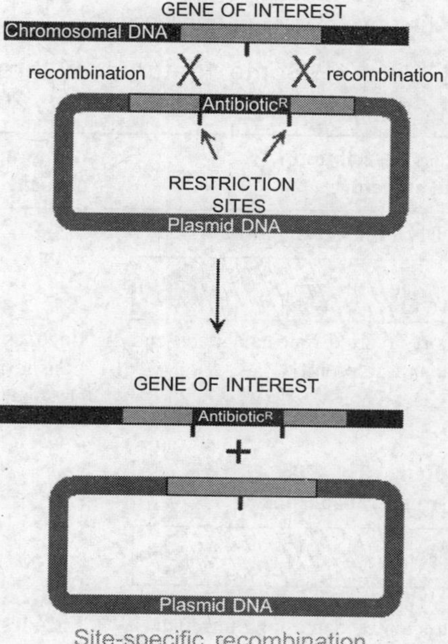

Site-specific recombination.

Unlike general recombination, site - specific recombination is guided by recombination enzymes that recognise short, specific nucleotide sequences present on one or both ends of the recombining DNA molecules. Each type of mobile element generally encode the enzyme that mediates its own movement and contains special sites upon which the enzyme acts. Many elements also carry other genes. For example, virus encode coat proteins that enable them to exist outside cells, as well as essential viral enzymes. The spread of mobile elements that carry antibiotic resistance genes is a major factor underlying the widespread distribution of antibiotic resistance in bacterial population (e.g., spread of resistance for antibiotics such as ampicillin and tetracycline). Site - specific recombination mainly occurs by **transposition** or **transpositional site - specific recombination** which involves breakage reactions at the ends of the mobile DNA segments embedded in chromosomes and the attachments of those ends at one of many different nonhomologous target DNA sites. It does not involve the formation of heteroduplex DNA.

In *transposition,* a specific enzyme, usually encoded by the transposon and called a **transposase,** acts on a specific DNA sequence at each end of the transport — first disconnecting it from the flanking DNA and then inserting it into a new target DNA site. There is no need for homology between the ends of the element and the insertion site. Most transposons move only very rarely (once in 10^5 cell generations for many elements in bacteria).

Types of Transposons

On the basis of their structure and transposition mechanisms, transposons can be grouped into following *three* large classes (Table 4.3) :

1. **DNA only transposon.** The mobile element exists as DNA throughout its life cycle — the translocating DNA segment is directly cut out of the donor DNA and joined to the target site by a transposase enzyme.

2. **Retroviral - like retrotransposons.** They move by a less direct mechanism. An **RNA polymerase enzyme** first transcribes the DNA sequence of the mobile element into RNA. The enzyme **reverse transcriptase** then transcribes this RNA molecule back into DNA using the RNA as a template, and it is this DNA copy that is finally inserted into a new site in the genome. For historical reasons, the transposase - like enzyme that catalyzes this insertion reaction is called an **integrase** rather than a transposase.

3. **Non retroviral retrotransposons.** Each of these transposons moves by making a DNA copy of an RNA molecule that is transcribed from it. However, the mechanism involved for these

nonretroviral retrotransposons is distinct from that in which the RNA molecule is directly involved in the transposition reaction.

All these three types of transposable elements range in length from 1000 to about 12000 nucleotide pairs; each family contains many members. In addition to transposable elements, there are selected viruses that can move in and out of host cell chromosomes; these viruses are related to the first two classes of transposons.

Table 4.3. Some features of three major classes of transposable elements (Source **Alberts** *et al.*, 2004).

Class description and structure	Genes in complete element	Mode of movement	Examples
1. DNA only transposons			
Short inverted repeats at each end of transposable unit	Encodes transposase enzyme	Moves as DNA, either excising or following a replicative pathway	1. Pelement (*Drosophila*) 2. *Ac - Ds* (maize) 3. *Tn3* and *IS1* (*E. coli*) 4. *Tam3* (snapdragon)
2. Retroviral - like retrotransposons			
Directed repeated long terminal repeats (LTRs) at ends	Encodes reverse transcriptase and resemble retrovirus	Moves via an RNA intermediate produced by promoter in LTR	1. *Copia* (*Drosophila*) 2. *Ty1* (yeast) 3. THE - 1 (human) 4. Bs1 (maize)
3. Nonretroviral retrotransposons (A A A A / T T T T)			
Poly A at 3′ end of RNA transcript; 5′ end is often truncated	Encodes reverse transcriptase	Moves via an RNA intermediate that is often produced from a neighboring promotor	1. F element (*Drosophila*) 2. *L1* (human) 3. *Cin4* (maize)

Mode of Transposition of DNA Transposons

DNA transposons may have the following *two* methods of transposition (Fig. 4.4) :

1. Replicative transposition. This transposition mechanism involves direct interaction between the donor transposon and the target site resulting in the copying of the donor element, *e.g.*, Shapiro's model.

2. Conservative transposition. This transposition mechanism involves excision of the element and re-integration at a new

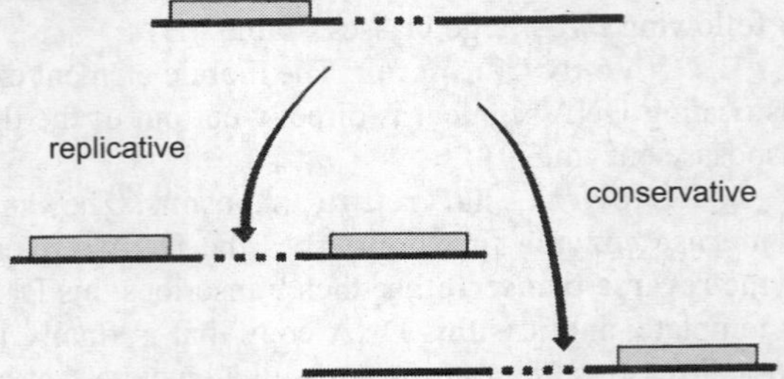

Fig. 4.4. Replicative and conservative transposition. Retroelements transpose replicatively via an RNA intermediate (after Brown, 2002).

site, *e.g.*, cut-and-paste model.

Both mechanisms require enzymes which are usually coded by genes within the transposons. DNA transposons use either the replicative or conservative pathway (some can use both).

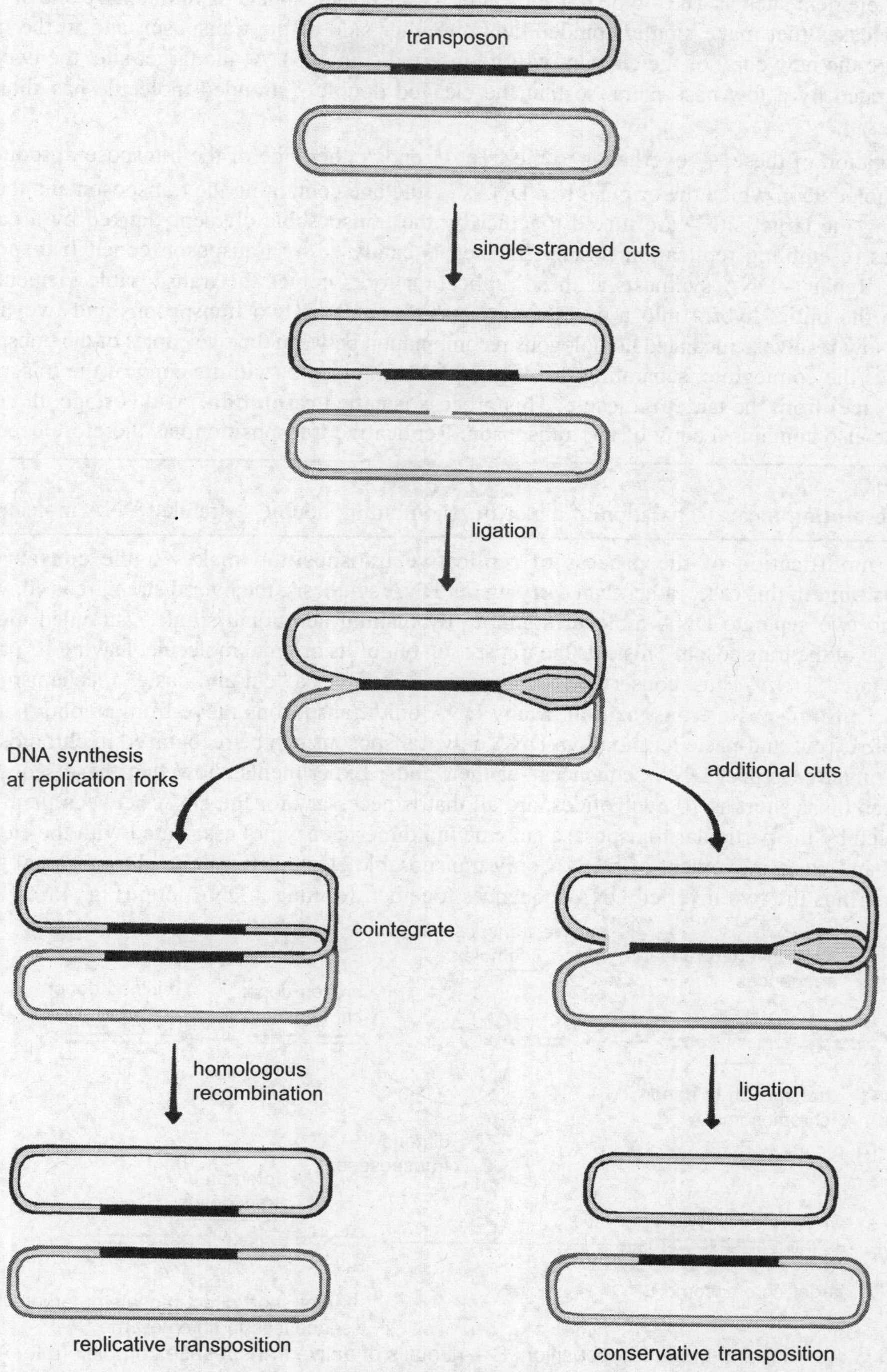

Fig. 4.5. A model for the process resulting in replicative and conservative transposition.

1. Shapiro's model of transposition : A number of models for replicative and conservative transposition have been proposed over the years but most are modifications of a scheme originally outlined by **J.A. Shapiro** (1979). According to Shapiro's model, the *replicative transposition* of a bacterial element such as Tn3 - type transposon or a transposable phage is initiated by one or more **endonucleases** that make single stranded cuts on either side of the transposon and in the target site where the new copy of the element will be inserted (Fig. 4.5). At the target site, the two cuts are separated by a few base pairs, so that the cleaved double - stranded molecule has short 5′ overhangs.

Ligation of these 5′ overhangs to the free 3′ ends either side of the transposon produces a hybrid molecule in which the original two DNAs — the one containing the transposon and the one containing the target site – are linked together by the transposable element flanked by a pair of structures resembling replication forks. This step is catalysed by transposon coded **transposase** enzyme. Repair - DNA synthesis at these replication forks copies the transposable element and converts the initial hybrid into a **co-integrate,** which contains two transposons and two linked DNAs. Now **resolvase** mediated homologous recombination between the two copies of the transposon uncouples the cointegrate, separating the original DNA molecule (with its copy of the transposon still in place) from the target molecule. This process is called **resolution.** At this stage, the target molecule also contains a copy of the transposon. Replicative transposition has therefore occurred.

Box 4.3.

Resolution means separation of a pair of recombining double - stranded DNA molecules.

A modification of the process of replicative transposition makes it the **conservative transposition.** In this case, rather than carrying out DNA synthesis, the hybrid structure is converted back into two separate DNA molecules simply by making additional single - stranded nicks at either side of the transposon. This cuts the transposon out of its original molecule, leaving it "pasted" into the target DNA. Thus, conservative transposition involves a "cut-and-paste" mechanism.

2. Cut-and-paste transposition. Many DNA only transposons move from a donor site to a target site by cut-and-paste termination. DNA only transposons can be recognised in chromosomes by the "inverted repeat DNA sequences" at their ends. Experiments show that these sequences, which can be as short as 20 nucleotides, are all that is necessary for the DNA between them to be transposed by the particular transposase enzyme (an dimeric enzyme) associated with the element. The cut-and-paste movement of a DNA only transposable element from one chromosomal site to another brings the two inverted DNA sequences together, forming a DNA loop (Fig. 4.6).

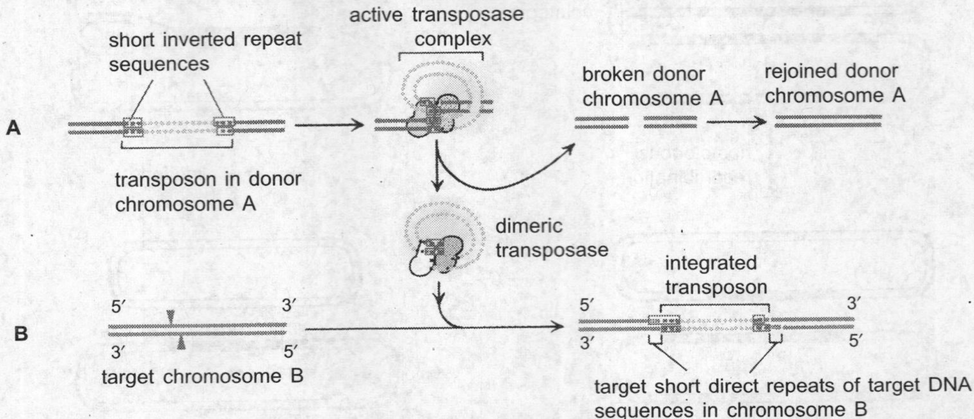

Fig. 4.6. A — cut-and-paste transposition; B — Details of active transposase complex (after Alberts *et al.*, 2004).

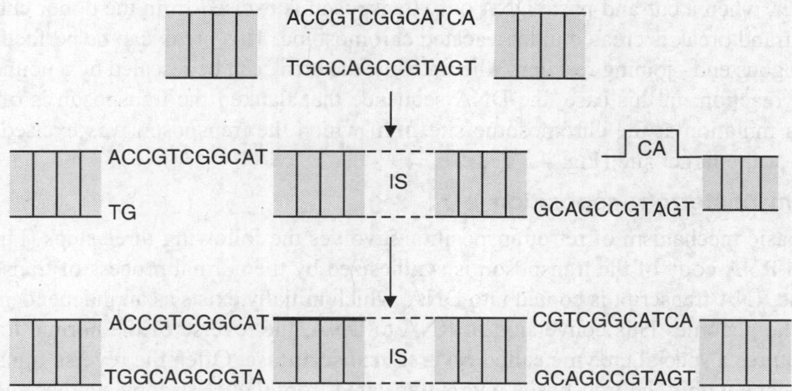

Fig. 4.7. Production of target site duplication by the insertion of IS element. The element causes breaks in the target DNA at the sites indicated by the arrows. After insertion, DNA synthesis fills in the gaps left by the staggered gaps, producing direct repeats on each side of the element (after Gardner *et al.,* 2001).

Insertion into the target chromosome, catalysed by the transposase enzyme, occurs at random site through the creation of staggered breaks in the target chromosome. As a result, the insertion site is marked by a short direct repeat of the target DNA sequence (Fig 4.6).

Since, the breaks made in two target DNA strands are staggered (Fig. 4.7), two short, single - stranded gaps are initially formed in the product DNA molecule, one at each end of the inserted transposon. These gaps are filled in by a host cell **DNA polymerase** and **DNA ligase** to complete the recombination process, producing a short duplication of the adjacent target DNA sequence. These flanking direct repeat sequences serve as convenient markers of a prior transpositional site - specific recombination event.

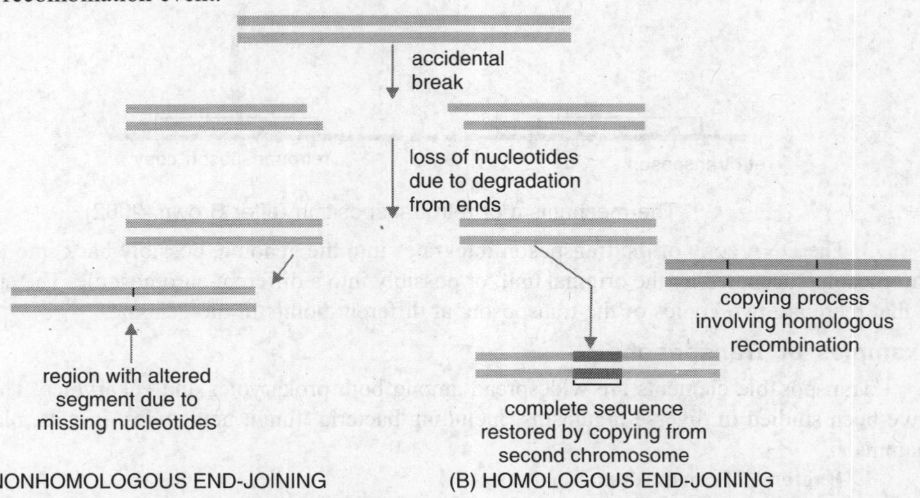

Fig. 4.8. Two different types of end - joining for repairing double - strand breaks. A nonhomologous end - joining alters the original DNA sequence when repairing broken chromosomes. B - Homologous end - joining. Here undamaged chromosome act as the template for transferring genetic information to the broken chromosome, repairing it with no change in DNA sequence (after Alberts *et al.,* 2004).

Further, when a cut-and-paste DNA only transposon is excised from the donor chromosome, a double - strand break is created in the vacated chromosome. This break can be perfectly "healed" by a homologous end - joining reaction. Alternatively, the break can be resealed by a nonhomologous end joining reaction; in this case, the DNA sequence that flanked the transposon is often altered producing a mutation at the chromosome site from which the transposon was excised. Mutation also occurs at the target site (Fig. 4.8).

Mechanism of Retrotransposition

The basic mechanism of retrotransposition involves the following three steps (Fig. 4.9) :

1. An RNA copy of the transposon is synthesized by the normal process of transcription.

2. The RNA transcript is copied into DNA, which initially exists as an independent molecule outside of the genome. This conversion of RNA to DNA, the reverse of the normal transcription process, requires a special enzyme called **reverse transcriptase.** Often the reverse transcriptase is coded by a gene within the transposon and is translated from the RNA copy synthesized in step 1.

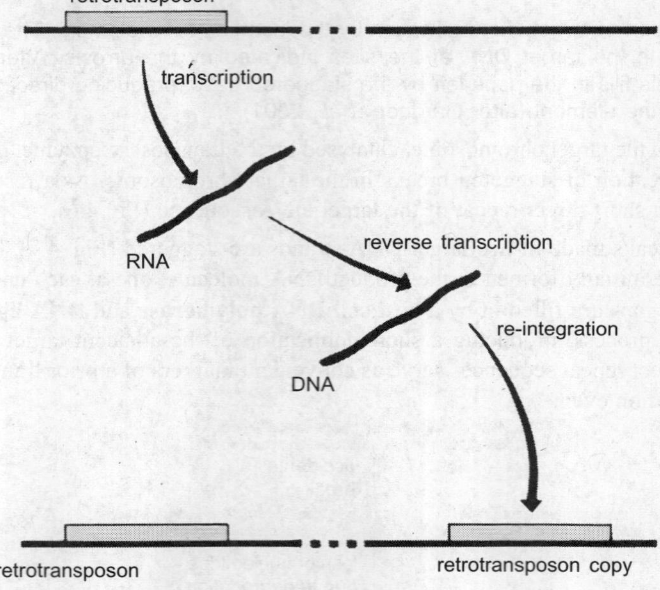

Fig. 4.9. The mechanism of retrotransposition (after Brown, 2002).

3. The DNA copy of the transposon integrates into the genome, possibly back into the same chromosome occupied by the original unit, or possibly into a different chromosome. The end result is that there are two copies of the transposon, at different points in the genome.

Examples of Transposons

Transposable elements are widespread among both prokaryotes and eukaryotes. Thus, they have been studied in diverse organisms, including bacteria, fungi, nematodes, insects, plants and mammals.

1. Bacterial Transposons

***(i)* Insertion sequence (IS).** During the 1960s geneticists were puzzled by a set of unusual mutations which were occasionally found in operons of *E.coli* and other bacteria. These are called **polar mutations** because they affect not only the gene in which the mutation occurs but also other genes downstream of the mutation (Fig. 4.10). Several types of polar mutations are known and often they are ones that *introduce a transcription termination signal into the operon.*

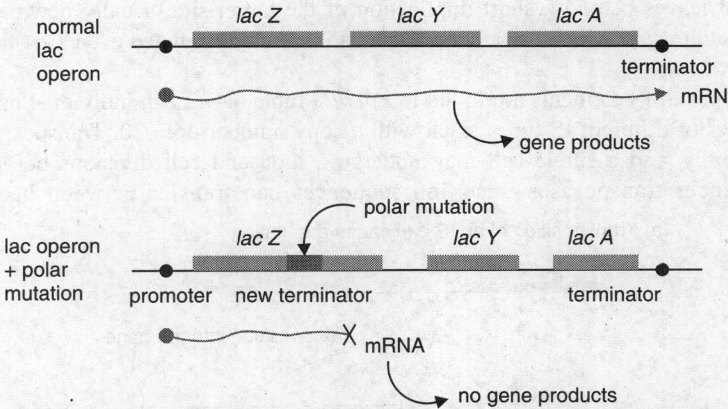

Fig. 4.10. The effect of a polar mutation in the *lac* operon. The polar mutation in *lacZ* prevents expression of *lacY* and *lacA*, even though the last two genes are not themselves altered (after Brown, 1998).

In contrast to other known mutations these polar mutations are always **irreversible.** Further, if a plasmid is introduced into a bacterium that carries a polar mutation of this type, then similar irreversible polar mutations suddenly appear in genes carried by the plasmid (Fig. 4.11). It happens as though the mutation is able to jump from one DNA molecule to another.

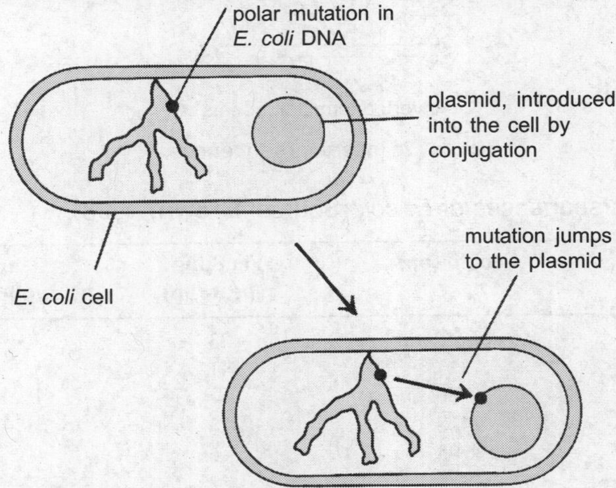

Fig. 4.11. The unusual type of polar mutation that is able to jump from one position to another on *E. coli* DNA molecule (after Brown, 1998).

The first hint of an explanation of this strange phenomenon appeared when the plasmids that had become mutated were studied. It was found that these plasmids had increased in size, generally a kilobase or more. When the DNA sequences of the mutated plasmids were examined it was discovered that the mutations were caused by the insertion into the target genes of new pieces of DNA, which were subsequently called **insertion sequences** or **ISs** (Fig. 4.12). These insertion sequences possess the remarkable capability of being able to move around the genome from site to site by a process dependent on recombination and transposition. Moreover, when an IS transposases

Box 4.4.

IS elements are the simplest forms of transposons. They are so called because they were discovered as spontaneous insertions within bacterial operons that prevented transcription or translation of concerned gene into which insertion has occurred.

to a new site it leaves behind a short duplication of the target site, usually between 5 to 15bp, which means that the gene which carried the IS probably remains mutated even after the IS element has departed.

Various types of IS elements are found in *E. coli* (Table 4.4). Each individual bacterium may contain a number of different IS types, each with a copy number upto 20. Transposition does not occur so frequently, and a single cell may undergo a thousand cell divisions before one of its insertion sequences transposases. Insertion sequences can transfer between bacteria during

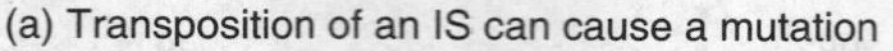

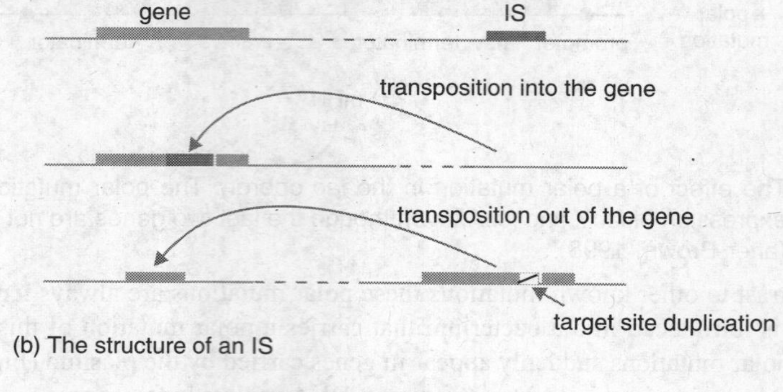

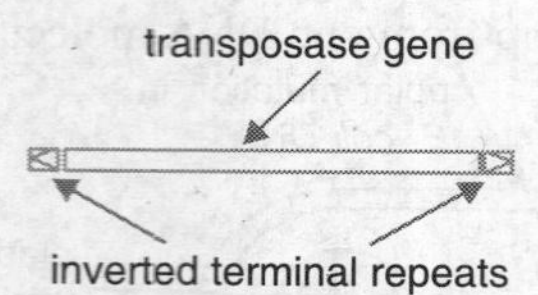

Fig. 4.12. Insertion sequences.

Table 4.4. Insertion sequences of *E. coli* (Source: **Brown,** 1998).

Insertion sequence	Length (bp)	Flanking repeat (bp)	Target site duplication (bp)
1. IS1	769	23	9
2. IS2	1327	41	5
3. IS3	1300	—	1
4. IS4	1428	18	11
5. IS5	1195	16	4
6. IS10	1329	22	9
7. IS50	1531	9	9

conjugation, and can also transfer between bacteria during conjugation and can also transfer between related species. The most part of each IS element is taken up by a single gene (except IS1 which has two) coding for a transposase enzyme that catalyses the transposition event. The IS element has at its ends a pair of **inverted repeats** (Box 4.5) between 9 and 41 bp in length, depending on the IS type (Fig. 4.12).

Box 4.5.

Inverted Repeats

Consider the following squence of a DNA molecule :

5′ — A G G C A T — 3′

3′ — T C C G T A — 5′

Inverted repeat of above DNA sequence will be :

5′ — A G G C A T A T G C C T — 3′

3′ — T C C G T A T A C G G A — 5′

Repeats may occur adjacent to one other, as shown here, or may flank a different DNA sequence, as with the repeats at the ends of insertion sequences and Tn - 3 type transposons.

(ii) **Composite transposons.** They consist two insertion sequences flanking an internal region which usually includes at least one gene, often coding for resistance to an antibiotic (Fig. 4.13A). Transposition occurs because of the transposase enzyme coded by IS elements. There are several different composite transposons (Table 4.5), some over 10kb in size. Often these types of element are carried by the plasmids.

Table 4.5. Composite and Tn3 - type transposons of *E. coli*

Transposons	Size (bp)	Flanking element	Gene carried
(a) Composite transposons			
(i) Tn 5	5700	IS50	Kanamycin resistance
(ii) Tn 9	2500	IS1	Chloramphenicol resistance
(iii) Tn10	9300	IS10	Tetracycline resistance
(b) Tn3 - type transposons			
(i) Tn 3	~ 5000	38 bp inverted repeats	Ampicillin resistance
(ii) Tn 501	~ 5000	38 bp inverted repeats	Mercury resistance

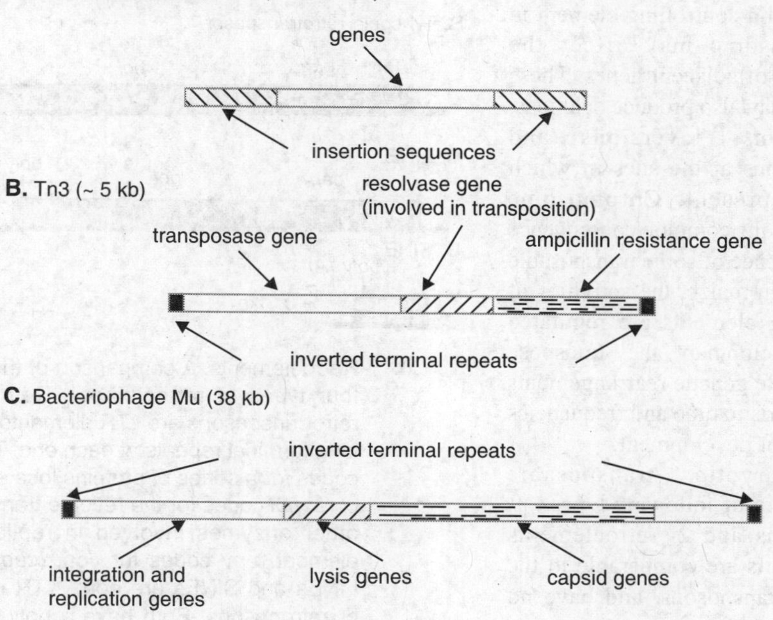

Fig. 4.13. Three types of bacterial transposable elements.

(iii) **Tn3 - type transposons.** They are not flanked by IS elements but are still able to transpose because they carry a different transposase enzyme (Fig. 4.13, Table 4.5).

(iv) **Transposable phages.** Certain phages such as Mn and D108 transpose as part of their normal infection cycle (Fig 4.13C). This means that besides inserting themselves into the host DNA (like phage λ), the genomes of these phages are able to move about from place to place in the host DNA (unlike phase λ). Transposable phages may be transposons that have evolved genes for protective proteins, enabling them to exist outside the cell, or they may be ordinary phages that have learned to transpose (see **Brown,** 1998).

Significance of bacterial transposable element. Transposons have no apparent function. Many evolutionary geneticists suggested that transposons are an evolutionary accident that once created, are self - maintaining. Since they may exist without an noticeable benefit to the host's phenotype, transposons have been referred to as **selfish DNA,** *i.e.*, they exist only to further their own existence (propagation), contributing nothing to the cell as a whole. However, it has been established that bacteria which carry a large number of transposons have a **selective advantage** over those with few transposons, possibly because transposons increase the mutation rate of the bacterium to adapt to changing environmental conditions.

In higher organisms, the retroviruses, which include some viruses implicated in cancer as well as the AIDS virus, resemble transposable elements. In the laboratory, transposable elements are used as **mutagens** because they inactivate genes that they transpose into, or possibly modulate the expression of a gene downstream from an insertion site (*e.g.*, in lac operon of *E. coli*) in both cases, a phenotypic change indistinguishable from a mutation.

2. Eukaryotic Transposons and Retroelements

Eukaryotes contain a variety of transposable elements that were originally discovered in maize and described as **controlling elements.** Insertion of a controlling element in a chromosome may affect the expression of adjacent genes. These elements may also produce deletions, duplications, inversions and translocations at the sites in which they are present. Chromosome breakage is the common consequence of the presence of some transposable elements. In maize, the activities of controlling elements are regulated during development; they transpose and promote genetic rearrangements at characteristic time and frequencies during plant development.

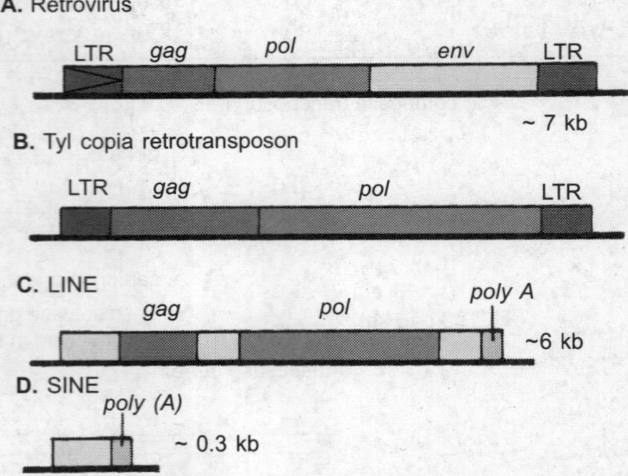

Fig. 4.14. Retroelements. A comparison of the structure of four types of retroelement. Retroviruses and retrotransposons are LTR elements that possess long terminal repeats at each end. The *gag* gene codes for a series of proteins located in the virus core; *pol* codes for the reverse transcriptase and other enzymes involved in replication of the element. *env* codes for coat (capsid) proteins. LINEs and SINEs are non - LTR retroelements or retroposons. Both have a poly (A) region (a long series of A nucleotide) at one end (after Brown, 2002).

Eukaryotic transposable elements are of following types : 1. transposons and 2. retroelements **Transposons** are comparable to the bacterial transposons, and have no life outside the genome, *e.g.,* controlling elements of maize. **RNA transposons** or **retroelements** are features of eukaryotic genomes but have not so far been discovered in prokaryotes (see **Brown** 2002). They have attracted great attention because of their close similarity with free - living viruses, called **retroviruses.** These viruses include both benign forms and also virulent types such as the human immunodeficiency viruses (HIV) that cause AIDS. These structural relationships are illustrated in Figure 4.14 and are of following types :

1. Retroviruses. These are exogenous viruses whose genomes are made of RNA. Their genetic organisation have been shown in (Fig. 4.14). They infect many types of vertebrates. Once inside a cell, the RNA genome is copied into DNA by the reverse transcriptase enzyme specified by the viral *pol* gene, and the DNA copy integrates into the host genome. New viruses can be produced by copying the integrated DNA into RNA and packaging this into virus coat proteins, the latter coded by the *env* genes on the virus genome.

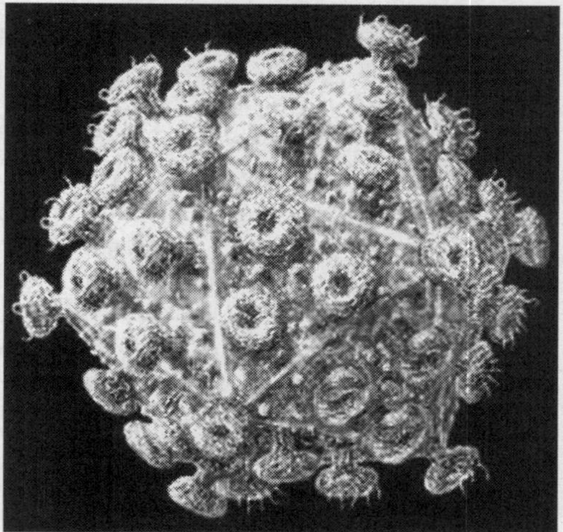
Retroviruses.

2. Endogenous retroviruses (ERVs). These are retroviral genomes integrated into vertebrate chromosomes. Some are still active and might, at some stage in a cell's lifetime, direct the production of exogenous viruses, but most are decayed relics that have lost the capacity to form viruses (**Patience** *et al.,* 1997). These inactive sequences are interspersed repeats of the genome and they are not capable of additional proliferation.

3. Retrotransposons. These elements have sequences similar to ERVs but are features of non - vertebrate eukaryotic genomes (*i.e.*, plants, fungi, invertebrate and microbial eukaryotes) rather than vertebrates. Retrotransposons are of many types and each has very high copy numbers in some genomes. Most of the interspersed repeats in maize are retrotransposons; in maize these elements form almost half of the genome (**San Mignel** *et al.,* 1996). There are two types of retrotransposons :

(*i*) *Ty3 / gypsy* family. The members of this family have the same set of genes as an ERV, *e.g.*, T3 genes of yeast, *gypsy* gene of *Drosophila.*

(*ii*) *Ty1 copia* family. The members of this family lack the *env* gene.

Both of these types of retrotransposons are able to transpose but the absence of the *env* gene means that the *Ty1/copia* group cannot form infectious virus particles. In fact, despite the presence of *env* in the *Ty3/gypsy* genome, some of these elements can form viruses and hence should be regarded as invertebrate retroviruses (**Song** *et al.,* 1994; **Peterson - Burch** *et al.,* 2000). Although technically they are interspersed elements, retrotransposons are sometimes found in clusters in a genome sequence as a result of the presence of preferred integration sites for transposing elements.

The three types of retroelements described so far are **LTR elements**, as they have long terminal repeats at either end which play a role in the transposition process (Fig. 4.14). Other retroelements do not have LTRs. These are called **retroposons** and in mammals include the following (Table 4.6) :

1. LINEs (Long Interspersed Nuclear Elements)

They contain a reverse - transcriptase - like gene probably involved in the retrotransposition process (Fig 4.14C). An example is the human mobile element LINE - 1 which is 6.1 kb long and has a copy number of 516000 in the human genome (Table 4.7).

2. SINEs (Short Interspersed Nuclear Elements)

They do not have a reversed transcriptase gene but can still transpose, probably by "borrowing" reverse transcriptase enzymes that have been synthesized by other retroelements (Fig. 4.14D). The commonest SINE in the human genome is *Alu*, which has a copy number of over one million (Table 4.8). *Alu* seems to be derived from the gene for the 7SL RNA, a non - conding ScRNA (= small cytoplasmic RNA) involved in transport of proteins around the cell. The first *Alu* element may have arisen by the accidental reverse transcription of a 7SL RNA molecule and integration of the DNA copy into the human genome. Some *Alu* elements are copied into RNA, providing the opportunity for proliferation of the element.

Examples of Eukaryotic Transposons

A. Transposons in Maize

In maize at least 14 different transposable systems are known, *e.g.*, *En, As-Ds, mu,* etc. Each of these elements is present in multiple copies in maize genome. These have the following general characteristics :

Phos mμ transposons in maize.

1. Autonomous and nonautonomous controlling elements. The controlling elements of maize may be grouped into two classes : 1. autonomous controlling elements; and 2. nonautonomous controlling elements. **Autonomous controlling elements** have the ability to excise and transpose themselves and corresponding nonautonomous elements. **Nonautonomous controlling elements** do not have the ability to transpose; they transpose only when an autonomous member of same family is present elsewhere in the genome. The nonautonomous controlling elements may be derived from autonomous controlling elements by the loss of *trans*-acting functions needed for transposition. The autonomous controlling elements of maize are *Ac* ('activator'; V9 copies/genome), *MuI* (mutator; > 10 copies/genome), *Spm* ('suppressor mutator') and *Dd* ('dotted'). The nonautonomous controlling elements are *Ds* (dissociator, released to *Ac*), *dspm* (defective *Spm*; related to *Spm*) and *I* ('inhibitor'; related to *En*). The nonautonomous controlling elements possess the sites such as terminii, on which transpoases act. They also have the defective *cis*-acting functions which make the elements permanently stable (since they can no longer be activated to transpose).

2. Change of phase. Autonomously controlling elements may be reversibly inactivated during certain phases of plant development; this is called '**change of phase**'. The phenomenon of change of phase is heritable but relatively unstable so that a controlling element cycles between active and inactive states during development. The inactivation of both *Ac* and *Mu* seems to be due to cytosine **methylation** in the target sequence CAG/GTC. Both *Ac* and *Mu* elements have several such target sites in each element. There may be self - regulative controls of transposition; an increase in the frequency of its transposition.

Example of transposons of maize are the following :

1. En controlling element. The *En* element (*En* = enhancer) codes for a single enzyme called **transposase,** which is believed to induce its transposition. A deletion of the *En* element during coding for this enzyme results in the element *I* ('inhibitor'); *I* element is present in several copies in each maize line but it is nontransposable in the absence of *En* which acts as the signalling for *I*. A similar situation exists for the *Ac* - *Ds* system maize chromosome, it may generate three - base duplication, *e.g.*, ATA, ATA, etc., in the gene from which it moves out; produce an inversion for a segment of the gene; or even cause duplication of entire gene. In view of these genetical effects, transposable elements are often referred to as **genome reorganizers.**

2. *Ac-Ds* controlling elements. *Ac* (activator) controlling elements are 4,563 bp long, encode a single mRNA of 807 codons (which encode transposase enzyme) from a gene having 5 exons.

They have 11 bp inverted repeats at their ends. *Ac* elements generate repeats of 8bp at the sites of their insertion. The nonautonomous counterparts, of *Ac* elements occur in a variety of forms, *e.g., Ds1, Ds2d1, Ds2d2, Ds6, Ds9*, etc., all of which contain the same 11 bp inverted repeats at their ends. *Ds* (= dissociation) elements are shorter than *Ac*, and the length of deletion varies from merely 194 bp in *Ds9* to over 2.5 kb in *Ds6* (Fig. 4.15). Evidently, non-autonomous controlling elements (*Ds* elements) are derived from autonomous elements by deletions or other changes that inactivates their transposase enzyme coding regions, but the sites on which this enzyme acts remain intact. Members of *Ds1* family, however, are different their only resemblance to *Ac* is that they possess the same 11 bp terminal inverted repeats. *Ds1* controlling elements are short and were possibly not derived directly from *Ac* elements.

Like the bacterial composite transposon such as *Tn10*, transposition of *Ds* elements occur by a nonreplicative mechanism. The *recipient site* is frequently on the same chromosome as the *donor site* and often quite close to it. In them, transposition almost always occurs after DNA replication, and usually only of the two copies produced by replication undergoes transposition. The basis for this may be similar to that for *Tn10* and could involve asymmetrical **methylation sites** as C* GTG / G* CAC. After replication, the daughter DNA duplexes will have different methylated sequences, *viz.*, C* GTG / GCAC and CGTG / G* CAC; one of these sequences may be recognised by the transposase enzyme more efficiently than the other.

The donor site suffers a break, which could lead to a simple loss of the chromosome fragment distal to the centromere (hence lacking a centromere) or produce a breakage - fusion - bridge cycle. A loss of acentric fragment allows the expression of recessive allele present on the corresponding region of the homologous chromosome. The breakage - fusion - bridge cycle, however, continues through the future generations causing changes (deletion, duplication) in the affected chromosome.

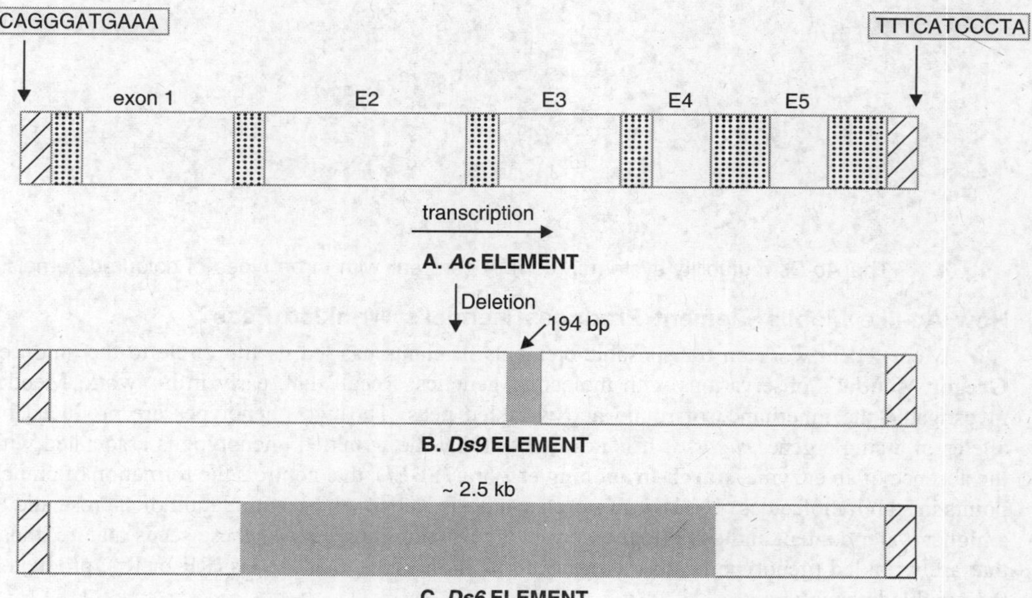

Fig. 4.15. Organization of maize *Ac* and *Ds* elements. A — an *Ac* element produces a single mRNA from a gene having 5 exons (depicted as Exon 1, E2, E3, E4 and E5) and ends in 11 bp inverted repeats. B— and C— Ds elements are usually derived by deletion of internal sequences from the *Ac* element; the region of deletion is depicted as solid blue area).

How purple spots are produced in maize kernels?

Purple spots in maize kernel as a result of dominant functioning alleles.

Ds element affects the phenotype (in maize) by blocking expression of the genes it transposes into as well as causing the loss of alleles in acentric chromosome fragments lost when *Ds* breaks its chromosome. In coloured figure, we see *three* kinds of kernels in the ear of corn - *purple, bronze without purple spots* and *bronze with purple spots.* The purple kernels result from the presence of dominant functioning alleles providing enzymes in the pathway for purple pigment. In those kernels that are bronze without spots, *Ds* elements have transposed into both copies of the B_z^2 locus, disrupting the pigment pathway. Without the *Ac* element present, the *Ds* elements remain in place and the kernels are a uniform bronze colour. In those bronze kernels with purple spots, the *Ac* element has entered the genome in the genetic cross. In the presence of *Ac, Ds* leaves its site in some of the cells, restoring the activity to the B_z^2 locus. This restored activity results in purple spots in those cells and their progeny with the functioning B_z^2 allele (**Tamarin**, 1996). Purple spots difer in size based on time of excision of the *Ds* element during the development of the kernel : early yields large spots; late yields small spots.

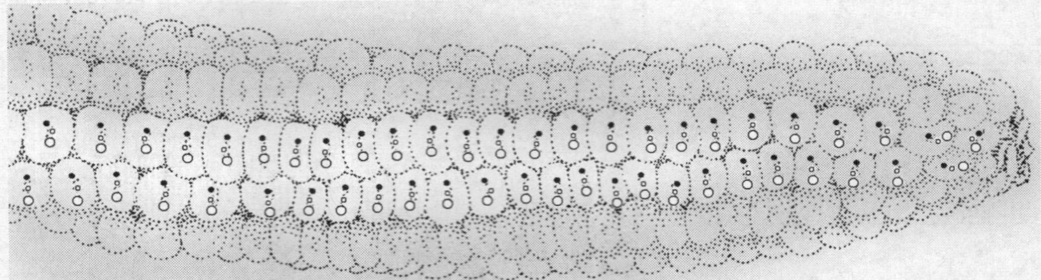

Fig. 4.16. The *Ac-Ds* mutability system in corn. A corn ear with three types of coloured kernels.

How Ac-like Mobile Element Produces Mendel's Wrinkled Peas?

More recent work on transposable elements in plants has led us full circle to the union of **Gregor Mendel's** observations with molecular genetics. Recall that, early in his work, Mendel investigated the inheritance of round and wrinkled peas. The two phenotypes are produced by alleles of a single gene, *rugosus.* It is now known that the wrinkled phenotype is associated with the absence of an enzyme, **starch-branching enzyme (SBEI),** that controls the formation of branch points in starch molecules. The lack of starch synthesis leads to the accumulation of sucrose and to a higher water content and osmotic pressure in the developing seeds. As these seeds mature, those that are wrinkled (genotype *rr*) lose more water than do the smooth seeds (RR or Rr), producing the wrinkled phenotype.

The structural gene for SBEI has been cloned and characterized in both wild type and mutant genotypes. In the *rr* genotype, the SBEI protein is non-functional, presumably because the SBEI gene is interrupted by a 0.8 - kb insertion, resulting in the production of an abnormal RNA transcript. The inserted DNA has 12 - bp inverted repeats at each end that are highly homologous to the terminal sequences in the transposable element *Ac* from maize and to other *Ac* - like elements from snapdragons and parsley. Terminal repeated sequences and the genetic information encoding a transposase enzyme appear to be universal components of transposons in all organisms studied.

3. *Spm-dSpm* controlling elements. The *Spm* (suppressor mutator) elements have 13bp inverted repeats at their termini and have two genes. Gene *tnpA* has 11 exons that are spliced to produce a 2,500 base mRNA, which encodes TnpA protein. Protein TnpA binds to a 12bp consensus regions of *Spm.* The action of protein TnpA is required for excision, but the excision of itself is effected by a hypothetical protein called TnpB. The protein TnpB is presumably coded by a 6000 base mRNA produce by alternative splicing of *tnpA* transcript; this mRNA contains the two open - reading frames, ORF - 1 and ORF - 2, both are present in the first large intron of gene *tnpA* (Fig. 4.17). ORF - 1 and ORF - 2 are present in an alternatively spliced RNA of 6,000 bases; this RNA is present at ~ 1% of *tnpA* mRNA. Protein TnB protein presumably binds the 13 bp terminal inverted repeats, and cleaves the termini for transcription.

A group of *Spm* derivatives are weakly autonomous; they are denoted as *Spm-w.* These controlling elements may show deletion in ORF - 1 and ORF - 2 regions; this indicates that *TnpB* function during transposition may be bypassed or substituted. The *dSpm* controlling elements show deletion of *tnpA* exon sequences and the length of deleted sequences varies greatly.

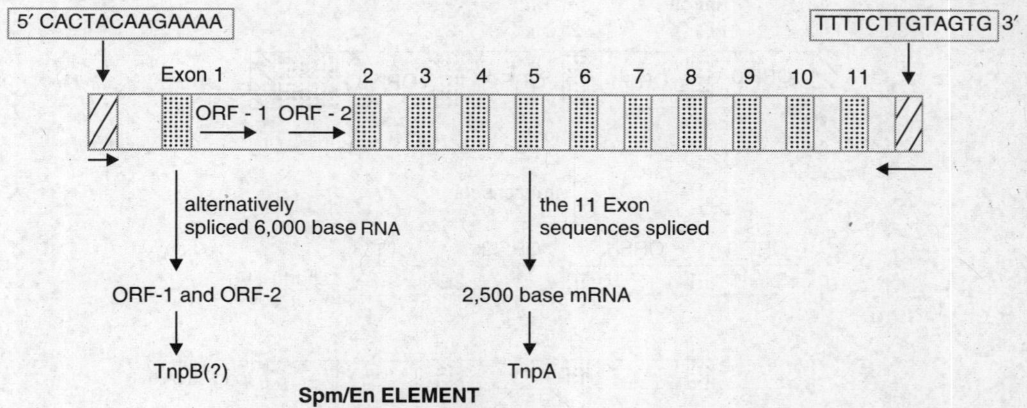

Fig. 4.17. The organization and function of *spm/En* controlling elements. The *Spm* and *En* elements differ at < 10 positions.

4. Dotted (*dt*) element. Dotted (*Dt*) element was discovered by **Marcus Rhoades** in 1938, when from a maize's ear of a pure breeding pigmented genotype of maize among kernels a dihybrid segregation ratio of 13 : 3 : 1 (*i.e.*, 13 pigmented : 3 dotted : 1 colourless) was obtained. This was surprising because no segregation was expected. The genetic analysis suggested that the original phenotype (full colour) was controlled by two genes with the genotype A_1, A_1, *dt dt* and that two mutations ($A_1 \rightarrow a_1$ and $dt \rightarrow Dt$) gave the genotype A_1 a_1 Dt *dt* which gave the following phenotypic and abbreviated genotypic ratio (Fig. 4.18) : 12/16 pigmented (A_1Dt = 9/16; A_1dt = 3/16) : 3/16 dotted (a_1Dt) : 1/16 colourless (a_1 dt).

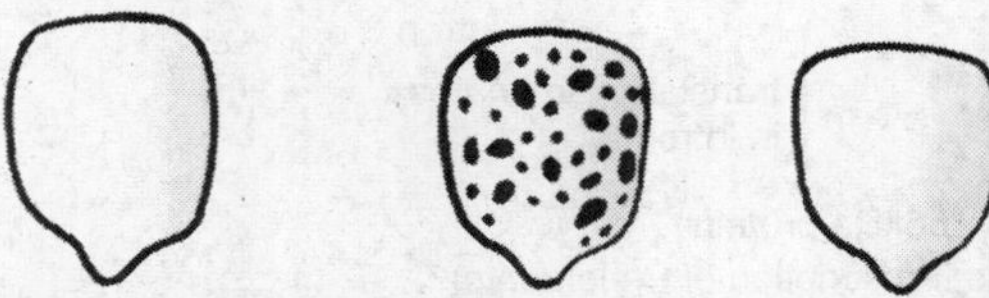

Fig. 4.18. Appearance of dotted phenotype in maize, due to two simultaneous mutations ($A_1 \rightarrow a_1$) and $dt \rightarrow Dt$).

The dotted phenotype (appears in kernels and leaves of maize) in a_1Dt, would probably result from the reverse mutation ($a_1 \rightarrow A_1$) in some of the somatic cells, and the frequency of this mutation should be unusual high. This led to discovery of a_1 as an unstable mutant allele (its instability is dependent on *Dt*) showing high rate of reverse mutation. It was also shown that once reverse mutations have occurred, they are stable, reverted A_1 being transmitted to the progeny in a stable form. It means that *Dt* causes mutations in a_1 to A_1, which remain stable.

B. Transposons in *Drosophila*

The genome of *Drosophila melanogaster* has several transposable elements, *e.g.*, P factor.

P factor element. In *D. melanogaster,* F_1 hybrids derived from certain strains show a series of defects, including high mutation rate, high frequency of chromosomal aberrations, non - disjunctions and sterility. The phenomenon by which such abnormal hybrids are produced is called **hybrid dysgenesis.** For example when P (paternal contributing) type of males are mated to M (maternal contributed) type females, the F_1 progeny are sterile, while those from the reciprocal cross (P female × M male) are normal fertile. The P type flies have a large number (30 to 50 copies) of **P factors** (the transposable elements) which become active in the cytoplasm of M flies. The active P factors transpose readily creating chromosome breaks and new mutations (*i.e.,* hybrid dysgenesis).

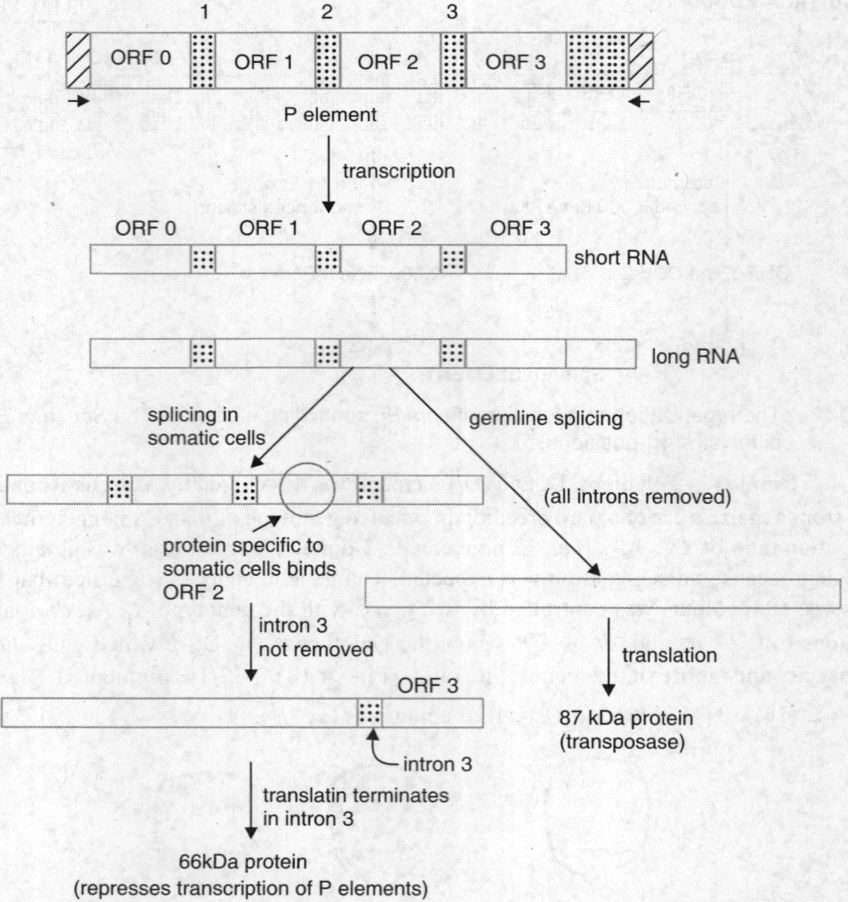

Fig. 4.19. Tissue - specific splicing of P element transcript leads to transposition in germline cells of F_1 individuals from crosses between P males and M females.

The P elements vary in length, the longest being ~ 2.9 kb. They all have inverted terminal repeats of 31 bp (Fig. 4.19) and generate direct repeats of 8 bp in target DNA. The full length P factors have 4 ORFs (= open reading frames), while their shorter types arise by internal deletions in them. The shorter P elements arise more frequently so that about 2/3 of the elements present in P males are of this type. Transposition of P elements is nonreplicative like that of *Tn10* (of *E. coli*), but the breaks in donor chromosomes are often repaired by using the corresponding sequence from homologous chromosomes. Transposition requires ~ 150 bp of terminal DNA and transposase enzyme binds to 10bp sequences that are adjacent to the 31bp inverted terminal repeats. Further, transposition of P elements is tissue - specific in that it occcurs only in the germline cells; it begins at the stage when rapid cell division starts in the germ line. This tissue specificity occurs due to the tissue specific alternative splicing of P elements transcripts as follows :

1. **Germline splicing.** It is a defective splicing pattern in which the 3 introns separating the 4 ORFs are spliced out generating an mRNA that encodes 87 kDa transposase enzyme. This enzyme induces P element transposition (Fig. 4.19).

2. **Somatic splicing.** It is induced by a protein, present in somatic cells of *Drosophila.* This protein binds to sequences in ORF-2 and prevents the removal of 3rd intron, which has a termination codon. The mRNA generated by this splicing produces a 66 kDa protein that does not contain information from ORF-3. This protein represses transcription of P elements, thus, it stops transposition and forms the basis for reciprocal differences seen in P × M crosses of *Drosophila* (Fig. 4.19).

P cytotype and M cytotype. *Drosophila* of P strain contain P elements in their genome (*P chromosomes*) and their cytoplasm has the 66 kDa protein; the cytoplasm of such flies is called **P cytoplasm** and they are said to have **P cytotype.** In contrast, flies belonging to M strain do not have P elements in their chromosomes (*M chromosomes*) and lack the 66 kDa protein in their cytoplasm. These flies are said to have **M cytoplasm** that gives rise to the **M cytotype.** The 66 kDa protein is a highly stable protein and is evidently present in very large quantities in the eggs so that the P cytotype effect on transposition of P elements may continue for more than one generation.

Thus, when P males are mated to P females, or M males are mated to P females, the eggs contain the 66 kDa repressor protein. The transcription of P elements, therefore, remains suppressed and there is no transposition either in somatic or germline cells. But when M females are crossed to P males, the eggs lack the 66 kDa protein. P elements are, therefore, transcribed to produce the 66 kDa protein in somatic cells which suppresses further transcription of P elements. But in germline cells, defective splicing of P element transcripts leads to the production of transposase; this helps in P element transposition and hybrid dysgenesis.

The P element is highly invasive when placed into a new population. The source of P element seems to be another species of *Drosophila.* Since hybrid dysgenesis reduces inbreeding, it helps in specification by creating breeding barriers. Thus, multiple hybrid dysgenesis system may develop in different population; this would lead to inability to intermate and therefore to specification.

4.3. EXAMPLES OF RETROPOSONS (RETROELEMENTS)

1. **Copia elements in *Drosophila*.** The *copia* elements are 5000 bp long, have direct terminal repeats of 276 bp, they are always intact, sometimes they occur in circular form and they have a single long (4,227 bp) reading frame. *Copia* is present in 20 – 60 copies per genome depending on the strain of the fly and is represented by a large number of closely related sequences (the copia family; Table 4.6). These sequences code for about abundant poly $(A)^+$ mRNAs. The name *Copia* is based on the presence of this sequence in copius amounts in cellular RNA of *Drosophila.*

In cultured cells, the number of *copia* elements per genome increases upto 2 to 3 times the number found *in vivo.* In cultured cells, the rate of transposition increases to 10^{-3} to 10^{-4} events per generation. A direct repeat of 5bp of target DNA is generated by insertion of a *Copia* element. Each of the terminal direct repeats of *Copia* elements are identical, indicating their relationship with retroposons. The single *Copia* ORF has homologies with *gag* and *pol* sequences of retroviruses.

Table 4.6. Properties of some families of Copia like elements

Family	Number/ genome	Length (kb)	Length of LTR (bp)	Length of inverted terminal repeat (bp)
1. Copia	60	5.0	276	13/17
2. 412	40	7.5	481	8/10
3. 297	30	7.0	412/415	3
4. *mdg*1	25	7.2	442/444	13/16
5. *mdg*3	15	5.5	283/268	15/18
6. B 104	80	8.7	429	3
7. 17.6	40	6.9	512	0
8. *mdg*4 /gypsy	10	7.3	479	415

The mRNAs from *Copia* sequences are translated into a variety of proteins *in vitro,* ranging from 18 to 51 kDa, whose functions remain unknown. The *Copia* elements must have originated from retroviruses and they are able to transpose (but the mode of transposition is unknown). Integration of *Copia* element produces mutations, a classical example being white-apricot (w^a) mutation for eye colour caused by insertion of a copia-like element.

2. FB elements of *Drosophila*. The fold - back or FB elements range from 500 to 5,000 bp (base pair) and occur in about 30 copies per genome. They carry long (250 to 1,250 bp) inverted repeats at the termini. The repeats may be adjoining (FB1, FB5) or may be separated (FB3, FB4, FB7, FB8, FB9), but can always fold back to form stem and loop structures. FB elements cause mutations by insertion or by their effect on gene expression (due to their presence near a control region). FB elements are also found flanking another class of transposon called *TE* (*e.g., TE1, TE28, TE77, TE98*). TE1 was first identified due to transposition of white eye locus from X-chromosome to second chromosome.

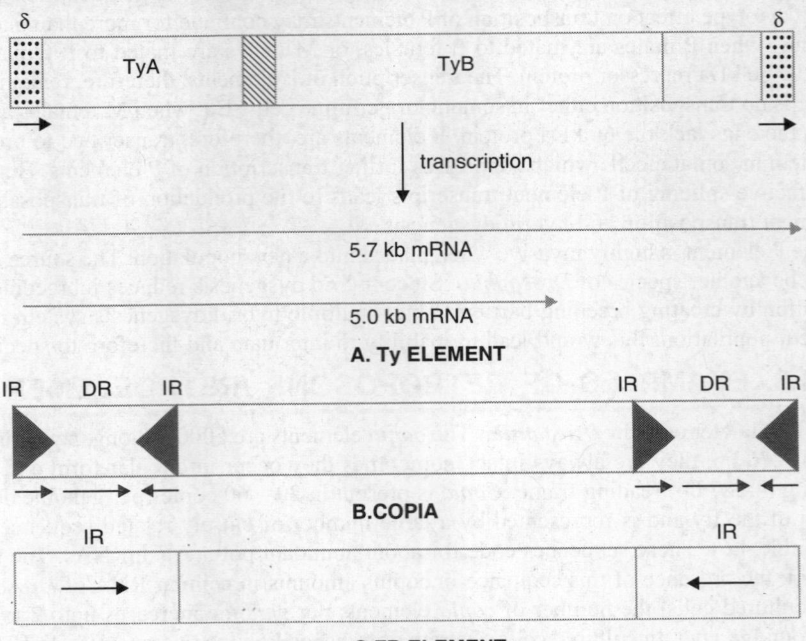

Fig. 4.20. A — T_y element of yeast; B — *Copia* element of *D. melanogaster* and C — *FB* element of *D. melanogaster.* IR, inverted repeat; DR, direct repeat.

3. T_y elements in yeast. Yeast T_y (= transposon *y*east) elements are an example of retroposons. T_y elements are very divergent. There are two main types of T_y elements, *viz.*, T_y1 and T_y917. They transpose at a frequency of ~ 10^{-7} to 10^{-8} through an RNA intermediate. This RNA is reversed transcribed to yield a DNA duplex that becomes inserted into the yeast genome generating a 5bp long direct repeats in yeast DNA. Such a transposition causes mutation in target gene due to insertion. Transpositions of T_y elements is controlled by genes within the T_y elements; they form noninfective virus - like particles, which accumulate within such cells in which transposition has been induced. Only some of the yeast T_y elements are active, while most of them have lost the ability to transpose.

Each T_y element is 6.3 kb long and has 330 bp direct repeats at its ends (no inverted repeats or IRs are found in bacteria); these repeats are called **delta (δ) repeats.** All T_y elements share δ repeats, a long region at the left end, another region in the centre and short region adjacent to the right δ repeat (Fig. 4.20). In a typical yeast genome, there are ~ 30 copies of T_y1 type and ~ 6 copies of the T_y 917 type element; in addition, there are ~ 100 copies of independent δ elements; these are called **solo deltas.** The sequences of δ elements very greatly, but the two δ elements of a single T_y element are identical. Thus, δ sequences seem to be similar to the LTR sequences of retroviruses. Although, the two δ sequences of a T_y element are identical, yet a *promoter* is active in the δ repeat at the right end; this feature is also found in the retroviruses. The T_y elements can be excised by homologous recombination between the δ sequences, leaving behind a solo δ sequence in the host DNA.

The T_y elements are transcribed into two species of poly $(A)^+$ mRNAs of which one is 5 kb long and the other is 5.7 kb long. These mRNAs form > 75% of the total mRNA of a haploid yeast cell. Transcription for both mRNA species begins ~ 95 bp from the left end of T_y element. Further, the T_y elements have two open - reading frames (ORFs), which are read in different phases and overlapped by 13 codons. One ORF is T_yA which encodes a DNA - binding protein. The other ORF is T_yB which contains regions that have homologies with reverse transcriptase, protease and integrase sequences of retroviruses. The organization and functions of T_yA and T_yB are analogous to the behaviour of retroviral *gag* and *pol* functions. Thus, T_yB is expressed only when a frameshift event allows the termination codon at the end of T_yA to be bypassed; as a result, T_yB product is fused with that of T_yA.

Moreover, the T_y elements comprise regions of homology within and between chromosomes. Recombination between T_y elements usually damages the chromosome either by deletion, inversion or other more effective structural changes in chromosomes, *e.g.*, dicentric chromosomes. Recombination between T_y elements seem to occur in bursts and gene conversion occurs between T_y elements present at different locations. Quite interestingly, although both the δ repeats of an active T_y element have th same sequence, a promoter is active in one δ elements, while a terminator is functional in the other. The T_y element behaves like a retrovirus that has lost its *env* gene so that it is unable to properly package its genome into virus particles.

4. Retrotransposons in plants. Retrotransposons are ubiquitous in plants. They make upto 50% to 80% of the genomic DNA (*e.g.,* in maize), thus, playing a major role in gene and genome evolution in plants. For example, LTR – retrotransposons are 100 bp to 13.5 kbp in size (Table 4.7) and possess the following coding sequences : 1. *gag,* encoding proteins involved in maturation and packaging of RNA, and proteins necessary for integration of retrotransposons into the genome; (*ii*) *pol,* encoding **reverse transcriptase (RT)** and **Rnase H** activities, needed for replication/ transposition to a new position on a chromosome. The *T_y1-copia* and *T_y3-gypsy* types of LTR-retrotransposons differ in order of the above coding sequences, particularly the *int* and *Rnase* sequences. Among non-LTR retrotransposons, LINEs, though simpler and smaller in size (1000 bp to 6 kbp), still encode many of the proteins encoded by the LTR retrotransposons. For instance, they have *gag* and *pol,* but not *int* and instead have endonuclease (*en*). SINEs do not encode any of these genes and are derived from RNA polymerase III products such as tRNA. In plant's chromosome, the retrotransposons are either dispersed on all chromosomes or can be clustered in some regions such as telomeres, centromeres, pericentromeric regions and also in the interstitial heterochromatic regions. An individual retrotransposon can be present as a single copy or it may be represented by upto as many as 50,000 copies per genome. There can be dozens of different types of retrotransposons within a species (*e.g.*, maize). Some of known plant retrotransposons have been listed in Table 4.7.

Table 4.7. Some selected retrotransposons in plants (Source: **Gupta,** 2004)

Types of retrotransposons	Size (in kb)	Number of copies	Plant / Crop
I.LTR retrotransposons			
1. T_y–1 Copia group			
(*i*) BARE - 1	12.1	> 50,000	Barley
(*ii*) Opie - 1	8.7	> 30,000	Maize
(*iii*) PREM - 2	9.5	> 10,000	Maize
(*iv*) SIRE - 1	11.0	> 7,200	Maize
(*v*) Tnt - 1	5.3	> 100	Maize
2. T_y– gypsy group			
(1) Cerebra	~ 10.0	1,500	Barley
(2) Cinful - 1	8.6	20,000	Maize
(3) Grand - 1	13.5	> 1,300	Maize
(4) Zeon - 1	7.3	> 1,250	Maize
(5) Cyclops - 2	12.5	1,000	Pea
(6) RIRE3	~ 10.5	10	Rice
II. Non-LTR retrotransposons			
1. *LINES*			
(*i*) Cin 4	1 – 6.8	50 – 100	Maize
(*ii*) Del 2	4.5	25,000	*Lilium*
2. *SINEs*			
(*i*) p - SINE1	0.125	> 100	Rice
(*ii*) $S1_{Bn}$	0.170	500	*Brassica napus*

5. Tam 1 element in Snapdragon (*Antirrhinum majus*)

In snapdragon, Tam1 element is responsible for pattern of colour seen on petals of some variegated strains and is 17 kb long. *Tam1* element is found in an unstable mutant, *Nivea recurrens,* where *Tam1* is present in the region of *chalcone synthase* gene and causes 3bp target site duplication. *Tam1* element also has 13bp inverted repeats at its termini, which are identical to those in Spm18 of maize. Other transposable elements of this series are numbered *Tam2* to *Tam10.* Putative transposable elements are also suggested to be present in soya bean.

6. Interspersed Repeats and Transposable Elements of Human Genome

The draft sequences have shown that approximately 62% of the human genome comprises **intergenic regions,** the parts of the genome that lie between genes and which have no known function. These sequences used to be called **junk DNA** but the term is falling out of favour of modern molecular biologists. Recently, bulk of the intergenic DNA is found to be made up of repeated sequences of one type or another : 1. **genome wide** or **interspersed repeats,** whose individual repeat units are distributed around the genome in an apparently random fashion; 2. **tandemly repeated DNA,** whose repeat units are placed next to each other in an array (Fig. 4.21).

Interspersed repeats include four types of mobile elements : SINEs, LINEs, LTR elements and DNA transposons. An interesting feature of these interspersed repeats is that each type appears to be derived from a **transposable element**, a mobile segment of DNA which is able to move around the genome from one place to another. Many of these mobile elements leave copies of themselves when they move, which explains how they propagate and become common throughout the genome. There are two main classes of transposable elements : those that transpose via an RNA intermediate and those that do not. LINEs, SINEs and LTR elements are examples of the first class and DNA transposons are examples of the second class. Characteristic features, types and subtypes of different genome - wide repeats have been tabulated in Table 4.8.

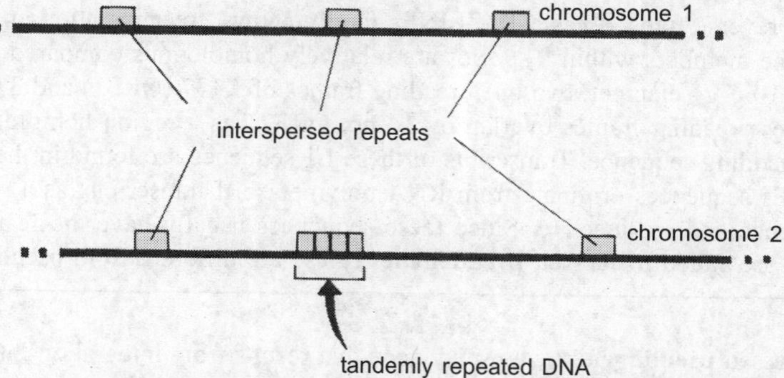

Fig. 4.21. The two types of repetitive DNA : interspersed repeats and tandemly repeated DNA.

Just for example, SINEs form most numerous genome - wide repeats of human genome. They comprise three subtypes : *Alu* element (approximately 1,090,000 copies in the genome), *MIR* (393,000 copies) and *MIR3* (75,000). Altogether genome wide repeats make up 44% of the draft genome sequences, some 1400 Mb of DNA (**Brown**, 2002).

Table 4.8. The types of genome - wide repeats in the human genome (Source: **Brown**, 2002).

Type of repeat	Subtype	Approximate number of copies in the human genome
1. SINEs		**1,558,000**
	1. Alu	1,090,000
	2. MIR	393,000
	3. MIR3	75,000
2. LINEs		**868,000**
	1. LINE-1	516,000
	2. LINE-2	315,000
	3. LINE-3	37,000
3. LTR elements		**443,000**
	1. ERV class I	112,000
	2. ERV (k) class II	8,000
	3. ERV (L) class III	83,000
	4. MaLR	240,000
4. DNA transposons		**294,000**
	1. hAT	195,000
	2. Tc-1	75,000
	3. PiggyBac	2000
	4. Unclassified	22,000

7. Mammalian Retroposons

Mammals contain a variety of retroposons; most of these belong to the following two families :

1. LINEs (Long Interspersed Elements). They are derived from transcripts of RNA polymerase II.

2. SINEs (Short Interspersed Elements). They are derived from transcripts of RNA polymerase III.

L1 retroposon is a good example of LINEs sequence. It is found in 20 to 50 thousand copies per genome. A typical L1 element is ~ 6,500 bp long, terminating in an A-rich region. It appears that L1 element is originated as a mobile gene coding for its own transposition as its reading frame homologous to reverse transcriptase. The LINEs family shows great variation among individual members, but the members within a species are relatively homologous compared to those within species. In 6,581 bp L1 element, two open reading frames of 1,137 (ORF1) and 3,900 (ORF2) bp are present.; these reading frames overlap by 14 bp. ORF2 has a region homologous to reverse transcriptase encoding sequence. Transcripts of these L1 sequence are found in the cells.

The LINEs sequences originate from RNA polymerase II transcripts; as a result, they lack promoter sequence and are inactive. Since these sequences usually have the features of mature transcript, they are called **processed pseudogenes** (these are different from pseudogenes). These

Box 4.6.

A **processed pseudogenes** is a pseudogene that results from integration into the genome of a reverse transcribed copy of mRNA (**Brown**, 2002).

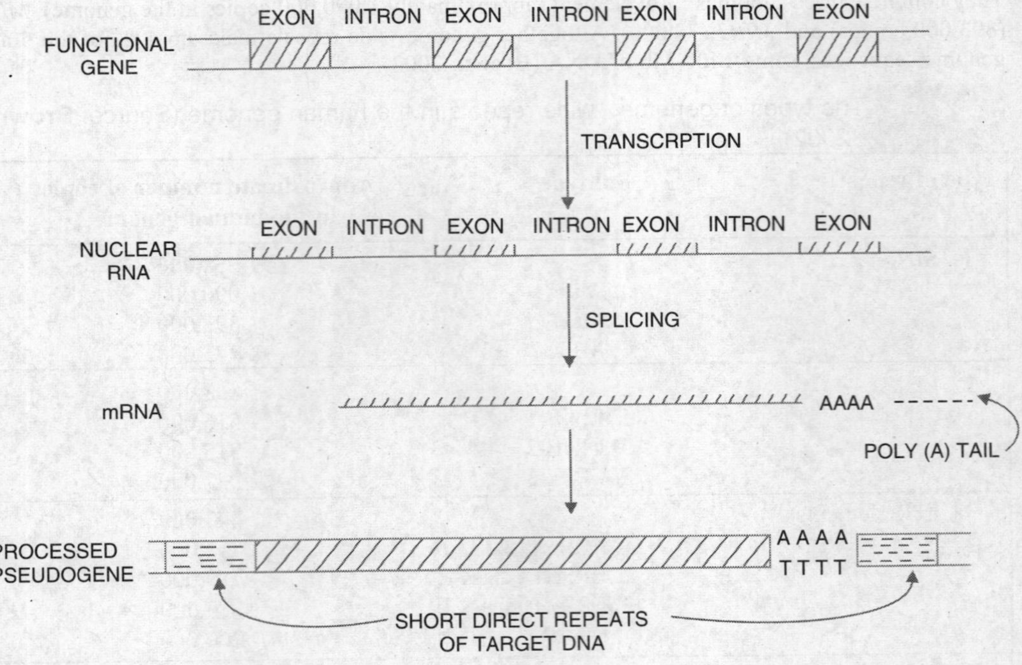

Fig. 4.22. Mode of origin and features of processed pseudogenes (*i.e.*, LINEs sequences). They have 3′ AT sequence, their 5′-end corresponds exactly with mRNA and they are flanked by short direct repeats.

pseudogenes have their 5′ ends having exact similarity with mRNA, many of them have exactly joined exon sequences and they usually end in a short 3′ stretch of AT base pairs. This 3′ stretch is probably derived from the poly (A) tail of the mRNA (Fig. 4.22). Presumably these pseudogenes have been originated from a mobile gene ecncoding its own transposition. It indicates that L1 elements belong to the viral superfamily of retroposons. The processed pseudogenes are flanked on either side by a short direct repeat, each of which is presumed to have been generated by a transposon - like insertion event.

The ***Alu* family** is an good example of SINEs sequence of human genome. These sequences are so named because they all are cleaved at a single site located at 170bp by the restriction enzyme *Alu1*. The sequence of *Alu family* are ~ 300 bp long, occur in ~ 300,000 copies per human genome and are widely dispersed in her genome. The human *Alu* family seems to have originated by a 130bp tandom duplication with an unrelated sequence of 31bp inserted into the right half of the dimer.

The *Alu* sequence is related to 7SL RNA, which is a component of the signal recognition particle. The 5′ terminal 90 bases of 7SL RNA are homologous to the left end of the left half (130 bp) of an *Alu* element, while the 40 bases at the 3′ -end of 7 SL RNA are homologous to the right end of the left half of *Alu* sequence; the central 160 bases of 7 SL RNA are absent from the *Alu* sequence. The gens for 7 SL RNA are transcribed by RNA polymerase III, and the SINEs are derived from transcripts of this RNA polymerase III. It is likely that the *Alu* sequences originated from the transcripts of 7 SL RNA genes by a deletion of the central 160 bases of this RNA.

The presence of multiple copies of *Alu* sequence in a single RNA molecule can produce secondary structure in the RNA. The presence of *Alu* sequences in the form of inverted repeats seems to be responsible for most of the secondary structure found in mammalian nuclear RNAs.

Members of *Alu* family resemble transposons in being flanked by short repeats of target DNA; the lengths of these DNA repeats are different for individual members of the family. *Alu* sequences have a 14 bp region that is almost identical with a sequence present at the origin of replication of papova virus such as SV40 and hepatitis B virus.

Other examples of SINEs sequence include ***B1 family*** which occurs in mice and ***Alu - equivalent family*** of chinese hamster.

Questions And Problems

1. What a plasmid is and provide examples of the different types of plasmid that are known.
2. Contrast the various transposable genetic elements in bacteria, maize, *Drosophila,* and humans. What properties do they share?
3. The initial discovery of IS units in bacteria involved their presence upstream (5′) to three genes controlling galactose metabolism. All three genes were affected despite not being contiguous with the IS unit. Offer an explanation as to why this occurs.
4. In maize, the recessive allele *bz* produces a lighter colour in the aleurone than does the dominant allele, B_z. Ears on a homozygous b_z/b_z plant were fertilised by pollen from a homozygous B_z/B_z plant. The resulting cobs contained kernels that were uniformly dark except for a few on which light spots occurred. Suggest an explanation.
5. Sometimes solo copies of the LTR of a retrotransposon, such as *gypsy,* are found in *Drosophila* chromosomes. How might these solo LTRs originate ?
6. *Escherichia coli* cells that are resistant to kanamycin are infected with bacteriophage lambda. Phage particles that were purified from the lysed cells were used to infect a kanamycin sensitive strain called C107 and cells that were lysogenised were picked from turbid plaques. These cells were subsequently plated on a medium containing kanamycin and several kanamycin resistant colonies appeared. Suggest an explanation for the transfer of kanamycin resistance to the C107 strain.

Answers to Problems

3. It is probable that the IS occupied or interrupted normal function of a controlling regions related to the galactose genes.
4. Transposon - induced mutation or loss of a B_z allele in the endosperm.
5. Through crossing - over between the LTRs of a gypsy element.
6. A transposon carrying the kanamycin resistance gene transposed from the original cells to some of the infected phage and subsequently conferred resistance on some of the lysogenised C107 cell, produced by the second infection.

Multiple Choice Questions

Choose the correct answer from the four alternatives given.

1. Certain genetic elements that regularly "jump" to new locations, often inactivating the gene into which they become inserted, are called
   (*a*) episomes
   (*b*) heterochromatin
   (*c*) transposons
   (*d*) overlapping genes
2. The jumping genes in maize were discovered by
   (*a*) Beadle and Tatum
   (*b*) Barbara McClintok
   (*c*) H. G. Khorana
   (*d*) T. H. Morgan
3. Genes which confer antibiotic resistance on bacteria are located on
   (*a*) chromosomal DNA
   (*b*) plasmid
   (*c*) RNA
   (*d*) polysome
4. Plasmids found in cells of bacteria are molecules of
   (*a*) DNA
   (*b*) RNA
   (*c*) proteins
   (*d*) none of these
5. Which of the following is called resistance transfer factor?
   (*a*) Col - factor
   (*b*) R - factor
   (*c*) F - factor
   (*d*) none of these
6. Plasmid is a
   (*a*) double - stranded circular DNA
   (*b*) single - stranded DNA
   (*c*) extrachromosomal linear DNA
   (*d*) none of the above

Answers

1. (*c*) 2. (*b*) 3. (*b*) 4. (*a*) 5. (*b*) 6. (*a*).

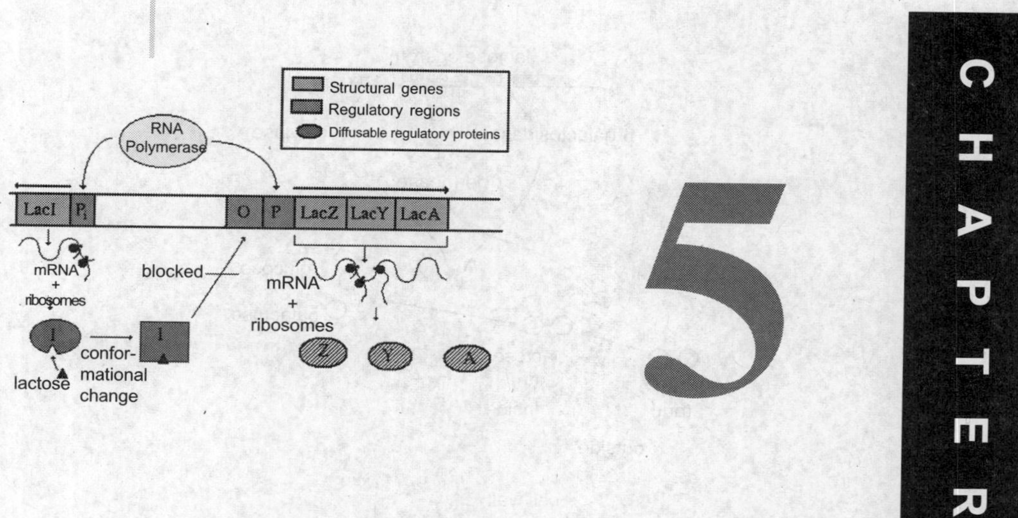

CHAPTER 5

Multigene Families

Chapter Contents

The majority of genes are spaced out more or less randomly along the length of DNA molecule. In some cases, however, they are grouped into distinct clusters. Sometimes the individual genes in a cluster are unrelated and there is no apparent reason or advantage to having them organized in this way. More commonly, clusters are made of genes that contain related units of biological information, e.g., operons and multigene family.

5.1. OPERONS

Operons are quite common aspects of the organization of genes in bacteria. A operon is a cluster of genes coding for a series of enzymes that work in concert in an integrated biochemical pathway (Fig. 5.1). The first operon to be discovered was the lactose or lac operon of *E.coli* which is a cluster of three genes each coding for one of the three enzymes involved in conversion (catabolism) of the disaccharide lactose into its monosaccharide units glucose and galactose. These enzymes are not required all the time, only when lactose is available. But when needed they are required together, *i.e.*, the three units of biological information carried by *lac* must be read at the same time. A

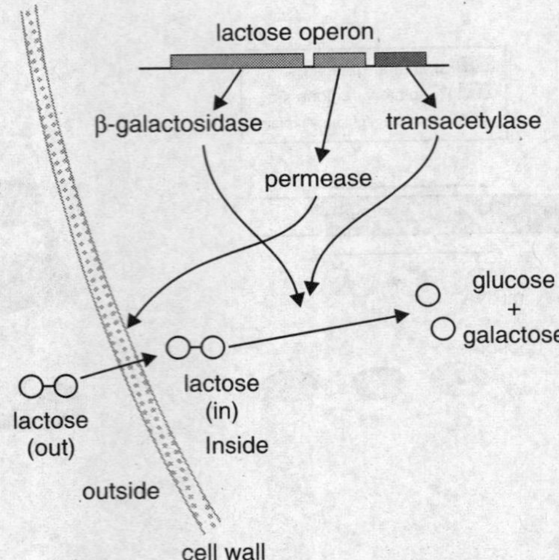

Fig. 5.1. The lac (lactose) operon of *E.coli*. The three genes specify three enzymes involved in utilization of lactose as a carbon and energy source. The permease transports lactose into the bacterial cell, and B-galactosidase and transacetylase split lactose into glucose and galactose (after Brown, 1998).

sophisticated system enabling the *lac* genes to be expressed together and only when needed has evolved round the fact that the genes are clustered.

5.2. MULTIGENE FAMILIES

Operons have no direct counterparts in organisms other than bacteria. In contrast, multigene families are found in many organisms, mainly eukaryotic organisms. A multigene family or gene family is also a cluster of related genes, but in multigene families the individual genes have identical or similar nucleotide sequences, and therefore contain identical or similar pieces of biological information.

Why do Multigene Families are Required ?

Some proteins in the cell are required in small quantities, but other proteins may be required in large quantities, so that the demands can not be easily met. These demands can be met in either of the following two ways: (1) In some cases, a solitary gene is present in the cell but is transcribed repeatedly to produce large number of messenger RNA molecules and these mRNA molecules act as templates for many rounds of protein synthesis. For example, although the gene coding for silk worm protein **fibroin** is present in only one copy per haploid complement, but it keeps on making silk actively. Similarly, in a cell (*e.g.*, erythroblasts) engaged in exclusive synthesis of haemoglobin, a solitary copy of the gene produce large quantities of α and β polypeptide chains of haemoglobin. However, although a whole family of related α globin genes are found in each cell, different but solitary α globin genes function at different stages of development. In other cases, multiple copies of a gene are found in the same cell, all taking part in transcription within a cell, giving rise to large quantities of the gene product as in case of ribosomal RNA (rRNA) genes, histone genes, tRNA genes, heat shock protein genes, etc.

Types of Multigenes

Members of multigene families share DNA sequence homology, descend from a single ancestral gene and their gene products frequently have similar functions. Members of multigene families are often, but not always, found together in a single location along a chromosome (Box 5.1).

Box 5.1.

Dispersed family of gene

In some multigene families, the individual members are clustered, as with the globin genes, but in others the genes are dispersed around the genome. An example of a *dispersed family* is the five human genes for *aldolase*, an enzyme involved in energy generation, which is located on chromosomes 3, 9,10, 16 and 17. The important point is that, even though dispersed, the members of the multigene family have sequence similarities that point to a common evolutionary origin.

Multigene families are of following main types:

1. Simple multigene families

In these gene families, all of the genes are exactly or virtually the same. For example, most higher organisms have multiple copies of the gene for the 5S rRNA (a component of the ribosome), probably because large quantities of 5S rRNA molecules must be synthesized at certain times, creating a demand that just one of a few genes would be unable to satisfy. Rather than being spread randomly throughout the chromosomes, the 5S rRNA genes are clustered into a multigene family. In humans, for example, there is a family of about two thousand 5S rRNA genes in chromosome 1.

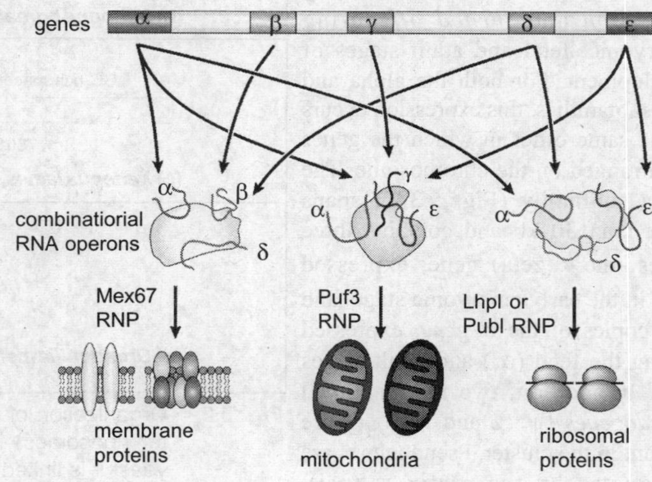

Dispersed family of Gene.

The 5S rRNA genes are present independently of the rDNA which is localized at the NOR in eukaryotes. However, in prokaryotes and yeast, 5S rRNA genes are present in close vicinity of rDNA (Fig. 5.2). The 5S rRNA genes are also organized in tandem repeats, each repeat consisting of a gene 120 bp long and a spacer region. The length of complete repeat is 375 bp in *Drosophila*. In wheat, there are two loci for 5S rRNA having repeats of different lengths, 480 bp and 500 bp. These repeat units may sometimes carry pseudogenes (instead of functional genes), which are not used for synthesis of any RNA.

2. Complex Multigene Families

A multigene family is a group of similar, but not necessarily identical sequences, each sequence representing a gene, so that the gene is present in multiple copies. Complex multigene families are of following types:

A. Multigene families with divergent members

(i) Globin gene family in human genome

The human α-globin and β-globin genes are subfamilies of the globin gene *superfamily* and are two of the most intensively studied regions of human genome. There is a cluster of three α -

globin genes on the short arm of chromosome 16 and another with five β - globin genes on the short arm of chromosome 11. Members of both subfamilies share nucleotide sequence similarity, but members of the same subfamily have the greatest amount of sequence similarity.

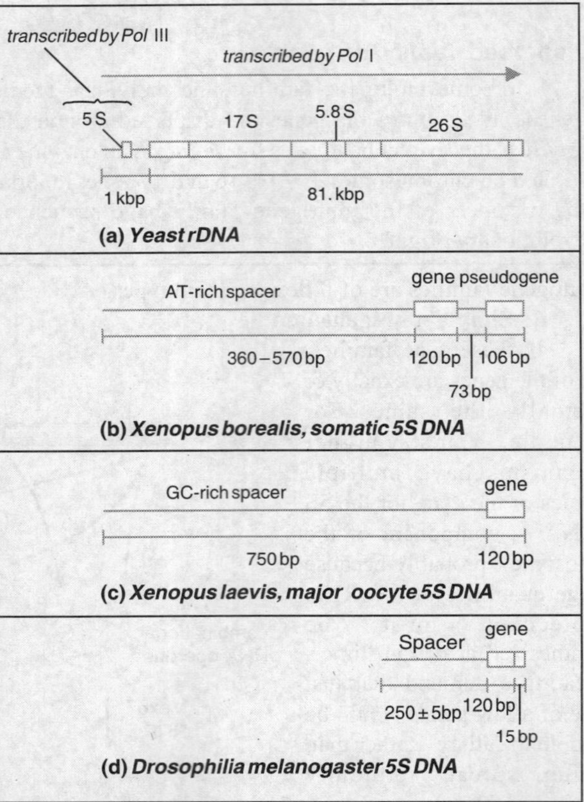

Fig. 5.2. Organization of 5S DNA (produces 5S r RNA for ribosomes) in several organisms; only in yeast it is linked with repeat units of r DNA; in other cases it is independent on a separate chromosome.

Haemoglobin is a tetramer, containing two α - and two β - polypeptides. Each polypeptide incorporates a haem group that reversibly binds oxygen. Within each subfamily, genes are coordinately *turned on* and *turned off* during embryonic, fetal and adult stages of development. For both the alpha and beta subfamilies, this expression occurs in the same order in which the genes are arranged on the chromosome. The alpha subfamily (Fig. 5.3A), spans more than 30 kb and contains three genes: the ζ (zeta) gene, expressed only in the early embryonic stage, and two copies of the α *gene,* expressed during the fetal (α_1) and adult stages (α_2). In addition, two nonfunctional *pseudogenes* (ψ z and ψ α_1) are present in the cluster. Pseudogenes are designated by the prefix ψ (psi), followed by the symbol of the gene that they most resemble. Thus, the designation ψ α_1 indicates a pseudogene of the adult α_1 gene. *Pseudogenes are non-functional versions of genes that resemble other gene sequences, but contain significant nucleotide substitutions, deletions and duplications that prevent their expression.*

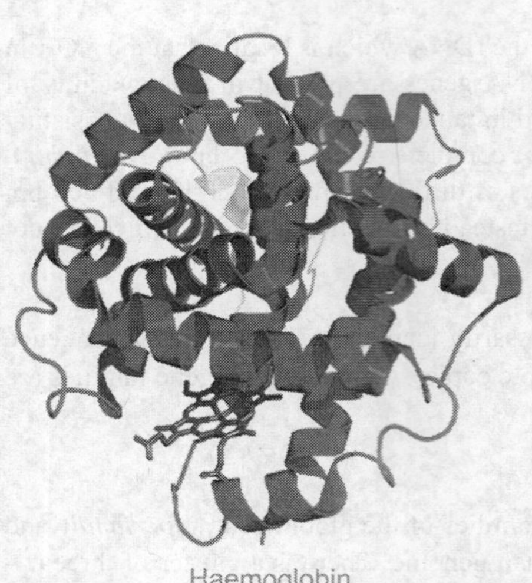

Haemoglobin.

The organization of the alpha subfamily members and the location of their introns and exons reveal several interesting features. First,as is common in eukaryotes, the DNA encoding the three functional α genes occupies only a small portion of the region containing the subfamily. Most of the DNA in this region is intergenic spacer. Second, each functional gene in this subfamily contains two *introns* at precisely the same positions. Third, the nucleotide sequences within corresponding *exons* are nearly identical in the *z* and α genes. Both genes encode polypeptide chains of 141 amino acids. However, their intron sequences are highly divergent, even though they are about the same size. Significantly, much of the nucleotide sequence of each gene is contained in these noncoding introns.

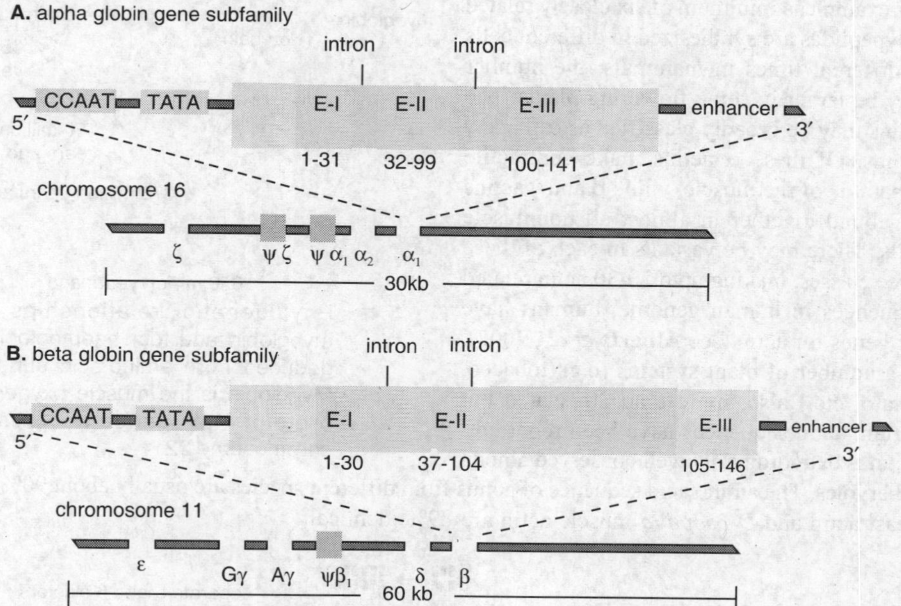

Fig. 5.3. Organization of the alpha-globin gene subfamily (A) on chromosome 16 and the β - globin gene subfamily on chromosome 11 (B). Also shown the internal organization of the α_1 gene and the β gene. Each gene contains three exons (E-I, E-II, E-III) and two introns. The numbers below the exons in dictate the amino acids in the gene product encoded by each exon (after Klug and Cummings, 2003).

The human β - globin gene cluster is longer than the α - globin cluster and contains five genes spaced over 60 kb of DNA (Fig. 5.3B). As with the alpha subfamily, the order of genes on the chromosome parallels their order of expression during development. Of the five genes, three are expressed prior to birth. The ε (epsilon) gene is expressed only during embryogenesis, while the two nearly identical γ (gamma) gene (Gγ and Aγ) are expressed only during fetal development. The polypeptide products of the two γ-genes differ only by a single amino acid. The two remaining genes, δ (delta) and β (beta), are expressed following birth. Finally a single pseudogene $\psi\beta_1$ is present within the subfamily. All five functional genes encode proteins with 146 amino acids and have two similarly sized introns at exactly the same positions. The second intron in the β gene is significantly larger than its counterpart in the functional α genes. These similarities reflect the *evolutionary history* of each subfamily and the events such as gene duplication, nucleotide substitution and chromosome translocations that produced the present day globin superfamily.

It has been shown that in these sequences (α cluster genes and β cluster genes), the coding regions show little divergene, but spacers in between the genes show considerable diversity as in the case in many other gene families. The only explanation for this is that the protein function puts a constraint on the evolution of coding sequence, while no such constraint is exercised in spacer region which is fast evolving. Based on the degree of similarity in amino acid sequences of globin of different mammals, evolutionary trees have been prepared in the past (Fig. 5.4 and Fig. 5.5).

(ii) Actin gene family in eukaryotes

Actin is a cytoskeleton protein. Actin genes represent a multigene family, where the members are homologous, but non-identical, giving rise to slightly different variants so that different members function either at different times in the same tissue or in different tissues at the same time. For example, a minimum of six closely related polypeptides are synthesized in different cells at different times in mammals; the number may be larger in some flowering plants. The actins may be broadly classified as α,β and γ chains. Of these α actins make contractile apparatus of the muscle, while β and γ actins are found together in almost all nonmuscle cells. There may be variants in each of these three classes, making upto 20-30 actin related sequences in human genome; humans have six genes for actin (see **Alberts** *et al.*, 2002). In a number of plant systems (e.g., tobacco, potato, etc.) also, more than 20 related but variant actin sequences have been reported. Actin is extraordinarily well conserved among eukaryotes. The amino acid sequence of actins from different species are usually about 90% identical. Yeast actin and *Drosophila* muscle actin are 89% identical.

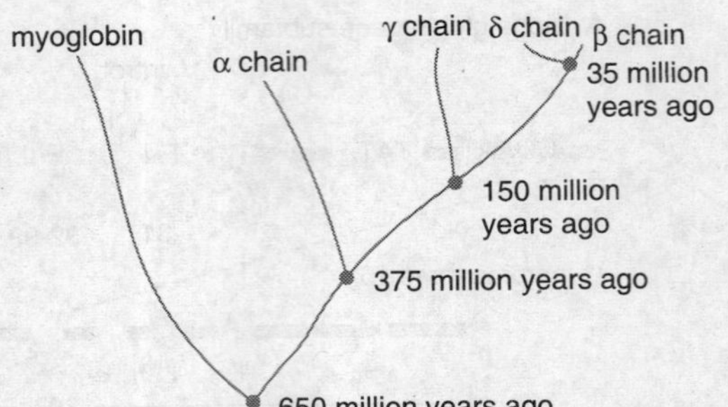

Fig. 5.4. Phylogenetic relationships among myoglobin and four haemoglobin chains deduced from amino acid substitutions. Myoglobin is the muscle oxygen-binding protein; its gene is present on chromosome 22.

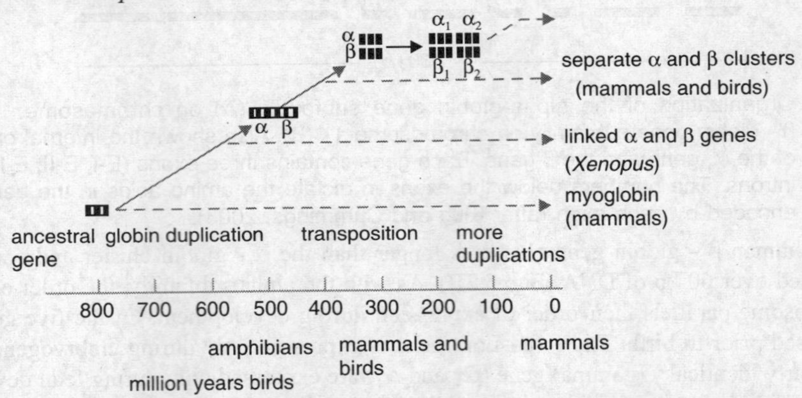

Fig. 5.5. Myoglobin differentiation into α and β chains from earliest period in the history.

(*iii*) *Other multigene families with divergent members.* Other multigene families have divergent members; *e.g., (i)* albumin, ; α feroprotein (major components of blood plasma), serine proteases, the interferons (for defence against viral infection) and the immunoglobulins; (*ii*) chorion proteins making the major components of egg shells in insects.

B. Multigene families with identical genes.

In some cases, for producing large quantities of a gene product, many copies of identical genes are required. These genes may occur as repeat units, each of which comprises (i) a *coding region,* which is highly conserved to produce exactly the same gene product and (ii) a *spacer region,* which may show divergence. Examples of these multigene families are the following:

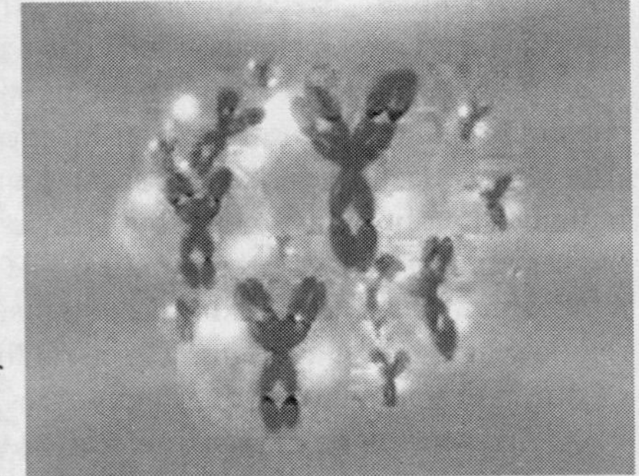

Immunoglobulins.
(Multigene Families)

(i) **Multiple copies of histone genes.** Organization of the histone genes is a variation on the theme established in the globin gene family. Here a cluster of five related, but nonidentical genes, separated from each other by highly divergent intergenic spacer regions, is tandemly repeated many times.

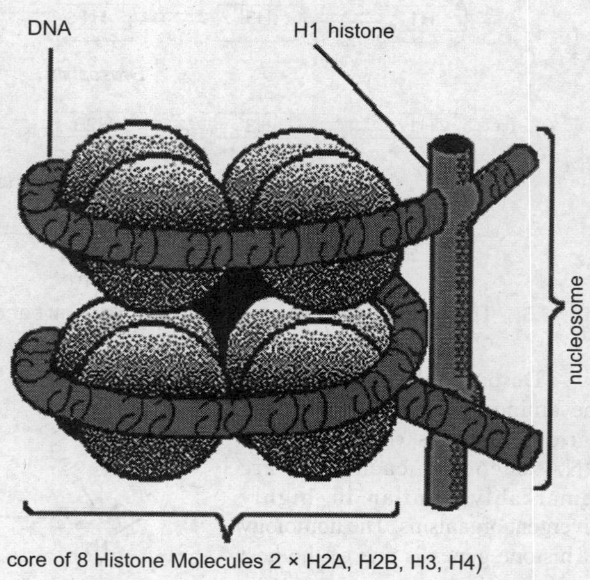

Histones.

As we already know that histones are positively charged (basic) proteins which interacts with the negatively charged phosphate groups of DNA to form nucleosomes, the chromatin units. In rapidly diving cells (e.g., cleaving fertilized egg of frog), histone synthesis must keep pace with DNA replication and the necessary quantity of histones must appear each time, as one cell cycle runs into the next. Multiple copies of the histone cluster allow the coordinate synthesis of DNA and histone during S phase of cell cycle.

Many interesting correlation support this idea. Yeast cells, with much less DNA than other eukaryotes, have only two clusters of four histone genes(they lack histone H1). In contrast, clusters of five histone genes are tandemly repeated from 10 to 800 times in many complex organisms (Table 5.1). Mitotic cell division is very rapid during development in sea urchins and newts and these organisms have one of the largest set of histone gene clusters.

Table 5.1. Numbers of histone gene clusters in selected eukaryotes (Source: **Klug** and **Cummings,** 2003).

Organism	Repeats
1. Yeast	2
2. Chickens	10
3. Mammals	20
4. *Xenopus*	40
5. *Drosophila*	100
6. Sea urchin	300-600
7. Newt (amphibian)	600-800

Polarity differences of transcription. Beyond its tandem arrangement, the histone gene family shows several other differences from other families.

First, almost all histone genes lack introns. Second, individual genes within clusters of a given species often orient in opposite directions with respect to transcription. As shown in Figure 5.6, the arrangement of genes and their polarity of transcription, shown by the direction of arrow, vary in the sea urchin and *Drosophila*, two organisms where this cluster has been extensively studied. Studies also show polarity differences in other organisms.

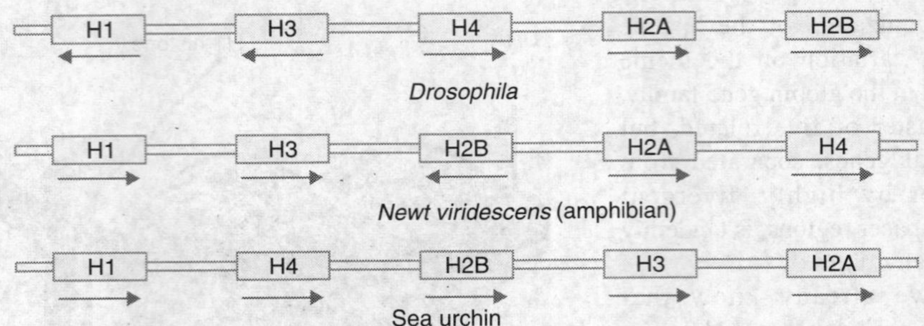

Fig. 5.6. The histone gene cluster in *Drosophila, Newt* and sea urchin. Arrows indicate the direction of transcription.

Despite polarity differences the amino acid sequences of the various histones (reflecting the DNA encoding each one) are remarkably similar in highly divergent organisms. The homology of histone genes is one of the best examples of sequence conservation through evolution.

Another interesting feature of the multigene family of histone genes observed in *Drosophila* is the presence of an '*attachment site*' per repeat unit (Fig. 5.7); 100 such repeat units being present in each haploid genome. With the help of attachment site, the genes (DNA) remain attached to *nuclear scaffold*. These attachment sites help in transcription.

In mice and humans, members of the histone gene family are clustered, but do not form tandem arrays. A region at the distal tip of the long arm of human chromosome 7 contains all copies of the histone gene family, interspersed with other nonhistone genes. Therefore, no single pattern of histone gene organization applies to all organism (see **Klug** and **Cummings,** 2003).

(*ii*) Ribosomal RNA (rRNA) genes in tandem arrays. For the production of about 10 million ribosomes per eukaryotic cell, rRNA genes are present in multiple

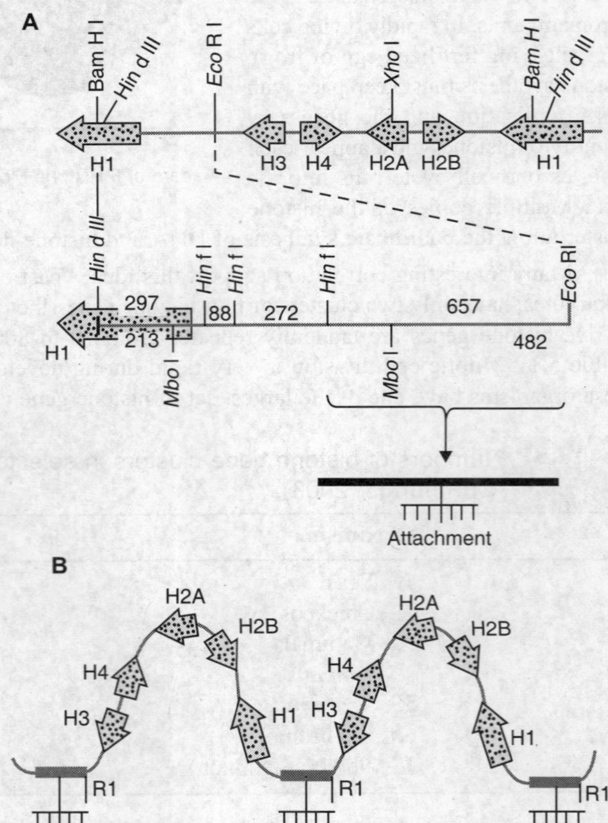

Fig. 5.7. A – A restriction map of major histone gene cluster in *Drosophila,* each repeat unit has one specific attachment site to the nuclear scaffold, this attachment site is Hin1/ECO RI DNA segment, 657 bp long; B – Schematic representation showing how *Drosophila* histone repeats might be attached to the nuclear scaffold.

copies at **nucleolar organizing regions(NOR)** of specific satellited chromosomes. The number of these genes may vary from 50 to 30,00 in a cell and this number may be unequally distributed on NOR's, if more than one such loci are present. The DNA comprising these genes is called **rDNA** (*ribosomal DNA*), which is repetitive in nature. Each repeat unit has (i) a coding region with genes that specify 18S, 5.8S and 28S rRNA molecules; (ii) a spacer region called *intergenic spacer (IGS)* and (iii) *internal transcribed spacers (ITS)* one each between 18S and 5.8S genes and between 5.8S and 28S genes. Because parts of intergenic spacer (IGS) adjoining region, known as *external transcribed spacer (ETS),* is also transcribed, the use of the term *nontranscribed spacer (NTS)* for the whole spacer region has been considered to be inappropriate. In fact, the NTS makes only a part of IGS (Fig. 5.8), the remaining part of IGS being ETS.

The IGS region in its turn has a region consisting of a tandem array of variable number of *subrepeats* ranging from 100 – 300 bp (bp = base pairs) in length. The variation in number and size of the subrepeats in IGS is responsible for variation in length of the repeat units (IGS + coding region).

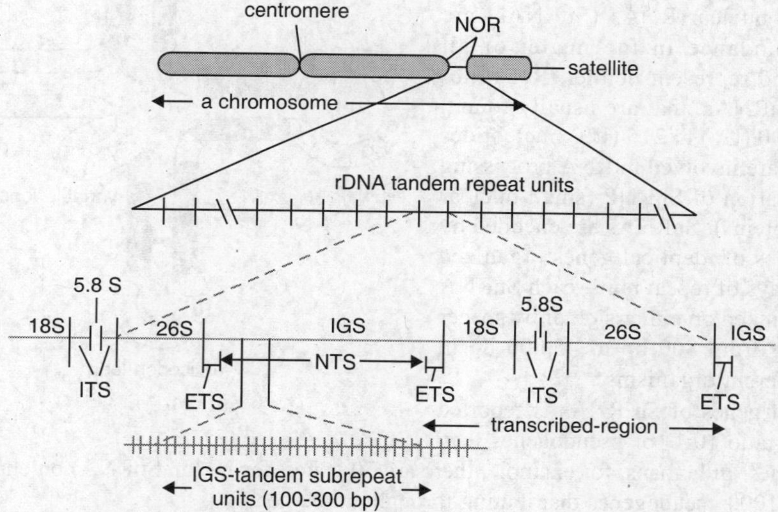

Fig. 5.8. Organization of ribosomal DNA (rDNA) in nucleolar organizing region (NOR) of a eukaryotic chromosome. Abbreviations: IGS = intergenic spacer; NTS = non-trascribed spacer; ETS = external transcribed spacer; ITS = internal transcribed spacer; 18S, 5.8S, 26S = different genes responsible for noncoding RNAs such as 18S rRNA, 5.8S rRNA and 26S rRNA.

The rRNA repeat units are usually looped off from the main chromosome fibres, in the form of extended threads at nucleolar organizing region. These loops, in association with specific proteins form the nucleoli, where rRNA synthesis and processing actually takes place. The number of nucleoli and the corresponding number of nucleolar organizing chromosomes in an organism may vary from one to several. At each locus (NOR), rDNA repeat units may evolve independently both in length and also in the nucleotide sequence of the spacer region. As a result of this, the length of the repeat unit including coding and spacer regions (nontranscribed spacer or NTS; external transcribed spacer or ETS and internal transcribed spacers or ITS) varies from about 7 kb (kilobase) pairs to 14 kb. In wheat and barley, it is usually 9 kb to 10 kb (Fig. 5.9).

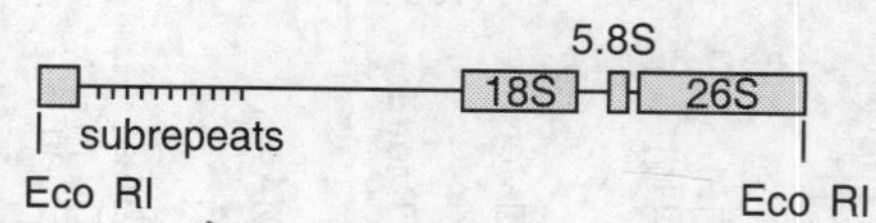

Fig. 5.9. An outline of the structure of rRNA in wheat and barley.

***(iii)* Transfer RNA (tRNA) multigene families.** The genes for each of the different tRNAs are also found in multiple copies to meet the heavy demands of the cell for the production of tRNA molecules. Ten to several hundred genes for each tRNA are present in each haploid genome. While in some cases with fewer copies of a gene, these copies are dispersed, in other cases such as in *Xenopus* (a toad, an amphibian), tandem repeats of long sequences (each repeat having genes for several different tRNAs) are found. In *Xenopus,* for example, a 3.2 kb repeat length has 8 genes, 2 for $tRNA_f^{met}$ and six other, for six different tRNA molecules (Fig. 5.10). In still other cases (e.g., *Drosophila)* genes for completely different tRNA species are clustered over a length of several thousand base pairs. Sometimes, tRNA pseudogenes are also found.

***(iv)* Small nuclear RNA (Sn RNA) genes.** Small nuclear RNAs (SnRNAs) are found in abundance in the nuclei of all eukaryotes and represent neither tRNA, nor rRNA. Six SnRNAs, that are usually found, range from 100 (U6) to 215 (U3) nucleotides in length and are involved in RNA processing through formation of SnRNP (small nuclear ribonucleoproteins). SnRNAs are encoded by multiple copies of identical genes organized in tandem arrays of repeat units, each SnRNA gene being flanked on either side of by spacer DNA ranging from 800 bp to 45,000 bp in length in different organisms.

Structure of tRNA.

DNA of genes of Sn RNAs is reported to have high ratio(10:1) of pseudogenes than functional genes. In humans, for example, there are 30 genes, for Ul (an SnRNA) on chromosome 1 and 500 to 1000 pseudogenes distributing throughout the genome.

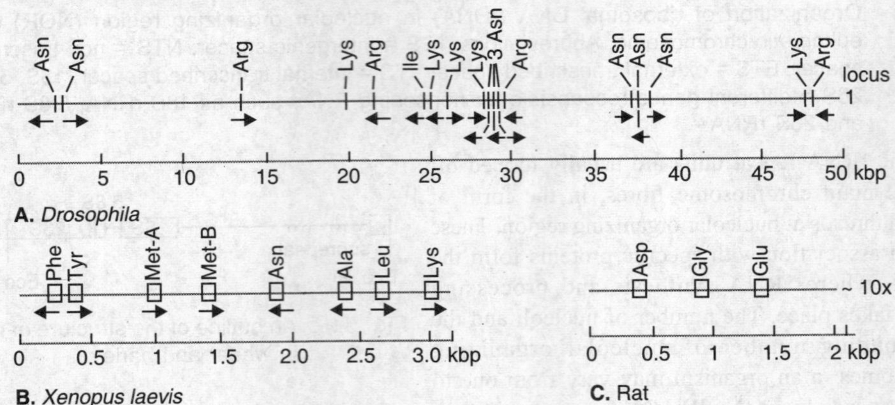

Fig. 5.10. Arrangement of tRNA genes in higher eukaryotes. A – One of the 50 – 60 sites on *Drosophila* chromosomes where tRNA genes are clustered. B – A 3.18 kilobase – pairs unit from *Xenopus* DNA, that is repeated about 150 times; C – Rat DNA has about 10 repeat units per haploid genome, here only one of them is shown.

C. Multigene families for storage proteins in crop plants. Storage proteins of several crop plants have been reported to be encoded by multigene families and are represented mainly as *prolamins* (major storage proteins in cereals; soluble in aqueous alcohol) or *globulins* (major storage proteins in legumes; soluble in salt). Three to ten genes for each of the ten prolamin families in wheat; and 18 genes belonging to three families of globulin genes are known in pea. Homologies in genes coding for prolamins and globulins in different crop plants have been reported. Their multiplicity appropriately meets the demand for rapid synthesis of these proteins in the developing seeds.

5.3. CONCERTED EVOLUTION OF MULTIGENE FAMILIES

Multigene families are found to have mechanisms that prevents the individual copies from accumulating mutations and hence diverging away from the functional sequence. This is called *concerted evolution.* Thus, if one copy of the family aquires an advantageous mutation then it is possible for that mutation to spread throughout the family until all members possess it. Following genetic mechanisms may be involved in such homogenization of concerted evolution:

1. Gene conversion. Gene conversion can result in the sequence of one copy of a gene being replaced with all or part of the sequences of a second copy. Multiple gene conversion events could therefore maintain identity among the sequences of the individual members of a multigene family (see **Brown,** 2002).

2. Unequal crossing over. Unequal crossing over may lead to *duplications or deletions* leading to homogenization, *i.e.*, either mutant member will be deleted or multiply and spread throughout the gene family. In yeast, it has been reported that the rates of gene conversion and unequal crossing over are higher than the rate of mutations so that homogeneity can be maintained. Further, certain specific sequences are also located in the multiple gene families which stimulate higher rates of recombination in a multigene family. In rDNA, NTS (nontranscribed spacer) region has been shown to be hot spot of recombination thus facilitating homogenization.

QUESTIONS AND PROBLEMS

1. What are multigene families and how are they identified? Discuss their significance in eukaryotic systems.
2. A multigene family may consist of identical or non-identical genes. Discuss the significance of each of these two cases using suitable examples.
3. Describe briefly the organization of sequences in rRNA, 5S rRNA and histone gene families, pointed out the major differences.
4. What aspects of gene structure and organization are displayed by the human globin gene families?
5. The β - globin gene family consists of 60 kb of DNA, yet only 5 percent of the DNA encodes β - globin gene products. Account for as much of the remaining 95 per cent of DNA as you can.

ANSWERS TO PROBLEMS

5. While the β - gene family is a relatively large (60 kb) sequence and restriction analyses show that it is composed of six genes, one is a pseudogene and therefore does not produce a product. The five functional genes each contain, two similarly sized intron, which when included with non-coding blanking region (5' and 3'), and spacer DNA between gene, accounts for the 95% mentioned in the question.

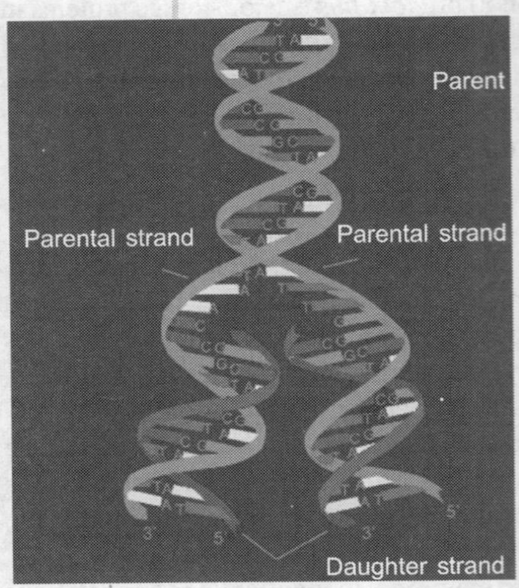

CHAPTER

DNA Replication

Chapter Contents

As a carrier of genetic information DNA has the following two important functions: **1. Heterocatalytic function** – When DNA directs the synthesis of chemical molecules other than itself (*e.g.,* synthesis of RNA, proteins, etc.), then such functions of DNA are called heterocatalytic functions. **2. Autocatalytic functions** – The functions of DNA which directs the synthesis of DNA itself, are called autocatalytic functions. Here, we are concerned only with autocatalytic function of DNA.

6.1. WATSON AND CRICK'S MODEL FOR DNA REPLICATION

The double helix model of DNA molecule of Watson and Crick is beautifully embodied a built-in template system for self-replication or autocatalytic function. Because of the specificity of base pairing, the sequence of bases along one chain automatically determines the base sequence along the other. Thus, each chain of the double helix can serve as **template** for the synthesis of the other. For the replication of **DNA** molecule, **Watson** and **Crick** proposed that replication involved the disruption of hydrogen bonds followed by a rotation and separation of the two polynucleotide strands. Each purine and pyrimidine base of each polynucleotide strand is thought to attract a complementary free nucleotide available for polymerization in cell and to hold it in place by means of the specific hydrogen bonds.

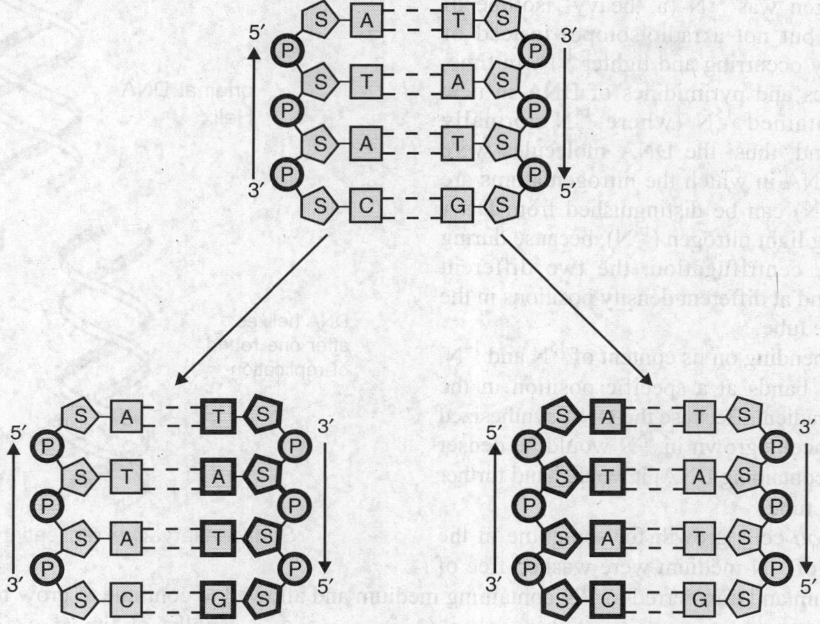

Fig. 6.1. Semi-conservative model of DNA replication.

Once held in place on the parent template chain, the free nucleotides were sewn together by formation of the phosphate diester bonds that linked adjacent deoxyribose residues, forming a new polynucleotide molecule of predetermined base sequence. Thus, two double helical molecules, identical with each other, are formed (Fig. 6.1).

1. Experimental Evidence for Semiconservative DNA Replication in *E.coli*

The Watson – Crick's model of DNA structure and replication suggested that once DNA replication is initiated, the two original polynucleotide strands of the duplex of helix will unwind, at least locally, so that each can serve as a template for a new strand.

An immediate prediction follows from this proposal: both duplexes that result from replication should be **hybrid** in nature, each containing an old strand derived from the original molecule and a new strand which has been formed during the replication process. Since each of the two double helices or duplexes conserves only one of the parent polynucleotide strands, the process is said to be **semiconservative**. This prediction is diagrammed in Fig. 6.2 in which old DNA is shown in black and new DNA in white. Figure 6.2 also outlines what would be predicted if these two hybrid duplexes went on to replicate themselves. Four duplexes would result, two of which would contain a single strand derived from the original chromosome and two of which would contain totally new DNA strands.

Besides the semiconservative mode of DNA replication, the following two methods of DNA replication were deemed equally feasible: (i) **conservative replication**, in which both strands of parent double helix would be conserved and the new DNA molecule would consist of two newly synthesized strands; and (ii) **dispersive replication**, in which replication would involve fragmentaion of the parent double helix and the intermixing of pieces of the parent strands with newly sythesized pieces, thereby forming the two new double helices.

***(i)* Meselson and Stahl's experiment.** **M. Meselson** and **F.W. Stahl** (1958) verified the semiconservative nature of DNA replication in a series of elegant experiments using isotopically labelled DNA and a form of isopycnic density gradient centrifugation. They cultured *Escherichia coli* cells in a medium in which the nitrogen was ^{15}N (a 'heavy' isotope of nitrogen, but not a radioisotope) instead of commonly occurring and lighter ^{14}N. In time, the purines and pyrimidines of DNA in new cells contained ^{15}N (where ^{14}N normally occurs) and, thus, the DNA molecules were denser. DNA in which the nitrogen atoms are heavy (^{15}N) can be distinguished from DNA containing light nitrogen (^{14}N), because during isopycnic centrifugation, the two different DNAs band at different density positions in the centrifuge tube.

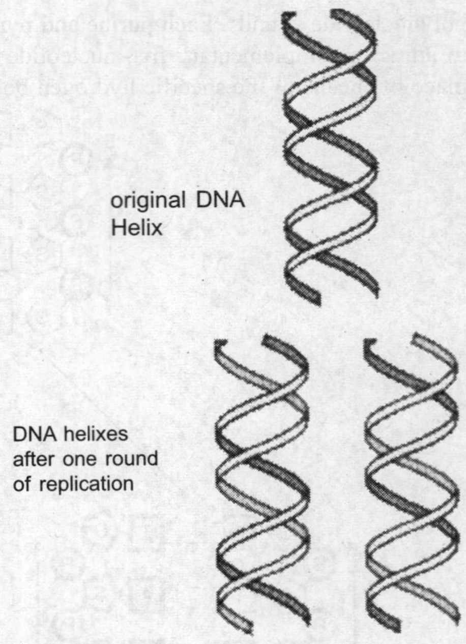

Semiconservative replication.

Depending on its content of ^{15}N and ^{14}N, the DNA bands at a specific position in the density gradient. Because the DNA synthesized by *E.coli* cells grown in ^{15}N would be denser than ^{14}N containing DNA, it would band further down the tube.

E.coli cells growth for sometime in the presence of ^{15}N-medium were washed free of the medium and transferred to ^{14}N-containing medium and allowed to continue to grow for specific lengths of time(*i.e.,* for various numbers of generation time). DNA isolated from cells grown for one generation of time in the ^{14}N medium had a density intermediate to that of the DNA from cells grown only in ^{15}N-containing medium (identified as *generation O*; Fig. 6.3) and that of DNA from cells grown only in ^{14}N-containing medium (*the controls*). Such a result immediately ruled out the possibility that DNA replication was conservative, because the conservative replication would have yielded two DNA bands in the density gradient for **generation 1** (*i.e.,* F_1 cells). The single band of intermediate density (identified as **"hybrid DNA"**) consisted of DNA molecules in which one strand contained ^{15}N and the other contained ^{14}N. When the incubation in the ^{14}N-medium was

original parent molecule

first generation daughter molecules

second generation daughter molecules

Fig. 6.2. Semiconservative DNA replication.

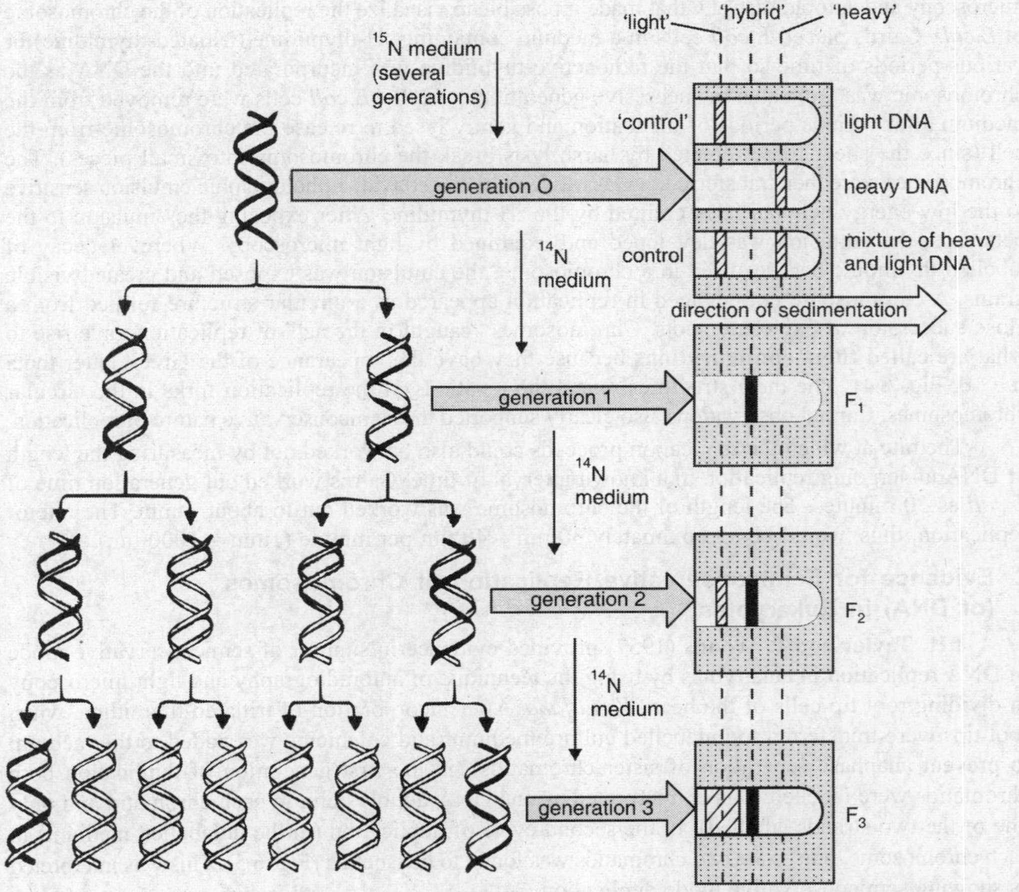

Fig. 6.3. Results of the Meselson-Stahl experiments (right) and their interpretation (left) (after Sheeler and Bianchi, 1987).

carried out for two generations of time (*i.e.,* **generation 2**), two DNA bands were formed – one at the same density position as the DNA from cells grown exclusively in ^{14}N medium (*i.e.,* light controls) and the other of intermediate density. Subsequent generations produced greater numbers of DNA molecules that banded at the "light" (^{14}N-containing DNA) position in the density gradient. These results are consistent only with the model of semiconservative replication. Studies using other prokaryotes as well as eukaryotes indicates that semi-conservative replication of DNA is probably a universal mechanism.

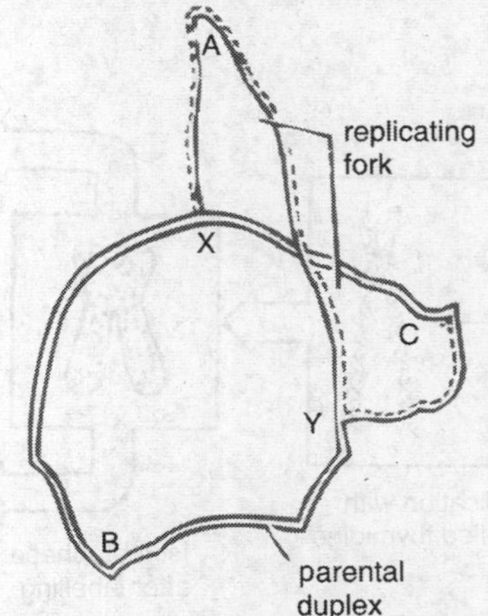

Fig. 6.4. Theta (θ) configuration of replication *E.coli* chromosome obtained by Cairns. Loops A and B have completed replication, whereas loop C remains to replicated. X and Y are the replication forks.

(*ii*) Visualization of replication in *E.coli.* In 1963, **J.Cairns** developed a technique employing a combination of microscopy and autoradiography that made it possible to visualize the replication of the chromosome of *E.coli.* Cairns placed *E.coli* cells in a medium containing ^{3}H-thymidine (tritiated-thymidine)for various periods of time so that the radioactive thymidine was incorporated into the DNA as the chromosome was replicated in successive generations of cells. *E.coli* cells were removed from the medium after various periods of incubation and gently lysed to release the chromosome from the cell (since the shear forces created by harsh lysis break the chromosome into small pieces). The chromosomes were then transferred to glass slides and coated with a photographic emulsion sensitive to the low energy beta-particles emitted by the ^{3}H-thymidine. After exposing the emulsion to the beta rays, the emulsion was developed and examined by light microscopy. Wherever decay of labelled thymidine had occurred in a chromosome, the emulsion was exposed and created visible grains. A chromosome not engaged in replication appeared as a circular structure formed from a close succession of explosure spots. Chromosomes "caught in the act" of replication gave rise to what are called **theta configurations** because they have the appearance of the Greek letter theta (*i.e.,* θ, Fig. 6.4) The theta structures reveal the positions of the replication forks in the circular chromosomes. Cairns' observations also clearly supported the semiconservative nature of replication.

The rate at which the replication proceeds could also be worked out by measuring the length of DNA undergoing replication in a known interval of time. Cairns worked out generation time of *E.coli* as 30 minutes. The length of the chromosome was worked out to about 1 mm. The rate of replication, thus, would be approximately 30 μm – 40 μm per minute (1mm = 1000 μm).

2. Evidence for Semiconservative Replication of Chromosomes (or DNA) in Eukaryotes

J.H. Taylor and **P. Woods** (1957) provided evidence in support of semiconservative mode of DNA replication in eukaryotes by using the technique of autoradiography and light microscopy in dividing root tip cells of the bean, *Vicia faba.* After incorporation of tritiated thymidine, when root tips were transferred to unlabelled culture medium (and colchicine was added to the medium to prevent anaphase separation of sister chromatids), in the first generation of duplication both chromatids were labelled (this is interpreted as one DNA double helix in each chromatid and only one of the two strands labelled). In the second cycle of duplication (in the unlabelled medium) in each chromosome, one of the two chromatids was found to be labelled (Fig. 6.5). This was interpreted as showing semiconservative mode duplication.

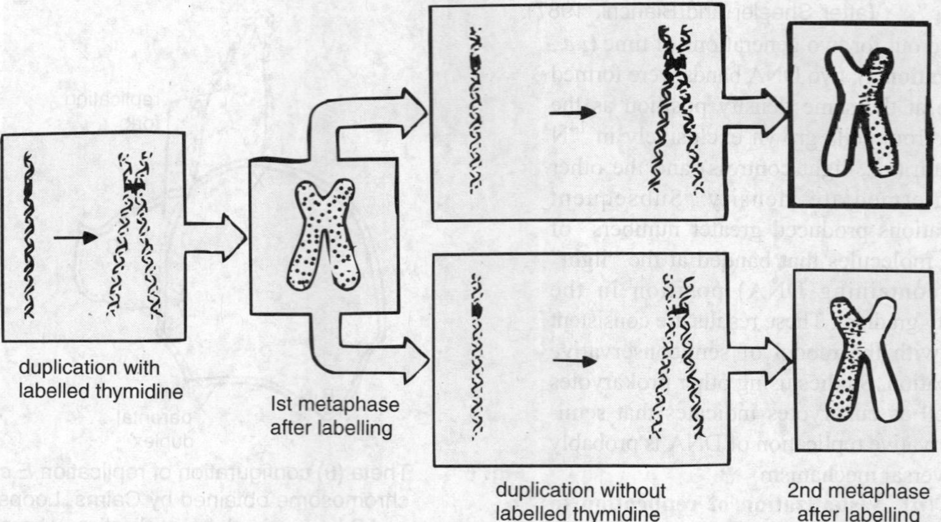

Fig. 6.5. Taylor and Wood's experiment on *Vicia faba* root tips using autoradiography technique (after Sheeler and Bianchi, 1987).

3. Semidiscontinuous DNA Replication

Various experimental evidences have suggested that DNA synthesis is continuous on one strand (3′ to 5′ strand), called **leading strand** and discontinuous on the other strand (5' to 3' strand), called **lagging strand**. Since DNA synthesis always proceeds in 5′ to 3′ direction, so, on the lagging strand, synthesis takes place discontinuously in pieces, called **Okazaki fragments** (after the name of discoverer **R. Okazaki**, 1968). Okazaki fragments average about 1,500 nucleotides in prokaryotes and 150 in eukaryotes (see **Tamarin**, 2002). Later on, these pieces are fused with the help of ligase enzyme to form an intact lagging strand. Such a DNA replication, where the leading strand is synthesized continuously and the lagging strand is synthesized discontinuously, is called **semidiscontinuous replication**. Once initiated, continuous DNA replication can proceed indefinitely. DNA polymerase III on the leading – strand template has what is called **processivity**: once it is attaches, it does not release until the entire strand is replicated. Discontinuous replication, however, requires the repetition of four steps: primer synthesis, elongation, primer removal with gap filling and ligation.

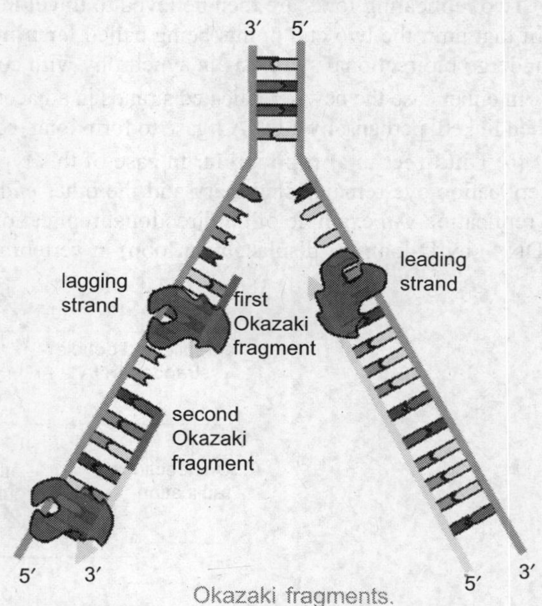

Okazaki fragments.

4. Unidirectional and Bidirectional DNA Replication

Regarding the direction of replication, DNA replication may be of the following two types:

(*a*) **Bidirectional replication.** All known DNA molecules, with only few exceptions, replicate as circles (or bubbles/eyes) and, hence, initiate within the helix. In the electron microscope, eukaryotic

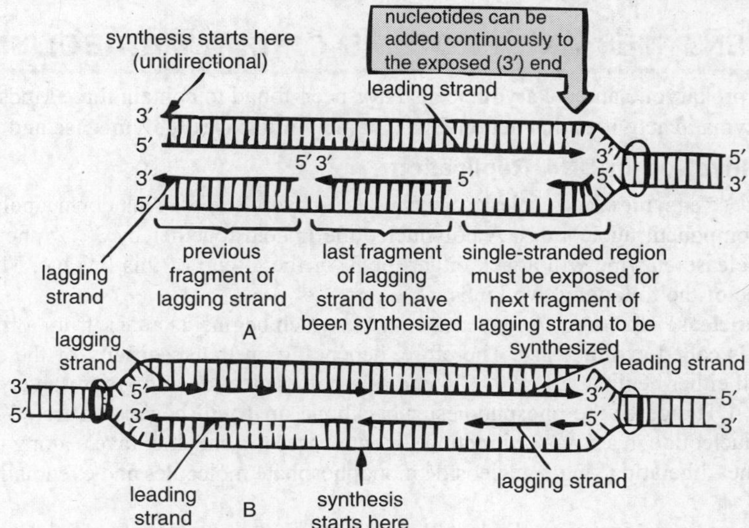

Fig. 6.6. A – Unidirectional DNA replication; B – Bidirectional DNA replication.

chromosomes are found to contain multiple expanding replication eyes, in contrast to a single eye of the prokaryotic DNA. Further DNA synthesis within a given replication unit eye is initiated some where at or near the midpoint of the unit at a site termed as the origin (O); prokaryotes contain a solitary origin, while eukaryotes have multiple (up to several thousands) origins for DNA replication. Both ends of the eye are moving and serve as **replication forks**(Fig. 6.6).

Two replicating forks are then believed to travel in opposite directions until they reach either end of that unit, the two end points being called **termini(T)**. A given replication unit may or may not undergo bidirectional synthesis in synchrony with continuous units.

In either case the newly replicated strands in adjacent units will eventually meet. These strands are then linked, perhaps, by a DNA ligase to form long, continuous daughter DNA strands (Fig. 6.8).

(*b*) Unidirectional replication. In case of this type of DNA replication, one of two ends of the replication eye remains stationary and the other end serves as the replication fork and moves with replication. An example of unidirectional replication is the replication of mitochondrial DNA (mt DNA) by D-loop (or displacement loop) in vertebrates.

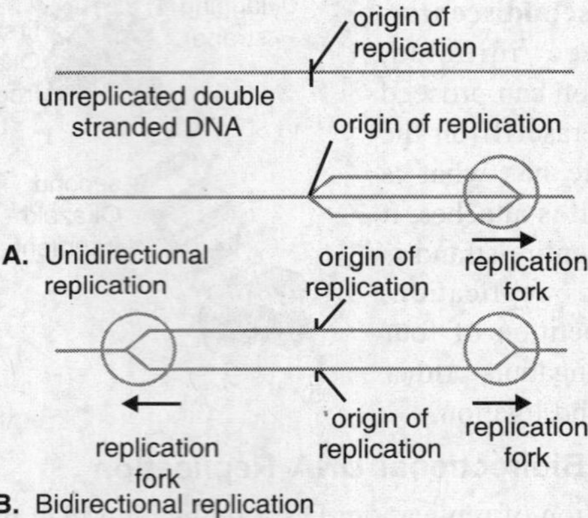

Fig. 6.7. The difference between unidirectional and bidirectional replications in the movement of replication forks.

6.2. ENZYMES AND PROTEINS OF DNA METABOLISM

Different prokaryotic and eukaryotic cells have been found to contain three kinds of nuclear enzymes or enzymatic activities that act on DNA, namely **nucleases, polymerase** and **ligases**.

A. Enzymes involved in DNA Replication

(I) Nuclease enzymes. The nuclease enzymes act to hydrolyze or break down a polynucleotide chain into its component nucleotides. A polynucleotide is held together by 3′, 5′ phosphodiester bonds and a nuclease enzyme will attack either the 3′ or the 5′ end of this linkage. The nuclease enzymes may be of the following two kinds.

(*a*) Exonuclease enzymes. A nuclease enzyme which begins its attack from a free end of a polynucleotide is called **exonuclease**. Therefore, depending on the specificity of the enzyme, an exonuclease will either begin at a free 3′-OH end of a polynucleotide and progressively cleave the bonds on the 3′-OH side of the phosphodiester backbone or it will begin at a free 5′-P end and digest the polynucleotide in a $5' \rightarrow 3'$ direction. In both cases the enzyme travels along the chain in a stepwise manner, liberating single nucleoside monophosphate molecules and eventually digesting entire polymer.

(*b*) Endonuclease enzyme. Endonuclease enzyme also attacks one of the two sides of phosphodiester linkages, but they react with those bonds that occur within the interior of a

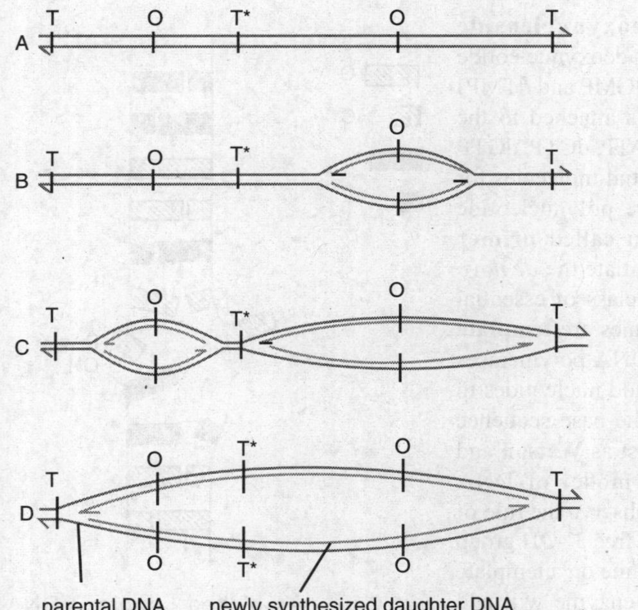

Fig. 6.8. Bidirectional mode for mammalian chromosomal DNA replication. Origins are indicated by O and termini by T. A – DNA prior to replication; B – replication started in right-hand replication unit; C – replication started in left-hand replication unit and completed at termini of right-hand unit; D – replication completed, sister duplexes joined at common terminus T.

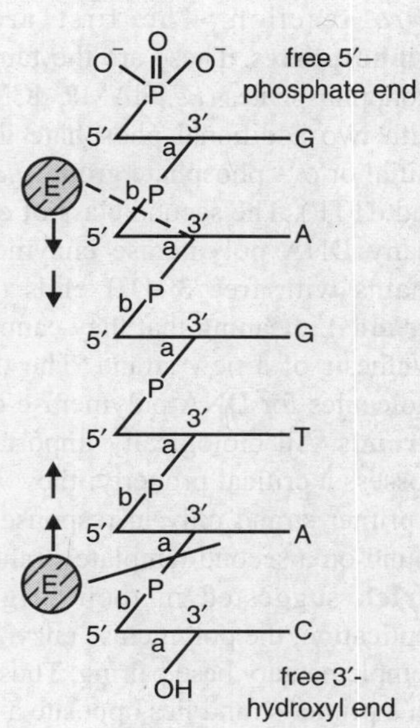

Fig. 6.9. Exonuclease action on a nucleotide chain. The 3′ side of each phosphodiester linkage is labelled a, the 5′ side is labelled b. An exonuclease that cleaves a linkages, starting from a free 3'-OH end (*e.g.,* snake venom phosphodiesterase) is shown in check. An exonuclease that cleaves b linkages, starting from a free 5′-P end (*e.g.,* bovine spleen phosphodiesterase) is shown in line shade (after Lehninger, 1970).

polynucleotide chain. If the polynucleotide chain is single-stranded (*e.g.,* viral DNA), such an attack will obviously cut the chain into two pieces. If, however, the polynucleotide strand is a member of a DNA helix (*e.g.,* prokaryotic and eukaryotic DNA), a single endonucleolytic cut will create a **nick** in the helix; the helix remains in one piece but it now possesses a gap that contains two free ends, which can serve as substrates for exonucleases. A nicked double helix suffers a localized disruption of its secondary structure. The molecules becomes free to bend or rotate around its intact strand and the two broken ends are free to dangle. The increased molecular motion in the region of the nick will very likely interrupt hydrogen bonding between bases in the vicinity of the nick, thus, effecting a limited "untravelling" of the helix.

(II) Polymerase or replicase enzymes. A polymerase enzyme catalyses the formation of a polymer and cellular polymerase enzymes of genetic interest are those that bring about the synthsis of one polynucleotide chain that is a copy of another. A polymerase enzyme is called **replicase** enzyme when the copy of polynucleotide chain so produced is inherited by daughter cells or viruses, that is, when the enzyme brings about chromosome replication (see **Goodenough** and **Levine**, 1974).

***In vitro* DNA polymerization.** To understand the mechanism of DNA-replication inside the living cell (*in vivo*), molecular biologists tried *in vitro* polymerization of DNA and found that in addition to DNA polymerase enzyme, three classes of organic molecules are essential for an *in* *vitro* reaction. The first are **deoxynucleoside triphosphates**. These are the familiar deoxynucleotide monophosphates (*i.e.,* dAMP, dCMP, dGMP and dTMP) with two additional phosphate groups attached to the initial or α – phosphate group (*i.e.,* dATP, dCTP, dGTP and dTTP). The second class of essential molecules for many DNA polymerase enzymes are polynucleotide chains with free **3′-OH** ends (often called **primer strands**), meaning that they cannot initiate the *de novo* synthesis of a new strand. The third class of essential molecules for DNA polymerase enzymes are **template strands**. All biologically important DNA polymerases possess a critical property: they will add nucleotides to a primer strand only in response to the base sequence found on a second template strand. Just as **Watson** and **Crick** suggested in their original model of DNA replication, the polymerase enzymes observe the rule of complementary base pairing. Thus, if a free 3'-OH group on a primer strand lies opposite a thymine on a template strand (see Fig. 6.11), a polymerase enzyme will add only an adenine group to the primer, even when dCTP, dTTP, dGTP are also present in the reaction mixture.

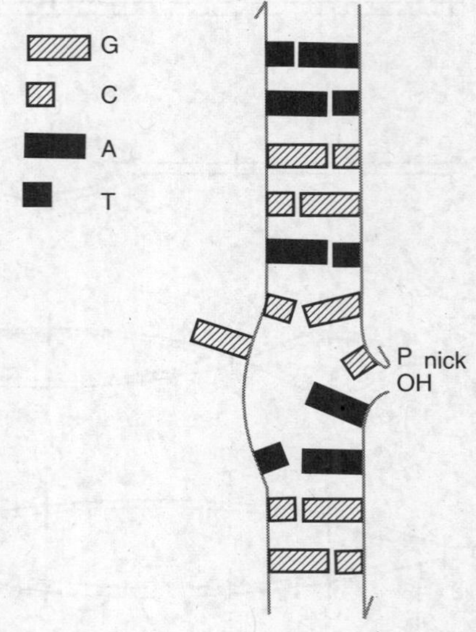

Fig. 6.10. A nicked duplex DNA molecule such as is formed by an endonuclease enzyme.

Once the appropriate molecules are present and the appropriate ionic conditions are maintained, *in vitro* polymerization reaction summarized in Fig. 6.11 will occur. The α-phosphate of a nucleoside triphosphate molecule forms a 3′, 5′ phosphodiester bond with a free 3′–OH in the growing polynucleotide chain, and a molecule of **pyrophosphate** (P ~ P) is simultaneously released. Pyrophosphate containes a "high-energy" or " ~ " bond, meaning that when the released pyrophosphate is hydrolyzed into two phosphate molecules, energy is liberated which drives the polymerization process forward.The resultant polymerization will always proceed in a net 5′ → 3′ direction, meaning that the nucleotide at the 3′ end is always the most recently added to the chain.

(1) Prokaryotic DNA polymerases. Three different DNA polymerases are known in *E.coli* and other prokaryotes (see Table 6.1), of which **DNA polymerase I** and **II** are meant for DNA repair and **DNA polymerase III** is meant for actual DNA replication.

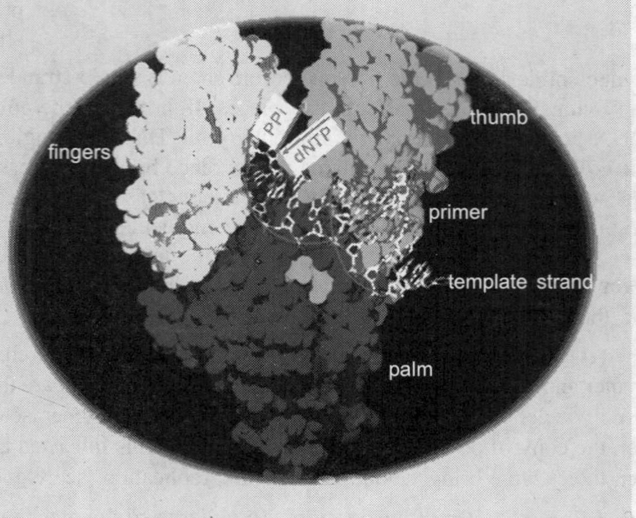

DNA polymerase.

1. DNA polymerase I. This enzyme was isolated around 1960 by **Arthur Kornberg** and was the first enzyme suggested to be involved in DNA replication. It is also called **Kornberg**

enzyme. DNA polymerase I enzyme is now considered to be a DNA repair enzyme rather than a replication enzyme. This enzyme is known to have five active sites, namely **template site, primer site, 5′ → 3′ cleavage** or **exonuclease site, nucleoside triphosphate site** and **3’ → 5’ cleavage site** (or **3′ → 5′ exonuclease site**) (Fig. 6.13). DNA polymerase I is mainly involved in removing RNA primers from okazaki or precursor fragments and filling the resultant gaps due to its 5′ → 3′ polymerizing capacity. DNA polymerase I enzyme can also remove thymine dimers produced due to UV – irradiation and fill the gap due to excision. Both polymerization (= chain elongation) and exonuclease activity of DNA polymerase I have been shown in Fig. 6.16. This is called **proof reading** or **editing function** of this enzyme.

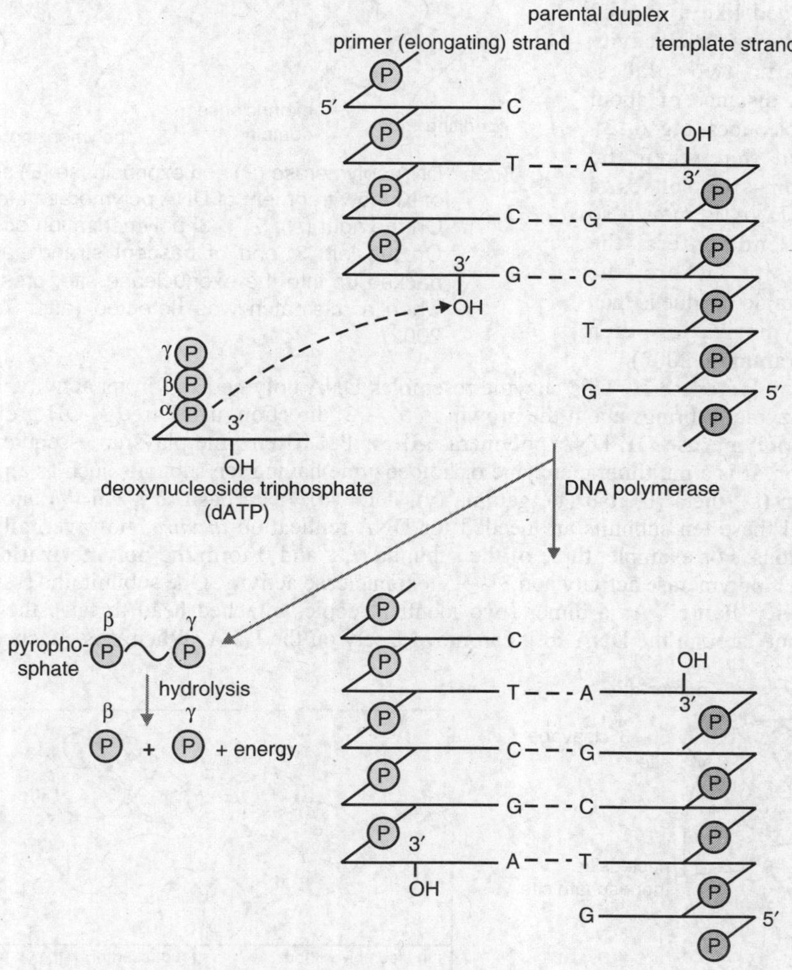

Fig. 6.11. DNA polymerization. A dATP molecule is shown being added to a parental DNA duplex or helix at its 3′ –OH end. As DNA polymerase catalyzes the formation of the phosphodiester bond, pyrophosphate (P ~ P) is hydrolyzed into two phosphate molecules.

Klenow fragment/polymerase. T. Steitz and coworkers have done much X-ray crystallography work that has given us a good look at the structure of a DNA polymerase. In fact, most of their work has been done on a fragment of DNA polymerase I called **Klenow fragment.**

Klenow fragment/polymerase was initially prepared by cutting the natural *E.coli* DNA polymerase I enzyme into two segments with a protease. One of these segments retained the polymerase and 3′ → 5′ exonuclease activity, but lacked the 5′ → 3′ exonuclease. Klenow enzyme is shaped like a cupped right hand with enzymatic activity taking place in two places, separated by a distance of about two or three nucleotides (Fig. 6.12). It is proposed that when the polymerization site senses a mismatch, the DNA is moved so that the 3′ end enters the exonuclease site, where the incorrect nucleotide residue is then cleaved. Polymerization then continue (see **Tamarin,** 2002).

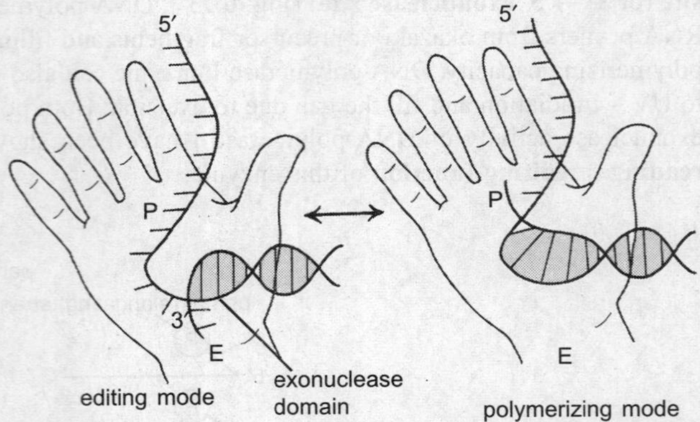

Fig. 6.12. DNA polymerase (P) and exonuclease (E) activities of Klenow fragment of DNA polymerase I in *E.coli*. On the right (A), 5′ → 3′ polymerization occurring. On the left, 3′ end of nascent strand has been backed up into the exonuclease site, presumably when a mismatch was detected (after Tamarin, 2002).

2. **DNA polymerase-II.** This enzyme resembles DNA polymerase-I in its activity, but is a DNA repair enzyme. It brings about the growth in 5′ → 3′ direction, using free 3′-OH groups.

3. **DNA polymerase-III.** DNA polymerase-III or Pol III enzyme plays an essential role in DNA replication. It is a multimeric enzyme or holoenzyme having ten subunits such as alpha (∝), beta(β), epsilon (ε), theta (θ), tau (τ), gamma (γ), delta (δ), delta dash (δ′), chi (χ) and psi(Ψ) (Table 6.2). All these ten subunits are needed for DNA replication *in vitro;* however, all having different functions. For example, three of the subunits α, ε and θ form the **polymerization core,** with both 5′ →3′ polymerase activity and 3′ → 5′ exonuclease activity. One subunit, the **β subunit,** is a "**processivity clamp**". As a dimer (two identical copies attached head to tail), the protein forms "doughnut" around the DNA so it can move freely on the DNA. When it is attached to the

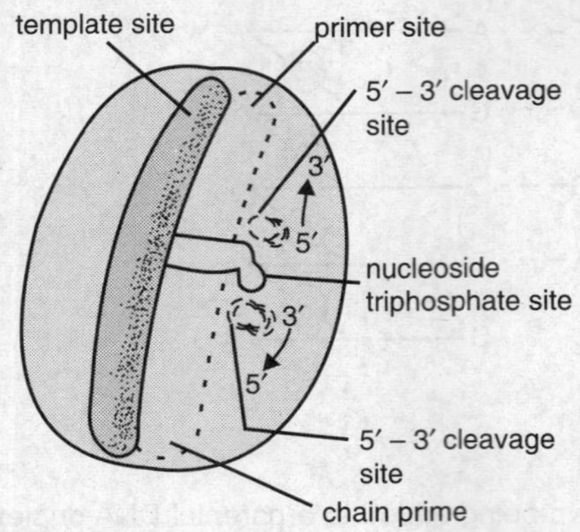

Fig. 6.13. A model of DNA polymerase-I enzyme, showing its five-different active sites

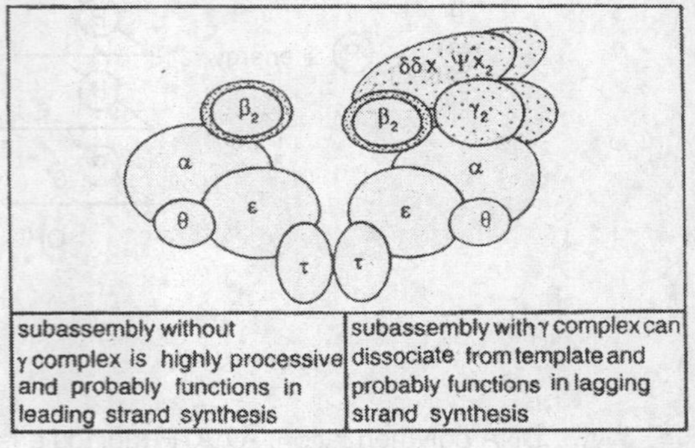

Fig. 6.14. Heterodimeric assembly of *E.coli* polymerase III holoenzyme to facilitate simultaneous leading-and lagging strand synthesis.

core enzyme, the polymerase is held tightly to DNA and shows high processivity: the leading strand is usually synthesized entirely without the enzyme leaving the template (Fig. 6.15). The remaining subunits are involved in processivity control and replisome formation. They allow polymerase to move off and on DNA of lagging-strand template as Okazaki fragments are completed (a process known as **polymerase cycling**).

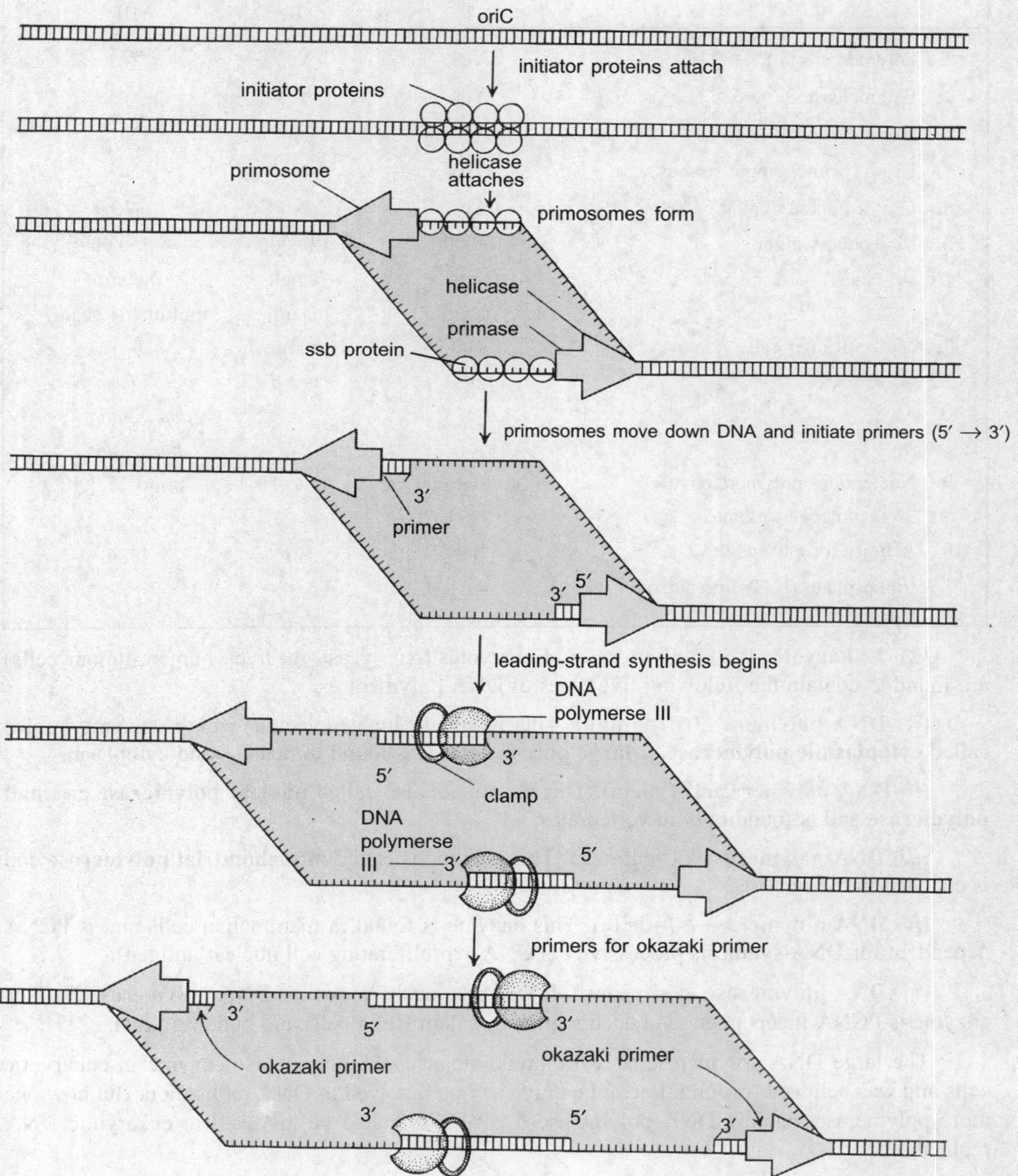

Fig. 6.15. Events at the origin of DNA replication in *E.coli*. The DNA opens up at *oric* to create two moving Y-junctions (replication forks). Initiator proteins attach and then bind helicase. The helicase then binds primase, forming a primosome. After the primer forms and two copies of DNA polymerase III are bound, the polymerization begins(after Tamarin, 2002).

Table 6.1. Comparison of different characters of three types of DNA polymerases of *E.coli.*

Character	DNA polymerase		
	I	II	III
1. Polymerization 5′ → 3′	Yes	Yes	Yes
2. Exonuclease 3′ → 5′	Yes	Yes	Yes
3. Exonuclease 5′ → 3′	Yes	No	No
4. Use of primer single strands	Yes	No	No
5. Use of nicked duplex or helix	Yes	No	No
6. Molecular weight	109,000 (single chain)	120,000 (single chain)	> 250,000 (hetero-multimeric chain)
7. Molecules per cell	400	Not known	10–12
8. Gene	*pol A*	*pol B*	*pol C* (*dna E*), *dna N, dna X, dna Q, dna T*
9. Nucleotides polymerized at 37°C molecules/minute	up to 1,000	up to 50	up to 15,000
10. Affinity for nucleoside triphosphate (NTPs) precursors	low	low	high

(2) Eukaryotic DNA polymerases. Eukaryotes (*e.g.,* yeast, rat liver, human tumour cells) are found to contain the following five types of DNA polymerases:

(i) **DNA polymerase α (=alpha).** This relatively high molecular weight enzyme is also called **cytoplasmic polymerase** or **large polymerase**. It is found in nucleus and cytoplasm.

(ii) **DNA polymerase β (=beta).** This enzyme is also called **nuclear polymerase** or **small polymerase** and is found only in vertebrates.

(iii) **DNA polymerase γ (=gamma).** This enzyme is called **mitochondrial polymerase** and is encoded in the nucleus.

(iv) **DNA polymerase δ (=delta).** This enzyme is found in mammalian cells and is PCNA dependent for DNA-synthesis processivity (PCNA = proliferating cell nuclear antigen).

(v) **DNA polymerase ε (=epsilon).** It was previously known as **DNA polymerase II**. This enzyme is PCNA independent and occurs in mammalian HeLa cells and budding yeast.

The large DNA polymerase α is the predominant DNA polymerase enzyme in eukaryotic cells and was believed for long time to be only enzyme involved in DNA replication. But now one more polymerase, namely DNA polymerase δ is also found to be involved in eukaryotic DNA replication.

(III) DNA ligases. DNA ligase enzymes are capable of catalyzing phosphodiester bond formation between free 3′–OH and free 5′–P groups of a nick of DNA which is created by endonuclease enzyme, thereby restoring an intact DNA duplex. Many DNA ligases have already

Table 6.2. Summary of the enzymes and proteins involved in DNA replication in *E.coli* (Source: **Tamarin,** 2002).

Enzyme or **Protein**	**Genetic locus**	**Function**
1. DNA polymerase I	*polA*	Gap filling and primer removal
2. DNA polymerase II	*polB*	Replicating damaged templates
3. DNA polymerase III (Holoenzyme)		
(i) α subunit	*dnaE*	Polymerization core; 5′ → 3′ polymerase
(ii) ε subunit	*dnaQ*	Polymerization core; 3′ → 5′ exonuclease
(iii) θ subunit	*holE*	Polymerization core
(iv) β subunit	*dnaN*	Processivity clamp (as a dimer)
(v) τ subunit	*dnaX*	Preinitiation complex; dimerization of core
(vi) γ subunit	*dnaX*	Preinitiation complex: loads clamp
(vii) δ subunit	*holA*	Processivity core
(viii) δ′ subunit	*holB*	Processivity core
(ix) χ subunit	*holC*	Processivity core
(x) Ψ subunit	*holD*	Processivity core
4. Helicase	*dnaB*	Primosome; unwinds DNA
5. Primase	*dnaG*	Primosome; creates Okazaki fragment primers
6. Initiator protein	*dnaA*	Binds at origin of replication
7. DNA ligase	*lig*	Closes Okazaki fragment
8. Ssb protein	*ssb*	Binds single stranded DNA
9. DNA topoisomerase I	*topA*	Relaxes supercoiled DNA
10. DNA topoisomerase type II (DNA gyrase)		
(i) α subunit	*gyrA*	Relaxes supercoiled DNA; ATPase
(ii) β subunit	*gyrB*	Relaxes supercoiled DNA
11. Topoisomerase IV	*parE*	Unconcatenates DNA circles
12. Termination protein	*tns*	Binds at termination sites

been discovered. The ligase enzyme from *E.coli* requires the presence of oxidized nicotinamide adenine dinucleotide (NAD^+) a cofactor, whereas the ligase enzyme specified by T_4 bacteriophage requires ATP to bring about the joining reaction.

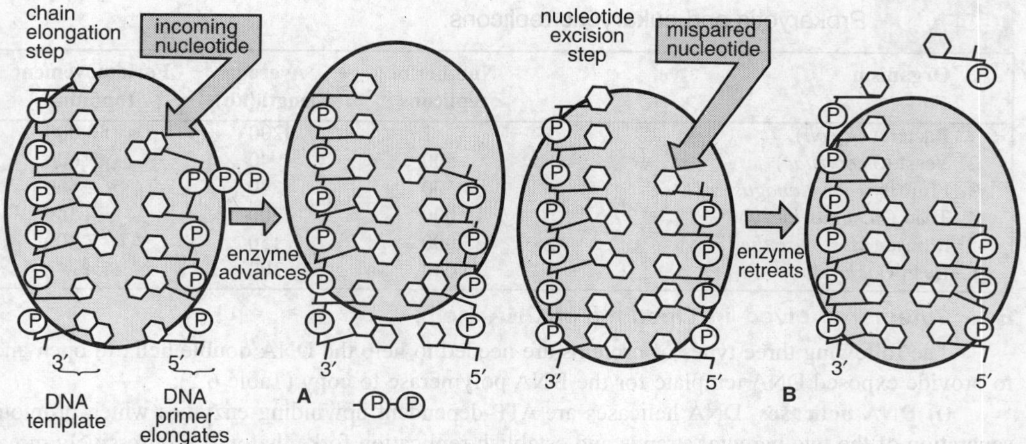

Fig. 6.16. Functions of DNA polymerase-I enzyme. A-Mode of polymerization of nucleotides, on the primer site of the enzyme, B-Removal of mispaired nucleotide activity of the enzyme.

Roles of DNA Primers in DNA Replication

No known DNA polymerase can initiate synthesis of DNA without the availability of a **primer RNA strand**. So, before actual DNA replication starts, short RNA oligonucleotide segments, called **RNA primers** or simply the **primers,** have to be synthesized by **DNA primase** enzyme utilizing ribonucleoside triphosphates. This RNA primer is synthesized by copying a particular base sequence from one DNA strand and differs from a typical RNA molecule in that after the synthesis the primer remains hydrogen-bonded to the DNA template(**Freifelder**, 1985). The primers are about 10 nucleotides long in eukaryotes and they are made at intervals on the lagging strand where they are elongated by the DNA polymerase enzyme to begin each Okazaki fragment. These RNA primers are later excised and filled with DNA with the help of DNA repair system in eukaryotes (or DNA polymerase I in *E.coli*).

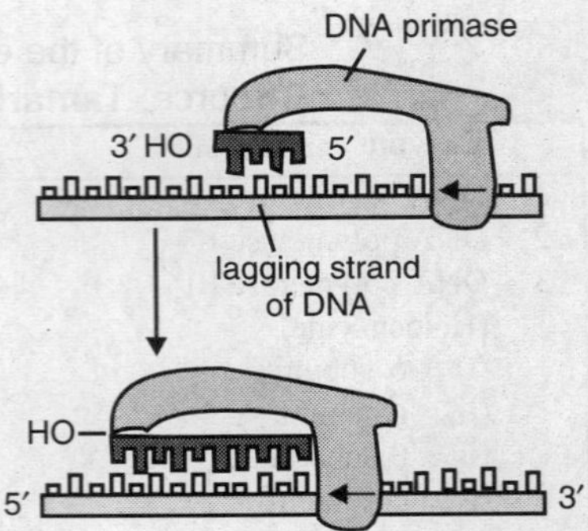

Fig. 6.17. Formation of RNA primer on lagging strand by DNA primase enzyme. Unlike DNA polymerase, this enzyme can start a new polynucleotide chain by joining two nucleotide triphosphates together (after Alberts *et al.*,1989).

In bacteria, two different enzymes are known to synthesize primer RNA oligonucleotides – **RNA polymerase** (on the leading strand) and **DNA primase** (on the lagging strand).

Replicons

DNA replication in prokaryotes and eukaryotes is attained in discrete units, called **replicons**. The number of replicons may vary in a genome from one in bacteria (*E.coli*) and 500 in yeast to several thousands in plants and animals (Table 6.3). For example, in the *E.coli* there is a single replication with the **origin**, identified as a genetic locus *ori C* (245 bp). The origin is A: T rich, a feature that is related to unwinding of DNA to initiate replication. In *E.coli*, there are also termination sites (*ter A-F*), each consisting of ~ 23 bp. The process of termination of DNA replication requires the product of *tus* gene (**Tus protein** or *TBP*, *i.e.*, ter binding protein) which recognizes *ter* or termination sites.

Table 6.3. Prokaryotic and eukaryotic replicons.

Organism	Number of replicons	Average length(kb)	Fork movement (bp/min)
1. Bacteria (*E.coli*)	1	4200	50,000
2. Yeast (*S.cervisiae*)	500	40	3,600
3. Fruit fly (*D.melanogaster*)	3,500	40	2,600
4. Toad (*Xenopus laevis*)	15,000	200	500
5. Mouse (*Mus musculus*)	25,000	150	2,200
6. Bean (*Vicia faba*)	35,000	300	Not known

B. Proteins Involved in Opening of DNA Helix

The following three types of proteins are needed to help the DNA double helix to open and to provide exposed DNA template for the DNA polymerase to copy (Table 6.2):

(*i*) **DNA helicases**. DNA helicases are ATP-dependent upwinding enzymes which promote separation of the two parental strands and establish replication forks that will progressively move away from the origin. DNA helicase hydrolyze ATP when they are bound to single strands of DNA. Hydrolysis of ATP can change the shape of a protein molecule in a cyclic manner that allows the protein to perform mechanical work. DNA helicases utilize this principle to move rapidly

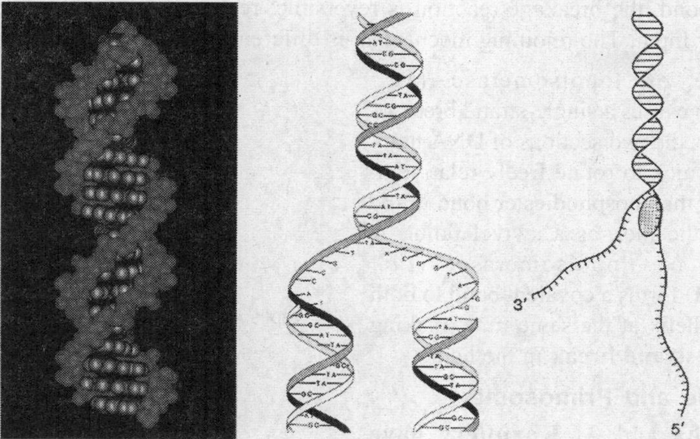

along a DNA single strand; when they encounter a region of double helix, they continue to move along their strand, thereby unwinding the helix (Fig. 6.18). Unwinding of the template DNA helix at a replication fork could in principle be catalyzed by two DNA helicases, acting in concert, one running along the leading strand and the other along the lagging strand.

(*ii*) **Helix-destabilizing strand.** (also called **single strand DNA-binding proteins** or **SSBPs**). Behind the replication fork, the single DNA strands are prevented from rewinding about one another (or forming double-stranded hair-pin loops in each single strands) by the action of SSB proteins. SSB proteins bind to exposed DNA strands without covering the bases, which, therefore, remain available for the templating process.

(*iii*) **Topoisomerases (DNA gyrases).** The action of a helicase introduces a positive supercoil into the duplex DNA ahead of the replication fork. Enzymes, called **topoisomerases**, relax the supercoil by attaching to the transiently supercoil duplex, nicking one of the strands and rotating it through the unbroken strand. the nick is then released. Thus, a DNA topoisomerase can be viewed as a "reversible nuclease" that adds itself covalently to a DNA phosphate, thereby breaking a phosphodiester bond in the DNA strand. because the covalent linkage that joins a topoisomerase to DNA phosphate retains the energy of the broken

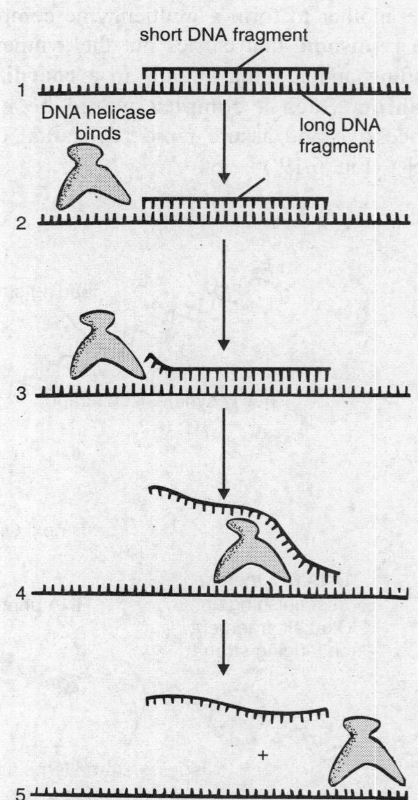

Fig. 6.18. Mode of action of DNA helicase (after Alberts *et al.*, 1989).

phosphodiester bond, the breakage reaction is reversible; resealing is rapid and does not require additional energy input. The rejoining mechanism is different from DNA ligase enzyme.

One type of topoisomerase (*i.e.*, **topoisomerase I**) causes a single-strand break or nick which allows the two sections of DNA helix on either side of nick to rotate freely relative to each other, using the phosphodiester bond in the strand opposite the nick as a swivel point. A second type of topoisomerase (*i.e.*, **topoisomerase II**) forms a covalent bond to both strands of DNA helix at the same time, making transient **double-strand break** in the helix.

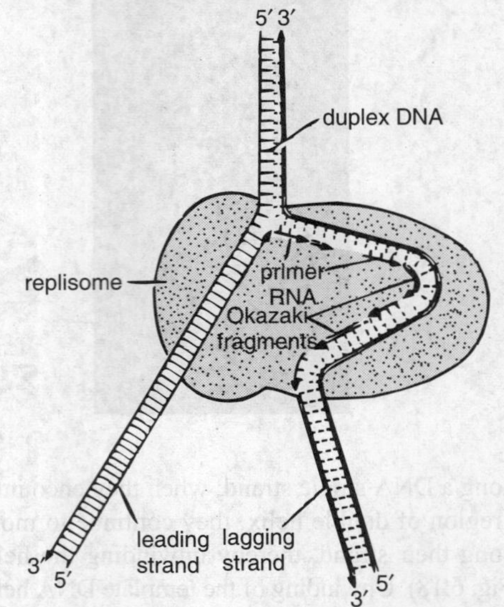

Fig. 6.19. Model of the replisome (after Sheeler and Bianchi, 1987)

Replisome and Primosome

N.K. Sinha and **A. Kornberg** have suggested that the DNA polymerases, RNA primases and helicases may be associated with one another to form a multienzyme complex – the **replisome** that carries out the synthesis of leading and lagging strands in a **coordinated fashion**. Such a complex would be highly processive and assure rapid replication of the DNA (Fig. 6.19.)

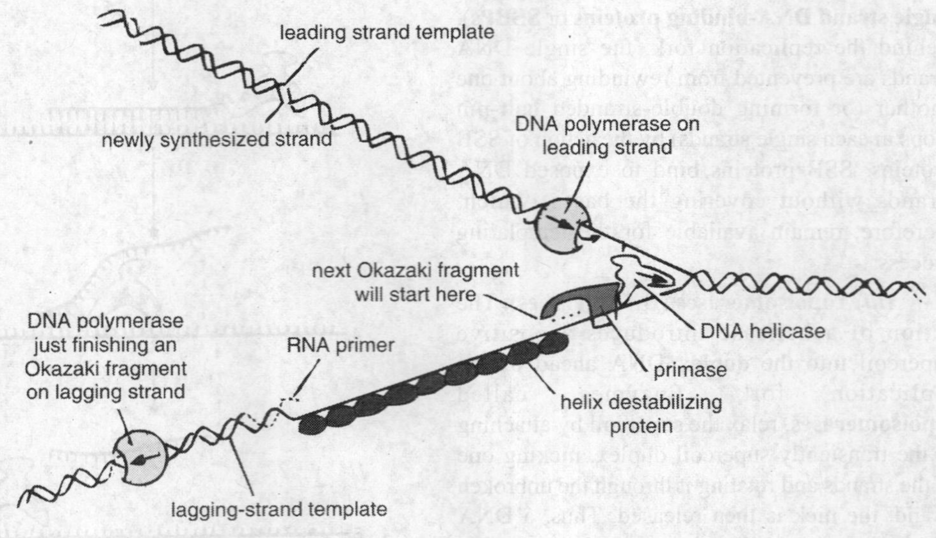

Fig. 6.20. A primosome in action (after Alberts *et al.*, 1989).

Likewise, the proteins at a replication fork cooperate to form replication machine, *i.e.*, the primase molecule is linked directly to the helicase to form a unit on the lagging strand called a **primosome** which moves with the fork, synthesizing RNA primers as it moves (Fig. 6.20).

6.3. MECHANISM OF DNA REPLICATION IN PROKARYOTES

In vitro DNA replication has been extensively studied in *E.coli* and in the phages and plasmids of *E.coli*. In *E.coli,* the process of DNA replication involves the following three steps.

1. Initiation of DNA replication. This process comprises three steps: (i) recognition of the origin (O), (ii) opening of DNA duplex to generate a region of single stranded DNA, and (iii) capture of **Dna B protein** (*i.e.,* 5′ → 3′ helicase; also acts as the activator of primase). Thus, Dna-A (or initiator protein)-ATP complex binds at 9 bp inverted repeat regions(R_1, R_2, R_3, R_4) of ori C of *E.coli* and promotes opening of the DNA duplex in a region of three direct repeats of 13-bp sequence (called **13-mers**). The opening occurs from right 13-mer leftwards and requires negatively supercoiled DNA and HU or IHF initiator proteins (Fig. 6.21). Dna B (=helicase) is transferred to exposed single stranded DNA (Fig. 6.22) and causes unwinding of the DNA in the presence of ATP, SSB protein and **DNA gyrase** (a topoisomerase). This results in unwinding of DNA duplex and the replication from ori proceeds in both directions (bidirectional); SSB binding occurs on single stranded regions and two Dna B complexes (=primosomes) are loaded one on each strand.

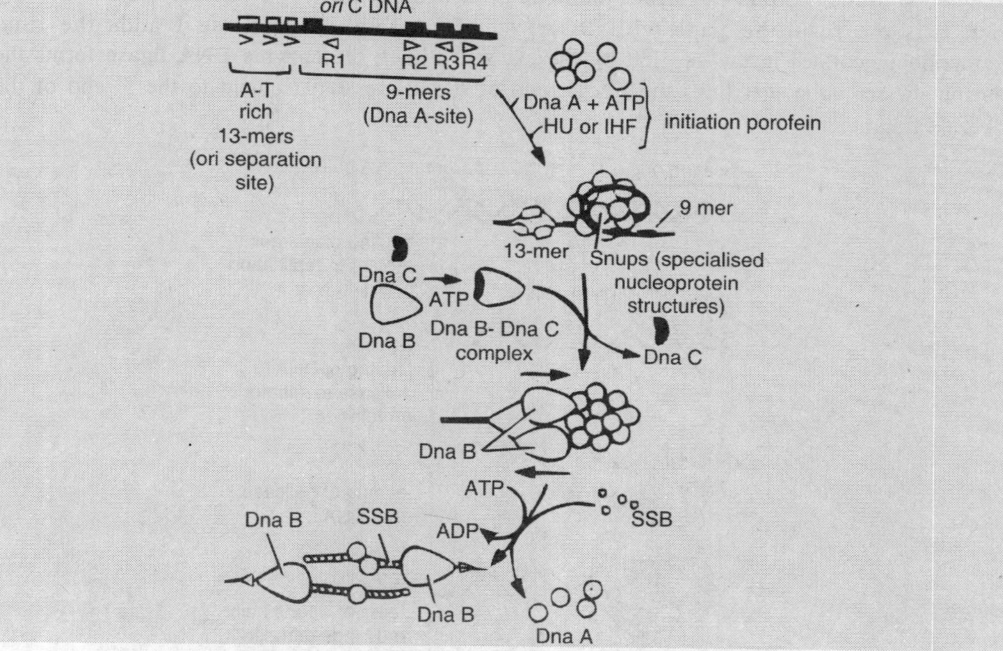

Fig 6.21. Mode of the initiation of DNA replication at *ori C* of *E.coli.*

2. Elongation of DNA chain. This step requires the presence of the following enzymes and factors: 1. Dna B or helicase (also called **mobile promoter**); 2. Primase (Dna G); 3. DNA polymerase holoenzyme (or DNA pol III HE); 4. SSB proteins; 5. RNAase H which removes RNA primers; 6. DNA polymerase I which is used for filling the gap created due to RNA primers and 7. DNA ligase (which converts primerless Okazaki fragments into continuous strand). During initiation to elongation transition, the following events occur:

1. As helicase (or Dna B) travels in 5′ → 3' direction, it generates a replication fork by opening the DNA duplex.

2. The DNA strand having helicase becomes the lagging strand. DNA primase associates with Dna B helicase, forming the primosome which synthesizes multiple primers for lagging strand and single RNA primer for the leading strand.

3. For the synthesis of lagging strand, the DNA pol III HE has to work on the same strand to which Dna B helicase is bound, but it travels in opposite direction.

4. Dna B helicase, Dna G primase and DNA pol III HE work together in strand elongation. Helicase and DNA polymerase assembly remains *processive, i.e.,* they remain tightly bound to the fork and stay bound through the reaction.

Synthesis (=elongation) of lagging and leading strands takes place by somewhat different methods; it is far more complex for lagging strand than for the leading strand:

A. Discontinuous synthesis on lagging strand. 1. Primase is taken up from solution and is activated by helicase (Dna B) to synthesize a RNA primer (10 to 20nt or nucleotides long) on the lagging strand. 2. The RNA primers are recognized by DNA pol III HE on the lagging strand and are utilized for synthesis of precursor or Okazaki fragments. In fact, each new RNA primer is recognized by the gamma (γ) subunit of DNA pol III HE and loaded with â subunit of the same polymerase. This preloaded β subunit may then capture the *core* of DNA poly III HE when it becomes available after finishing its synthetic job on the preceding Okazaki fragments. 3. On completion of the Okazaki fragments, the RNA primers are excised by DNA polymerase I, which then fills the resulting gaps with DNA. 4. After DNA **polymerase I** adds the final deoxyribonucleotides in the gap left by the excised primer, the enzyme **DNA ligase** forms the phosphodiester bond that links the free 3′ end of the primer replacement to the 5′ end of the Okazaki fragment.

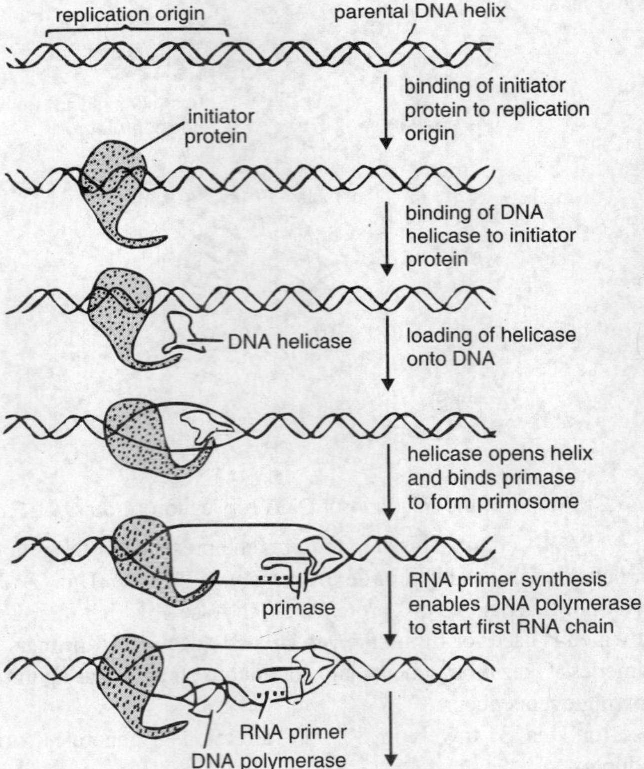

Fig. 6.22. Initial steps leading to formation of replication forks at the *E.coli* (after Alberts *et al.*, 1989).

B. Continuous synthesis on leading strand. (1) In bidirectional DNA replication, the leading strand is primed once on each of the parental strands. (2) The RNA primer of the leading strand is synthesized by RNA polymerase enzyme. (3) DNA pol III HE causes elongation of the leading strand and finally DNA pol I and ligase enzymes give final touch to the leading strand as in case of the lagging strand (Fig. 6.24).

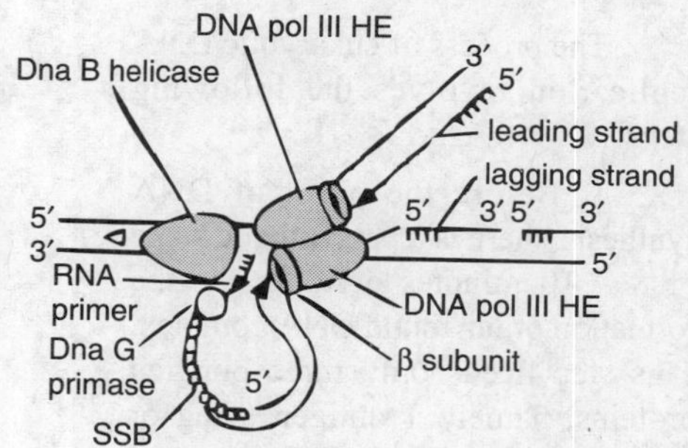

Fig. 6.23. Model of a DNA replication fork in prokaryotes during elongation.

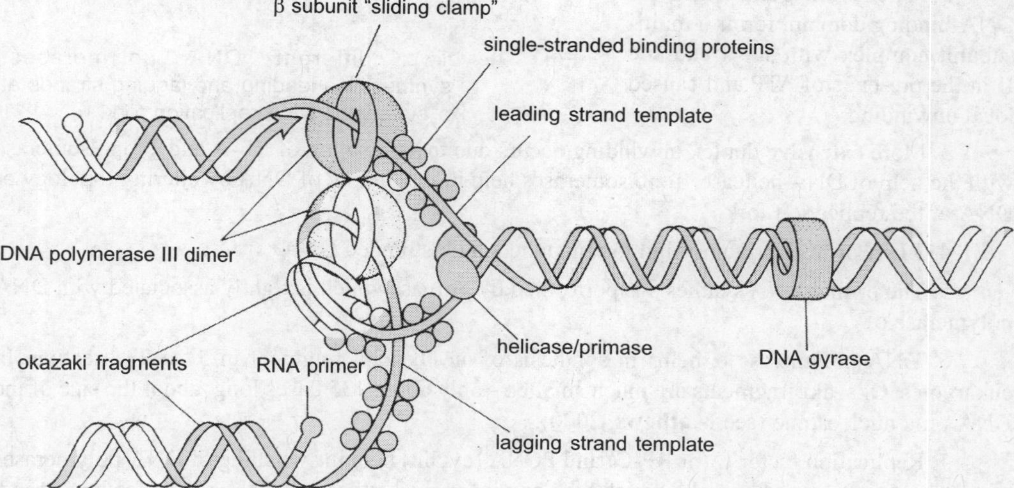

Fig. 6.24. Summary of prokaryotic DNA synthesis at a single replication fork. Various enzymes and proteins essential to the process are shown (after Klug and Cummings, 2003).

6.4. DNA REPLICATION IN EUKARYOTES

Eukaryotic DNA replication requires two different DNA polymerase enzymes, namely **DNA polymerase** α and **DNA polymerase** δ. DNA polymerase δ synthesizes the DNA on the leading strand (continuous synthesis), whereas DNA polymerase α synthesizes the DNA on the lagging strand (discontinuous DNA synthesis). Besides these two enzymes, six more factors are involved in eukaryotic DNA replication: (1) T antigen; (2) replication factor A or RF-A (also called RP-A or eukaryotic SSB); (3) topoisomerase I; (4) topoisomerase II; (5) proliferating – cell nuclear antigen (PCNA), also called **cyclin**), and (σ) replication factor C or RF-C.

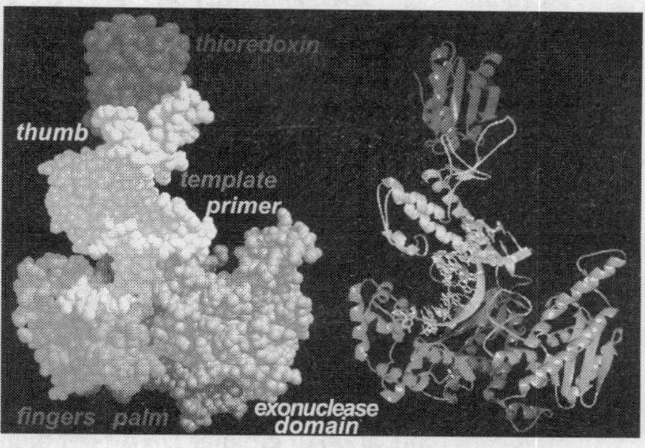

3-D Structure of DNA polymerase.

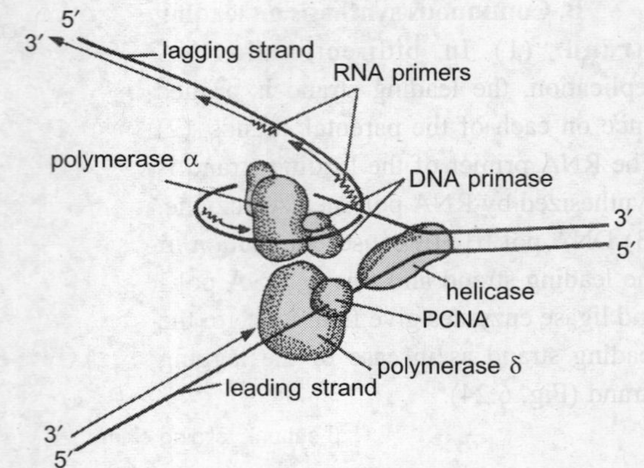

Fig. 6.25. Two different DNA polymerases, synthesizing leading and lagging strands at a eukaryotic DNA replication fork.

The process of eukaryotic DNA replication involves the following steps:

1. Before the onset of DNA synthesis, there is a presynthetic stage of 8 – 10 minutes duration for the formation of unwound DNA complex. This step needs only three purified proteins, namely T antigen (T-ag or tumour antigen), RF-A and topoisomerase I and II.

2. The T-antigen, using its DNA-binding domain, forms a multi-subunit complex with site I and site II in the presence of ATP and caused local unwinding.

3. More extensive duplex unwinding occurs due to association of RF-A and a topoisomerase with the help of DNA helicase. Topoisomerases help in unwinding of DNA by altering topology of DNA at the replication fork.

4. RF-A or SSB proteins bind to unwound single stranded DNA.

5. The primer RNA synthesis is performed by primase which is tightly associated with DNA polymerase α.

6. DNA polymerase α helps in synthesis of an Okazaki fragment in 5′ to 3′ direction. In eukaryotes, Okazaki fragments are much smaller – only about 135 bases long, about the size of the DNA on a nucleosome (see **Mathews**, 2000).

7. Replication factor C (or RF-C) and PCNA (cyclin) help in switching of DNA polymerases so that pol α is replaced by pol δ which then continuously synthesized DNA on the leading strand.

8. Another Okazaki fragment is then synthesized from the replication fork on the lagging strand by pol α-primase complex and this step is repeated again and again, till the entire DNA molecule is covered.

9. The RNA primers are removed and the gaps are filled as in prokaryotic DNA replication.

Recently, role of DNA polymerase ε in DNA replication has been stressed upon, so that three DNA polymerases (α, δ and ε) are now known to be involved in eukaryotic DNA replication. **A. Sugino** and coworkers have proposed that DNA polymerase α might function at both the leading and lagging strands (since polymerase α has primase activity), whereas polymerase ε and polymerase δ are involved in elongation of the leading and lagging-strands respectively.

6.5. MODELS OF DNA REPLICATION

The following three models have been proposed for DNA replication in different organisms:

1. Replication Fork Model

It occurs both in linear and circular DNA molecules and involves the formation of **replication forks** which either move in one direction in unidirectional replication or both directions in bidirectional replication.

In *E.coli*, replication of circular DNA is of bidirectional type involving intermediate theta (θ) type structures.

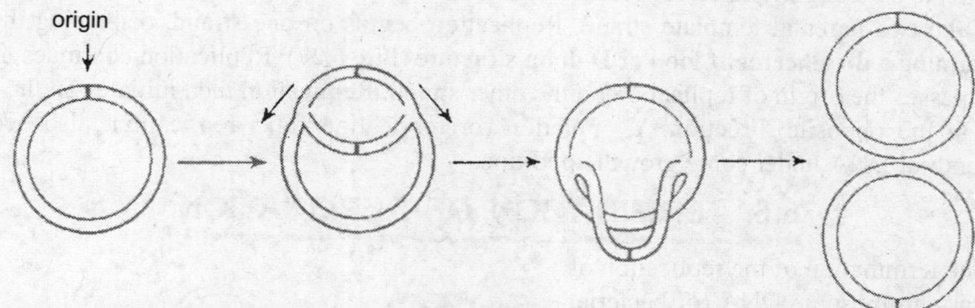

Fig. 6.26. Observable stages in the DNA replication of a circular chromosome, assuming bidirectional DNA synthesis. The intermediate figures are called theta structures (after Tamarin, 2002).

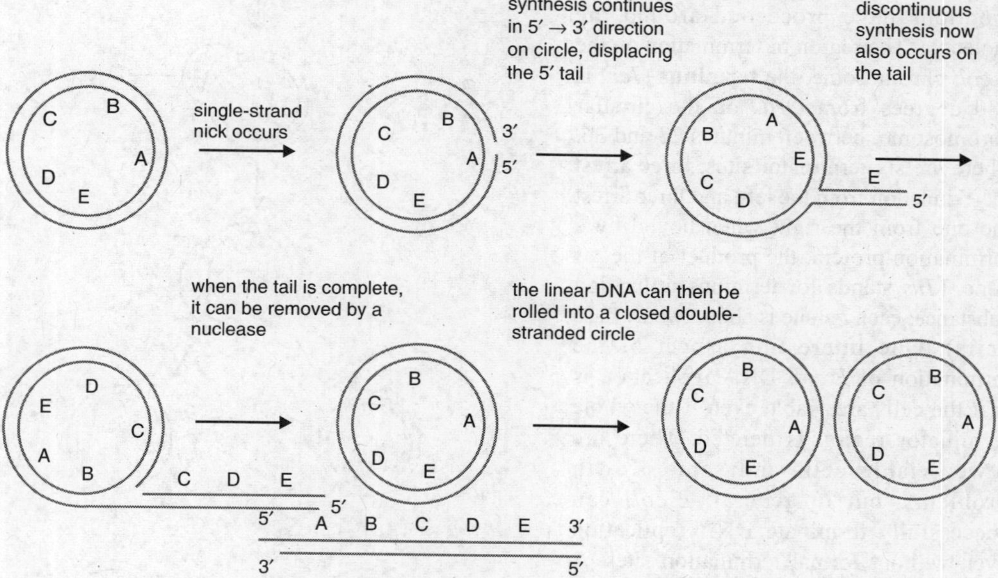

Fig. 6.27. Rolling circle model of DNA replication. The letters A to E provide landmarks on the chromosome (after Tamarin, 2002).

2. Rolling Circle Model (Rolling Circle Replication)

In the rolling circle mode of DNA replication, a **nick** (a break in one of the phosphodiester bonds) is made in one of the strands of the circular DNA and this nick has 3′ – OH and 5′ – P termini. Due to a helicase and *Ssb* protein α, Y–fork is generated. Ultimate result is replication of a circle and a tail (Fig. 6.26). This form of replication occurs in F plasmid or *E.coli* Hfr chromosome during conjugation. The F^+ or Hfr bacterial cell retains the circular daughter while passing the linear tail into the F^- cell, where replication of the tail takes place. Several phages (including l phage) also use this DNA replication method, filling their heads (protein coats) with linear DNA replicated from the circular molecule. Rolling circle model is also called **sigma (s) replication** and it is an efficient mechanism for rapid synthesis of multiple copies of a circular genome (**Novick**, 1998).

3. D-loop Model (Displacement Replication)

Chloroplasts and mitochondria (in eukaryotic cells) have their own circular DNA molecules that replicate by a slightly different mechanism. The **origin** of replication is at a different point on each of the two parental template strand. Replication begins on one strand, displacing the other while forming a **displacement loop** or **D–loop** structure (Fig. 6.27). Replication continues until the process passes the origin of replication on the other strand. Replication then initiates on the second strand, in the opposite direction. Y– junction (or replication fork) replication, also occurs in mitochondrial DNA under some growth condition.

6.6. TERMINATION OF REPLICATION

The termination of the replication of a circular chromosome DNA (of bacteria, mitochondria and chloroplasts) presents no major topological problems. At the end of the theta structure replication, both Y-junctions have proceeded around the molecule. The region of termination on the *E.coli* chromosome, the **terminus** (*Ter*) is 180 degrees from *OriC* on the circular chromosome, between minutes 28 and 36. There are six terminator sites; three arrest the Y-junction from the left, and three arrest the one from the right when bound by a termination protein, the product of the *tus* gene. (*Tus* stands for terminus utilization substance; each *ter* site is about twenty base pairs). One interesting aspect of the termination of *E.coli* DNA replication is that the cells are viable even if the whole terminator region is deleted. There are fewer viable cells and some growth problems, but in general, *E.coli* can successfully terminate DNA replication even without formal termination sites. A topoisomerase, (**topoisomerase IV**) then releases the two circles and DNA polymerase I and ligase enzymes close them up (Fig. 6.29).

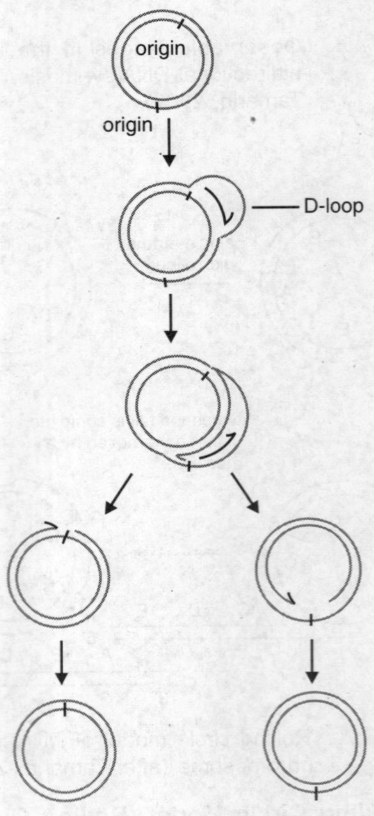

Fig. 6.28. D–loop form during mitochondrial and chloroplast DNA replication because the origins of replication are located at different places on the two strands of the double helix (after Tamarin, 2002).

DNA Partitioning in *E.coli*

Until very recently, geneticists believed that the partitioning of the *E.coli* chromosome was a passive process, unlike that in eukaryotes. Now, however, we know that *E.coli* DNA partitioning is indeed a very complex process. When DNA replication begins, the newly replicated origins of replication are segregated to opposite ends of the bacterial cell, acting as centromeres do. A ring of FtsZ protein, the products of *ftsZ* gene, form a ring at the middle of the already elongated cell and begins to create septum that will divide the already elongated cell into two daughter cells.

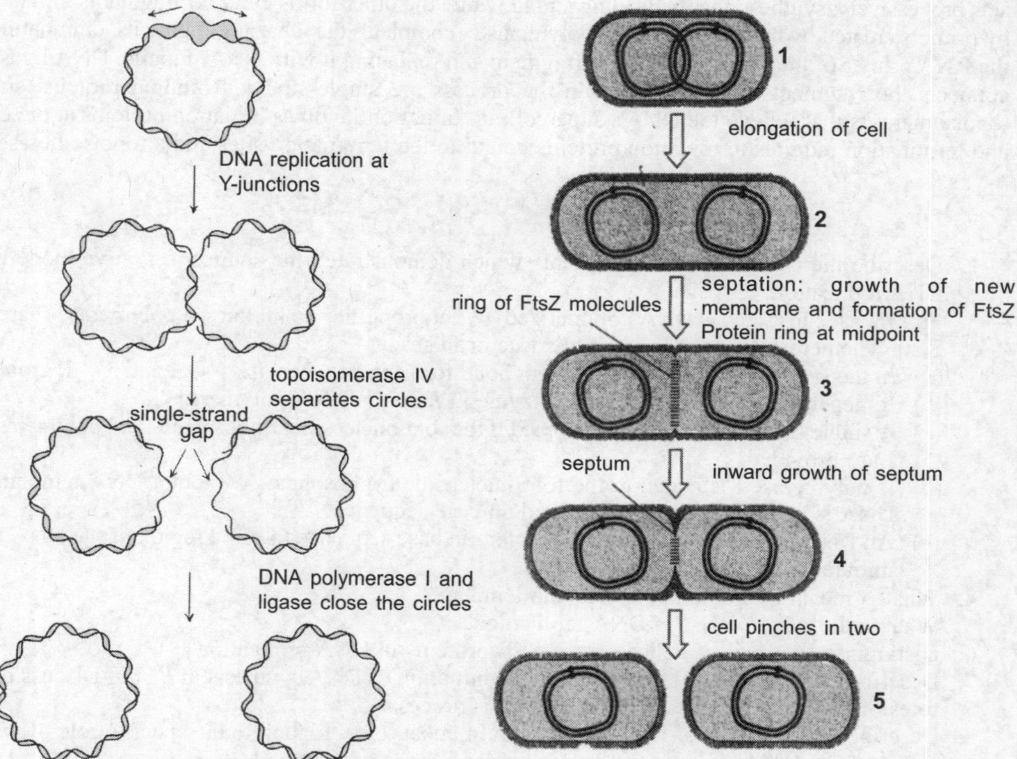

Fig. 6.29. The replication of circular DNA terminates when topoisomerase separates the circles and DNA polymerase I and ligase close the gap in each circle (after Tamarin, 2002).

Fig. 6.30. Binary fission in *E.coli.* After the DNA is replicated, the cell elongates and the DNA is partitioned in the cell. Separation then begins, in which new cell membrane material begins to grow and form a septum at approximately the midpoint of the cell. A protein molecule, called FtsZ, (It is a tubulin-like protein) facilitates this processs, *i.e.*, FtzZ forms a ring at approximately the midpoint of the cell. When the septum is complete, the cell pinches in two (after Raven *et al.,* 2005).

SOLVED PROBLEMS

Problem 1. Four distinct DNA polymerases: α, β, γ and δ, have been characterized in several higher eukaryotes. What are intracellular locations and functions of these polymerases?

Answer. DNA polymerases α, β and γ are located in the nuclei of cells; polymerase γ is located in mitochondria and chloroplasts. Current evidence suggests that polymerases α and δ are both required for the replication of nuclear DNA. Polymerase δ is thought to catalyze the continuous synthesis of the leading strand and polymerase α is believed to catalyze discontinuous synthesis of the lagging strand since it contains primase activity required for the repeated initiation of "Okazaki fragments". Polymerase γ presumbly catalyzes replication of organellar chromosomes. The function of polymerase β remains unknown, perhaps it functions in DNA repair pathways like polymerase I of *E.coli.*

Problem 2. What enzymes are involved in DNA replication in *E.coli*?

Answer. A replisome, consisting of a primosome (a primase and a helicase) and two polymerase III holoenzymes, forms at a Y-junction (*i.e.,* replication fork) on DNA. One polymerase acts processively, synthesizing the leading strand, while the other forms Okazaki fragments initiated by primers created by the primase. DNA polymerase I completes the Okazaki fragments, eliminating the RNA primer of the previous Okazaki fragment and replacing it with DNA. Finally, DNA ligase connects the fragments. Also involved in the process are single-stranded binding proteins and topoisomerases that relieve the DNA's supercoiling. Initiation involves initiation proteins at *ori C*, and termination requires termination proteins bound to the termination sites and a topoisomerase.

QUESTIONS AND PROBLEMS

1. Describe the two classical experiments which demonstrated the semiconservative mode of DNA replication.
2. From what substrates is DNA polymerized? What properties do all known polymerase share?
3. State whether each of the following is true or false:
   (*a*) In the synthesis of DNA the covalent bond forms in between a 3' – OH and 5' – P group.
   (*b*) In general, the DNA replicating enzyme in *E.coli* is DNA polymerase I.
   (*c*) A single strand of DNA can be copied if the four nucleoside triphosphates and polymerase I are provided.
   (*d*) If polymerase I is added to the four nucleoside triphosphates without a DNA template, DNA is synthesized but with a random base sequence.
   (*e*) An RNA primer must be complementary in base sequence to some region of the DNA to initiate DNA synthesis.
4. What is meant by the terms primer and template?
5. What is the role of RNA in DNA replication?
6. Distinguish the roles of helicases and SSB proteins in DNA replication?
7. Describe the different steps involved in the initiation of DNA synthesis in *E.coli.* Discuss the roles of different enzymes or proteins in this process.
8. Describe various steps of DNA replication in eukaryotes. Elaborate the specific role of two different DNA polymerases for the leading and the lagging strands.
9. Describe rolling circles and D-loops for DNA replication.
10. If one strand of DNA is found to have the sequence 5' A A C G T A C T G C 3', what is the sequence of nucleotides on the 3', 5' strand?
11. For a molecule of n-deoxyribonucleotide pairs, give a mathematical expression that can be used to calculate the number of possible sequences of those nucleotide pairs if only the "usual" bases are present.
12. Develop a formula for determining the length in micrometers of a DNA molecule whose number of deoxyribonucleotide pairs is known.
13. Phage T2 DNA is estimated to consist of about 200,000 deoxyribonucleotide pairs. What is the length in micrometers of its DNA complement?
14. Write short notes on the following:
    (*i*) Density centrifugation; (*ii*) Meselson and Stahl experiment;
    (*iii*) DNA polymerases; (*iv*) Unwinding proteins;
    (*v*) Replications; (*vi*) FtsZ protein
    (*vii*) Theta configuration; and (*viii*) Okazaki fragments.
15. Distinguish between the terms "dispersive", 'semiconservative' and 'conservative', as applied to DNA replication.
16. Explain why DNA replication must be primed and describe how the priming problem is solved by *Escherichia coli* and by eukaryotes.
17. Give a detailed description of the events occurring at the Y-fork or replication fork in *Escherichia coli.*

ANSWERS TO PROBLEMS

2. Deoxynucleoside triphosphates. Addition to 3′ –OH group; requirement for a template.
3. (*a*) and (*e*) are true.

4. Primer means nucleotide bound to DNA and having a 3′ –OH group; template means a polynucleotide strand whose base sequence can be copied.
5. RNA serves as a primer.
6. A helicase unwinds a helix. The SSB proteins prevent the helix from rewinding and prevent intramolecular base pairing from occurring.
10. 3′ T T G C A T G A C G 5′ because of pairing qualities of deoxyribonucleotides.
11. 4^n.
12. $L\,\mu\,m = 3.4 \times 10^{-4}$ P, where $L\,\mu\,m$ represents the length in micrometers, and P represents the number of nucleotide pairs.
13. 68 $\mu\,m$.

Multiple Choice Questions

Choose the correct answer from the four alternatives given.

1. DNA replication occurs in
   (*a*) G_1 phase (*b*) G_2 phase
   (*c*) S phase (*d*) interphase
2. Duplication of DNA is called
   (*a*) transduction (b) transcription
   (*c*) replication (*d*) translation
3. The process of DNA replication is
   (*a*) dispersive
   (*b*) conservative
   (*c*) semi-conservative
   (*d*) non-conservative
4. Semiconservative DNA replication was first demonstrated by
   (*a*) Taylor
   (*b*) Watson and Crick
   (*c*) Meselson and Stahl
   (*d*) Khorana
5. Taylor demonstrated that DNA replication is semi-conservative in
   (*a*) *Vicia faba* (*b*) mouse liver cells
   (*c*) peanut (*d*) HeLa cells
6. Isotopes used for proving semiconservative replication are
   (*a*) N^{14} and N^{31} (*b*) N^{14} and C^{14}
   (*c*) H^3 and N^{15} (*d*) C^{14} and P^{31}
7. The unwinding of DNA duplex is performed by an enzyme called
   (*a*) gyrase (*b*) lactase
   (*c*) maltase (*d*) ligase
8. The enzyme that reduces the tension during the unwinding of DNA helix in front of the replication form is
   (*a*) topoisomerase (*b*) helicase
   (*c*) ligase (*d*) polymerase
9. During DNA replication the strands separate by
   (*a*) DNA polymerase (*b*) gyrase
   (*c*) DNA helicase (*d*) DNA primase
10. During DNA replication, the term leading strand is applied to the one which replicates in
    (*a*) 5′ – 3′ direction continuously
    (*b*) 3′ – 5′ direction continuously
    (*c*) 5′ – 3′ direction discontinuously
    (*d*) 3′ – 5′ direction discontinuously
11. Which of the following enzymes makes short RNA chains using a DNA template?
    (a) DNA polymerase I
    (*b*) DNA helicase
    (*c*) DNA primase
    (*d*) DNA gyrase
12. Okazaki fragments are formed in
    (*a*) the splicing of mRNA
    (*b*) the synthesis of the lagging strand of DNA
    (*c*) the neurons of vertebrates
    (*d*) the oocytes of amphibians
13. The enzyme that removes the RNA primer after DNA replication is
    (*a*) primase
    (*b*) DNA polymerase I
    (*c*) DNA polymerase II
    (*d*) ligase
14. The enzyme that joins DNA fragments is
    (*a*) restriction endonuclease
    (*b*) lipase
    (*c*) ligase
    (*d*) peroxidase

Answers

1. (*c*) 2. (*c*) 3. (*c*) 4. (*c*) 5. (*a*) 6. (*c*) 7. (*a*)
8. (*a*) 9. (*c*) 10. (*a*) 11. (*c*) 12. (*b*) 13. (*b*) 14. (*c*).

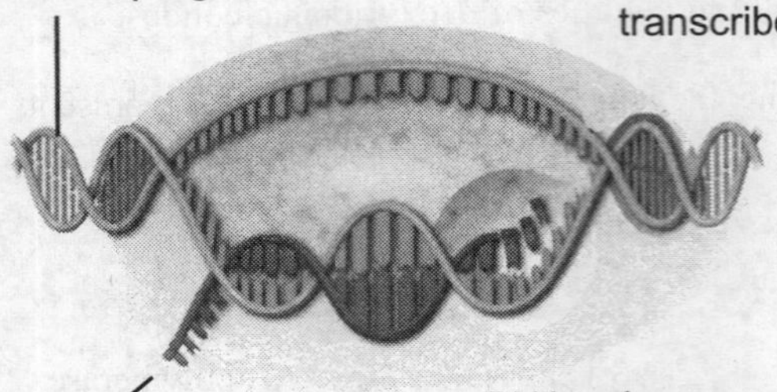

CHAPTER 7

Transcription

Chapter Contents

In the organisms (viz., prokaryotes and eukaryotes), where coded genetic informations are contained in the DNA molecule, different genetically controlled functions of their cells are performed by a different kind of nucleic acid, called **non-genetic ribonucleic acid (RNA)**. In the cells of such organisms, the DNA molecule occurs in the chromosomes of nucleus or nucleoids of eukaryotes or prokaryotes respectively, and the process of protein synthesis occurs in the cytoplasm. It is investigated that a DNA molecule does not leave the nucleus to participate directly in the process of protein synthesis but employs different types of non-genetic ribonucleic acid molecules for carrying its genetic informations from the nucleus to the site of protein synthesis, *i.e.*, ribosomes. Thus, gene expression accomplished by the transfer of genetic information from DNA to RNA molecules and then from RNA to protein molecules. RNA molecules are synthesized by using the base sequence (triplet codons) of one strand of DNA as a template in a polymerization reaction that is catalized by enzymes called **DNA-dependent RNA polymerases** or simply **RNA polymerases.** The process by which RNA molecules are initiated, elongated, and terminated is called **transcription.**

7.1. CHEMICAL COMPOSITION OF NON-GENETIC RIBONUCLEIC ACID (RNA)

Chemically, the non-genetic RNA is closely related with DNA. However, the non-genetic RNA is a single stranded polymer of ribonucleotide units and is composed of phosphoric acid, ribose (pentose) sugar and nitrogen bases, which are purines (adenine and guanine) and pyrimidines (cytosine and uracil). The molecular weights of the various RNA molecules vary from about 25,000 to over one million. Its base composition does not follow the base composition of DNA, although, in general, the proportion of A+C roughly equals G+U, as has been shown in Table 7.1.

Table 7.1. Base ratios of RNA from various sources (as molar percentages) (From, **Hall** *et al.*, 1974).

Source	Adenine	Guanine	Cytosine	Uracil
1. *Allium cepa* (onion seed)	24.9	29.8	24.7	20.6
2. *Phaseolus vulgaris* (bean seed)	24.9	31.4	24.1	19.6
3. *Cucurbita pepo* (pumpkin seed)	25.2	30.6	24.8	19.4
4. Ox liver	17.1	27.3	33.9	21.7
5. Yeast	25.4	24.6	22.6	27.4
6. *E. coli*	25.3	28.8	24.7	21.2

Further, some RNA molecules contain significant proportions of some methylated and an unusual nucleoside known as pseudouridine (ψU), in which the glycosidic bond is associated with position 5 of uracil rather than position 3.

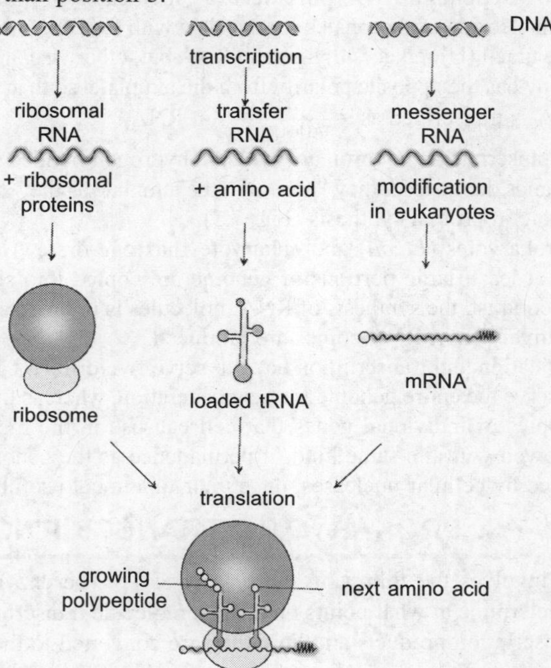

Fig. 7.1. Relationships among the three types of RNA – ribosomal, transfer and messenger – during protein synthesis. All three types are found together at the ribosome during protein synthesis (after Tamarin, 2002).

Types of RNA

In a cell, there are three different kinds of RNA which serve in three different roles in the process of protein synthesis. The first type is **messenger RNA (mRNA)** which carries the DNA sequence information to particles in the cytoplasm known as *ribosomes*, where the messenger RNA is translated. The second type is *transfer RNA* (tRNA), which brings the amino acids to the ribosomes, where protein synthesis takes place. The third type of RNA is a structural and functional part of the ribosome called ribosonal RNA (rRNA). The general relationship of the roles of these three types of RNA is diagrammed in Figure 7.1. In addition, small RNAs play other roles in cellular metabolism.

Further, DNA does not take part directly in protein synthesis because, in eukaryotes, translation occurs in the cytoplasm, whereas DNA remains in the nucleus. Molecular biologists suspected for a long time that genetic intermediate in bacteria and eukaryotes was RNA because the cytoplasmic RNA concentration increases with increasing protein synthesis, and the cytoplasmic RNAs carry nucleotide sequences complementary to the cells DNA. Proof of an RNA intermediate came when it was shown that mRNA directs protein synthesis.

7.2. COMPARISON BETWEEN DNA REPLICATION AND TRANSCRIPTION

The replication and transcription are two chief activities of DNA and they should be compared at the outset in the following manner:

1. In DNA replication an enzyme (or enzyme complex) matches up complementary deoxyribonucleoside triphosphates with a DNA template according to Watson-Crick pairing rules and the bases are polymerized to form a daughter DNA molecule. In DNA transcription a different enzyme, called **DNA-depenent RNA polymerase** medicates a similar process, except that complementary ribonucleoside-triphosphates are matched with the DNA template and RNA polymers are formed. The base uracil (U) replaces thymine in RNA but, otherwise, the DNA strand is faithfully copied. The RNA copy has the opposite polarity from the template, so that, for example, the sequence $\xrightarrow[\text{TACAAC}]{}$ in DNA is transcribed as $\xleftarrow[\text{AUGUUG}]{}$ in RNA.

2. The RNA transcripts do not ordinarily remain hydrogen-bonded with their DNA templates as DNA daughter strands do. Instead, they "peel off" the template as they are formed, thus, becoming available to participate in protein synthesis (Fig. 7.1).

3. DNA of prokaryotes (*E.coli*) and eukaryotes has one or few initiation points for DNA replication, so that, at least, large portions of genome are copied into single, enormous daughter DNA molecules. In contrast, the synthesis of RNA molecules is initiated at close intervals along the DNA and, thus, relatively short RNA copies are produced.

4. Lastly, replication and transcription have to serve two different functions. The purpose of replication is to conserve the entire genome for next generation, whereas the purpose of transcription is to make RNA copies of individual genes that cell can use in the metabolism. The copies are themselves not endowed with the same kind of permanence as the genetic material; instead, they are typically degraded by cellular nucleases, once their functional usefulness has been spent.

7.3. PROKARYOTIC TRANSCRIPTION

Transcription involves the following three aspects: 1. The enzymatic synthesis of RNA; 2. The signals that determine at what points on a DNA molecule transcription starts and stops; and 3. The types of transcription products and how they are converted to the RNA molecules needed by the cell.

1. Enzymatic Synthesis of RNA

The essential chemical characteristics of the synthesis of RNA are following:

(1) The precursors in the synthesis of RNA are the four ribonucleotide 5'-triphosphates (rNTP)-ATP, GTP, CTP, and UTP. On the ribose portion of each NTP, there are two OH groups – one each on the 2'-and 3'-carbon atoms.

(2) In the polymerization reaction a 3'-OH group of one nucleotide reacts with the 5'-triphosphate of a second nucleotide; a pyrophosphate is removed and a phosphodiester bond results (Fig. 7.2).

(3) The sequence of bases in RNA molecule is determined by the base sequence of the DNA. Each base, added to the growing end of the RNA chain, is chosen by its ability to base-pair with the DNA strand used as template; thus, the bases C,T,G and A in a DNA strand cause G,A,C, and U respectively, to appear in the newly synthesized RNA molecule.

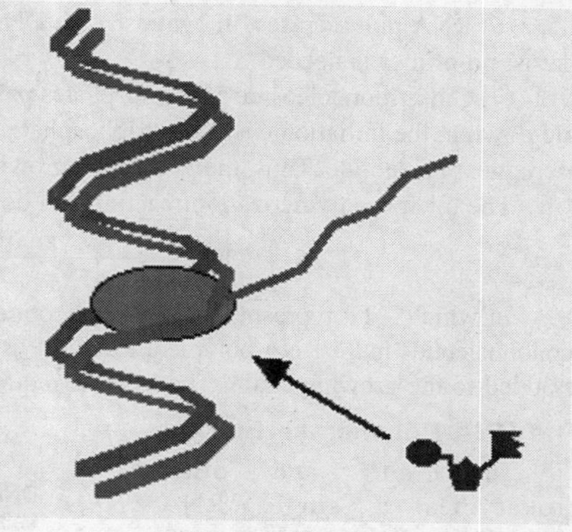

Transcription.

(4) The DNA molecule being transcribed is double-stranded, yet in many particular regions only one strand serves as a template (Fig.7.3).

(5) The RNA chain grows in 5′→ 3′ direction; that is, nucleotides are added only to the 3'-OH end of the growing chain. (This is the same as the direction of chain growth in DNA synthesis). The RNA molecule is terminated by a 5′-triphosphate at the non-growing end. The RNA molecule is antiparallel to the DNA strand being copied. Once intiated, RNA chains grow at a rapid rate – 40 to 50 nucleotides per second in *E.coli* at 37°C.

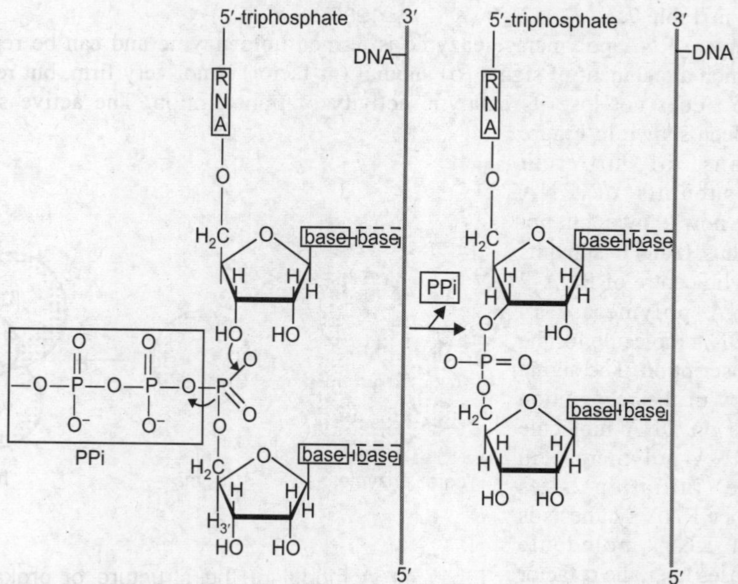

Fig. 7.2. Mechanism of the chain-elongation reaction catalyzed by RNA polymerase. The broken arrows join the reacting groups. The pyrophosphate group (PP) and the bold hydrogen do not appear in the RNA strand. The DNA template and the RNA strands are antiparallel as in double-stranded DNA (after Freifelder, 1985).

(6) RNA polymerases, in contrast with DNA polymerases, are able to initiate chain growth; that is, no primer is needed.

(7) Only ribonucleoside 5′-triphosphates participate in RNA synthesis and the first base to be laid down in the initiation event is a triphosphate. Its 3′-OH group is the point of attachment of the subsequent nucleotide. Thus, the 5′ end of a growing RNA molecule terminates with a triphosphate.

The overall polymerization reaction can be written as follows:

$$ {}_{n}NTP + XTP \xrightarrow[Mg^{2+}]{DNA,\ RNA\text{–}P} (NMP)_n\ XTP + {}_{n}PP_1 $$

In which XTP represnts the first nucleotide at the 5′ terminus of the RNA chain, NMP is a mononucleotide in the RNA polymerase, and PP_1 is the pyrophosphate released each time a nucleotide is added to the growing chain. The Mg^{2+} is required for all nucleic acid polymerization reactions.

The RNA Polymerase Enzyme

In *E.coli* and other prokaryotes a single RNA polymerase (RNA-P) enzyme is responsible for the synthesis of all kinds of RNAs (such as mRNA, tRNA and rRNA). RNA-P is, in fact, one of the largest enzymes known (MW 490,000). It consists of six subunits (*i.e.*, polypeptide chains) – two identical **alpha** (α) subunits and one chain of each of **beta** (β), **bete dash** (β′), **omega** (ω) and **sigma** (σ) subunits(Fig. 7.4). Some characteristics of each subunit of RNA-P has been tabulated in Table 7.2.

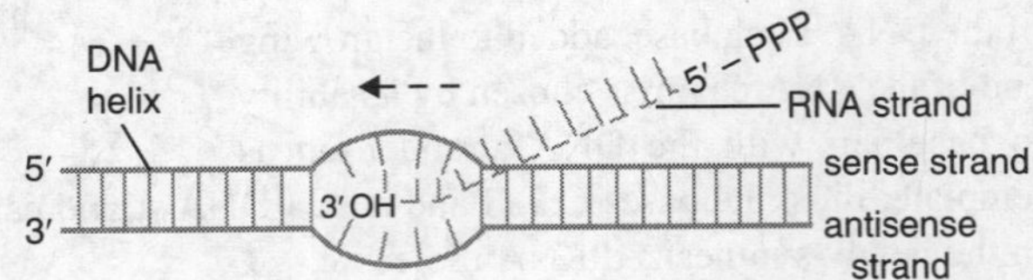

Fig.7.3. An RNA strand is copied only from strand A (sense strand) of a segment of a DNA molecule. No RNA is copied from strand B (antisense strand) in that region of the DNA molecule. However, elsewhere, for example, in a different gene, strand B might be copied; in that case strand A would not be copied in that region of the DNA. The arrow shows the direction of RNA chain growth (after Freifelder, 1985).

The complete RNA polymerase enzyme is termed **holoenzyme** and can be represented as $\alpha_2\beta\beta'\omega\sigma$ in which attachment of sigma (σ) subunit (or factor) is not very firm, but resultant **core enzyme** ($\alpha_2\beta\beta'$) does not lose its catalytic activity of transcription. The active sites of core enzyme have been shown in Figure 7.5. Functions of different polypeptide subunits of RNA polymerase are now known, but not in any detail. Thus, β and β′ subunits form the catalytic centre of RNA-P and help RNA polymerase in unwinding of DNA molecule for the purpose of transcription. The sigma (σ) factor helps in the recognition of start signals on DNA molecule and directs RNA polymerase in selecting the initiation sites (promotor). Once RNA synthesis is initiated and RNA moledcule becomes 8-9 bases long, the σ factor dissociates from the holoenzyme and then the core enzyme brings about elongation of mRNA (or any other RNA).

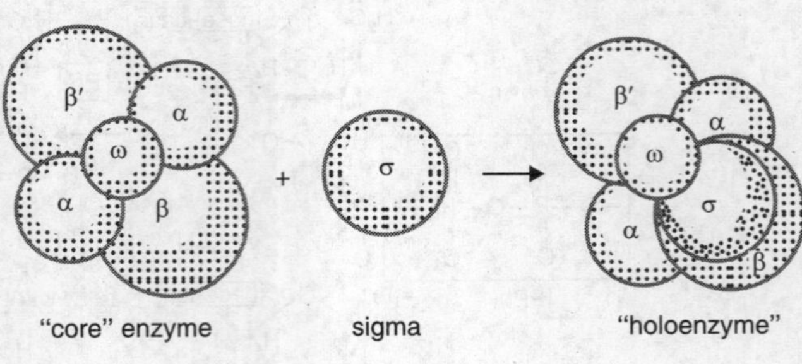

Fig. 7.4. A model of the structure of prokaryotic RNA polymerase showing association of 6 subunits or polypeptides (after Gardner *et al.*, 1991).

2. Binding of RNA Polymerase to Promoter and Initiation

Unlike replication, transcription does not progress along the entire length of a chromosome. Instead, certain parts of the chromosome are transcribed. Only one of the two strands of a DNA duplex is transcribed; this strand is called the **sense strand.** The other strand of DNA which is not transcribed at the moment, is called **anti-sense strand**. Like DNA, RNA is synthesized in the 5→3' direction from the single-stranded region of the DNA template. This localized unwinding moves along the molecule followed by single-recoiling of the helix behind the newly synthesized RNA. The region of sense strand of DNA which is actually transcribed into RNA, is called the **coding region**.

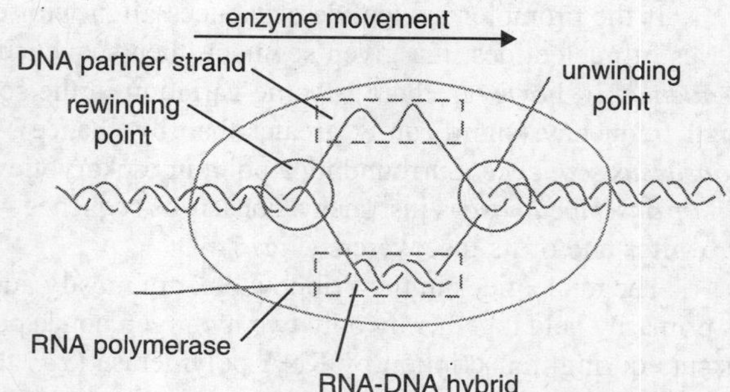

Fig. 7.5. Active centres in the core enzyme of bacterial RNA polymerase enzyme.

Table 7.2. Some basic characteristics of different subunits (=polypeptides) of RNA polymerase of *E.coli* (Source: **Gardner,** *et al.*, 1991).

	Subunit	Gene	Molecular weight (daltons)	Location	Function
1.	α_2	*rpo A*	41,000 each	Core enzyme	Promoter binding
2.	β	*rpo B*	155,000	Core enzyme	Nucleotide binding
3.	β'	*rpo C*	165,000	Core enzyme	DNA template binding
4.	ω	–	12,000	Core enzyme	–
5.	σ	*rpo D*	95,000	Sigma factor	Initiation.

The DNA region that RNA polymerase enzyme associates with immediately before beginning transcription is known as the *promoter.* The promoter is a significant part of gene expression in both prokaryotes and eukaryotes. Promoters contain the information for transcription initiation and are the major sites in which gene expression is regulated.

The molecule of enzyme of RNA polymerase covers a region of about sixty base pairs of DNA. This point was determined by causing the polymerase to bind to DNA and then digesting the mixture with nuclease enzymes, in a technique known as *footprinting* (Fig. 7.6). The polymerase "protects" or prevents degradation of the region it covers. The undigested DNA is then isolated and its size determined. Geneticists have gained much new information about the nature of recognition regions within promoters through recombinant DNA technology and nucleotide sequencing techniques. Sequencing of numerous promoters has revealed that they contain common sequences.

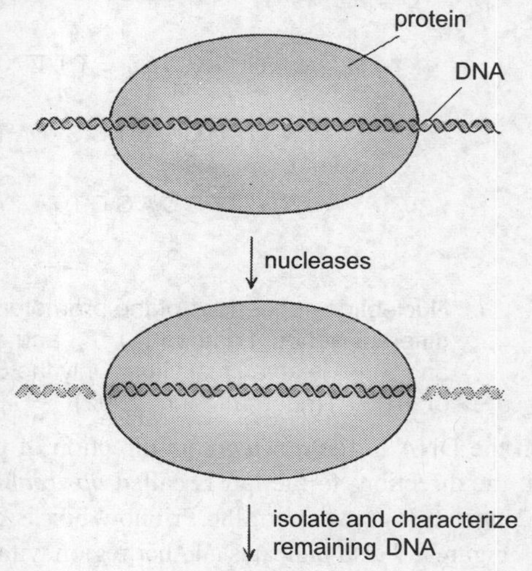

Fig. 7.6. Technique of footprinting. DNA in contact with a protein (e.g., RNA polymerase) is protected from degradation by nuclease enzyme. In next step the protected DNA is isolated and characterized (after Tamarin, 2002).

If the promoter nucleotide sequences align with each other, and each has exactly the same series of nucleotides in a given segment, then sequence of that segment is regarded as *conserved sequence.* If, however, there is some variation in the sequence, but certain nucleotides occur at a high frequency (significantly greater than by chance), then these nucleotides are said to form a *consensus sequence.* Surrounding a point in prokaryotic promoters about ten nucleotides before the *first transcribed base* is just such a consensus sequence – TATAAT. This sequence is called *Pribnow box* after one of its discoverers (Fig. 7.7).

The nucleotides in the Pribnow box are mostly adenines (A) and thymines (T), so the region is primarily held together by only two hydrogen bonds per base pairs. Since local DNA denaturation occurs during transcription by RNA polymerase (*i.e.*, the DNA is opened to allow transcription), fewer hydrogen bonds make this process easier energetically. When the RNA polymerase is bound at the promoter region (Fig. 7.7), it is in position to begin polymerization six to eight nucleotides down from the Pribnow box.

In Fig.7.7. have been shown the sequences which represent the *coding strand* of DNA. It is general convention to show the coding strand because both that strand and mRNA have the same sequences, substituting U for T in RNA, they are both complementary to the same *template strand* (which is also known as *anticoding strand* or *noncoding strand*; Fig. 7.8). Another convention is to indicate the first base transcribed by number +1 and to use positive numbers to count farther

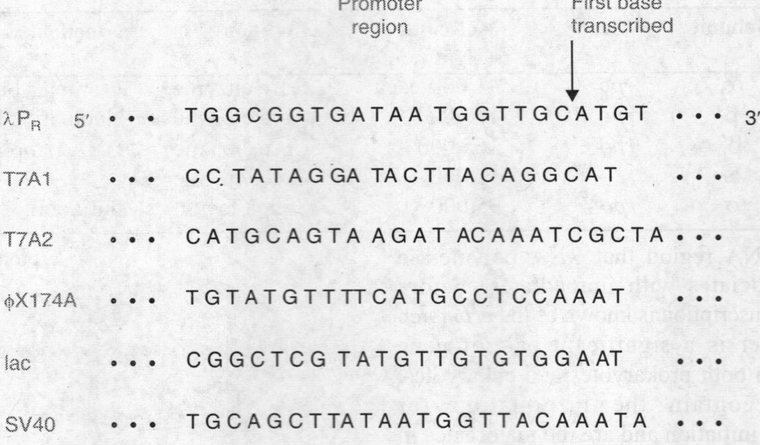

Fig. 7.7. Nucleotide sequences of the promoter region and the first base transcribed from several different genes. Lambda (λ), T_7, and ϕx174 are bacteriophages. *Lac* is an *E.coli* gene, and SV40 is an animal virus. Only the SV40 promoter has the actual consensus sequence of TATAAT(after Tamarin, 2002).

down the DNA in the *downstream* direction of transcription. If transcription is proceeding to the right, the direction to the left is called *upstream,* with bases indicated by negative numbers (Fig. 7.9) Under this convention, the Pribnow box is often referred to as the – 10 sequence.

Figure 7.9 also indicates another region with similar sequences among many promoters centred near – 35 and referred to as the *– 35 sequence.* The consensus sequence at – 35 is TTGTCA. Mutation studies have tried to determine the relative roles of the – 10 and – 35 sequences in transcription. In other words, mutations of bases in the – 10 and – 35 regions of promoter were examined to determine how they affected transcription initiation. The conclusion from these studies are that both regions contribute to the efficiency of binding of polymerase enzyme to promoter. In

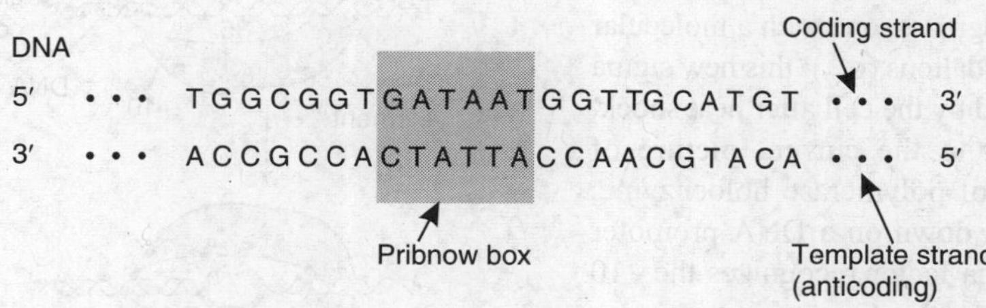

Fig. 7.8. The template (anticoding) strand of DNA is complementary to both the coding strand and the transcribed RNA. The sequences are from the promoter of the bacteriophage λP_R region (after Tamarin, 2002).

other words, the more each sequence differs from the consensus sequence, the less frequently that promoter initiates the transcription. The sigma factor recognizes both the –35 and the –10 sequences. The sigma factor is also sensitive to the spacing between these sequences, preferring (*i.e.*, being most efficient at seventeen base pairs.

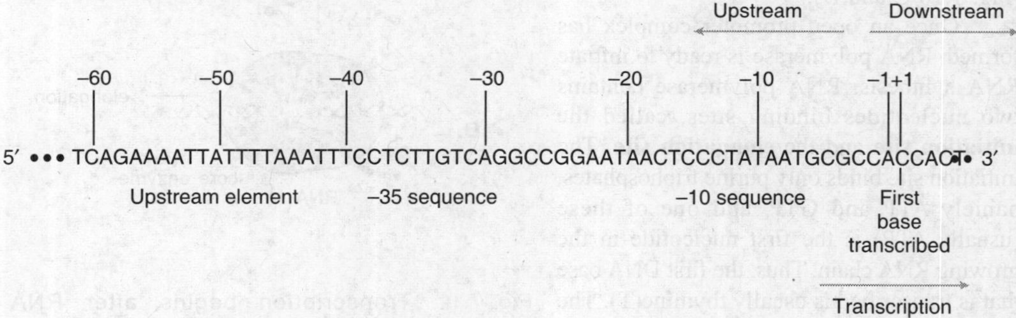

Fig. 7.9. Promoter of the *Escherichia coli* ribosomal RNA gene, rrnB. Note the – 10 and – 35 sequences and the upstream element. The first base transcribed (the transcription start site) is noted (+1), as well as, the upstream, downstream, and transcription direction (after Tamarin, 2002).

Moreover, in bacterial chromosomes, farther upstream from the – 35 sequence is a recognition element that are very strongly expressed, e.g., ribosomal RNA genes (Fig. 7.8). this *upstream element* or *UP element*, is about twenty base pairs long, is centered at – 50 and is rich in A and T. by mutational studies, it has been shown that adding UP element to promoters that do not normally have it greatly increases the rate of transcription. There are other recognition sites in prokaryotes, both upstream and downstream, at which various proteins attach that can enhance or inhibit transcription by direct contact with the polymerase (the α and σ subunits).

Because the holoenzyme (RNA polymerase) recognizes consensus sequences in a promoter, it is not surprising that some promoters are bound more efficiently than others or that different sigma factors exist within a cell. For example, in *E.coli*, the major sigma factor is a protein of 70,000 daltons, referred to as $ó_{70}$. There exist about five less common sigma factors, due to which a bacterial cell remains able to transcribe different genes under different circumstances. For instance, in an *E.coli* cell subjected to elevated temperatures, a group of new proteins, called *heat shock proteins*, which protect the cell to some extent against the elevated temperatures. These heat shock proteins all appear at once because they have promoters that a different sigma factor recognizes, one with a molecular weight of 32,000 daltons (σ_{32}); this new sigma factor is produced by the cell after heat shock.

According to the current picture of mode of action of polymerase holoenzyme, this enzyme sets down on a DNA promoter because the sigma factor recognizes the –10 and – 35 elements, the α proteins recognize the UP element and the α and σ subunits recognise proteins to bound to various other upstream elements, when present (Fig. 7.10B). This initiation complex is initially known as a *closed complex* because the DNA has not melted, which is the next step in transcription initiation(Fig. 7.10.B). After the transcription of 5 to 10 bases, the sigma factor is released (Fig. 7.10 C and D).

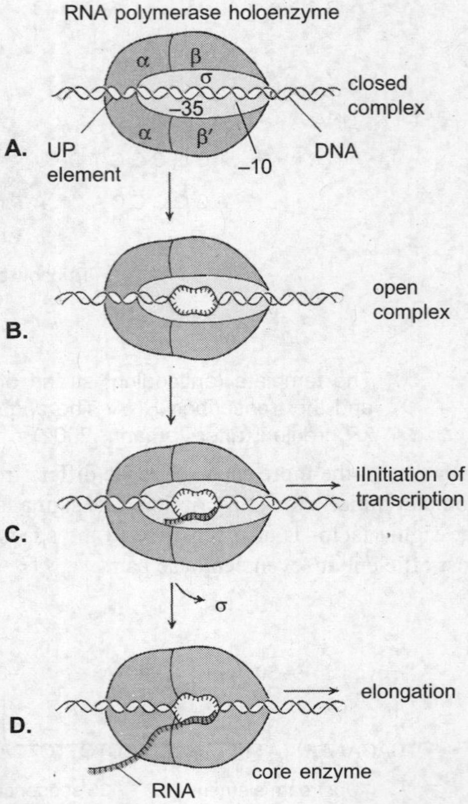

RNA polymerase holoenzyme
A. UP element; α; β; σ; –35; α; β′; –10; DNA; closed complex
B. open complex
C. iInitiation of transcription
σ
D. elongation; RNA; core enzyme

Fig. 7.10. Transcription begins after RNA polymerase attaches to the promoter, with specificity granted by the sigma factor. The DNA opens to form the open complex, transcription begins, the sigma factor leaves, and elongation starts (after Tamarin, 2002).

Once an open promoter complex has formed, RNA polymerase is ready to initiate RNA synthesis. RNA polymerase contains two nucleotides binding sites, called the **initiation site** and the **elongation site**. The initiation site binds only purine triphosphates, namely ATP and GTP, and one of these (usually ATP) is the first nucleotide in the growing RNA chain. Thus, the first DNA base that is transcribed is usually thymine(T). The initiating nucleoside triphophate binds to the enzyme in the open-promoter complex and forms a hydrogen bond with the complementary DNA base. The elongation site is then filled with a nucleoside triphosphate that is selected strictly by its ability to form a hydrogen bond with the next base in the DNA strand. The two nucleotides are then joined together, the first base is released from the initiation site, and initiation is completed. The dinucleotide remains hydrogen bonded to the DNA. The **elongation phase** begins when the polymerase releases the base and then moves along the DNA chain.

3. Elongation phase. Since there is just one bacterial RNA polymerase, the general mechanism of transcription is the same for all bacterial genes. The following description of elongation and termination, given in the context of mRNA synthesis, therefore apply equally well to the synthesis of non-coding RNA (*i.e.*, rRNA and tRNA).

During transcription, ribonucleotides are added one after another to the growing 3′ end of the RNA transcript, the identity of each nucleotide specified by the base-pairing rules: A base pairs with T or U; G base pairs with C. During each nucleotide addition, the β- and γ- phosphates are removed from the incoming nucleotide (Fig. 7.11) and the hydroxyl group is removed from the 3′ – carbon of the nucleotide present at the end of the chain.

Fig. 7.11. A triphosphate of ribonucleotide(a nucleotide).

At this stage of transcription, the bacerial RNA polymerase is in its core enzyme form, comprising four proteins, two relatively small (approximately 35 k Da) α subunits and one each of the related subunits β and β' (both approximately 150 k Da): the α subunit that has the key role in initiation has now left the complex. The RNA polymerase covers about 30 bp of the template DNA, including the *transcription bubble* of 12 – 14 bp, within which the growing transcript is held to the template strand of the DNA by approximately eight RNA – DNA base pairs (Fig. 7.12). The polymerase has to keep a tight grip on both the DNA template and the RNA that it is making

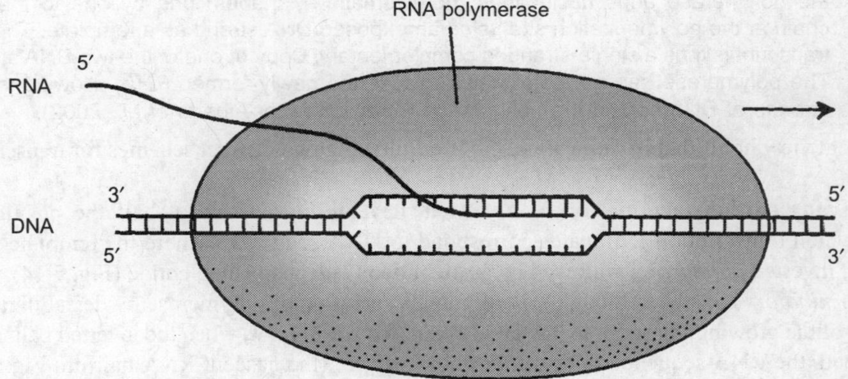

Fig. 7.12. Transcription elongation complex of *E. coli* (a schematic representation). The arrow indicates the direction in which the RNA polymerase moves along the DNA (after Brown, 2002).

in order to prevent the transcription complex from falling apart before the end of the gene is reached. However, this grip must not be so tight as to prevent the polymerase from moving along the DNA.

In the current model, which has been produced by X-ray crystallographic studies and crosslinking experiments, the DNA template lies between the β and β′ subunits, within a trough on the enclosed surface of β′. The active site for RNA synthesis also lies between these two subunits, with the non-template strand of DNA held within the β subunit and RNA transcript extruded from the complex via a channel formed partly by the β and partly by the β′ subunit (Fig. 7.13; **Korzheva** *et al.* 2000).

4. Termination phase. Exact mechanism involved in termination of transcription is still not known. However, molecular biologists believed that transcription is a stepwise addition process of nucleotide-by-nucleotide, with the polymerase pausing at each position and making a choice between continuing elongation by adding another ribonucleotide to the transcript, or terminating by dissociating from the template. Which choice is selected depends on which alternative is more favourable in thermodynamic terms (**Von Hippel**, 1998). This model emphasizes that in order for termination to occur, the polymerase has to reach a position on the template where dissociation is more favourable than continued RNA synthesis.

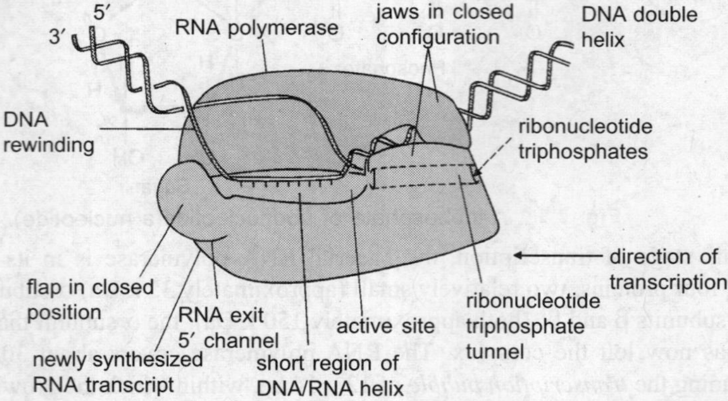

Fig. 7.13. DNA is transcribed by the enzyme RNA polymerase. The RNA polymerase moves stepwise along the DNA, unwinding the DNA helix at its active site. As it progresses, the polymerase adds nucleotides (here, small "T" shapes) one by one to the RNA chain at the polymerization site using an exposed DNA strand as a template. The RNA transcript is thus a single-stranded complementary. Copy of one of the two DNA strands. The polymerase has a rudder that displaces the newly formed RNA, allowing the two strands of DNA behind the polymerase to rewind (after Alberts *et al.,* 2002).

Current view holds that bacteria appear to use following two distinct schemes for transcription termination.

(*a*) Transcription termination at an intrinsic terminator. About half the positions in *E. coli* at which transcription terminates correspond to DNA sequences where the template strand contains an inverted *palindrome* followed by a run of deoxyadenosine nucleotide (Fig. 7.14). These *intrinsic terminators* have been thought to promote dissociation of polymerase by destabilizing the attachment of the growing transcript to the template, in two ways. *First*, when the inverted palindrome is transcribed, the RNA sequence folds into a stable *hairpin* (the nascent RNA hairpin; Fig. 7.14), this RNA – RNA base pairing being favoured over the DNA – RNA pairing that normally occur in the transcription bubble. This reduces the number of contacts made between the template and transcript, weakening the overall interaction and favouring dissociation. The interaction is further weakened when the run of As (= adenine bases) in the template is transcribed, because the resulting A-U base pairs have only two hydrogen bonds each, compared with three for each G – C pair. The net result is that termination is favoured over continued elongation (**Von Hippel**, 1998).

An alternative model which is based on crosslinking experiments has shown that the **RNA hairpins** makes contact with flap structure on the outer surface of the RNA polymerase β subunit, adjacent to the exit point of the channel through which the RNA emerges from the complex. Although the flap structure is quite distant (some 6.4 nm) from the active site of the polymerase, a direct connection is made between the two by a segment of β- sheet within the β-subunit. Movement of the flap could therefore affect the positioning of amino acids within the active site, possibly leading to breakage of the DNA-RNA base pairs and termination of transcription (**Korzheva** *et al.,* 2000; **Toulokhonov** *et al*., 2001). **Toulokhonov** *et al*., (2001) have demonstrated the presence of a protein called **NusA** which enhances termination at intrinsic promoters, interacts with the hairpin loops and flap structure and may stabilize the contact between the two.

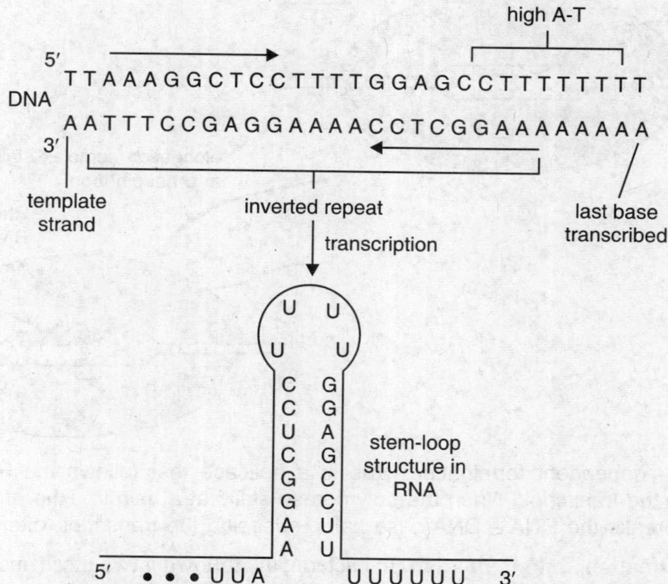

Fig. 7.14. An inverted-repeat base sequence characterizes terminator regions of DNA. Stem-loop (hair-pin) structures can occur as the RNA forms because of complementary sequences. The 3' poly-U tail indicates a rho-independent terminator (after Tamarin, 2002).

***(b)* Rho dependent transcription termination.** The second type of bacterial termination signal is **Rho dependent.** These signals usually retain the hairpin feature of intrinsic terminators, although the hairpin is less stable and there is no run of As in the template. Termination requires the activity of a protein called **Rho,** which attaches to the transcript and moves along the RNA towards the polymerase. If the polymerase continues to synthesize RNA then it keeps ahead of the pursuing Rho, but at the termination signal the polymerase stalls (Fig. 7.16). Rho is a helicase, which means that it actively breaks base pairs, in this case between the template and transcript, resulting in termination of transcription.

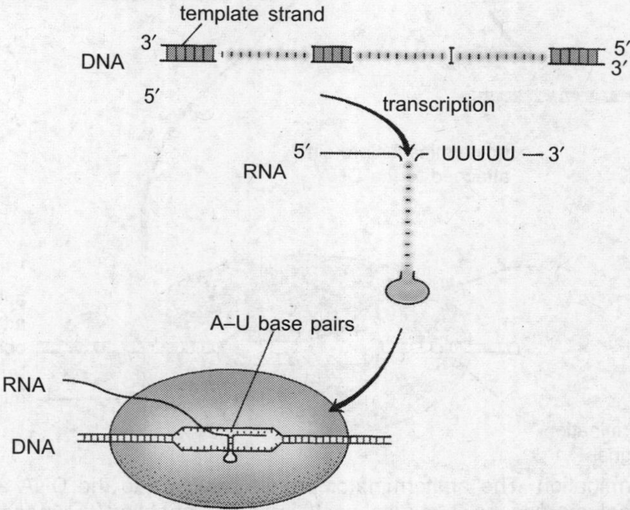

Fig. 7.15. Terminatiom at an intrinsic terminator. The presence of an inverted palindrome in the DNA sequence result in formation of a hairpin loop in the transcript (e.g., mRNA) (after Brown, 2002).

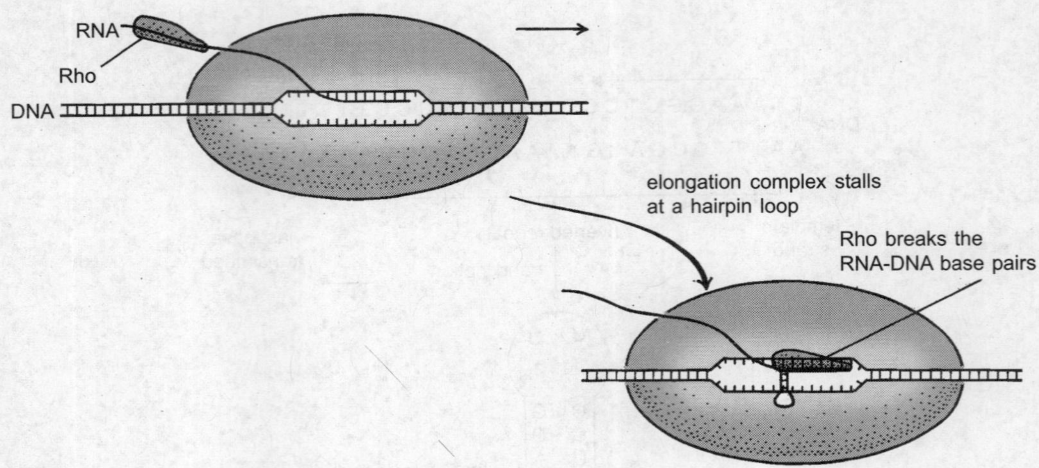

Fig. 7.16. Rho – dependent termination. Rho is a helicase that follows the RNA polymerase along the transcript. When the polymerase stalls at a hairpin, Rho protein catches up and breaks the RNA – DNA base pairs, releasing the transcript (after Brown, 2002).

5. Antitermination and attenuation. In bacteria, the following two mechanisms have evolved for influencing the repeated choice that the polymerase has to make between elongation and termination when copying a template (DNA).

A. Antitermination. This process occurs when the RNA polymerase ignores a termination signal and continues elongating its transcript until a second signal is reached (Fig. 7.17). Antitermination provides a mechanism whereby one or more of the genes at the end of an operon

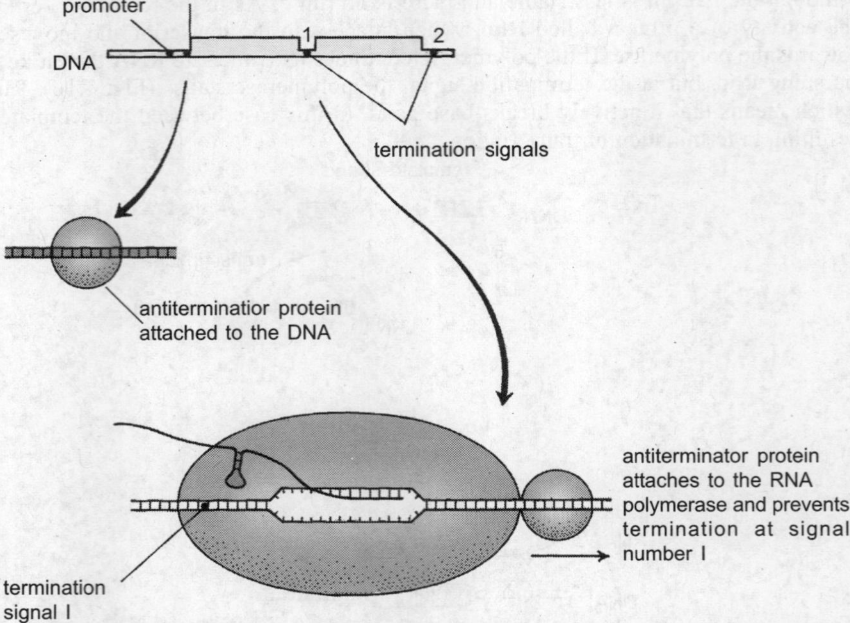

Fig. 7.17. Antitermination. The antiterminator protein attaches to the DNA and transfers to the RNA polymerase as it moves past; later on, enabling the polymerase to continue transcription through termination signal number 1, so the second of the pair of genes in this operon is transcribed (after Brown, 2002).

can be switched off or on by the polymerase recognizing or not recognizing a termination signal located upstream of those genes. Antitermination is controlled by an **antiterminator protein** (*e.g.*, N protein and Q protein of genome of bacteriophage λ of *E.coli*; **Friedman** *et al.*, 1987). This protein attaches to the DNA near the beginning of the operon and then transfer to the RNA polymerase as it moves past en route to the first termination signal. The presence of the antiterminator protein causes the enzyme to ignore the termination signal, presumbly by countering the destabilizing properties of an intrinsic terminator or by preventing stalling at a Rho-dependent terminator. Although the mechanics of the process are unclear, the impact that antitermination can have on gene expression has been described in detail especially during the infection cycle of bacteriophage λ (**Friedman** *et al.,* 1987).

B. Attenuation. The process of attenuation operates primarily with operons that code for enzymes involved in amino acid biosynthesis. The tryptophan operon of *E.coli* shows how it works. In tryptophan operon, two hairpin loops can form in the region between the start of the transcript and the beginning of *trpE.* The smaller of these loops acts as a termination signal, but the larger hairpin loop, which is closer to the start of the transcript, is more stable. The larger loop overlaps with the smaller loop (*i.e.*, termination hairpin), so only one of the two hairpins can form at any one time. Which loop forms depends on the relative positioning between the RNA polymerase and a ribosome which attaches to the 5′ end of the transcript as soon as it is synthesized in order to translate the genes into protein (Fig. 7.18)

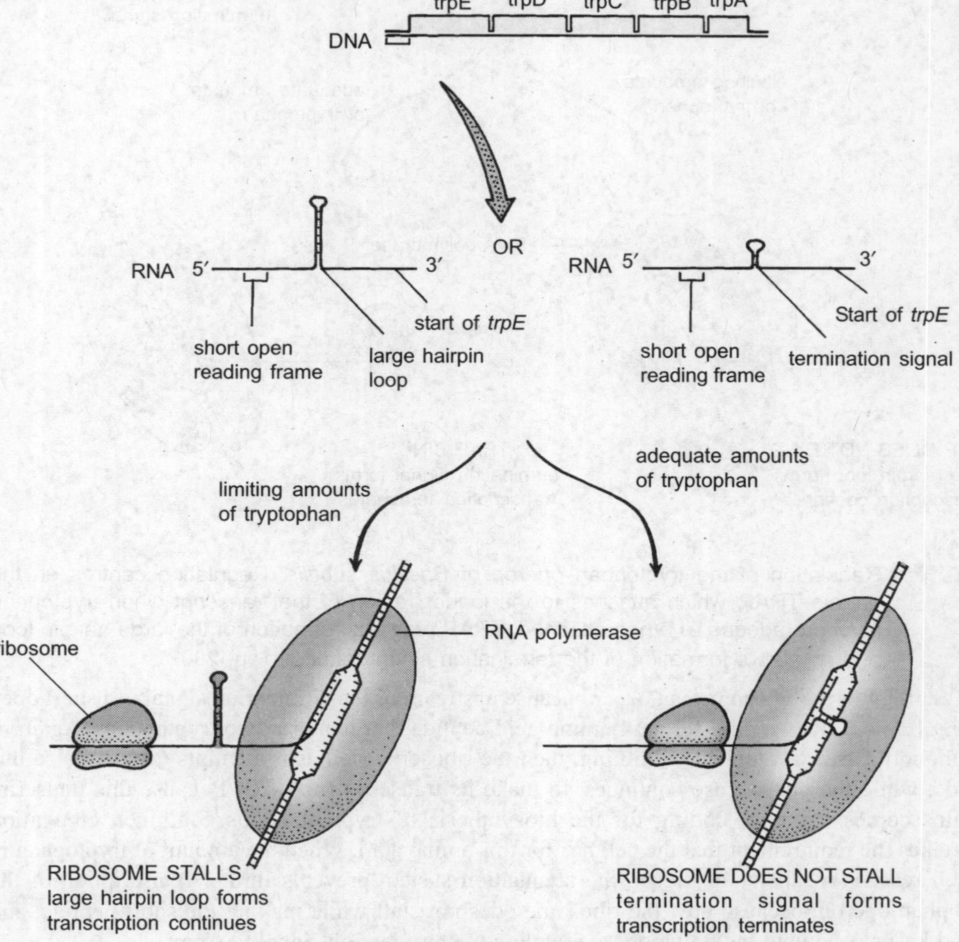

Fig. 7.18. Attenuation at the tryptophan operon of *E.coli.*

Thus, if the ribosome stalls so that it does not keep up with the polymerase, then the larger hairpin forms and transcription continues. However, if the ribosome keeps pace with the RNA polymerase then it dislocates the larger hairpin by attaching to the RNA that forms part of the stem of this hairpin. When this happens the termination hairpin is able to form and transcription stops.

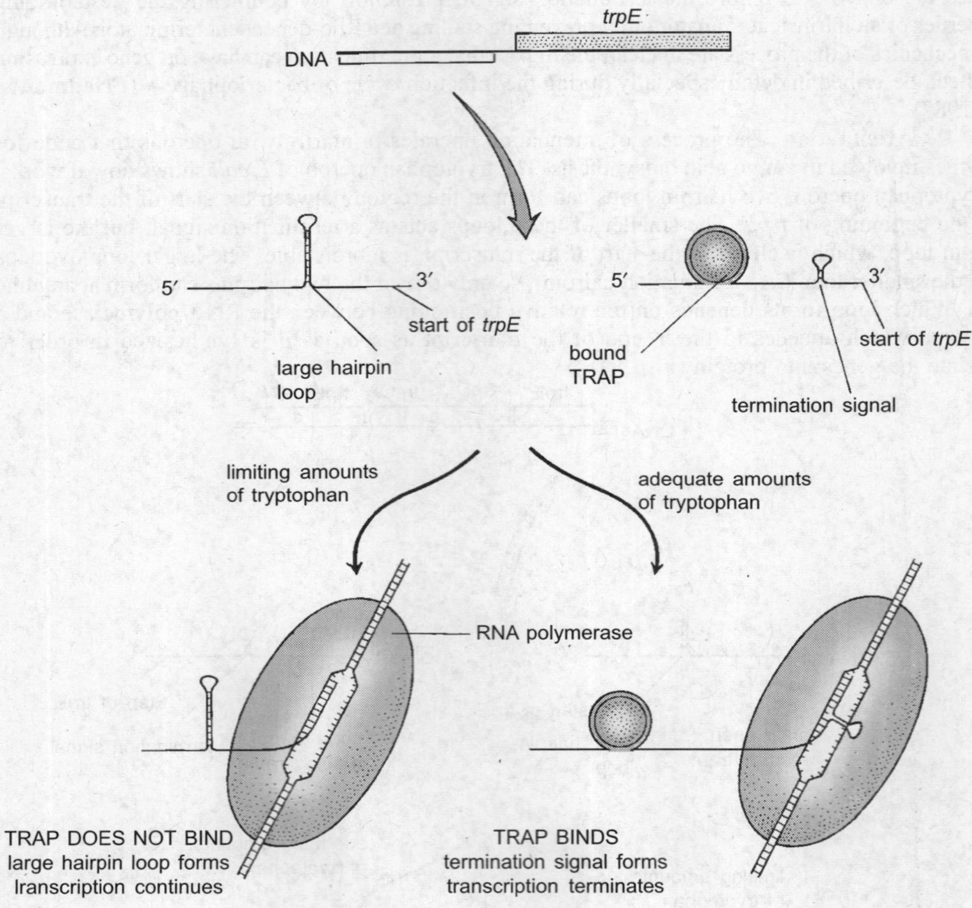

Fig. 7.19. Regulation of the tryptophan operon of *Bacillus subtilis.* Regulation centres on the protein TRAP, which attaches to the leader region of the transcript when tryptophan level are adequate. When attached, TRAP prevents formation of the large hairpin loop and so allows formation of the termination signal (after Brown, 2002).

Stalling of the ribosome can occur because upstream of the termination signal is a short open reading frame (ORF) coding for a 14-amino-acid peptide that includes two tryptophans. Again, if the amount of free tryptophan is limiting, then the ribosome stall as it attempts to synthesize this peptide, while the polymerase continues, to make its transcript (mRNA). Because this transcript contains copies of genes coding for the biosynthesis of tryptophan, its continued elongation undertakes the requirement that the cell has for this amino acid. When the amount of tryptophan in the cell reaches a satisfactory level, the attenuation system prevents further transcription of the tryptophan operon, because now the ribosome does not stall while making the short peptide, and instead keeps pace with the polymerase, allowing the termination signal to form.

In fact, tryptophan operon of *E.coli* is controlled not only by attenuation but also by a **repressor.** Exactly how attenuation and repression work together to regulate expression of the operon is not known, but it is believed that repression provides the basic on-off switch and attenuation modulates the exact level of gene expression that occurs. Other operons of *E.coli*, such as those for biosynthesis of histidine, leucine and threonine, are controlled entirely by attenuation. Interestingly, in some bacteria, including *Bacillus subtilis*, the tryptophan operon is one of those that regulated entirely by attenuation. In these bacteria, attenuation is mediated not by the speed at which the ribosome tracks along the mRNA, but by an RNA-binding protein called **trp RNA-binding attenuation protein (TRAP)** which in the presence of tryptophan, attaches to the mRNA in the region equivalent to the short ORF of *E.coli* transcript (Fig. 7.19). Attachment of TRAP leads to formation of the termination signal and stoppage of transcription (**Antson** *et al.*, 1999).

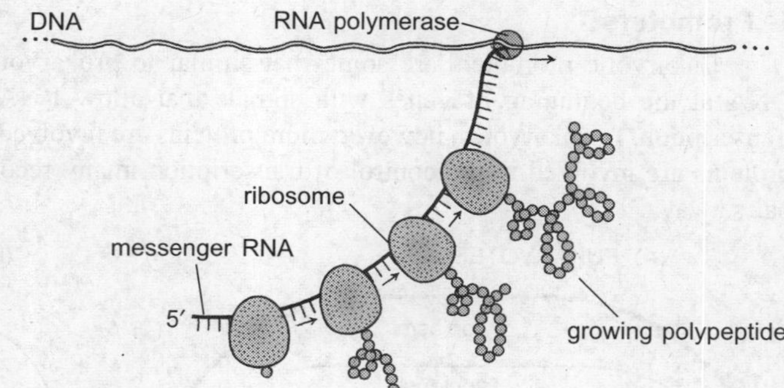

Fig. 7.20. In prokaryotes, translation of messenger RNA (*i.e.*, mRNA) by ribosomes begins before transcription is complete. Ribosomes attach to the growing mRNA strand when the 5′end becomes accessible. They then move along the RNA as it elongates. When the first ribosome moves from the 5′ end, a second ribosome can attached and so on (after Tamarin, 2002).

7.4. INITIATION OF TRANSCRIPTION IN EUKARYOTES

The mRNA molecules make up the **transcriptosome** (*i.e.,* the entire mRNA content of a cell) and specify the protein content of the cell. As the central players in genome expression, mRNAs have received the greatest attention from the molecular biologists. Transcriptional events in prokaryotes (bacteria) are different in many respects from those in eukaryotes. In bacteria, mRNA do not undergo any significant forms of processing: the primary transcript that is synthesized by the RNA polymerase in itself the nature mRNA and its translation usually begins before transcription is complete (Fig. 7.20). The messenger RNA is synthesized in the 5′ → 3′ direction, and it is near the 5′ end of mRNA that translation begins. As soon as the 5′ end of mRNA is available, a ribosome can attach to the mRNA and move along it in the 5′ → 3′ direction lengthening the growing polypeptide as it moves. When the first ribosome moves away from the 5′ end of the transcript, a second ribosome can attach and begin translation. These processes evidently are repetitive. In eukaryotes, however, messenger RNA is synthesized in the nucleus, but protein synthesis takes place in the cytoplasm. (Such a regional division of labour is not present in *E.coli* because the bacterium has no nucleus). Before a mRNA leaves the nucleus, it is highly modified by processes that generally do not occurs in prokaryotes. Eukaryotic processing of primary transcript includes 1. 5' capping; 2. RNA splicing and 3. 3' polyadenylation (Fig. 7.21). Here now we will describe initiation of transcription in the eukaryotes.

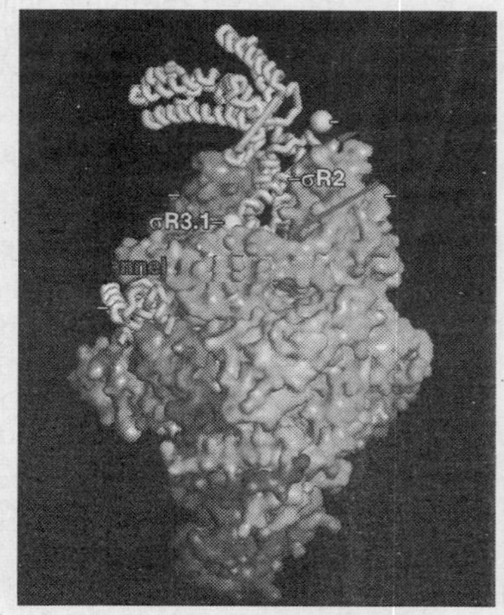

Model for the structural organisation of the bacterial RNA polymerase-promoter open complex.

1. Promoters

Eukaryotic promoters are somewhat similar to prokaryotic promoters; both are regions of DNA at the beginning of genes with signals that allow RNA polymerase to attach and begin transcription. In eukaryotes, however, more proteins are involved in promoter recognition and more proteins are involved in the control of transcription, many recognizing signals thousands of base pairs away.

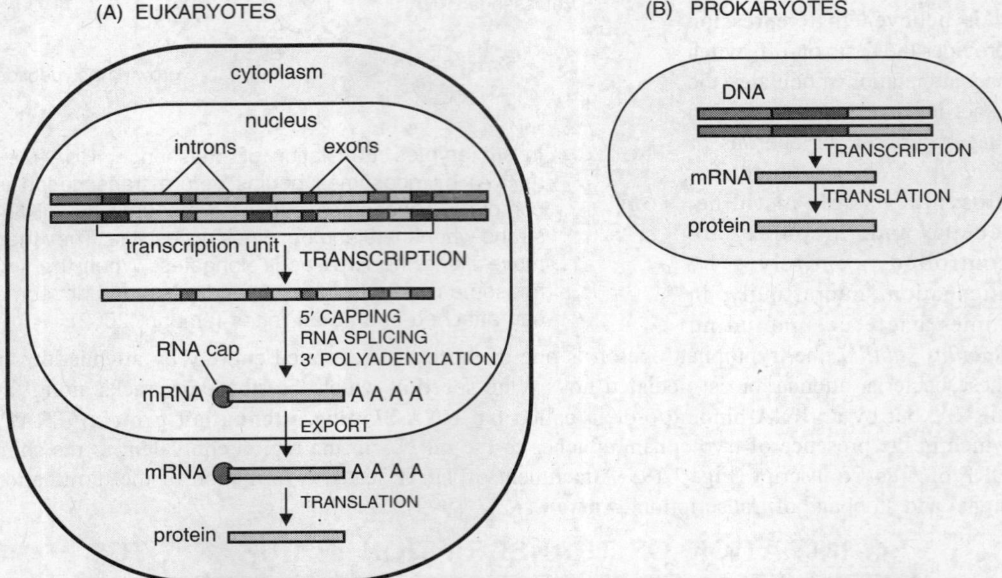

Fig. 7.21. Steps leading from gene to protein in eukaryotes. In eukaryotic cells the RNA molecule produced by transcription is called primary RNA transcript. It contains both coding (exon) and noncoding (intron) sequences. Before it is translated into protein, the two ends of the RNA are modified, the introns are removed by an enzymatically catalyzed RNA splicing reaction and resulting mRNA is transported from the nucleus to the cytoplasm (after Alberts *et al.*, 2002).

In eukaryotes, there are three major classes of RNA polymerases which are designated as I, II and III and are found in the nucleus. The three polymerases have different properties (see Table 7.3) and can be distinguished by the ions required for their activity, the optimal ion strength and their sensitivity to inhibition by various antibiotics (e.g., α- amanitin). Each of these enzymes is a large protein (~500,00 daltons), with two large and several (8 to 10) smaller subunits.

Table 7.3. Properties of three different eukaryotic nuclear RNA polymerases.

Enzymes	Location	Product and abudance	Sensitivity to α-amanitin
1. RNA polymerase I	Nucleolus	rRNA (50 – 70%) (except 5S rRNA)	Not sensitive
2. RNA polymerase II	Nucleoplasm	hnRNA (mRNA) (20 – 40%)	Sensitive
3. RNA polymerase III	Nucleoplasm	rRNA(~ 10%) (and 5S rRNA)	Inhibited in animals at high levels; not in yeast and insects

All three eukaryotic RNA polymerases (I, II and III) recognize a seven–base sequence, TATAAAA, located at about – 25 on the promoter DNA. it is similar to the – 10 sequence in prokaryotes and is called the **TATA box** (or **Hogness box** after its discover, D. Hogness). Since RNA polymerase II transcribes most of the genes in eukaryotes, so let us describe it in suffiecient detail.

Among the large number of promoters that have been sequenced, a few lack the TATA box, yet are still transcribed. Transcription initiation in these promoters appears to be controlled by a CT-rich area, called the **initiator element (INR),** at +1 of the transcript (close to the transcription start site), coupled with a **downstream promoter element (DPE)** at about +28 to the +34 of the

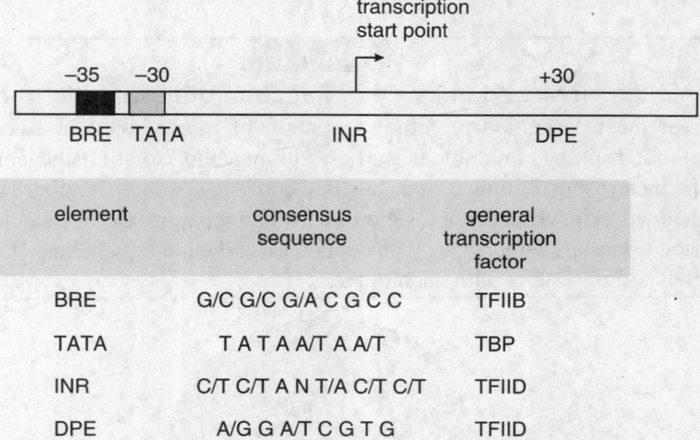

element	consensus sequence	general transcription factor
BRE	G/C G/C G/A C G C C	TFIIB
TATA	T A T A A/T A A/T	TBP
INR	C/T C/T A N T/A C/T C/T	TFIID
DPE	A/G G A/T C G T G	TFIID

Fig. 7.22. Consensus sequences found in the vicinity of eukaryotic RNA polymerase II start points. The name given to each consensus sequence(first column) and the general transcription factor that recognizes it (last column) are indicated. N indicates any nucleotide and two nucleotides separated by a slash indicate an equal probability of either nucleotide at the indicated position. Most polymerase II promoters have a TATA box sequence, and those do not typically have a "strong" INR sequence. Although most of the DNA sequences that influence transcription initiation are located "upstream" of the transcription start point, a few, such as the DPF shown in the figure, are located in the transcribed region (after Alberts *et al.,* 2002).

transcript. In TATA – less promoters, a protein called **TFIID** requires both these elements to bind. The initiator element has a consensus sequence of TCAGTT or TCATTC, and the downstream promoter element has the consensus sequence of AGACGTG or GGTCGTG.

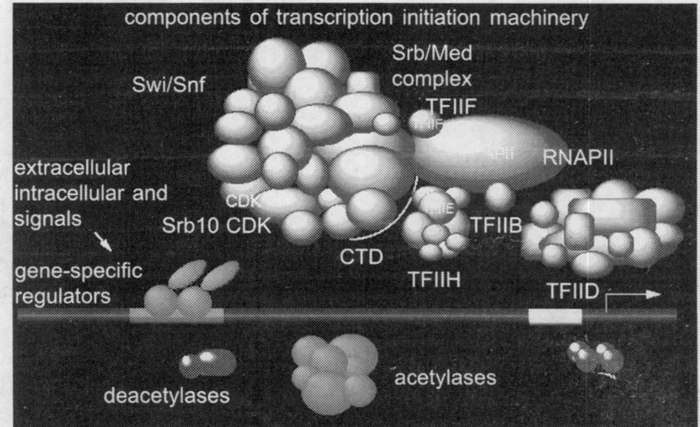

RNA Polymerase Holoenzyme.

2. General Transcription Factors

RNA polymerase II of yeast is a protein of twelve subunits. This enzyme cannot locate promoters or attach to DNA in a stable fashion. To attach at the beginnings of genes, RNA polymerase II must interact with several proteins called **general transcription factors.** The proteins are "general" because they assemble on all promoters used by RNA polymerase II. In eukaryotes, general transcription factors are named after the polymerase they work with. Thus, the transcription factor that recognizes the TATA box for polymerase II genes is called TFIID (D being the fourth letter of the alphabet for the fourth transcription factor so named). TFIID is composed of one subunit that recognizes the TATA sequence, called **TATA binding protein (TBP),** and upto a dozen other proteins called **TBP – associated factors (TAFs)**, which recognize the initiator element, when present and aid in regulating transcription. TFIID is virtually similar to the sigma factors of prokaryotic RNA polymerase. One interesting aspect of the binding of TBP is that it causes a significant bending and opening of the DNA. This bending may be an important signal for other binding proteins (Fig. 7.24).

Box 7.1.

Techniques employed in study of initiation of eukaryotic transcription

Much of the information regarding initiation of transcription in eukaryotes has been gathered by foot printing, mutational studies, cloning and isolating the genes and proteins involved, and then reconstituting various purified combinations in test tube. These studies have been combined with kinetic research to determine which arrangements are stable, immunological research to isolate various components with antibodies, and photocrosslinking studies to determine which moieties are in contact with each other.

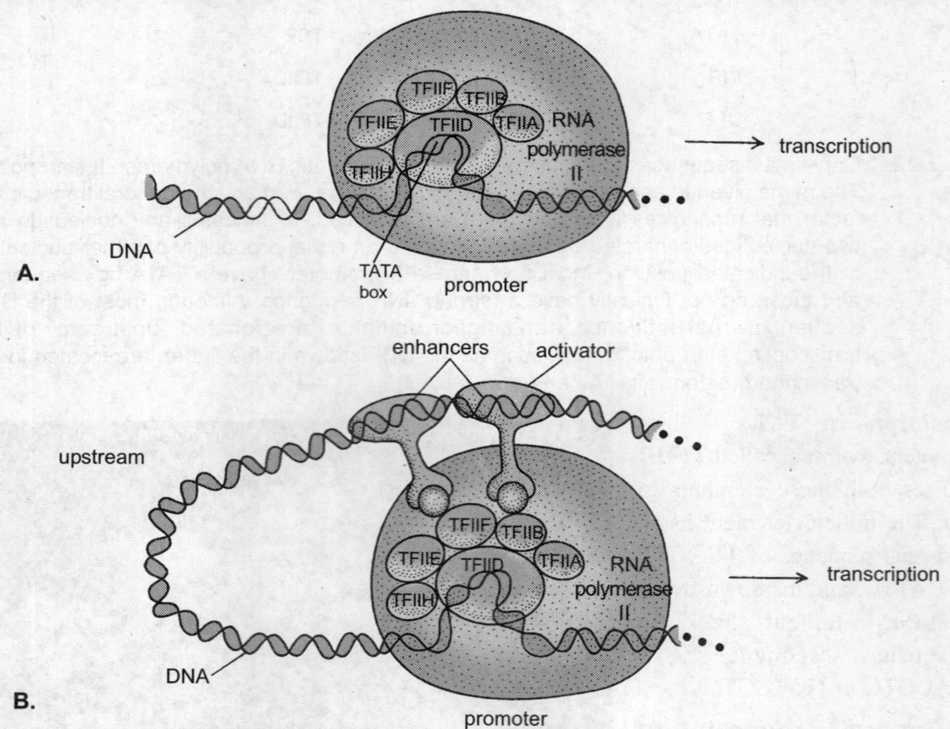

Fig. 7.23. A – Formation of an RNA polymerase II preinitiation complex at a promoter. TFIID binds to the TATA box. The other transcription factors are then recruited with the polymerase. B – Two activators are shown bound at one end (their DNA domains) to enhancers upstream on the DNA. The activators are bound at their other ends (their transcriptional activation domains) to other protein associated with the polymerase machinery. Phosphorylation of the polymerase initiates activated transcription (after Tamarin, 2002).

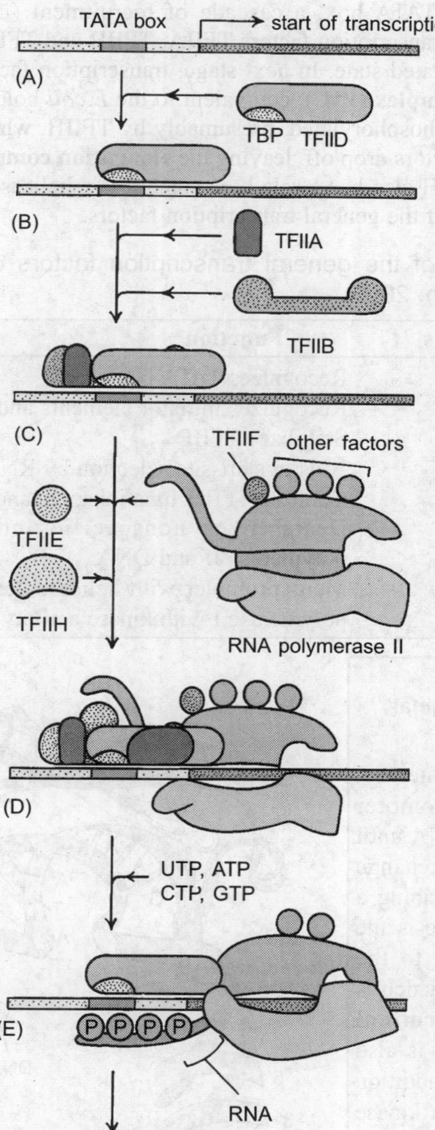

Fig. 7.24. Initiation of transcription of a eukaryotic gene by RNA polymerase II. To begin transcription, RNA polymerase requires a number of general transcription factors (called TFIIA, TFIIB and so on). A – The promoter contains a DNA sequence called the TATA box, which is located 25 nucleotide away from the site at which transcription is initiated. B – The TATA box is recognized and bound by transcription factor TFIID, which then enables the adjacent binding of TFIIB (C); for simplicity the DNA distortion produced by the binding of TFIID is not shown. D – The rest of the general transcription factors, as well as the RNA polymerase itself, assemble at the promoter. E – TFIIH then uses ATP to pry apart the DNA double helix at the transcription start point, allowing transcription to begin. TFIIH also phosphorylate RNA polymerase II, changing its conformation so that polymerase is released from the general factors and can begin the elongation phase of transcription. As shown, the site of phosphorylation is a long C – terminal polypeptide tail that extends from the polymerase molecule (after Alberts *et al.,* 2002).

Once TFIID binds to the TATA box, a cascade of recruitment (*i.e.,* binding) of other transcription factors take place. Transcription factors TFIIA, TFIIB and TFIIF bind, as does RNA polymerase II in an unphosphorylated state. In next stage, transcription factors TFIIE and TFIIH bind, forming a **pre-initiation complex (PIC),** equivalent to the *E.coli* holoenzyme (Fig. 7.24A). The RNA polymerase II is then phosphorylated, presumably by TFIIH, which is a kinase; at this point, most of the transcription factors drop off, leaving the **elongation complex,** which carries out a basal rate of transcription. TFIIH also has a role here since it is a helicase (Box 7.2). Table 7.4 summarizes the postulated roles of the general transcription factors.

Table 7.4. Postulated roles of the general transcription factors of RNA polymerase II (source **Tamarin,** 2002).

General transcription Factors	Function
1. TFIID, TBP	Recognizes TATA box
2. TFIID, TAFs	Recognizes initiator elements and regulatory proteins
3. TFIIA	Stabilizes TFIID
4. TFIIB	Aids in start-site selection by RNA polymerase II
5. TFIIE	Controls TFIIH functions; enhances promoter melting
6. TFIIF	Destabilizes nonspecific interactions of RNA polymerase II and DNA
7. TFIIH	Melts promoter with helicase activity; activates RNA polymerase I with kinase activity

Box 7.2.

Phosphorylation of C–terminal domain

Like the bacterial polymerase, polymerase II remains at promoter synthesizing short lengths of RNA until it undergoes a conformational change and is released to begin transcribing a gene. A key step in this release is the addition of phosphate groups to the "**tail**" of the RNA polymerse (which is known as the **CTD** or **C-terminal domain**). This phosphorylation is also catalysed by TFIIH, which in addition to a helicase contains a protein kinase as one of its subunits (Fig. 7.24). The polymerase can then disengage from the cluster of general transcription factors, undergoing a series of conformational changes that tighten its interaction with DNA and acquiring new proteins that allow it to transcribe for long distances without dissociating.

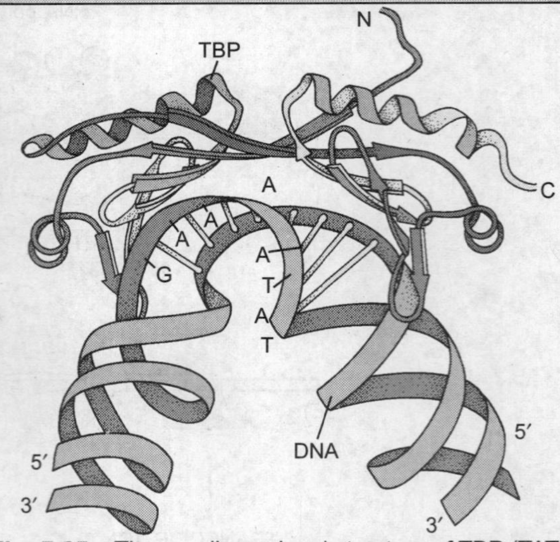

Fig. 7.25. Three – dimensional structure of TBP (TATA – binding protein) bound to DNA. The TBP is the subunit of the general transcription factor TFIID that is responsible for recognizing and binding to the TATA box sequence in the DNA. The unique DNA bending caused by TBP, i.e., two kinks in the double helix separated by partly unwound DNA, may serve as a land mark that helps to attract the other general transcription factors. TBP is a single polypeptide chain that is folded into two very similar domains.

3. Specific Transcription Factors. For activated transcription, a high level of transcription, to take place, other factors are needed that are involved in controlling which promoters are actively transcribed. These other factors are **activators** or **specific transcription factors** that can bind to DNA sequence, called **enhancers**. Enhancers are often hundreds or thousands of base pairs upstream from the promoter (Fig. 7.23B)

These specific transcriptional activators have domains (regions) that recognize their specific enhancer sequences, regions that recognize proteins associated with the polymerse (general transcription factors) and regions that allow the joint attachment of other transcription factors (Fig. 7. 23B). Similar to activators and enhancers, repressors can bind to silencer regions of DNA, often far upstream of the promoters, to repress transcription. Thus, many genes are associated with numerous and complex arrangements of transcription factors, providing intricate control of transcription.

For specific transcription factors to attach to both enhancers and the polymerse machinery, possibly thousands of base pairs apart, the DNA evidently bends to allow them to come into the range of polymerse.

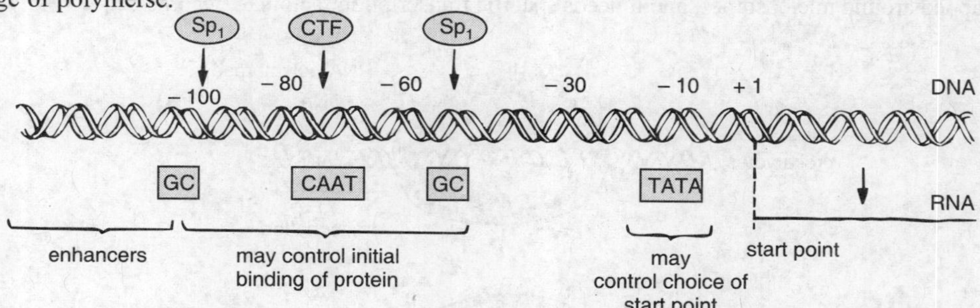

Fig. 7.26. An eukayotic DNA segment showing promoter sites (TATA, CAT, GC boxes and enhancer sites).

Although RNA polymerase I and II seem to have termination signals similar to rho-independent promoters in prokaryotes, termination of transcription of RNA polymerse II genes is more complex, coupled with further processing of the mRNA.

Further unlike prokaryotic RNA polymerases, eukaryotic RNA polymerases do proof reading (showing $3' \rightarrow 5'$ exonuclease activity). Moreover, eukaryotic DNA is complexed with histone proteins that can interfere with transcription. In turn, part of the RNA polymerase II complex is made up of proteins that can disrupt the histones bound to the DNA. In addition, the RNA polymerase II complex

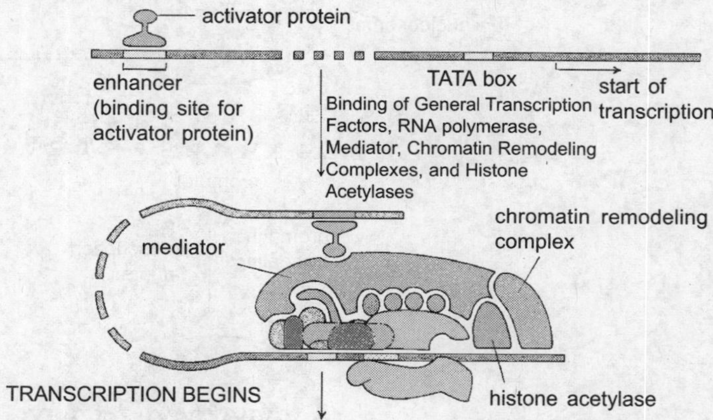

Fig. 7.27. Initiation of transcription by RNA polymerase II in a eukaryotic cell. Transcription initiation *in vivo* requires the presence of transcription activator proteins. These proteins bind activator proteins. These proteins bind to specific short sequences in DNA. a typical eukaryotic gene has many activator proteins (here only one is shown), which together determine its rate and pattern of transcription. Sometimes acting from a distance of several thousand nucleotide pairs (indicated by dashed DNA molecule), these gene regulatory proteins, help RNA polymerase, the general factors, and the mediator all to assemble at the promoter. In addition, activators attract ATP – dependent chromation remodeling complexes and histone acetylases (after Alberts *et al.,* 2002).

contains proteins that act as **mediators** between activators and the polymerase holoenzyme (Fig. 7.27). This complex coordination of the initiation of transcription in eukaryotes has been termed **combinatorial control**, the huge initiation complex may contain 100 or more different polypeptides (see **Alberts** *et al.,* 2002). Finally, transcription in archaea (archaebacteria), although under much simpler control than in eukaryotes, resembles transcription in eukaryotes rather prokaryotes. The study of details of the transcription process – its **initiation, control** and **termination** – is one of the most active and exciting areas in modern genetics.

Control of Transcription in Eukaryotes

In prokaryotes, an RNA polymerase holoenzyme with its promoter-recognizing sigma factor is generally active, transcribing at high levels; repressors are needed to prevent transcription. In eukaryotes, an RNA polymerase holoenzyme (*e.g.*, RNA polymerase II), with its promoter-recognizing TFIID, is generally not transcribing; it needs access to the promoter, which is usually wrapped around nucleosomes, and it needs specific transcription factors to become active (Fig.7.28).

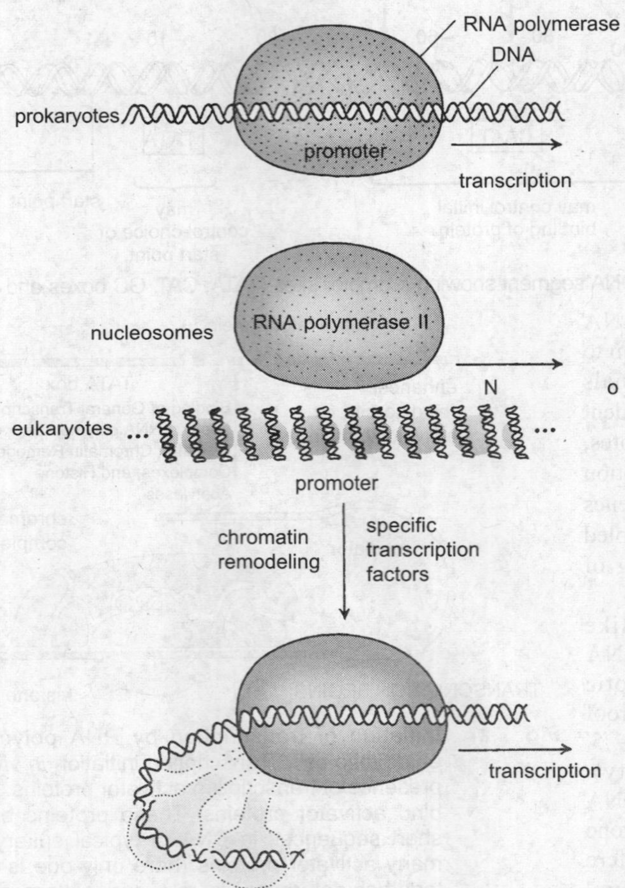

Fig. 7.28. In prokaryotes, the default condition is active transcription. In eukaryotes, the default condition is no transcription since the DNA of promoters is usually wrapped around nucleosomes and specific transcription factors are needed to recruit the polymerase holoenzyme. Transcription in eukaryotes is generally initiated when specific transcription factors bind to enhancer sequences near the promoter, and chromatin is remodeled at the promoter (after Tamarin, 2002).

Thus, although the parts of the transcribing machinery of prokaryotes and eukaryotes are generally similar, the essence of prokaryotic transcription is activity, whereas the essence of eukaryotic transcription is inactivity. In addition, eukaryotes generally do not have operons; however, groups of eukaryotic genes involved in the same pathway or function can be induced simultaneously by having common enhancers that respond to the same specific transcription factors. Such a group of genes is called a **synexpression group.**

1. Chromatin remodeling. For transcription to take place in eukaryotes the DNA must be available for the preinitiation complex to form, with its RNA polymerase and general transcription factors. It appears that DNA wrapped around nucleosomes is often not accessible for the formation of the preinitiation complex, but is available for recognition by transcription-activating proteins, also called *specific transcription factors* (as compared to general transcription factors of the RNA polymerase machinery). One model of initiation of transcription by genes whose promoters are wrapped around nucleosomes is for specific transcription factors recruit chromatin-remodeling proteins. There are two general classes of proteins that remodel nucleosomes: *histone acetyl transferases* and ATP-dependent chromatin remodeling proteins such as the *SWI/SNF complex* in yeast. Thus, the presence of one or more specific transcription factors can begin the process of transcription by recruiting chromatin-remodeling proteins that allow the RNA polymerase access to the promoter.

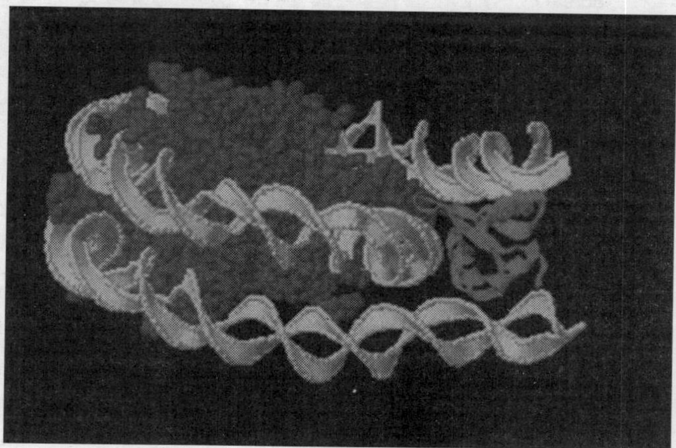

Chromatin remodeling.

2. Specific transcription factors. Eukaryotic transcription is started with the formation of a preinitiation complex formed by the amalgamation of a group of general transcription factors (such as TFIID in RNA polymerase II formation). Proteins that exert control over transcription at specific promoters are the specific transcription factors (Fig. 7.23). These proteins generally have two domains : a domain that recognizes specific DNA sequence, and a domain that recognizes another protein such as a protein in the preinitiation complex. Thus, these proteins recognize signals in the vicinity of the promoter of a gene, bind there, and initiate transcription. Currently, we believe that the majority of specific transcription factors act by recruiting the components of the DNA polymerase holoenzyme. Thus, the binding of a specific transcription factor at a promoter is the first step in the formation of a pre-initiation complex at the promoter of a gene. Some transcription activating proteins also recruit chromatin-remodeling proteins.

An example of a specific transcription factor is Dorsal protein, the product of the *dorsal* gene in fruit flies, active in development. Dorsal protein controls the transcription of several genes and at several different levels of protein concentration. The ability to have different effects at different concentrations is extremely important, allowing gradients of the same protein to control the expression of different genes. One gene Dorsal protein controls is the *rhomboid*, which has three sites in its promoter that Dorsal protein binds to, initiating transcription. Another gene, *twist*, also has three sites in its promoter that bind Dorsal protein, also initiating transcription. However, the *rhomboid* sites are more efficient in binding Dorsal protein; thus, *rhomboid* is transcribed at lower concentrations of Dorsal protein than *twist* is (Fig.7.29). One other signal in the control of transcription that is of current interest is methylation.

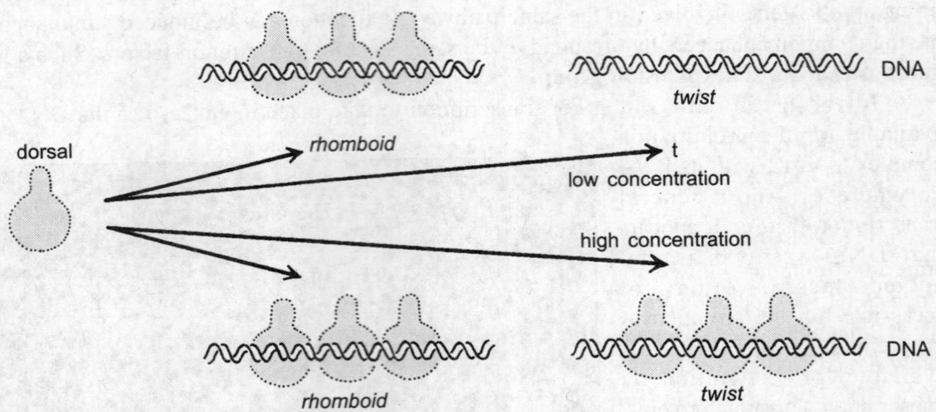

Fig.7.29. In fruit fly the *rhomboid* and *twist* genes each have three enhancer sequences that are recognized by the Dorsal protein or transcription factor. However, the recognition sequences of the *rhomboid* gene are more efficient at binding Dorsal protein than the recognition sequences of the *twist* gene. Thus, *rhomboid* is induced at both low/and high concentrations of Dorsal protein whereas *twist* is induced only at high concentration (after Tamarin, 2002).

3. Methylation of DNA. The importance of methylation in DNA-protein interactions is well known. Molecular biologists have showed that a particular DNA sequence could be protected from restriction endonucleases if it was methylated. A small percentage of cytosine residues are methylated in many eukaryotic organisms, mainly in CpG sequences; 80% of the cytosines in CpG sequences in human DNA are methylated. (**Note.** Often when we refer to a sequence of two bases on the same strand of DNA, we put a "p" between them –CpG to indicate that they are on the same strand connected by a phosphodiester bond and not on two different strands as a hydrogen-bonded base pair).

The *degree* of methylation of DNA is related to the silencing of gene. Genes that are dormant at one stage of development but active in another, are usually less methylated when active and more fully methylated when inactive. For example, adenovirus, a cancer-causing virus, has been observed in many eukaryotic cell lines. In most lines in which the adenovirus DNA has integrated into the host chromosome, late viral genes are turned off. These genes are highly methylated at their CCGC or GCGC sites.

Arabidopsis thaliana.

In addition, chemicals that prevent methylation frequently activate previously dormant genes. For example, a chemical compound, called *5-azacytidine*, inhibits methylation; X chromosomal genes, which are normally deactivated, can be reactivated by treatment with 5-azacytidine. There are numerous other examples of the activation of genes after treatment with this chemical. The activated genes lack methylated cytosines that were previously methylated. Finally, the possibility exists that DNA methylation can affect chromatin structure.

Recent work has also indicated that the methylation itself may not prevent transcription, but rather may be signal for transcriptional inactivity. In the thale cress plant, *Arabidopsis thaliana* a protein named Mom (for *Mo*rpheus molecule) has been discovered that , when mutated, results in genes that have heavy methylation level but are actively transcribed. Thus, the methylation level can be separated from the transcriptional activity of genes, although the two usually occur together. *Arabidopsis* is providing to be a good model in the study of the role of methylation in transcriptional activation because other common model organisms, namely fruit flies, yeast and the nematode, *Caenorhabditis elegans*, do not have methylation of their DNA.

Further interest has been generated in the role of methylation in *regulating* gene expression by the discovery of Z DNA, and the fact that Z DNA can be stabilized by methylation. This observation has led to a model of transcriptional regulation based on alternative DNA structures. Sequences (such as CpG repititions) that could exist as Z DNA exist as B DNA when being transcribed. If the gene is to be silenced (turned off), the CpG sequences are converted to stable Z DNA by methylation, which then blocks transcription. This feasibility has gained some interest because of the recent discovery of an enzyme, double stranded *RNA adenosine deaminase(ADAR1)*, that binds to Z DNA sequences.

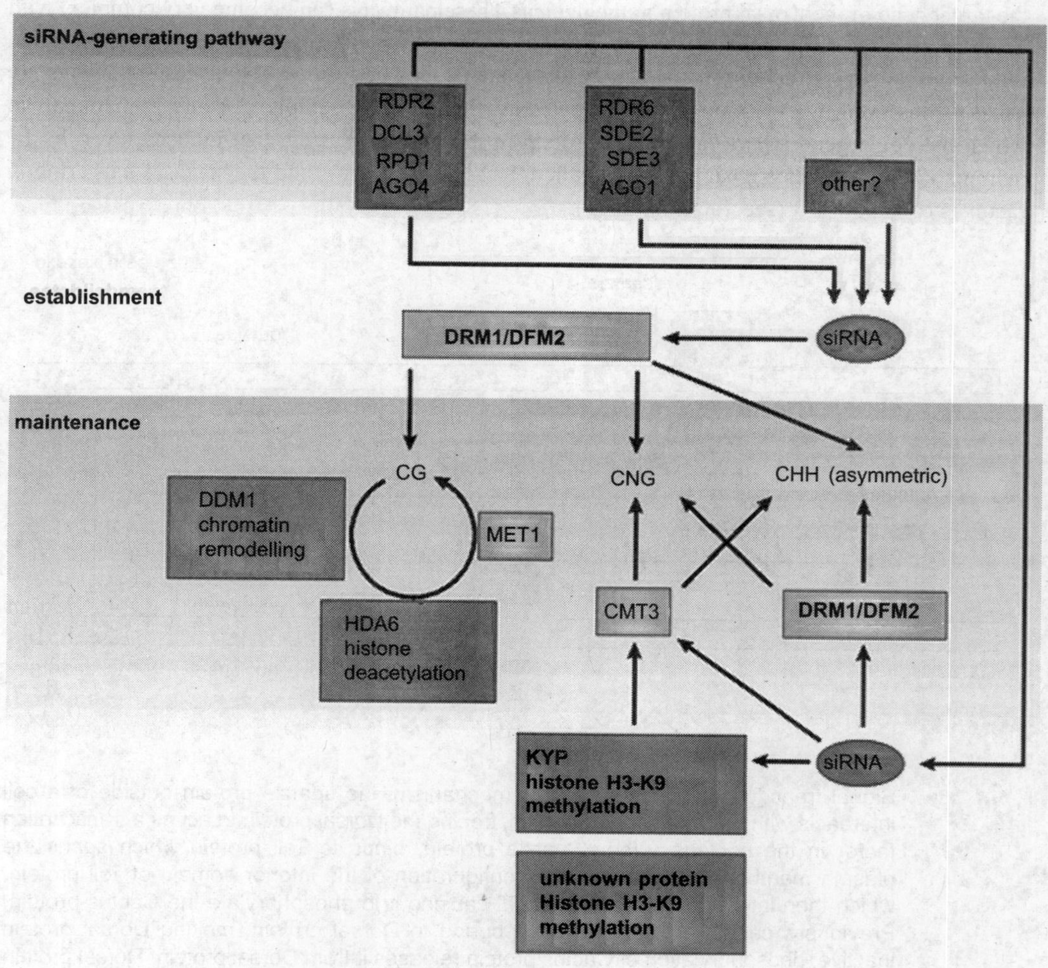

DNA Methylation.

4. Signal transduction. Let us see, how specific transcription activation factors appear at specific times? During development of animals, control of gene expression requires that genes be expressed at specific times and under specific circumstances. If transcription is usually controlled by specific transcription factors, what determines the appearance of these factors at the appropriate times and places? For both of these requirements, there exists one common mechanism, called **signal transduction pathway**, in which signals pass from the external environment through the cytoplasm, into the nucleus.

For example, in a signal transduction pathway involved in development of the fruit fly, the *Toll protein* spans the plasma membrane (Fig. 7.30). It acts as a receptor for the *Spatzle protein,* which when detected, causes a change in the cytoplasmic end of Toll protein, activating it. Activated Toll protein activates *Pelle*, a protein kinase that phosphorylates the *Cactus* protein, causing it to dissociate from Dorsal protein. Once Dorsal protein dissociates from Cactus protein (which acts to repress Dorsal protein), Dorsal protein becomes an active specific transcription factor that can cross the nuclear membrane and activates its target genes (Fig.7.30). We thus see that Spatzle protein attaching to its receptor protein (*i.e.*, Toll protein) on the cell surface results in activation of the target gene of the Dorsal protein in the nucleus.These pathways can become very complex, with many protein elements. More elements mean more sensitive control of various processes, often requiring that several conditions be met before gene is activated. In addition, these pathways are usually conserved by evolution. A similar pathway, though more complex, occurs in mammals in which the target gene is *interleukin-1,* a protein in the immune system that induces fever. In the mammalian protein, the signal protein is called Toll-like receptor-4 and the specific transcription factor is called NF-kB (see Tamarin, 2002).

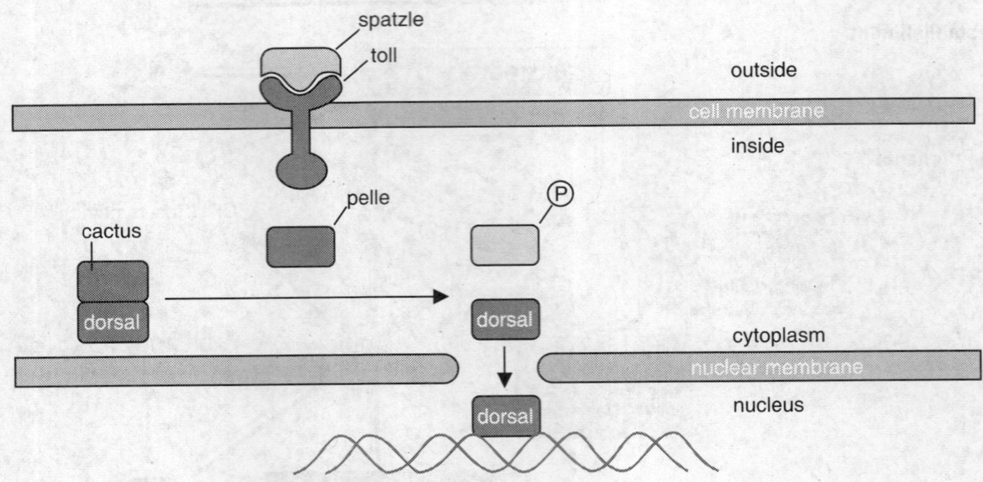

Fig.7.30. Signal transduction pathway. In this mechanism, the spatzle protein outside of a cell intereacts with the Toll receptor protein, freeing the Dorsal protein to act as a transcription factor in the nucleus. When Spatzle protein, binds to Toll protein which spans the plasma membrane, it changes the configuration of the interior domain of Toll protein, which then interacts with pelle protein, causing it to phosphorylate the Cactus protein. Previously, cactus protein had been bound to Dorsal protein, making Dorsal protein inactive; phosphorylation of Cactus protein releases it from Dorsal protein. Dorsal protein is then free to cross the nuclear membrane and act as a transcription factor (after Tamarin, 2002).

5. Transcription elongation produces super helical tension of DNA. Once it has initiated transcription, RNA polymerase does not proceed smoothly along a DNA molecule, rather it moves jerkily, pausing at some sequences and rapidly transcribing through others. RNA polymerases involved in *RNA-elongation* both bacterial and eukaryotic, are associated with a series of *elongation factors,* proteins that decrease the likelihood that RNA polymerase will dissociate before it reaches the end of a gene. These factors typically associate with RNA polymerase shortly after initiation has occurred and help polymerases to move through the wide variety of different DNA sequences that are found in genes. Eukaryotic RNA polymerases must also cope with chromatin structure as they move along a DNA template. Experiments have shown that bacterial RNA polymerases, which never encounter nucleosomes in vivo, can nonetheless transcribe through them *in vitro,* suggesting that a nucleosome in easily traverses. However, eukaryotic polymerases have to move through forms of chromatin that are more compact than a simple nucleosome. It therefore seems likely that they transcribe with the aid of *chromatin remodeling complexes.* These complexes may move with the RNA polymerases or may simply seek out and rescue the occasional stalled RNA polymerase. In addition , some elongation factors associated with eukaryotic RNA polymerase facilitate transcription through nucleosomes without requiring additional energy. It is not yet understood how this is accomplished, but these proteins may help to dislodge parts of the nucleosome core as the RNA polymerase transcribes the DNA of a nucleosome.

Elongating RNA polymerases also have to phase out some another type of barrier, both bacterial and eukaryotic. To discuss this issue, we need first to consider a refined property inherent in the DNA double helix called *DNA supercoiling.* DNA supercoiling represents a conformation that DNA will adopt in response to superhelical tension; conversely, creating various loops or coils in the helix which can create such tension. A simple way of visualizing the topological constraints that cause DNA supercoiling is illustrated in Figure 7.31A. There are approximately 10 nucleotide pairs for every helical turn in a DNA double helix. In a gene a helix whose two ends are fixed (Fig. 7.31B) with respect to each other (as they are in a DNA circle, such as bacterial chromosome, or in a tightly clamped loop, as thought to exist in eukaryotic chromosomes). In this case, one large DNA supercoil will form to compensate for each 10 nucleotide pairs that are opened (unwound). The formation of this super coil is energetically favourable because it restores a normal helical twist to the base paired regions that remain which woned otherwise need to be overwound because of the fixed ends.

Superhelical tension is also created as RNA polymerase moves along a stretch of DNA that is anchored at its both ends (Fig. 7.31). As long as the polymerase is not free to rotate rapidly (and such rotation is unlikely given the size of DNA polymerases and their attached transcripts), a moving polymerase generates *positive* superhelical tension in the DNA in front of it and *negative* helical tension behind it. For eukaryotes, this situation is thought to provide a bonus : the positive superhelical tension ahead of the polymerase makes the DNA helix more difficult to open but this tension should facilitate the unwrapping of DNA in nucleosomes, as the release of DNA from the histone core helps to relax positive superhelical tension.

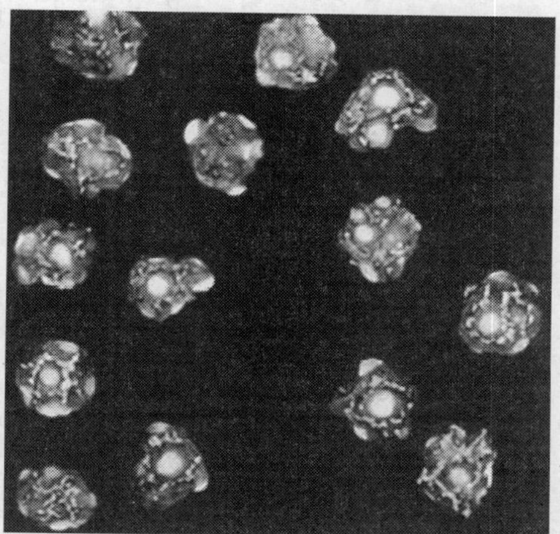

Chromatin remodelling complexes.

Any protein that propels itself along a DNA strand of a double helix tends to generate superhelical tension. In eukaryotes, DNA topoisomerase enzyme rapidly removes this superhelical tension. But, in bacteria, a specialized topoisomerase called *DNA gyrase* uses the energy of ATP hydrolysis to pump super coils continuously into the DNA, thereby maintaining the DNA under constant tension. These are *negative supercoils,* having the opposite handedness from the *positive supercoils* that form when a region of DNA helix opens (see Fig. 7.31B). These negative super coils are removed from bacterial DNA whenever a region of helix opens, reducing the superhelical tension. DNA gyrase therefore makes the opening of the DNA helix energetically favourable compared with helix opening in DNA that is not supercoiled. For this reason, it usually facilitates those genetic processes in bacteria, including the initiation of transcription by bacterial RNA polymerase, that require helix opening.

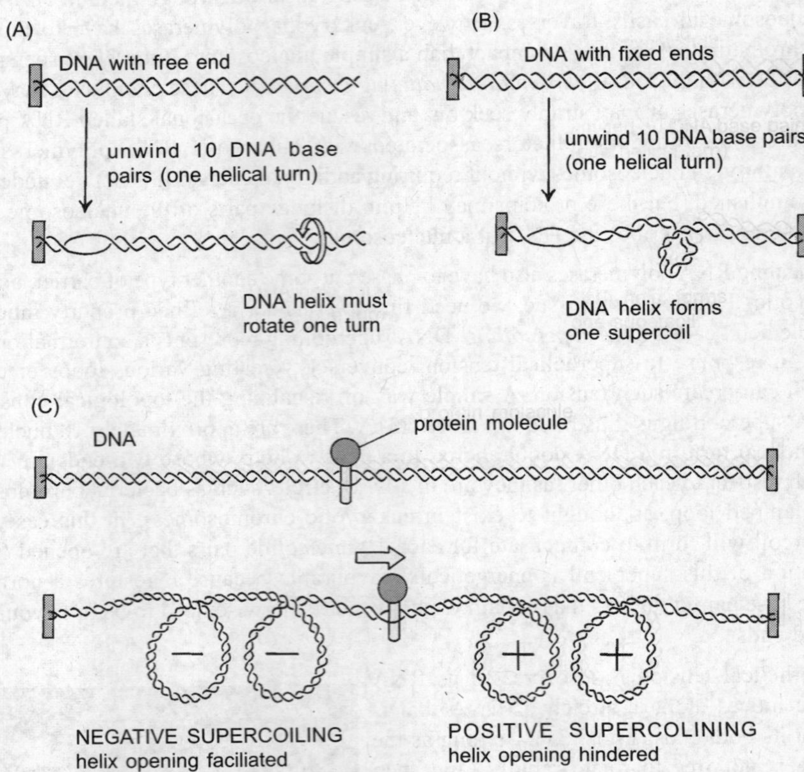

Fig. 7.31. Superhelical tension in DNA causes DNA super coiling A-For a DNA molecule with one free end (or a nick in one strand that serves as a swivel), the double helix rotates by one turn for every 10 nucleotide pairs opend; B- If rotation is prevented, superhelical tension is introduced into the DNA by helix opening. One way of accommodation this tension would be to increase the helical twist from 10 to 11 nucleotide pairs per turn in the double helix that remains in this example; the DNA helix, however, resists such a deformation in a spring-like fashion, preferring to relieve the superhelical tension by bending into super coiled loops. As a result, one DNA supercoil forms in that DNA double helix for every 10 nucleotide pairs opend. The super coil forms in this case is positive super coil. C- Supercoiling of DNA is induced by a protein tracking through the DNA helix. The two ends of the DNA show here are unable to rotate freely relative to each other, and the protein molecule is assumed also to be prevented from rotating freely as it moves. Under these conditions, the movement of the protein causes an excess of helical turns to accumulate in the DNA helix ahead of the protein and a deficit of helical turns to arise in the DNA behind the protein as shown (after Alberts *et al.,* 2002).

Questions And Problems

1. (*a*) What genetic attributes does RNA share with DNA?
   (*b*) RNA is of three types in eukaryotes and bacteria. What are they? Where are they located in the cell? Where are they produced? What are their characteristics and functions?
2. Given a single strand of DNA ... 3′ TACCGAGTAC 5′...., construct
   (a) the complementary DNA chain,
   (*b*) the mRNA chain which would be made from strand.
3. (*a*) From what substrates is RNA made? (*b*) On what template?
   (*c*) By which enzyme ? (*d*) Is a primer required?
4. Describe the differences, if any, between the chemical reactions catalyzed by DNA polymerase and RNA polymerase.
5. Describe the structure of *E.coli* RNA polymerase and discuss the roles of different components of their enzyme in RNA synthesis on DNA template.
6. What are the functions of the core enzyme and the holoenzyme *in vivo.?*
7. What is a transcription unit? Is it the same thing as a gene?
8. In what way, relating to polycistronic mRNA, do eukaryotic and prokaryotic protein synthesis differ?
9. Describe the structure and functions of three RNA polymerases known in eukaryotes.
10. Describe the promoter sites for initiation of transcription in prokaryotes and eukaryotes.
11. What are transcription factors? Describe them for three different RNA polymerases in eukaryotes.
12. Describe the post-transcription modification of heterogeneous nuclear RNA in eukaryotes.
13. Describe specific transcription factors.
14. Describe the relationship of the three types of RNA at a ribosome which relationships make use of complementarity?
15. What is meant by the term “upstream” and “downstream”?
16. What is Pribnow box? Describe its evolutionary and biochemical significance.
17. An RNA molecule is isolated having 3′–OH terminus and 5′–P terminus. What information does this fact provide?
18. What is a consensus sequence? What is a conserved sequence?
19. Why do you think that most promoter regions are A-T rich?
20. What is the difference between a σ-70 and a σ-32?
21. What is footprinting? How did it help to define promoter?
22. Can one nucleotide be a conserved sequence?
23. What is a stem-loop or hair-pin structure? An inverted repeat? A tandem repeat? Draw a section of a DNA double helix with an inverted repeat of seven base pairs.

Answers To Problems

3. (*a*) Ribonucleoside 5’ triphosphates.
   (*b*) Double-stranded DNA
   (*c*) RNA polymerase.
   (*d*) No.
4. The reactions are identical, however, substrates are different, *i.e.*, DNA polymerase joins deoxynucleotides and RNA polymerase joins ribonucleotides.
7. A transcription unit is a section of DNA extending from a promoter to an RNA polymerase termination site. It is usually not a gene, but typically includes many genes.
8. In eukaryotes all mRNA is monocistronic.

15. Upstream and downstream usually refer to regions in the 5′and 3′ directions respectively, from a particular site that is being discussed.
17. Since the triphosphate of the primary transcript is absent, the molecule has been processed.
18. A *Consensus sequence* is made up of the nucleotides that appear in a significant proportion of cases when similar sequences are aligned. A *conserved sequence* consists of nucleotides found in all cases in when similar sequences are aligned . For example, the Pribnow box is the consensus sequence TATAAT.
19. A double helix of DNA must unwind in order for transcription to occur. A-T pairs, because they have only two H-bonds, are more easily disrupted than G-C pairs.
20. The superscripts of the sigma factors refers to their molecular weights (e.g., σ-70 is 70,000 daltons). Different sigma factors usually recognize different prokaryotic promoters.
21. **Footprinting** is a technique in which DNA in contact with a protein is exposed to nucleases; only DNA protected by the protein is undigested. Promoters could be isolated by protection with RNA polymerase in the absence of ribonucleotides- the polymerase will not move- and then sequenced.
22. Conserved sequences are invariant sequences of DNA or RNA recognizable to either a protein or a complementary sequence of DNA or RNA. However, in group II introns, an adenine (A) is needed near the 3′ end of the intron for lariat formation. Thus, this single nucleotide, given its relative position in the intron and possible surrounding bases, is a conserved sequence of one.
23. A stem-loop (or hair-pin) structure can form when a single strand of DNA or RNA has a double helical section. An *inverted repeat* is a sequence read outward on both strands of a double helix from a central point. A *tandem repeat* is a segment of nucleic and repeated consecutively; that is, the same sequence repeats in the same directions on the same strand:

    5′- TCCGGTCCGGTCCGG- 3′
    3′- AGGCCAGGCCAGGCC- 5′

    A DNA sequence with a *seven-base inverted repeat* is

    5′ – ATTACCGCGGTAAT-3′
    3′- TAATGGCGCCATTA-5′

MULTIPLE CHOICE QUESTIONS

Choose the correct answer from the four alternatives given.

1. Transcription is the transfer of genetic information from
   (*a*) chromosome to cytoplasm
   (*b*) tRNA to mRNA
   (*c*) DNA to mRNA
   (*d*) mRNA to rRNA
2. Nongenetic RNA is of
   (*a*) one type
   (*b*) two types
   (*c*) three types
   (*d*) non functional type
3. Regarding transcription which one of these is false?
   (*a*) uses RNA polymerase
   (*b*) one of the mother strands of DNA is the template
   (*c*) is bidirectional
   (*d*) reverse transcription used RNA dependent DNA polymerase
4. The enzyme required for transcription is
   (*a*) DNA polymerase
   (*b*) RNA polymerase
   (*c*) endonuclease
   (*d*) all the above
5. Pribnow box contribute to
   (*a*) protein synthesis
   (*b*) ATP synthesis
   (*c*) RNA synthesis
   (*d*) none of the above
6. The process involved in the RNA formation on DNA template is
   (*a*) translation
   (*b*) transduction

(*c*) transcription
(*d*) transformation

7. Transcription takes place in
(*a*) matrix
(*b*) nucleus
(*c*) cytosol
(*d*) cytoplasm

8. Essential components of eukaryotic cistron are
(*a*) intron
(*b*) exons
(*c*) operons
(*d*) operator and regulator genes

9. After initiation of transcription with core enzyme RNA polymerase the sigma factor is
(*a*) functionless
(*b*) released to take part again
(*c*) used during closing of chain
(*d*) retained and it performs special function

10. During transcription, the DNA site at which RNA polymerase binds is called
(*a*) promoter
(*b*) regulator
(*c*) receptor
(*d*) enhancer

ANSWERS

1. (*c*)	2. (*c*)	3. (*c*)	4. (*b*)	5. (*c*)
6. (*c*)	7. (*b*)	8. (*d*)	9. (*b*)	10. (*a*).

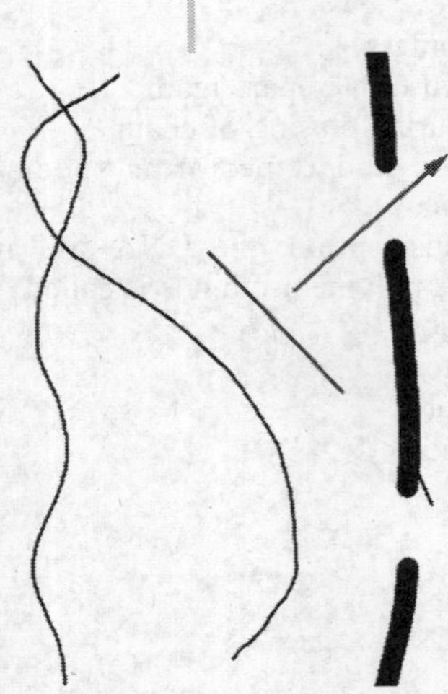

CHAPTER

Messenger RNA (Coding RNA)

Chapter Contents

Messenger RNA (mRNA) acts as the intermediate molecule between the gene and the polypeptide translation product. Its existence was postulated by **Crick** and his associates during the 1950s. Among the indirect evidence for mRNA at that time was the knowledge that *in eukaryotes the genes reside on the chromosomes in the nucleus, whereas protein synthesis occurs in ribosomes in the cytoplasm.* The physical separation of genes and ribosomes means that some sort of messenger molecule must carry the biological information from nucleus to cytoplasm. In bacteria the physical separation of DNA and ribosomes is less distinct but nevertheless a messenger molecule is still necessary.

Evidence that this messenger is in fact RNA came first from **Elliot Volkin** and **Lazarus Astrachan** in 1956, but more convincingly from **Sol Spiegelman** and **Benjamin D.Hall** in 1961. Both groups demonstrated that after infection of a culture of bacteria with a bacteriophage, the new RNA that is synthesized is related in sequence to the phage DNA, suggesting that the phage genes are copied into RNA before synthesis of the phage proteins occurs. Shortly afterward two independent groups-**Sydney Brenner, Francois Jacob** and **Mathew Meselson**, who did the crucial experiment at the California Institute of Technology) and

James Watson's group at Harvard- directly identified mRNA molecules in *E. coli* cells.

8.1. LIFE SPAN OF mRNA

Messenger RNA molecules are not generally long lived in the cell, *i.e.*, most of them are unstable. Most bacterial mRNAs have a **half life** of only a few minutes and so are turned over very rapidly. In eukaryotic cells, mRNA molecules have longer half-lives (*e.g.*, 6 hours for many mammalian mRNAs) but are still subject to turnover. Turnover of mRNA is important because it

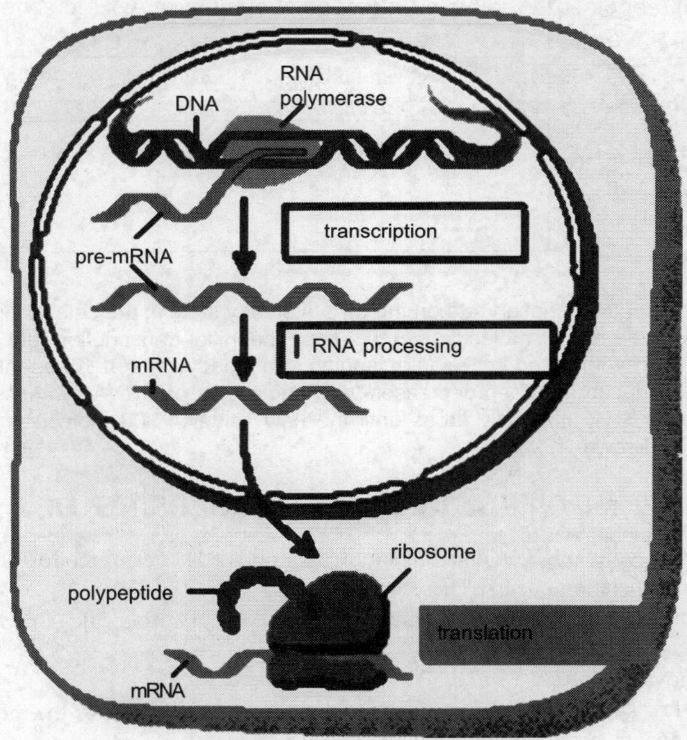

Functions of mRNA.

means that the absolute amount of a particular mRNA in the cell can be controlled by adjusting the rate at which the relevant gene is transcribed. If the transcription rate for the gene decreases then the level of the mRNA in the cell also decreases until a new steady state is reached (Fig. 8.1).

In fact, in both bacteria and eukaryotes mRNAs are known that are apparently resistant to cytoplasmic ribonuclease enzyme and survive for long period of time. For example, mRNA with life time of six hours has been detected in the bacterium *Bacillus cerus* at a time when the cells are induced to become long lived spores. Likewise, in differentiating eukaryotic cells mRNAs with a life time of days have been detected. For example, in the immature red blood cells (reticulocytes) of the mammals the mRNA is synthesized originally by the nucleus in early stages and expelled to the cytoplasm. In later stages, the nuclei of maturing reticulocytes degenerate but the mRNA exists up to 2 days for prolong utilization in the synthesis of globin protein of haemoglobin. Further, in extreme cases, such as in the state of dormancy adopted by many animal eggs and plant seeds, mRNA is maintained in a stable form for months or even years.

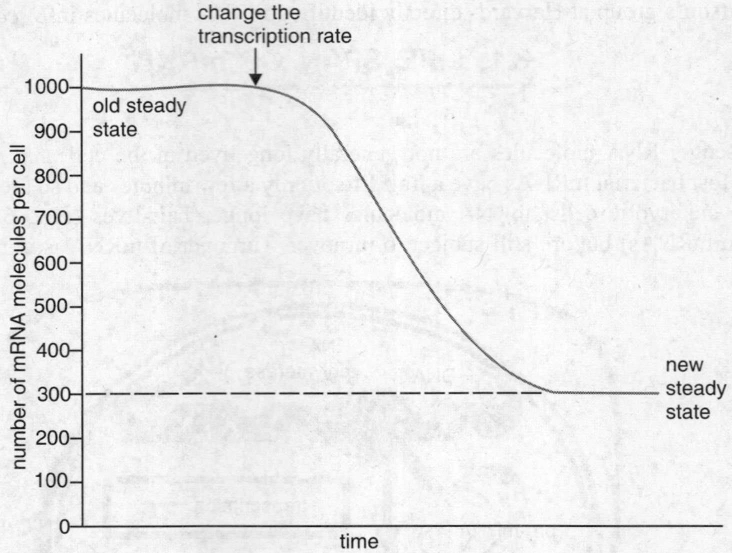

Fig. 8.1. Effect of transcription rate on the mRNA steady state in the cell. In this example, the half-life of the mRNA molecule is 5 minutes. The initial transcription rate is 100 molecules of mRNA synthesized per minute, leading to a steady state of 1000 molecules per cell. The transcription rate then decreases to 30 molecules of mRNA synthesized per minute; the steady state gradually drops until the new value of 300 molecules per cell is reached (after Brown, 1998).

8.2. MODIFICATION AND PROCESSING OF mRNA

Most eukaryotic mRNAs undergo a quite complicated series of modifications and processing events before translation occurs. These include:

1. Chemical modification to the two ends of the mRNA molecule (*Capping*).
2. Polyadenylation
3. Removal of introns (*Splicing*).
4. In a few special cases, alterations to the nucleotide sequence of the mRNA (*RNA editing*).

These events occur only in eukaryotes. The first clue that eukaryotic mRNAs undergo modification and processing came when RNA fractions present in the nucleus and cytoplasm were compared. The nucleus can be divided into two distinct regions : the *nucleolus,* in which rRNA genes are transcribed, and the *nucleoplasm,* where other genes, including those for mRNA, are transcribed. The nucleoplasmic RNA fraction is called *heterogenous nuclear RNA,* the name indicating that it is made up of a complex mixture of RNA molecules, some over 20 kb in length. The mRNA in the cytoplasm is also heterogenous, but its average length is only 2 kb. If the mRNA, in the cytoplasm is derived from hnRNA, then modification and processing events including a reduction in the length of the primary transcripts, must occur before the mRNA leaves the nucleus.

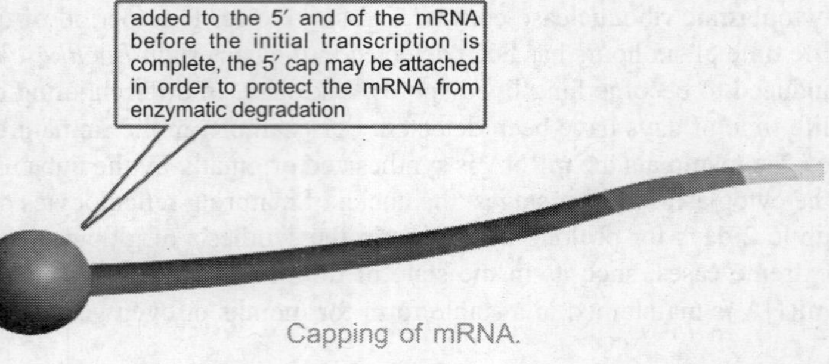

Capping of mRNA.

The hnRNA fraction includes mRNAs at various stages of modification and processing making it very difficult to study the exact order of events that occur within a single molecule. There is evidence that the end-modifications (cappings) are completed before all the introns are removed (splicing), but this may not be true for all mRNAs.

1. Role of CTD of RNA Polymerase II in mRNA Processing (RNA Factory)

RNA processing steps are tightly coupled to transcription elongation by an ingenious mechanism. As discussed previously, a key step of the *transition* of RNA polymerase II to the elongation mode of RNA synthesis is an extensive phosphorylation of the RNA polymerase II tail, called the CTD. This C-terminal domain of the largest subunit consists of a long tandem array of a repeated seven-amino acid sequence, containing two serines per repeat that can be phosphorylated. Because there are 52 repeats in CTD of human RNA polymerase II, its complete phosphorylation would add 104 negatively charged phosphate groups to the polymerase II. This phosphorylation step not only dissociates the RNA polymerase II from other proteins present at the start point of transcription, it also allows a new set of proteins to associate with the RNA polymerase tail onto the nascent RNA molecule to begin processing. As discussed next, some of these processing proteins seem to "jump"from the polymerase tail onto the nescent RNA molecule to begin processing it as emerges from RNA polymerase. Thus, RNA polymerase II in its elongation mode can be viewed as RNA factory that both transcribes DNA into RNA and process the RNA it produces (Fig. 8.2).

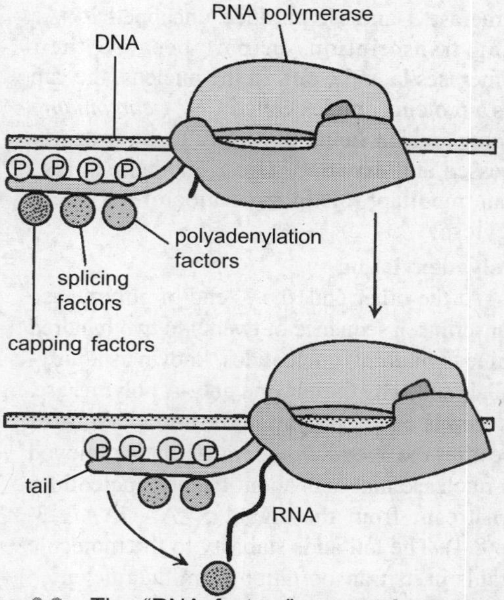

Fig. 8.2. The "RNA factory"concept for eukaryotic RNA polymerase II. Not only does the RNA polymerase II transcribe DNA into RNA, but it also carries pre-mRNA- processing proteins on its tail, which are then transferred to the nascent RNA at the appropriate time. There are many RNA – processing enzymes, and not all travel with the polymerase. For RNA splicing, for example, only a few critical components are carried on the tail, once transferred to an RNA molecule, they serve as a nucleation site for the remaining components. The RNA – processing proteins first bind to the RNA polymerase tail when it is phosphorylated late in the process of transcription initiation. Once RNA polymerase II finishes transcribing, it is released from DNA, the phosphates. Only this dephosphorylated form of RNA polymerase II is competent to start RNA synthesis at a promoter (after Alberts *et al.*, 2002).

2. RNA Capping

As soon as RNA polymerase II has produced about 25 nucleotides of RNA, the 5' end of the new RNA molecule is modified of a "cap"that consists of a modified guanine nucleotide (*i.e.*, 7 – methyl G). The capping reaction is performed by *three* enzymes acting in succession : One enzyme called *phosphatase* removes one phosphate from the 5′ end of the nascent RNA, another enzyme (a *guanyl transferase*) adds a GMP in a reverse linkage (5′ to 5′ instead of 5′ to 3′) and a third enzyme (a *methyl transferase*) adds a methyl group to the guanosine (Fig. 8.3). Because all three enzymes bind to the phosphorylated RNA polymerase tail, they are ready to modify the 5′ end of the nascent transcript as soon as it emerges from the polymerase.

The 5′-methyl cap signals the 5′ end of eukaryotic mRNA, and this landmark helps the cell to distinguish mRNA, from the other types of RNA molecules present in the cell. For example, RNA polymerase I and III produce uncapped RNAs during transcription, in part because their polymerases lack the tail. In the nucleus, the cap binds a protein complex called *CBC* (*cap binding complex*) which helps the RNA to be properly processed and exported. The 5′ methyl cap also has an important role in translation of mRNA in the cytosol.

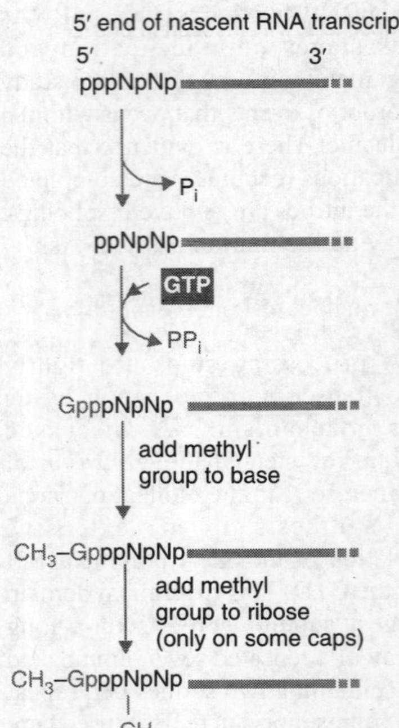

Fig. 8.3. The reactions that cap the 5′end of each RNA molecule synthesized by RNA polymerase II. The final cap contains a novel 5′- to – 5′linkage between the positively charged 7 – methyl G residue and the 5′end of the RNA transcript. The latter N represents any one of the four ribonucleotides, although the nucleotide that starts an RNA chain is usually a purine (an A or a G) (after Alberts *et al.*, 2002).

3. Polyadenylation

At the other end, the 3′ end of polymerase II transcripts, a sequence of twenty to two hundred adenine containing nucleotides, known as a **poly-A tail,** is added by the enzyme **poly-A polymerase.** This step is called **polyadenylation** which takes place after the 3′ end of the transcript is removed by a nuclease that cuts about twenty nucleotides downstream from the signal 5′–AAUAAA–3′ (Fig. 8.4). The tail adds stability to the molecule and aids in its transportation from the nucleus.

4. Introns and RNA Splicing

The protein coding sequences of eukaryotic genes are typically interrupted by noncoding intervening sequences (introns). Discovered in 1977, this feature of eukaryotic genes came as a surprise to molecular biologists, who had been until that time, familiar only with bacterial genes, which typically consists of a continuous stretch of coding DNA that is directly transcribed into mRNA. In marked contrast, eukaryotic genes were found to be broken up into small pieces of codon sequence (*expressed sequences or exons)* interspersed with much longer *intervening sequences or introns;* thus, the coding portion of a eukaryotic gene is often only a small fraction of the length of the gene (e.g., globin gene, see Fig. 8.5).

Types of introns. Molecular biologists have recognized seven distinct types of introns in eukaryotes, and additional forms in the archaea (Table 8.1). Two of these types, namely GV-AG and AU-AC introns, are found in eukaryotic protein coding genes.

Table 8.1. Types of introns

Intron type	Where found
1. GU-AG introns	Eukaryotic nuclear pre-mRNA
2. AU-AC introns	Eukaryotic nuclear pre-mRNA
3. Group I	Eukaryotic nuclear pre-rRNA, few bacterial RNAs.
4. Group II	Organelle RNAs, some prokaryotic RNAs
5. Group III	Organelle RNAs
6. Twintrons	Organelle RNAs
7. Pre-tRNA introns	Eukaryotic nuclear pre-tRNA
8. Archael introns	Various RNAs

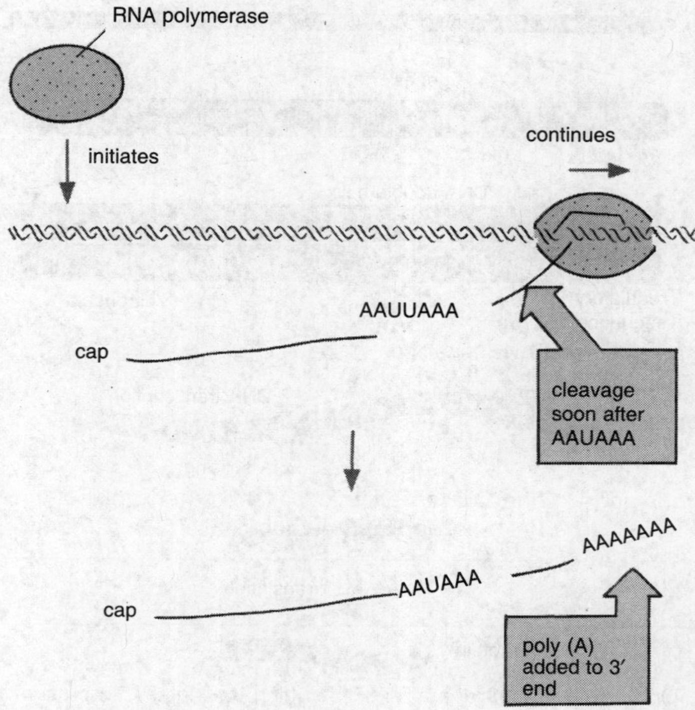

Fig. 8.4. Cleavage of signal AAUAAA sequence and polyadenylation at the cleaved 3′ end.

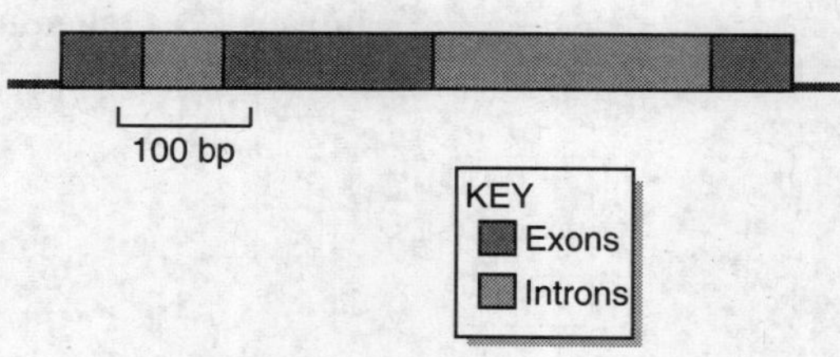

Fig. 8.5. Introns. Human ß-globin gene is 1423 bp in length and contains two introns, one of 131 bp and other of 851 bp, which together make up 69% of the length of the gene (after Brown, 2002).

In protein-coding genes, introns are less common in lower eukaryotes : the 6000 genes in the yeast genome contain only 239 introns in total, whereas many individual mammalian genes contain 50 or more introns. A eukaryotic pre-mRNA may contain many introns, perhaps over 100, taking up a considerable length of the transcript (Table 8.2). These introns must be excised and the exons joined together in the correct order before the transcript can function as a mature mRNA.

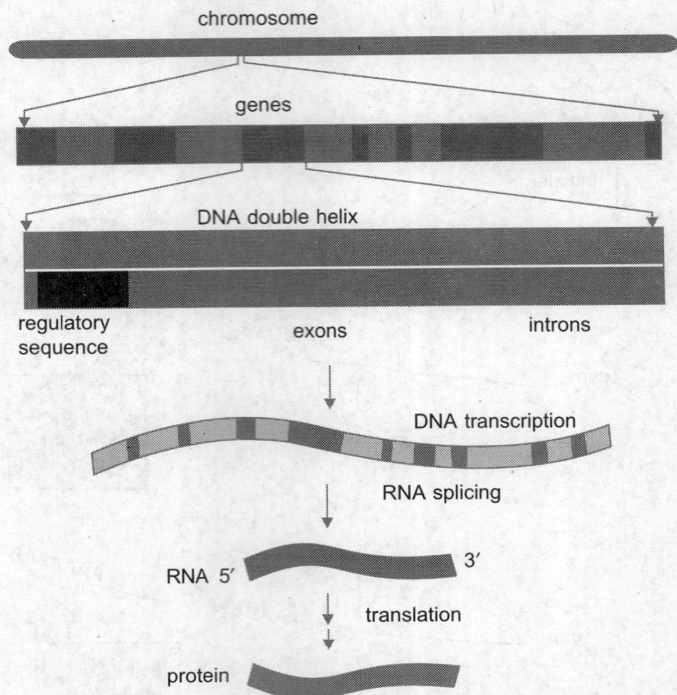

Only exons are included in mRNA which translates into protein.

Table 8.2. Introns in human genes (Source : **Brown,** 2002).

Gene	Length(kb)	Number of introns	Amount of the gene taken up by the introns(%)
1. Insulin	1.4	2	69
2. ß-globin (of haemoglobin)	1.6	2	61
3. Serum albumin	18	13	79
4. Type VII collagen	31	117	72
5. Factor VIII	186	25	95
6. Dystrophin*	2400	78	98

* A transmembrane glycoprotein complex of skeletal muscle cells and is absent in Duchenne muscular dystrophy and deficient in Becker muscular dystrophy.

Evolution of introns. When the same gene is compared in related species, molecular biologists find that some of the introns are in identical positions but that each species has one or more unique introns. This implies that some introns remain in place for millions of years,. Retaining their positions while species diversify, whereas others appear or disappear during this same period. This leads to two opposing hypothesis for the evolution of introns:

1. 'Intron late'hypothesis. This hypothesis holds that introns evolved relatively recently and are gradually accumulating in eukaryotic genomes.

2. 'Introns early'hypothesis. This hypothesis suggests that introns are very ancient and are gradually being lost from eukaryotic genomes.

A. GU-AG Introns and Protein-Mediated Splicing

In most of the pre-mRNA introns, the first two nucleotides of the intron sequence are 5′-GU-3′and the last two 5′-AG-3′. They are therefore called 'GU-AG' introns and all members of this class are spliced in the same way. The GU-AG motifs vary in different types of eukaryotes; in vertebrates they can be described as:

5' splice site → 5′–AG ↓ GUAAGU–3′

3' splice site → 5′–Py Py Py Py Py Py N C A G -3′

In these designations, 'Py' is one of the two pyrimidine nucleotides (U of C); 'N' is any nucleotide, and the arrow indicates the exon-intron boundary. The 5'splice site is also known as the **donor site** and the 3' splice site as the **acceptor site.**

There are certain **conserved sequences**. But they are present in some but not all eukaryotes. Introns in higher eukaryotes usually have a *polypyrimidine tract*, a pyrimidine- rich region located just upstream of the 3'end of the intron sequence (Fig.8.6).

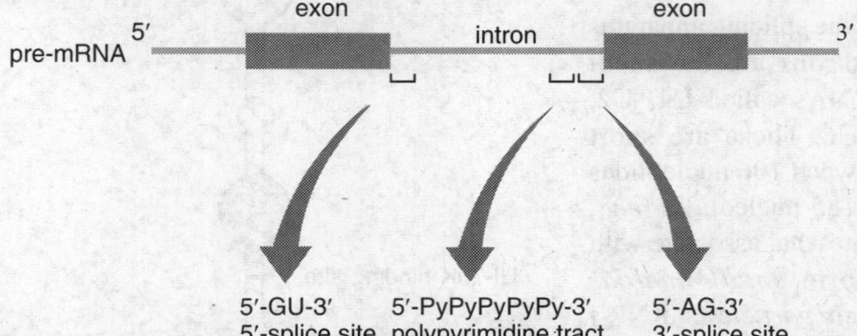

Fig. 8.6. Conserved sequences in vertebrates introns. Py, pyrimidine nucleotide (U or C).

This tract is not common in yeast. However, introns of yeast contain a 5′-UACUAAC-3′ sequence, located between 18 and 140 bp upstream of the 3′splice; these sites are absent in higher eukaryotes. The polypyrimidine tract and the 5′ –UACUAA-3′sequence have different functions.

Two steps of splicing-pathways

1. Cleavage of the 5′splice site. This step occurs by a **transesterification reaction** promoted by the hydroxyl group attached to the 2′carbon of an adenosine nucleotide located within the intron

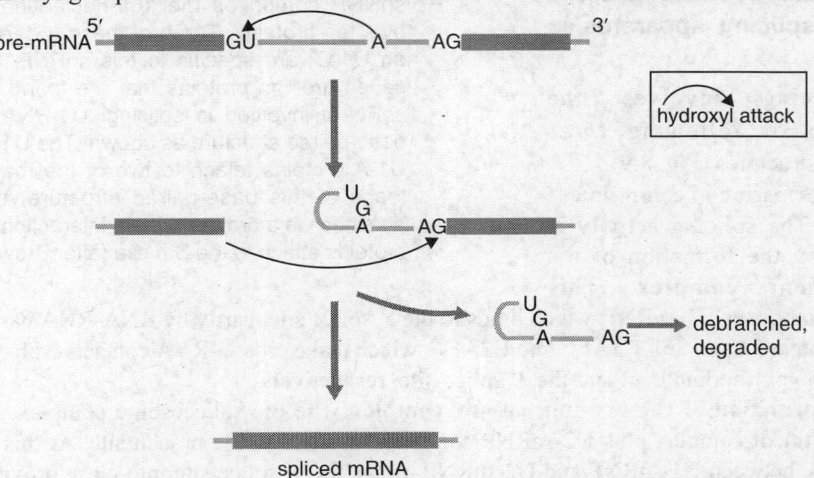

Fig. 8.7. Outline of splicing process. Cleavage of the 5'splice site is promoted by the hydroxyl (OH) group attached to the 2'-carbon of an adenosine nucleotide within the intron sequence. This results in the lariat structure and is followed by the 3'-OH group of the upstream exon inducing. Cleavage of the 3'splice site. This enables the two exons to be ligated, with the released intron debranched and degraded.

sequence. In yeast, this adenosine is the last one in the conserved 5′; -UACUAA-3′sequence. The result of the hydroxyl attack is cleavage of the phosphodiester bond at the 5′splice site, accompanied by formation of a new 5′-2′ phosphodiester bond linking the first nucleotide of the intron (the G of the 5′-GU-3′motif) with the internal adenosine. This means that the intron has now been looped back on itself to create a **lariat structure**. Lariat is a rope with running noose, for catching horses and cattle.

2. Cleavage of the 3′ splice site and joining of exons. These two reactions occur due to second transesterification reaction, which is promoted by the 3′-OH group attached to the end of the upstream exon. This hydroxyl group attacks the phosphodiester bond at the 3′ splice site, cleaving and so releasing the intron as the lariat structure. The free lariat structure is converted back to a linear RNA and degraded. At the same time the 3′end of the upstream exon joins to the newly formed 5′end of the downstream exon completing the splicing process.

The Splicing apparatus of GU-AG introns. The central components of the splicing apparatus for GU-AG introns are the small nuclear (sn) RNAs called U1, U2, U4, U5 and U6. These are short molecules [between 106 nucleotides (*e.g.*, U6) and 185 nucleotides (*e.g.*, U2) in vertebrates that associate with proteins to form *small nuclear ribonucleoprotein particles*(snRNPs) in the nucleus of eukaryotic cells (Fig. 8.8). The snRNPs, together with accessory proteins, attach to the transcript and form a series of complexes, the last one of which is the **spliceosome**, the structure within which the actual splicing reactions occur (**Smith** and **Valcarcel**, 2000).

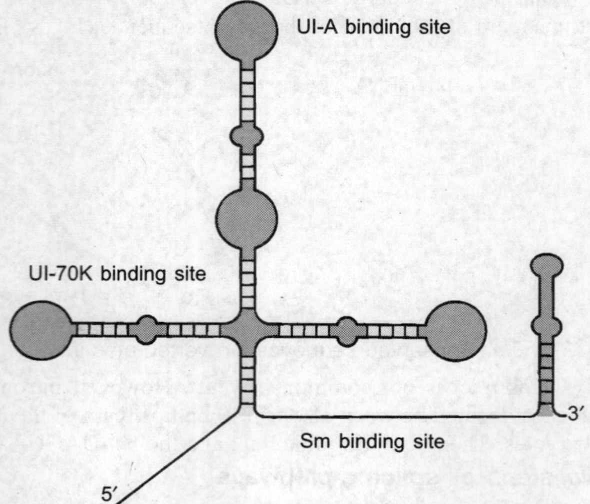

Fig. 8.8. Structure of U1-snRNP. The mammalian U1-snRNP comprises the 165-nucleotide UI-RNA plus ten proteins. Three of these (U1-70K, U1-A an U1-C) are specific to this snRNP, the other seven are Sm proteins that are found in all the snRNPs involved in splicing. U1-RNA forms a base-paired structure as shown. The U1-70K and U1-A proteins attach to two of the major stem-loops of this base-paired structure, the U1-C attaches via a protein-protein interaction. The Sm proteins attach to the Sm site (after Brown, 2002).

Role of splicing apparatus in splicing

Splicing involves the formation of following *three* transition structures (Fig. 8.9).

1. Formation of commitment complex. The splicing activity is initiated by the formation of the **commitment complex**. This complex comprises U1-snRNP, which binds to the 5′ splice site, partly by RNA-RNA base pairing, and the protein factors SFI, $U2AF^{35}$ and $U2AF^{65}$, which make protein-RNA contacts with the branch site, the polypyrimidine tract and the 3′ splice site, respectively.

2. Formation of the pre-spliceosome complex. The pre-spliceosome complex comprises the commitment complex plus U2-snRNP, the latter attached to the branch site. At this stage, an association, between U1-snRNP and U2-snRNP brings the 5′ splice site into close proximity with the branch point.

3. Formation of spliceosome. It is formed when U4/U6-snRNP (a single snRNP containing two snRNAs) and U5-snRNP attach to the pre-spliceosome complex. This results in addition interactions that brings the 3′ splice site close to the 5′ splice site and the branch point. All three key positions in the intron are now in proximity and the two transesterifications occur as a linked reaction, possibly catalyzed by U6-snRNP, completing the splicing process (Fig. 8.10).

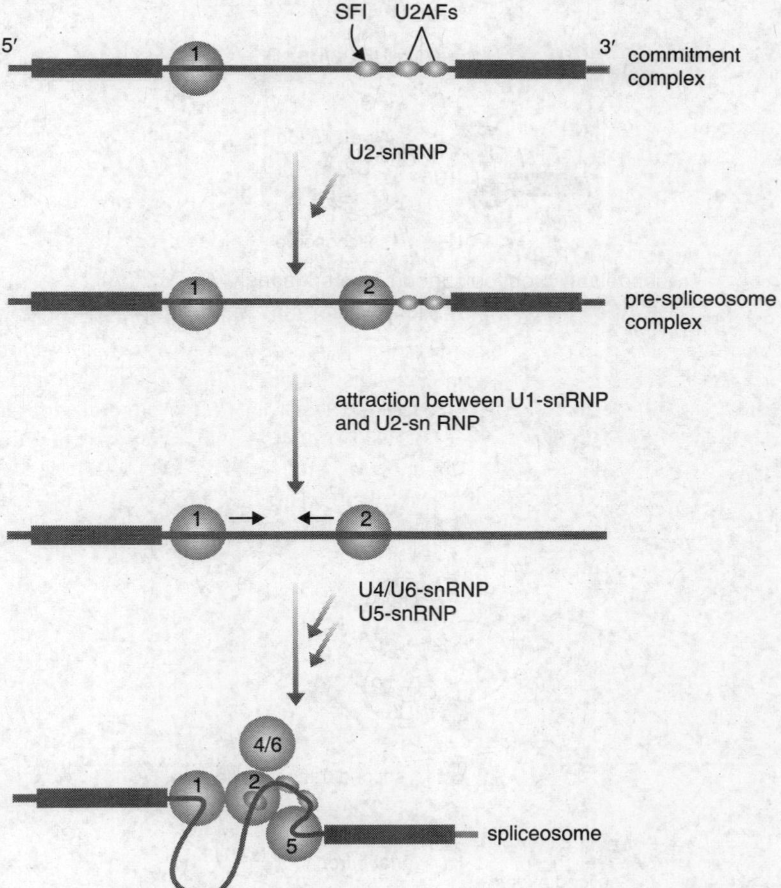

Fig. 8.9. The roles of snRNPs and associated proteins during splicing of pre-mRNA. During this little understood splicing process associations between the snRNPs are thought to bring the three critical parts of the introns– two splice sites and the branch point-into Close proximity (after Brown, 2002).

Selection of correct splice sites

A set of splicing factors called **SR Proteins** are found to be important in splice site selection. The SR proteins-so-called because their C-terminal domains contain a region rich in serine (abbreviations S) and arginine (R) were first implicated in splicing when it was discovered that they are components of the spliceosome. SR proteins have following functions:

1. They establish a connection between bound U1-snRNP and bound U2AF in the commitment complex (**Valcarcel** and **Green**, 1996). This indicates that SR proteins have a role in splice-site selection, since formation of the commitment complex being the critical stage of the splicing process, as this is the event that identifies which sites will be linked.

2. They interact with **exonic splicing enhancers (ESEs),** which are purine-rich sequences located in the exon regions of a transcript(**Blencowe** 2000). Our understanding about the role of ESEs and their counterparts, the **ESSs** or **Exonic splicing silencers (Del Gatto-Konczak** *et al.,* 1999) still is in a state of infancy, but their importance in controlling splicing is clear from the discovery that several human diseases, including one type of muscular dystrophy, are caused by mutations in ESE sequences.

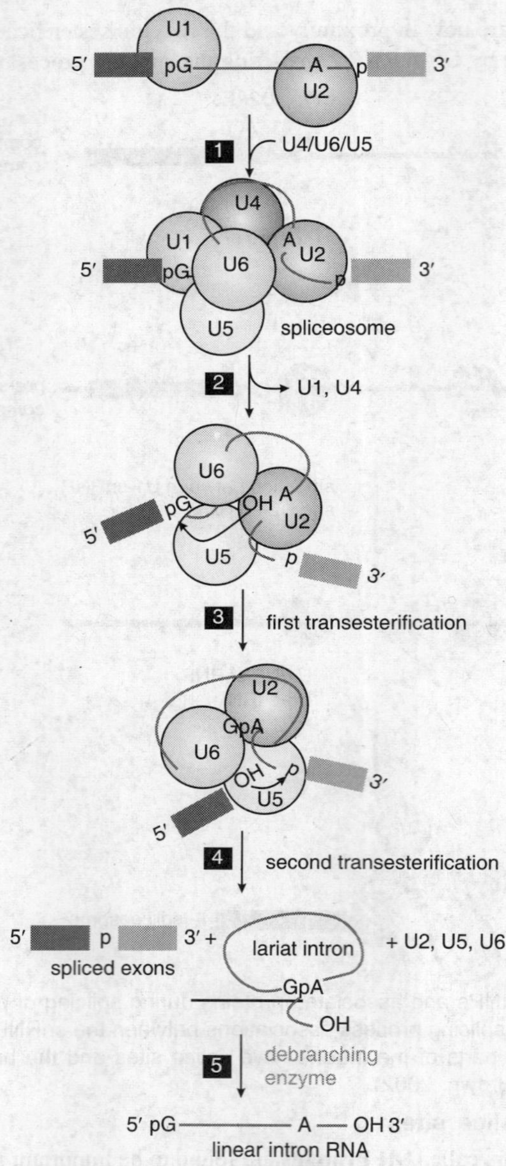

Fig. 8.10. Model of spliceosome-mediated splicing of pre-mRNA. 1.After U1 and U2 snRNPs associate with the pre-mRNA, via base-pairing interactions, a trimeric snRNP complex of U4, U5 and U6 joins the initial complex to form the spliceosome. 2. Rearrangements of base-pairing interactions between snRNAs converts the spliceosome into a catalytically active conformation and destabilizes the U1 and U4 snRNPs, which are released. 3. The catalytic core, thought to be formed by U6 and U2, then catalyzes the first transesterification reaction, forming the intermediate containing a 2',5'-phosphodiester bond. 4. Following further rearrangements between the snRNPs, the second transesterification reaction joins the two exons by a standard 3', 5'-phosphodiester bond and the remaining snRNPs. 5. The excised lariat intron is converted into a linear RNA by a debranching system (after Lodish *et al.,* 2004).

The sites of ESEs and ESSs indicates that assembly of the spliceosome is driven not simply by contacts within intron but also by interactions with adjacent exons. In fact, it is possible that an individual commitment complex is not assembled within an intron as shown in Figure 8.9, but initially bridges an exon (Fig. 8.11). This model hold two positive points: 1. it provides a means by which contact between an ESE or ESS and an SR protein could influence splicing; 2. it takes account of the large disparity between the lengths of exons and introns in vertebrates genes. In the human genome, for example, the exons have an average length of 145 bp compared with 3365 bp for introns (**IHGS**, 2001). Initial assembly of a commitment complex across an exon might therefore be less difficult task than assembly across longer introns.

3. Certain SR proteins, called **CASPs(=CTD-associated SR-like proteins**) or **SCAFs (=SR-like CTD-associated factors)** , form a physical connection between the spliceosome and the CTD of the RNA polymerase II transcription complex, and hence provide a link between transcript elongation and processing. As already discussed, like some of the polyadenylation proteins, it is probable that these **splicing factors** ride with the RNA polymerase II as this enzyme synthesizes the transcript, and are deposited at their appropriate positions at introns splice sites as soon as these are transcribed. Electron microscopy studies have shown-that transcription and splicing occur together.

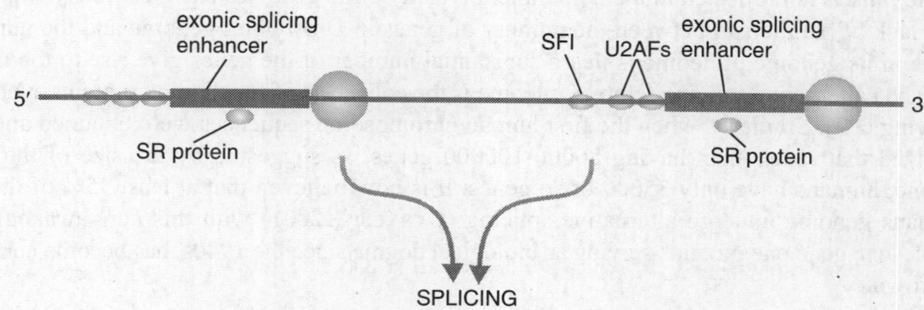

Fig. 8.11. An alternative model for assembly of the commitment complex. In this model, each individual commitment complex is built up across an exon, bringing the complex into close association with an exonic splicing enhancer or silencer and its attached SR proteins. Once complexes are formed over adjacent exons, splicing follows the pathway illustrated in Figure 8.8, the only difference being that the resulting spliceosome is made up of components of adjacent commitment complexes, rather than being derived from the single commitment complex shown in Figure 8.8. (after Brown, 2002).

Alternative splicing pathways

With the discovery of introns, it has been the prevailing concept that each gene always give rise to the same mRNA: in other words, there is a single **splicing pathway** for each primary transcripts (Fig. 8.12A). This assumption was found to be incorrect in the 1980s, when it was shown that the primary transcripts of some genes can follow two or more **alternative splicing pathways**, enabling a single transcript to be processed into related but different mRNAs and hence to direct synthesis of a range of proteins (Fig. 8.12 B). The alternative splicing is most common in higher animals. So while there are only three examples of alternative splicing in yeast *Saccharomyces cerevisiae,* there are much more examples in *Drosophila* and humans.

Presence of many alternative splicing pathways were became established when draft *Drosophila* sequence and sequence of microscopic nematode *Caenorhabditis elegans* were compared and found that fruit files have fewer genes than this nematode. Thus, the apparent greater physical complexity

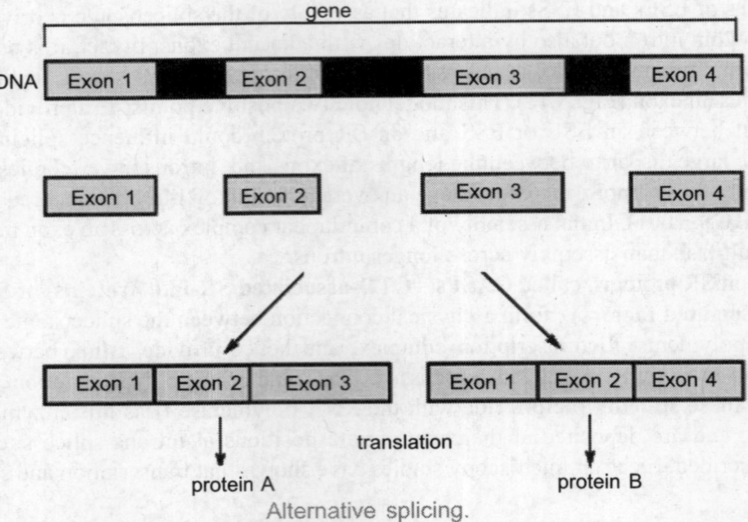

Alternative splicing.

of *Drosophila* is not reflected in the far simpler diversity of its genome. The most likely explanation for the lack of parallelism between the number of genes in *Drosophila* genome and the number of proteins in its genome proteome is that a substantial number of the genes give rise to the multiple proteins via alternative splicing (**Note.** Proteome is the collection of functioning proteins synthesized by a living cell). Similarly, when the first human chromosome sequences were obtained and it was recognized that rather than having 80000-100000 genes, as suggested by the size of the human proteome, humans have only 35000 or so genes. It is now believed that at least 35% of the genes in humans genome undergo alternative splicing (**Graveley**, 2001) with this new awakening, the principle 'one gene one protein', serving as biological dogma since the 1940s, has become completely overthrown.

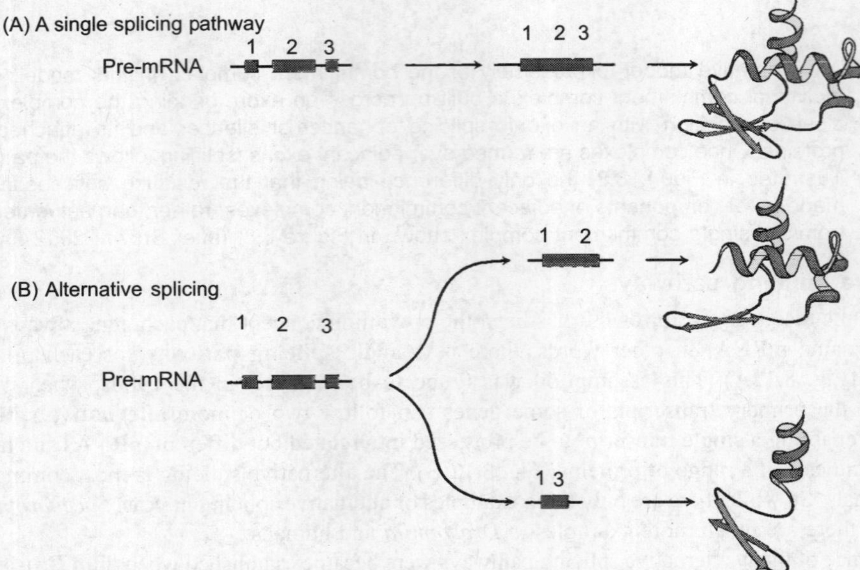

Fig. 8.12. Diagram explaining two modes of alternative splicing (after Brown, 2002).

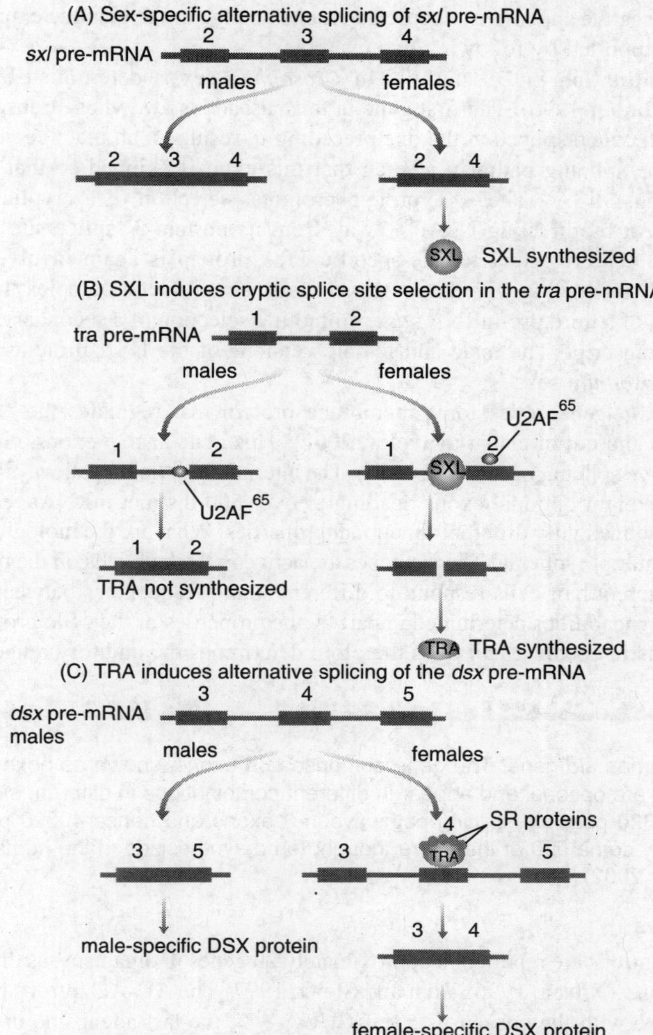

Fig. 8.13. Scheme of **Chabot** (1996) for expression of genes involved in sex determination of *Drosophila*. A–The cascade begins with sex-specific alternative splicing of the *sxl* pre-mRNA. In males all exons are present in the mRNA, but this means that a truncated protein is produced because exon 3 contains a termination codon. In females, exon3 is skipped, leading to full-length, functional SXL protein. B–In females, SXL blocks the 3'splice site in the first intron of the *tra* pre-mRNA. $U2AF^{65}$ is unable to locate this site and instead directs splicing to a cryptic site in exon2. This results in an mRNA that codes for a functional TRA protein. In males, there is no SXL protein so the 3'splice site is not blocked and a nonfunctional mRNA is produced. C–In males, exon4 of the dsx pre-mRNA is skiped. The resulting mRNA codes for a male-specific DSX protein. In females, TRA protein stabilizes the attachment of SR proteins to an exonic splicing enhancer located within exon 4, so this exon is not skipped, resulting in the mRNA that codes for the female specific DSX protein. The two versions of DSX protein are the primary sex determinants of the male and female physiologies. The female dsx mRNA ends with exon4 because the intron between exons 4 and 5 has not 5' splice site, meaning that exons 5 cannot be ligated to the end of exon4. Instead a polyadenylation site at the end of exon4 is recognized in females (after Brown, 2002).

Currently alternative splicing is regarded a crucial change in the genome expression pathway. This viewpoint is supported by following two examples:

1. **Sex determination in Drosophila.** In *Drosophila* sex is determined by an alternative splicing cascade (**Chabot** 1996). The first gene in this cascade is *sxl*, whose transcript contains an optional exon which, when spliced to the one preceding it, results in an inactive version of protein SXL. In females the splicing pathway is such that this exon is skipped so that functional SXL protein is made (Fig. 8.13A). The SXL protein promotes selection of a cryptic splice site in a second transcript, *tra*, by directing $U2AF^{65}$ away from its normal 3' splice site to a second site further downstream. The resulting female-specific TRA protein is again involved in alternative splicing, this time by interacting with SR proteins to form a multifactor complex that attaches to an ESE within an exon of a third pre-mRNA , *dsx*, promoting selection of a secondary, female-specific splice site in this transcript. The male and female versions of the DSX proteins are the primary determinants of *Drosophila* sex.

2. In human *slo* gene codes for a membrane protein that regulates the entry and exit of potassium ions into and out of cells (**Graveley,** 2001). This gene has 35 exons, eight of which are involved in alternative splicing events (Fig. 8.14). The alternative splicing pathways involve different combinations of the eight optional exons, leading to over 500 distinct mRNAs, each specifying a membrane protein with slightly different functional properties. What are the biological consequences of this example of multiple splicing. The *slo* genes are active in the hair cells on the basilar membrane of the cochlea. Different hair cells respond to different sound frequencies between 20 and 20 000 Hz, their individual capabilities determined in part by the properties of their Slo proteins. Alternative splicing of *slo* genes in cochlear hair cells therefore determines the auditory range of humans.

Fig. 8.14. The human *slo* gene. The gene comprises 35 exons, shown as boxes, eight of which (i.blue) are optional and appear in different combinations in different *slo* mRNAs. There are 40320 possible splicing pathways in 8 exons and hence 40320 possible mRNAs, but only some 500 of these are thought to be synthesized in the human cochlea (after Brown, 2002).

B. AU-AC Introns

The AU-AC introns are reported in approximately 20 genes in organisms as diverse as humans, plants and *Drosophila* (**Nilsen**, 1996; **Tarn** and **Steitz,** 1997) The AU-AC introns have a conserved branch site sequence with the consensus 5′-UCCUUAA-3′, the last adenosine in this motif being the one that participates in the first transesterification reaction. Thus, splicing pathway of AU-AC introns is very similar to that for GU-AG introns, but involves a different sets of splicing factors. Only the U5-snRNP is involved in the splicing mechanisms of both types of intron. The roles of U1-snRNP and U2–snRNP are taken by a previously discovered complex that had never been assigned a function, *e.g.*, U11/U12-snRNP, U4 atac/U6 atac-snRNP.

8.3. RNA EDITING

In the late 1980s, still another quite unexplected form of post transcriptional RNA processing was discovered in several organisms. In this form, referred to as **RNA editing, the nucleotide sequence of a pre-mRNA is actually changed prior to translation.** As a result, the ribonucleotide sequence of the mature RNA differ from the sequence encoded in the exons of the DNA from which the RNA was transcribed.

RNA editing is of following two types :

1. Substitution RNA editing;
2. Insertion or deletion RNA editing.

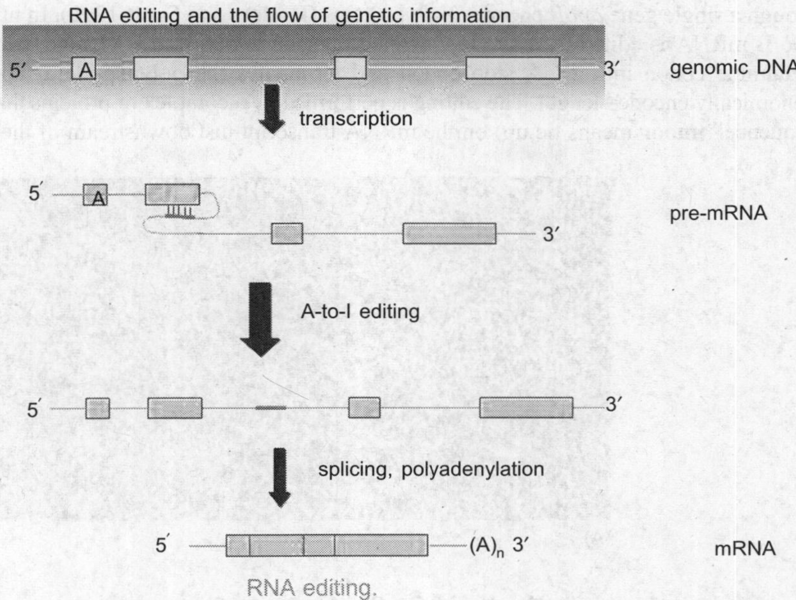

RNA editing.

1. Substitution RNA editing. In this process the identities of individual nucleotide bases are altered. Substitution editing is used in some nuclear-derived eukaryotic RNAs and is also very prevalent in mitochondrial and chloroplast RNAs transcribed in plants.

2.Insertion or deletion RNA editing. In this process nucleotides are added to or subtracted from the total number of bases.

Examples of RNA editing

1. The slime mold, *Physarum polycephalum* uses both substitution and insertion/deletion editing for its mitochondrial mRNAs.

2. RNA editing was particularly evident in mitochondrial proteins of a group of protozoan parasites, the trypanosomes (some of which cause African sleeping sickness); in one case, more than 50% of the nucleotides in the mRNA were added uridines (Us). Uridines were also **deleted** from the original sequence. These parasites had another curious trait- the existence of **minicircles** and **maxicircles** of DNA in specialized mitochondria, called **kinetoplasts**. In the average kinetoplasts, there are about 50 maxicircles and about 5000 minicircles, concentrated like chain links (Fig. 8.15A). The maxicircles contain genes for mitochondrial function. **Larry Simpson** and his colleagues showed in 1990, that both maxicircles and minicircles are templates for **guide RNA (gRNA),** RNA that guides the process of messenger editing.

The guide RNA forms a complement with the mRNA to be edited; however, the gRNA has the sequence complementary to that of the final mRNA, the one with bases added. Since the bases have not yet been added, a bulge occurs in the gRNA where the complement to be added is (Fig.8.15). The mRNA is then cleaved opposite the bulge by an enzyme **editing endonuclease**. A uridylate (U) is brought into the mRNA as complement to the adenine (A) with the enzyme **terminal-U-transferase.** An **RNA ligase** then closes the nick in the mRNA, which now has a uridylate added.

An exciting outcome of this research, aside from learning about a novel mechanism of messenger RNA processing, is the possibility of clinical rewards. Anytime there is a specialized pathway in a parasite not found in its host, it is possible to use that pathway to attack the parasite. Thus, this research might lead to new approaches of combating these trypanosome parasites.

3. The most well-studied examples of substitutional editing occur in mammalian nuclear encoded mRNA transcripts. For example, **apolipoprotein-B(apo-B)** exists in both a long and a short form,

though a single gene *apoB* encodes both proteins (**Hodges** and **Scott** 1992). In human intestinal cells, apo B mRNA is edited by a single C (called nucleotide 6666) to U change, which converts a CAA glutamine codon into a UA stop codon and terminates the polypeptide at approximately half its genomically encoded length. The editing is performed by a complex of proteins that bind to a "mooring sequence" (**moor** means tie up) on the mRNA transcript just downstream of the editing site.

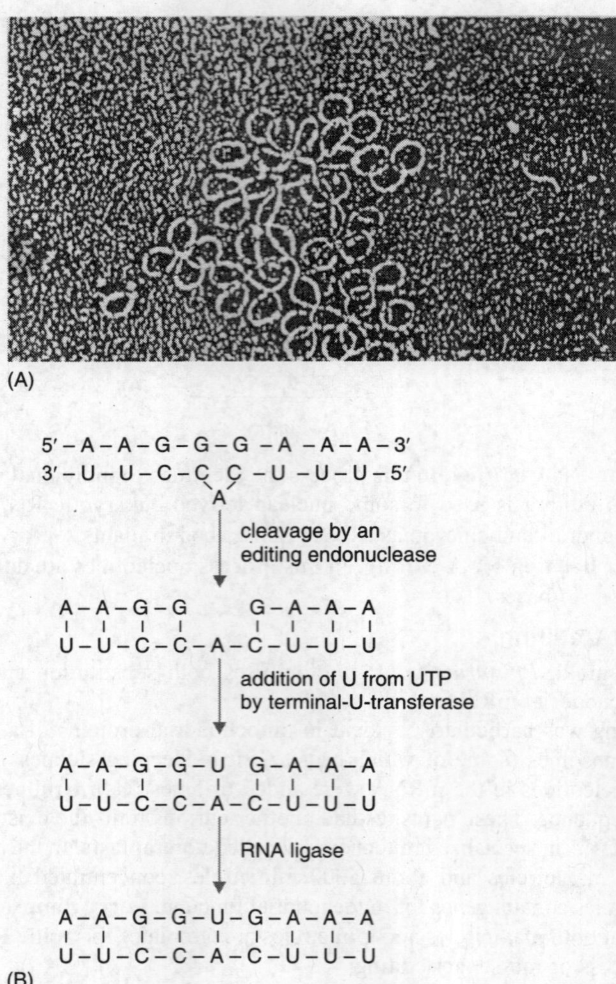

Fig. 8.15. RNA editing. A–870bp minicircles of DNA from *Leishmania tarentolae* (mitochondria). B– Mechanism by which a guide RNA is involved in the editing of a mRNA (after Tamarin, 2002).

The synthesis of subunits constituting the **glutamate receptor channels (Glu R)** in mammalian brain tissue is also affected by RNA editing. In this case, adenosine (A) to inosine (I) editing occurs in pre-mRNAs prior to their translation, during which I is read as guanosine (G). The enzymes produced by a family of three **ADAR** (Adenosine Deaminase Acting on RNA) genes are believed to be responsible for the editing of various sites within the glutamate channel subunits. The double-stranded RNAs required for editing by the **ADAR** enzymes are provided by intron-exon pairing of the GluR mRNA transcripts. The editing changes the physiological parameters (*i.e.*, solute permeability and desensitization response time) of the receptors containing the subunits.

The importance of RNA editing resulting from the action of ADARs is most apparent when one examine situations in which these enzymes have lost their functional capacity as a result of mutation. In several investigations, the loss of function was shown to have a lethal impact in mice. In one study, embryos heterozygous for a defective ADAR1 gene die during embryonic development as a result of a **defective haemopoietic system (Wang** *et al.,* 2000). In another study, mice with two defective copies of ADAR2 progress through development normally, but are prone to **epileptic seizures** and die while still in the weaning stage. Their tissues contain the unedited version of one of the Glu R products. The defect to death is believed to be in the brain. Heterozygote for the mutation are normal.

8.4. SELF-SPLICING AND RIBOZYMES

Sidney Altman in 1981 showed that RNA can have catalytic properties. In 1982, **Thomas Cech** and **Sidney Altman** discovered self-splicing by RNA ; both were awarded 1989 Nobel Prize in Chemistry. Working with an intron in the 35S ribosomal RNA precursor in the single-cell eukaryote (ciliated protozoan) *Tetrahymena thermophila*, Cech and his colleagues found that they could induce intron removal *in vitro* with no proteins present. A guanine-containing nucleotide (GMP; GDP or GTP) had to be present. Figure 8.16 diagram depicts how self-splicing occurs. The intron is acting as an enzyme; we call an RNA with catalytic or enzymatic properties a **ribozyme.**

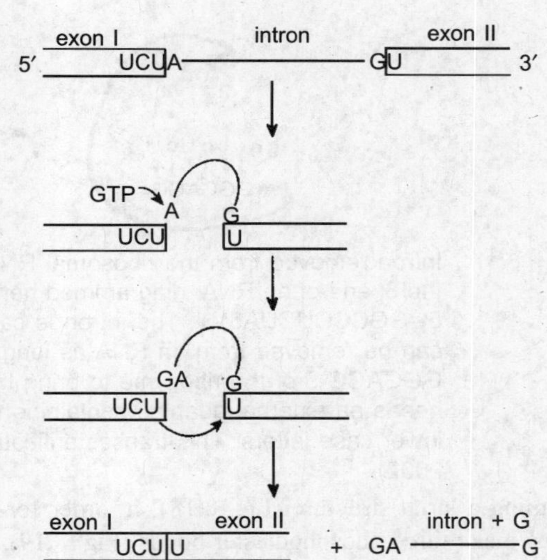

Fig. 8.16. Self-splicing of a ribsosomal RNA precursor in *Tetrahymena*. An external GTP is required. Two bonds transfer produce a shortened RNA and a free intron (after Tamarin, 2000).

During self-splicing, the U-A bond at the left (5′) side of the intron is transfered to the GTP. The U that is now unbounded displaces the G at the right (3′) side of the intron, reconnecting the RNA with a U-U connection and releasing the intron (Fig. 8.16). Since all bonds are reversible transfers (transesterifications) rather than new bonds, no external source of energy is required. Self-splicing introns of this type are called **group I introns**.

Although the first enzymatic activity of the ribozyme is its own removal, its secondary structure after removal gives it the ability to further catalyze reactions (Fig. 8.17). The reactions that ribozymes catalyze are transesterifications and the hydrolysis reaction of splitting an RNA molecule into two parts. Ribozymes can also perform other functions, including peptide bond formation. Currently, at least seven different classes of ribozymes are known, based on their enzymatic properties (Table 8.2). A ribozyme that can split other RNAs and that occurs in small plant pathogens (viruses) is called **hammerhead ribozyme** because of its shape.

Self-splicing has also been found in genes in the mitochondria of yeast. These introns are referred to as **group II introns** because they use a different mechanism of splicing that does not require an external nucleotide. Instead, the first bond is transferred within the intron to an adenosine,

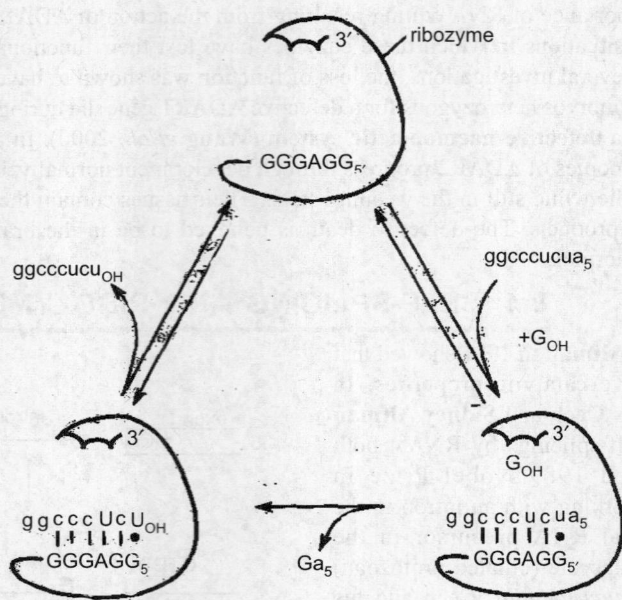

Fig. 8.17. Intron removed from the ribosomal RNA of *Tetrahymena* can catalyze the removal of the 3′ end of an RNA, diagrammed here as five AMP residues (a_5) from the sequence 5′–GGCCCUCUA5′-3′. The intron is called the *Tetrahymena* ribozyme. Any sequence can be removed from an RNA as long as there is a sequence complementary to the GGGAGG-5 of the ribozyme to bring the RNA into position. In reaction depicted here needs an external guanine containing nucleotide (G_{OH}) , substrate nucleotides are in lower case letters. This transesterification requires no external energy (after Tamarin, 2002).

forming a lariat structure (Fig. 8.18). In order for the lariat to form, the ribose of the adenosine must make three phosphodiester bonds (Fig. 8.19).

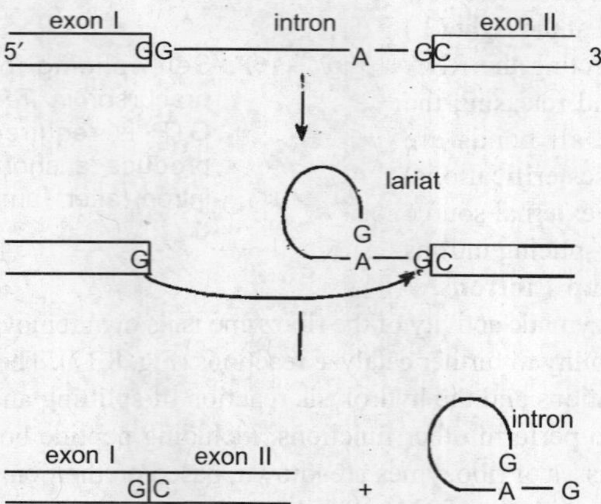

Fig. 8.18. Self-splicing of a group II intron results in a lariat configuration of the released intron. No external GTP is required since the first bond transfer takes place with an internal nucleotide, forming the loop of the lariat. A second bond transfer releases the lariat-shaped intron (after Tamarin, 2002).

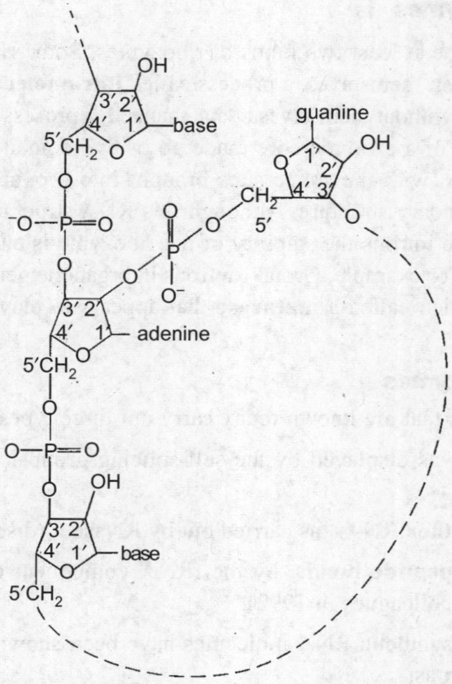

Fig. 8.19. The lariat branch point (see Fig. 8.18), formed during removal of a group II intron, occurs as three phosphodiester bonds form at the same ribose sugar. (The lariat loop is formed by the 2'- phosphodiester bond (after Tamarin, 2002).

Table 8.3. Examples of ribozymes.

Ribozyme	Description
1. Self-splicing introns	Some introns of Groups I, II and II splice themselves by an autocatalytic process. However, the splicing pathway of GU-AG introns includes at least some steps that are catalyzed by snRNAs (Newman, 2001)
2. Ribonuclease P	The enzyme that creates the 5′ ends of bacterial tRNAs consists of an RNA subunit and a protein subunit, with the catalytic activity residing in the RNA. (**Note.** The RNA catalyzed reactions are found to follow **Michaelis-Menten Kinetics,** which provides further evidence to the fact RNA component of ribonuclease P is acting like a true enzyme; see Becker *et al.,* 2007).
3. Ribozomal RNA	The peptidyl transferease activity required for peptide bond formation during protein synthesis is associated with the 23S rRNA of the large subunit of the ribosome.
4. $tRNA^{Phe}$	Undergoes self-catalyzed cleavage in the presence of divalent lead ions.
5. Virus genome	Replication of the RNA genomes of some viruses involves self-catlyzed cleavage of chains of newly synthesized genomes linked to head to tail. Examples are the plant viriods and virusoids and the animal hepatiis delta virus. These viruses form a diverage group with the self-cleaving activity specified by a variety of different base-paired structure called, e.g., hammerhead.

Structure of Ribozymes

There appear to be at least two kinds of ribozymes. Some ribozymes have folded structures and catalyze reactions on themselves, a process called **intramolecular catalysis**. Other ribozymes act on other molecules, without themselves being changed, a process called **intermolecular catalysis.** The ribozyme consists of a catalytic core made up of two domains, each one compirising two of the base-paired regions, with the splice sites brought into proximity by interaction between two other parts of the secondary structure. Although this RNA structure is sufficient for splicing, it is possible that with some introns the stability of the ribozyme is enhanced by non-catalytic protein factors that bind to it. For example, Group I introns in organelle genes, many of these containing an ORF coding for a protien called a **maturase** that appears to play a role in splicing (see **Brown,** 2002).

Functions of Ribozymes

Those ribozymes that are known today carry out three types of biochemical reactions:

1. Self-cleavage, as displayed by the self-splicing group I, II and III introns and by some virus genomes.

2. Cleavage of other RNAs, as carried out by RNase P (discovered by **Altman** *et al.,* 1983)

3. Synthesis of peptide bonds, by the rRNA component of the ribosome (discovered by **Harry Noller** and his colleagues in 1992).

In the test tube, synthetic RNA molecules have been shown to carry out other biologically relevant reactions, such as;

1. Synthesis of ribonucleotides (**Unran** and **Bartel,** 1998).

2. Synthesis and copying of RNA molecules (**Exland** and **Bartel,** 1996: **Johnston** *et al.,* 2001)

3. Transfer of an RNA-bound amino acid forming a dipeptide, in a manner analogous to the role of tRNA in protein synthesis (**Lohse** and **Szostak,** 1996).

The discovery of these catalytic properties solved the polynucleotide-polypeptide dilemma by showing that the **first biochemical systems could have been centered entirely** on RNA (**Bartel** and **Unran**, 1990).

Role of Ribozymes in the Origin of Life

Ideas about the RNA world during chemical evolution and origin of life have taken shape in recent years (**Robertson** and **Ellington,** 1998). Biologists now envisage that RNA molecules initially replicated in a slow and haphazard fashion simply by acting as templates for binding of complementary nucleotides which polymerized spontaneously. This process would have been very in accurate so a variety of RNA sequences would have been generated, eventually leading to one or more with incipient ribozyme properties that were able to direct their own, more accurately self-replication. It is possible that a form of natural selection operated so that the most efficient replicating systems began to predominate. A greater accuracy in replication would have enabled RNAs to increase in length without losing their sequence specificity, providing the potential for more sophisticated catalytic properties, possibly culminating in structures as complex as present day Group-I introns (e.g. *Tetrahymena* tRNA intron) and ribosomal RNAs (Fig. 8.2).

8.5. CHANCES OF ERRORS IN GENE SPLICING

There are certain topological problems which might pose chances of errors in gene splicing. Following two types of errors can commonly occur during splicing of a gene :

1. The first of these is the substancial distance that might lie between splice sites, possibly a few tens of kb, representing 100nm or more if the mRNA is in the form of linear chain. A means is therefore needed to bring the splice sites into proximity.

2. The second topological problem concerns **selection** of the correct splice site. All splice sites are similar, so if a pre-mRNA contains two or more introns then there is the possibility that the wrong splice sites could be joined, resulting in **exon skipping,** i.e., the loss of an exon from the mature mRNA (Fig. 8.20A). Equally unfortunate would be selction of a **criptic splice site,** a site within an intron or exon that has sequence similarity with consensus motifs of real splice sites (Fig. 37.20 B). Cryptic sites are present in most pre-mRNAs and must be ignored by splicing apparatus (see **Brown**, 2002).

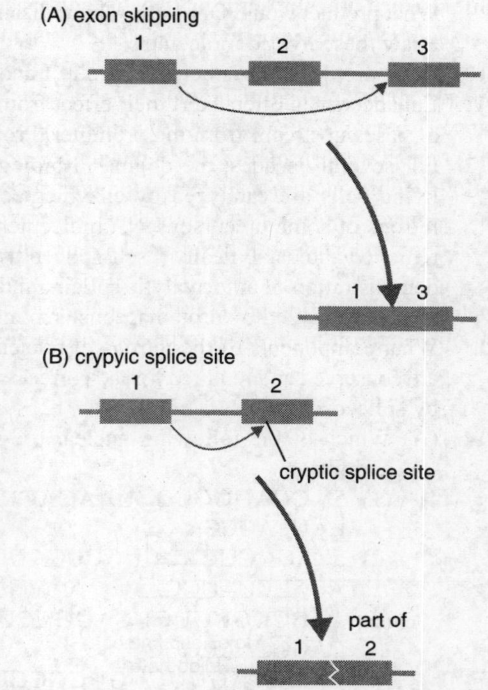

Fig. 8.20. Two aberrant forms of splicing. **A**-In exon skipping the aberrrant splicing results in an exon being lost from mRNA. **B**–When a cryptic splice site is selected, part of an exon might be lost from the mRNA (after Brown, 2002).

QUESTIONS AND PROBLEMS

1. These questions refer to eukaryotic RNA:
   (a) What is a cap? (b) At which end of mRNA is the poly (A)?
   (c) Are there eukaryotic mRNA molecules that do not contain either feature?
2. (a) What are intervening sequences or introns?
   (b) What is mRNA splicing?
3. What is heterogenous nuclear messenger RNA (hnRNA)? What are small nuclear ribonucleo-proteins?
4. What chemical groups are present at the origin and terminus of a molecule of mRNA that has just been synthesized?
5. (i) What is mRNA?
   (ii) What is polycistronic mRNA?
6. Many eukaryotic genes contain noncoding introns that separate the coding sequences or exons of these genes. At what stage during the expression of these mosaic genes are the noncoding intron sequences removed ?
7. In what ways are ribosomes and spliceosomes similar? Different?
8. What would be the effect on the final protein product if an intervening sequence were removed with an extra base? One base? Too few?

9. What product would DNA-RNA hybridization produce in a gene with five introns? No introns? Draw these hybrid molecules.
10. What are differences between group I and group II introns?
11. Enhancers can often exert their effect from a distance; some enhancers are located thousands of bases upstream from the promoter. Propose an explanation to account for this.
12. For several decades, the dogma in biology has been that molecular reactions occurring in living cells are catalyzed by enzymes composed of polypeptides. We now know that the introns of some precursor RNA molecules such as the rRNA precursors in *Tetrahymena* are removed autocatalytically ("self-spliced") with no involvement of any protein. What does the demonstration of autocatalytic splicing indicate about the dogma that biological reactions are always catalyzed by proteinaceous enzymes?
13. What components of the introns of nuclear genes that encode proteins in higher eukaryotes are conserved and required for the correct excision of intron sequences from primary transcripts by spliceosomes?
14. (A) Which of the following nuclear pre-mRNA nucleotide sequences potentially contains and intron?
    (1) 5′–CCAUGGCGCUAACACUGCCAAUUGGCCAAUACUGACCUGAAG-CAUCAGCCAA-3′
    (2) 5′-AGUCUCAUCUGUCCAUUGACUUCGAAACUGAAUCGUAACUCCU-ACGUCUAUGGA-3′.
    (3) 5′–GCUGUUUGUCAUGACUGACUGGUCACUAUCGUACUAACCUGU-CAUGCAAUGUC-3′
    (4) 5′–AGCAGUUCUGUCGCCUCGUGGUGCUGGCUGGCCCUUCGUCG-CUCGGGCUUAGCUA-3′
    (5) 5′-UUCGCAUGACGUACUUCUGAGACUACUACUACUACUAACGCAUCG-AGUCUCAA-3′
    (B) One of the five pre-mRNAs shown above is a likely candidate for splicing out an intron sequence. What mRNA nucleotide sequence would one expect to result from this splicing event?

Answers to Problems

1. (a) A terminal structure in which a methylated guanosine is in 5'-5'-triphosphate linkage at the 5'terminus of mRNA.
   (b) The 3' –OH end.
   (c) All mRNA molecules (except those of several viruses) are capped. Some mRNA molecules lack the poly (A) tail.
2. (a) Untranslated sequences that interrupt the coding sequence of a transcript and that are removed before translation begins.
   (b) Removal of introns.
6. The entire nucleotide-pair sequences-including the introns-of the genes are transcribed by RNA polymerase to produce primary transcript that still contain the intron sequences. The intron sequences are then "spliced out"off the primary transcripts to produce the mature, functional RNA molecules. In the case of protein encoding nuclear genes of higher eukaryotes, the introns are "spliced out "by complex ribosome-like macro-molecular structures called **spliceosomes.**
7. Ribosomes and spliceosomes both play essential roles in gene expression, and both are complex macromolecular structures composed of RNA and protein molecules. Ribosomes are located in cytoplasm, and spliceosomes in the nucleus. Ribosomes are larger and more complex than spliceosomes.

8. Removing one base too many or too few would result in shift in the reading frame during translation, thus radically altering the protein product.
9. Five introns

10. Group I introns are self-splicing introns that require a guanine (G)-containing nucleotide for splicing. Group II introns are similar but do not require an external nucleotide for splicing. Group I and II introns are released as linear and lariat-shaped molecules respectively.
11. Enhancers (E) bind activator proteins that also bind proteins of the polymerase at the promoter (P). In some eukaryotic genes, there are many enhancers, allowing numerous levels of control and forcing enhancers further and further upstream. For activators bound to enhancers to bind polymerase proteins, the DNA must lost around.

12. "Self splicing"of RNA precursors demonstrates that RNA molecules can also contain catalytic sites; this property is not totally restricted to proteins.
13. The introns of protein encoding nuclear genes of higher eukaryotes almost without exception begin (5′) with GT and end (3′) with AG. In addition, the 3′ subterminal A in the "TACTAAC box" is completely conserved; this A is involved in bond formation during intron excision.
14. (a) Sequence 5. It contains the conserved intron sequences: a 5′ GU, a 3′ AG, and UACUAAC internal sequence providing a potential bonding site for intron excision. Sequence 4 has a 5′GU and 3′ AG, but contains no internal A for the bonding site during intron excision.
    (b) 5′ – UAGUCUCAA-3′ ; the putative intron from the 5′ GU through the 3′AG has been removed.

MULTIPLE CHOICE QUESTIONS

Choose the correct answer from the four alternatives given.

1. Nuclear DNA sends information for protein synthesis through
   (*a*) tRNA
   (*b*) mRNA
   (*c*) rRNA
   (*d*) All the above
2. The RNA which carries the genetic message from the nucleus to the ribosome is
   (*a*) hn mRNA
   (*b*) mRNA
   (*c*) tRNA
   (*d*) rRNA
3. The type of RNA specifically responsible for directing the proper sequence of amino acids in protein synthesis is
   (*a*) chromosomal RNA

(*b*) rRNA
(*c*) tRNA
(*d*) mRNA

4. Maximum formation of mRNA occurs in
   (*a*) cytoplasm (b) nucleolus
   (*c*) ribosome (d) nucleoplasm
5. The ribozyme is a RNA
   (*a*) without sugar
   (*b*) with extra phosphate
   (*c*) with enzymatic property
   (*d*) without phosphate
6. mRNA is complimentary copy of
   (*a*) tRNA
   (*b*) rRNA
   (*c*) ribosomal DNA
   (*d*) a single strand of DNA
7. The cap structure of mRNA is of methyl GTP and tail of
   (*a*) methionine
   (*b*) formylmethionine
   (*c*) poly A
   (*d*) UAG codon
8. Synthesis of mRNA molecule is terminated by ____factor
   (*a*) alpha (*b*) beta
   (*c*) sigma (*d*) rho
9. Which RNA is having least age?
   (*a*) mRNA (*b*) tRNA
   (*c*) rRNA (*d*) none of these

Answers

1. (*b*)	2. (*b*)	3. (*d*)	4. (*d*)	5. (*c*)
6. (*d*)	7. (*c*)	8. (*d*)	9. (*a*).	

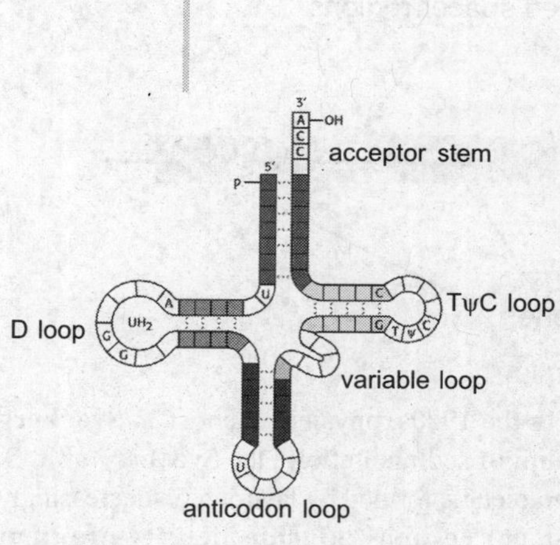

CHAPTER

Ribosomal RNA (rRNA) and Transfer RNA (tRNA)

Chapter Contents

9.1. RIBOSOMAL RNA (rRNA)

Ribosomes are cell organelles, which function as cell's protein factories. They provide for the sites of protein synthesis involving translation of the genetic information encoded in the messenger RNA (mRNA). When not engaged in protein synthesis, the ribosomes in a bacterial cell are always free, but in an eukaryotic cell, these may be found either free in the cytosol or attached to membrane of ER (called rough endoplasmic reticulum).

The initial progress in understanding the detailed structure of the ribosome came not from observing them with the electron microscope but by analyzing their components by **ultracentrifugation**. Ribosomes, as well as other small particles and molecules, are measured in units that describe their rate of sedimentation during density gradient centrifugation in sucrose. This technique gives information on size and shape (due to the speed of sedimentation) while simultaneously isolating the molecules. Isolation by centrifugation in sucrose is a relatively gentle isolation technique; the molecule still retain their biological

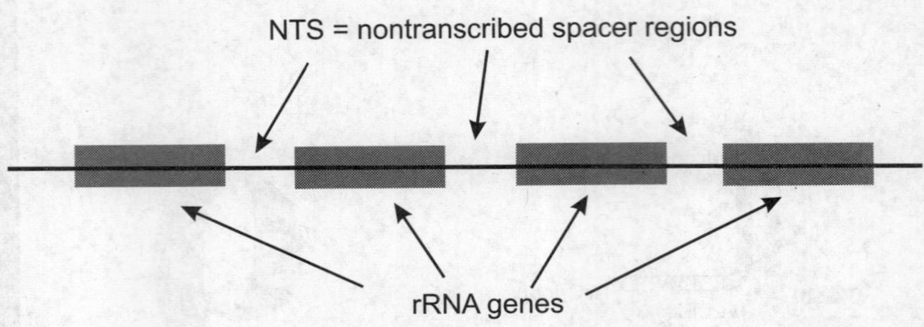

Ribosomal RNA genes.

properties and can be use for further experimentation. In the 1920s, physical chemist **T. Svedberg** developed ultracentrifugation, giving his name to the unit of sedimentation : the **Svedberg unit, S.**

In **sucrose density-gradient centrifugation,** the gradient is formed by layering on decreasingly concentrated sucrose solutions. In a related technique, **cesium chloride density-gradient centrifugation,** the gradient develops during centrifugation. The sucrose centrifugation is stopped after a fixed time, whereas in the cesium chloride technique, the system spins until it reaches equilibrium. The sucrose method tends to be more rapid. Samples can be isolated from a sucrose gradient by punching a hole in the bottom of the tube and collecting the drops in sequentially numbered containers. The first (lowest-numbered) containers will contain the heaviest molecules (with the highest S values).

Components of Ribosome

Intact ribosomes have sedimentation coefficients of 80S for eukaryotes and 70S for bacteria, and each can broken down into smaller components (Fig. 9.1):

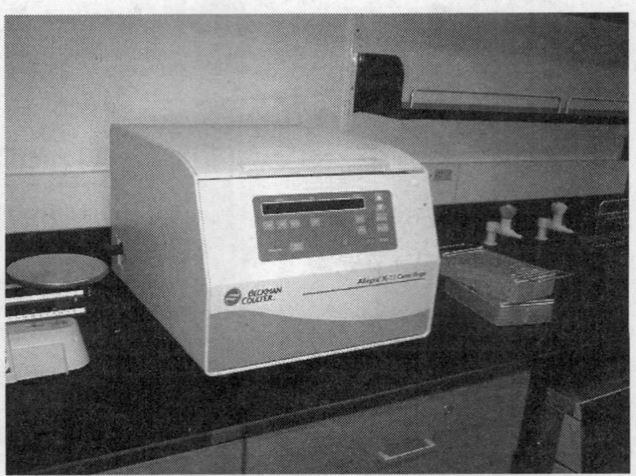

An analytical centrifuge.

1. Each ribosome comprises two subunits of unequal size. In eukaryotes these subunits are 60S and 40S, in bacteria they are 50S and 30S. Note that sedimentation coefficients are not additive because they depend on shape as well as mass; it is perfectly acceptable for the intact ribosome to have a S value less than the sum of its two subunits.

2. The large subunit contains three rRNAs in eukaryotes (the 28S, 5.8S and 5S rRNAs). In bacteria the equivalent of the eukaryotic 5.8 rRNA is contained within the 23S rRNA.

3. The small subunit contains a single r RNA in both types of organisms : an 18S rRNA in eukaryotes and a 16S rRNA in bacteria.

4. Both subunits contain a variety of **ribosomal proteins,** the numbers have been shown in figure 9.1. The ribosomal proteins of the small subunit are called S1, S2, S3, etc.; those of the large subunit are called L1, L2, etc. There is just one of each protein per ribosome, except for L7 and L12, which are present as dimers.

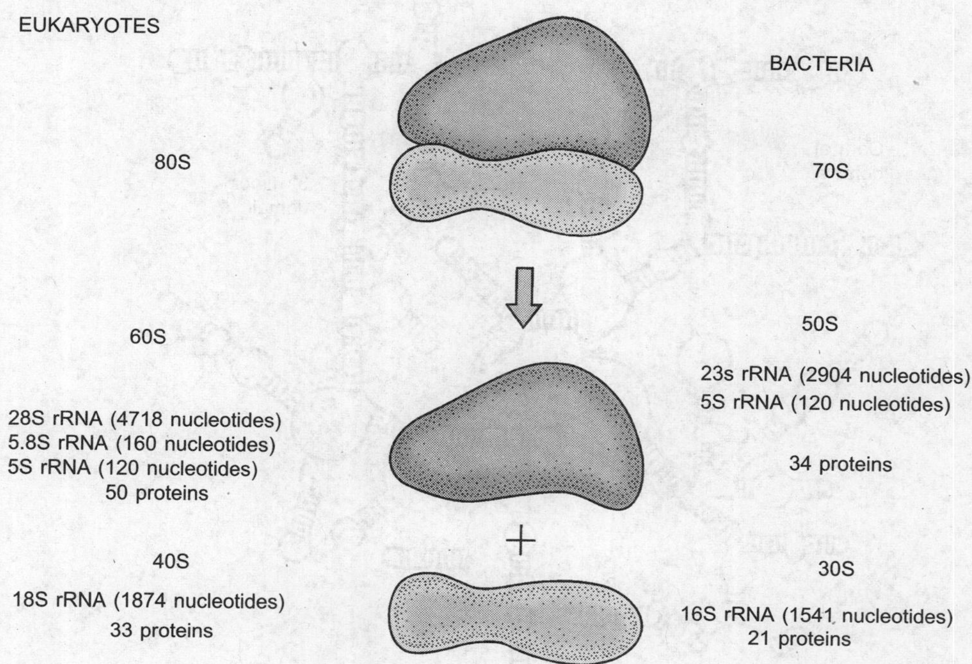

Fig. 9.1. The comparison of eukaryotic and bacterial ribosomes. The details refer to a 'typical' eukaryotic ribosome and the *Escherichia coli* ribosome. Variations between different species mainly concern the numbers of ribosomal proteins (after Brown, 2002).

Investigation of the Fine Structure of Ribosome

Once the basic composition of eukaryotic and bacterial ribosomes had been worked water, attention was focused on the way in which the various molecules of rRNAs and proteins fit together. Important information was provided by the first rRNA sequences, comparisons between these identifying conserved regions that can base-pair to form complex two-dimensional structures (Fig.9.2). This suggested that the rRNAs provide a scaffolding within the ribosome, to which the proteins are attached, an interpretation that under-emphasizes the active role that rRNAs play in protein synthesis.

Much of the subsequent research has concentrated on the bacterial ribosome, which is smaller than the eukaryotic version and available in large amounts from extracts of cells grown to high density in liquid cultures. Following technical approaches have been used to study the bacterial ribosome.

1. **Nuclease protection studies** enable contacts between rRNAs and proteins to be identified.

2. **Protein-protein crosslinking** identifies pairs or groups of proteins that are located close to one another in the ribosome.

3. **Electron microscopy** has gradually become more sophisticated, enabling the overall structure of the ribosome to be resolved in greater detail. Innovations such as **immunoelectron microscopy**, in which ribosomes are labeled with antibodies specific for individual ribosomal proteins before examination, have been used to locate the positions of these proteins on the surface of the ribosome.

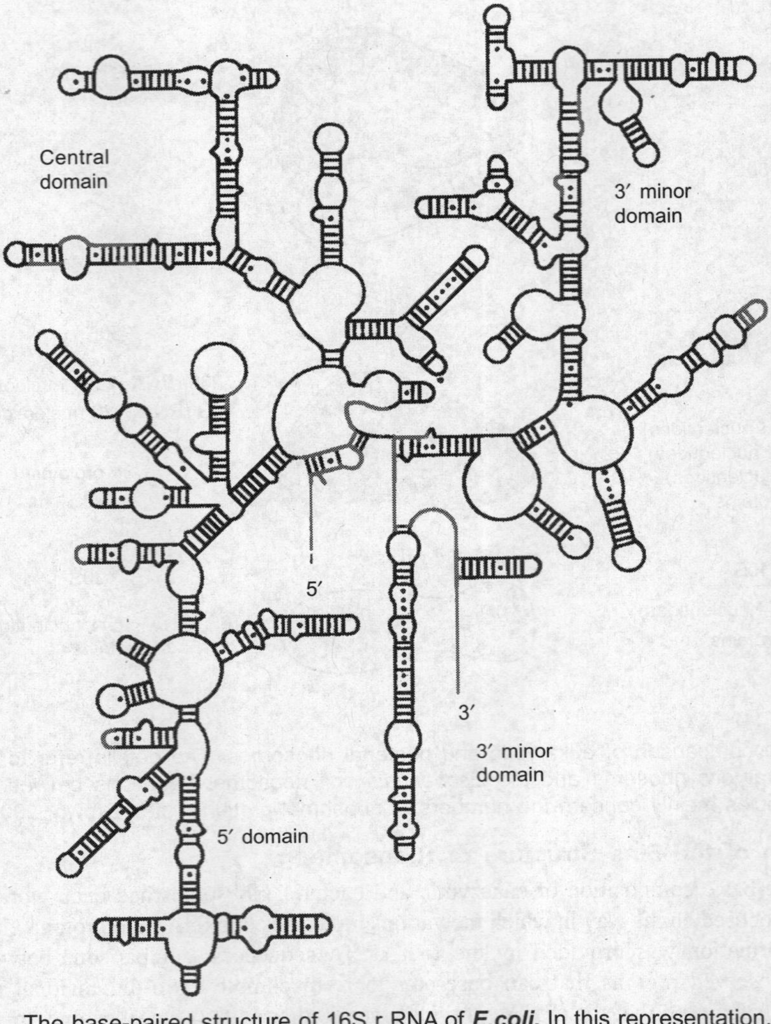

Central domain

3′ minor domain

5′

3′

3′ minor domain

5′ domain

Fig.9.2. The base-paired structure of 16S r RNA of ***E.coli***. In this representation, standard base pairs (G-C, A-U) are shown as bars; non-standard base pairs (e.g., G-U) are shown as dots (after Brown, 2002).

4. Site-directed hydroxyl radical probing makes use of the ability of Fe (II) ions to generate hydroxyl radicals that cleave RNA phosphodiester bonds located within 1nm of the site of radical production. This technique was used to determine the exact positioning of ribosomal protein S5 in the ribosome of *E.coli.* Different amino acids within S5 were labeled with Fe(II) and hydroxyl radicals induced in the reconstituted ribosomes. The positions at which 16S rRNA was cleaved were then used to infer the topology of the rRNA in the vicinity of S5 protein (Fig. 9.3; **Heilek** and **Noller**, 1996).

X-ray crystallographic apparatus.

5. X-ray crystallography. In recent years above discussed techniques have been increasingly supplemented by X-ray crystallography which has been responsible for the most exciting insights into ribosome structure. Analyzing the bulk amounts of X-ray diffraction data that are produced by crystals of an object as large as a ribosome is a enormous task, particularly at the level needed to obtain a structure that is detailed enough to be informative about the way in which the ribosome works (**Pennisi**, 1999).

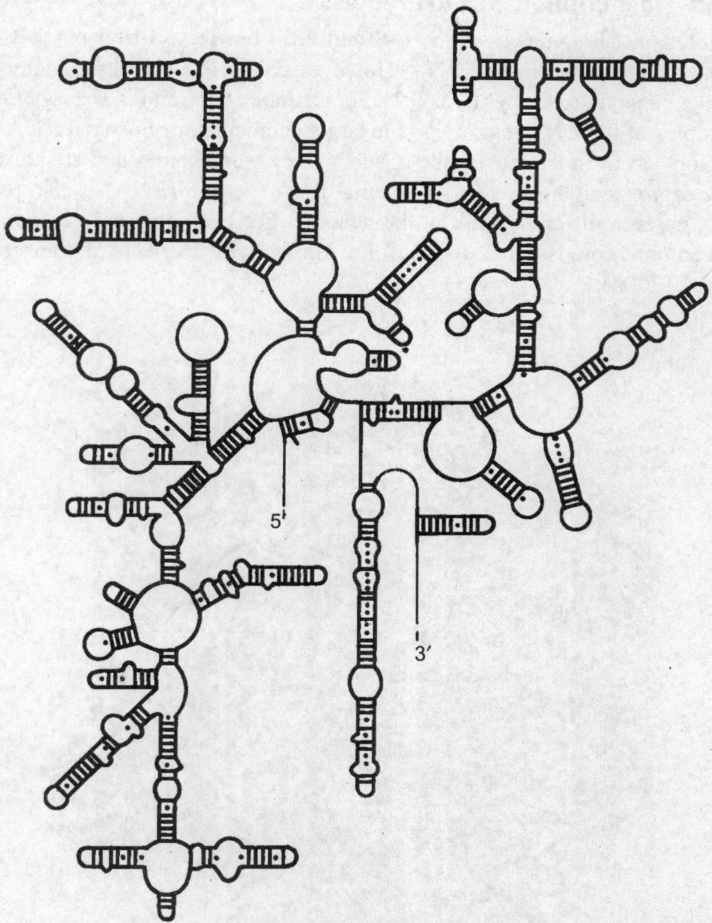

Fig. 9.3. Position within the 16S rRNA of *E.coli* that form contacts with the ribosomal protein S5. The distribution of the contact positions (shown in red) for this single ribosomal protein emphasizes the extent to which the base-paired secondary structure of the rRNA is further folded within the three-dimensional structure of the ribosome (after Brown, 2002).

Box 9.1.

By the help of X-ray diffraction, details of following ribosomal features have been revealed by different molecular biologists:

1. Binding of ribosomal proteins to their segments of r RNA : **Conn** *et al.,* 1999; **Agalarov** *et al.,* 2000).

2. Structure of the large and small subunits of ribosome : **Ban** *et al.*, 2000; **Wimberly** *et al.*, 2000.

3. Entire bacterial ribosome attached to mRNA and tRNA : **Yusupov** *et al.,* 2001.

In *E.coli,* all three ribosomal RNA segments are transcribed as a single long piece of RNA that is then cleaved and modified to form the final three pieces of RNA (*i.e.*, 16S, 23S and 5S). The region of DNA that contains the three ribosomal RNA molecules also contains genes for four tRNAs (Fig.9.5). There occur about ten copies of this region in each chromosome of *E.Coli* (see *Albert et al*., 2002). The occurrence of three r RNA segments on the same piece of RNA ensures a final ratio of 1:1:1, the ratio needed for ribosomal construction.

Nucleolus and Transcription in Eukaryotes

In eukaryotes, all rRNAs (*i.e.*, 5.8S, 18S and 28S) but the 5S ribosomal RNA section are transcribed as part of the same piece of RNA. However, eukaryotic cells have many copies of these rRNA genes, depending on the species (Box 9.1). For example, the fruit fly, *Drosophila melanogaster*, has about 130 copies of the DNA region that the larger segments of ribosomal RNA are transcribed from. These regions occur in tandem on the X and Y sex chromosomes and are known collectively as the *nucleolar organizer*. The smallest ribosomal RNA (i.e., 5S rRNA) is also produced from a duplicated gene, but at a different point in the genome. For example, in *D. melanogaster*, the 5S rRNA is produced on chromosome 2. It is still not known why 5S rRNA is transcribed separately (see **Alberts** *et al*., 2002)..

Fig. 9.4. This picture shows the molecular construction of 70S ribosome of the bacterium *Thermus thermophilus.* The small subunit is at the top, with the 16S rRNA in light blue and the small subunit ribosomal proteins in dark blue. The large subunit rRNAs are in grey and the proteins in purple. The golden area is the A site-the point at which aminoacylated tRNAs enter the ribosome during protein synthesis. This site and most of the region within which protein synthesis occurs, is located in the cleft between the two subunits (after Brown, 2002).

Fig. 9.5. The *E. coli* transcript that contains the three ribosomal RNA segments also contains four tRNAs and some spacer RNA that separates the tRNA are r RNA genes (after Tamarin, 2002).

Box 9.2.

Human cells contain about 200 rRNA gene copies per haploid genome. These are spread out in small clusters on five different chromosomes. Cells of the frog *Xenopus* also contain about 600 rRNA gene copies per haploid genome in a single cluster on one chromosome.

As we know that the eukaryotes have three RNA polymerases (enzymes). Eukaryotic RNA polymerase I (or polymerase A) transcribes only the nucleolar organizer DNA (*i.e.*, most of rRNAs except 5S). RNA polymerase II (or polymerase B) transcribes most genes (*i.e.*,genes for mRNAs). RNA polymerase III (or polymerase C) transcribes small genes, primarily the 5S rRNA gene and tRNA genes. Primase do primer synthesis during DNA replication. In addition, mitochondria, chloroplasts and some phages have other RNA polymerases.

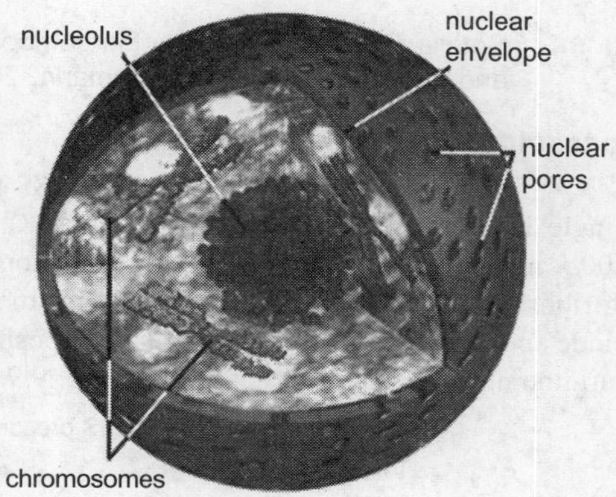

The Cell Nucleus.

At the nucleolar organizer, the *nucleolus* forms as dark globule found in eukaryotic nuclei. Nucleolus is the place where ribosomes are assembled. The various ribosomal proteins that has been synthesized in the cytoplasm migrate to the nucleus and eventually to the nucleolus, where, with the final forms of the rRNA, they are assembled into ribosomes.

In the nucleolar organizer, an untranscribed region of spacer DNA separates each repeat of the large ribosomal DNA gene. This has been shown in Figure 9.6 and diagrammed in Figure 9.7. In the electron micrograph (Fig. 9.6), the polarity of transcription is evident from the short RNA at one end of the transcribing segment and the long RNA at the other ends, with a uniform gradation between. Notice the many RNA polymerases are transcribing each region at the same time. The regions between the tranicribed DNA segments are the spacer DNA regions. Because each RNA gene has a fixed initiation site (promoter) and a fixed termination site, the transcripts adopt the characteristic Christmas tree' configuration (Fig. 9.6).

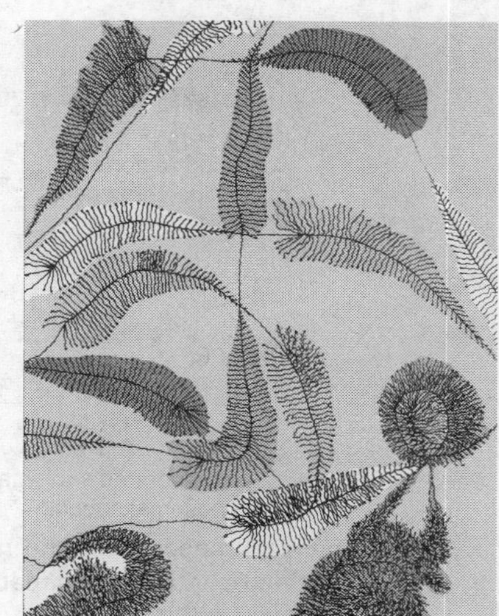

Fig. 9.6. Transcription of rRNA in the nucleolus of the newt, *Triturus*. Tandem repeats of the large rRNA genes are being transcribed. The polarity of the process (progressing from small to large transcripts), as well as the spacer DNA (thin lines between transcribing areas), is clearly visible.

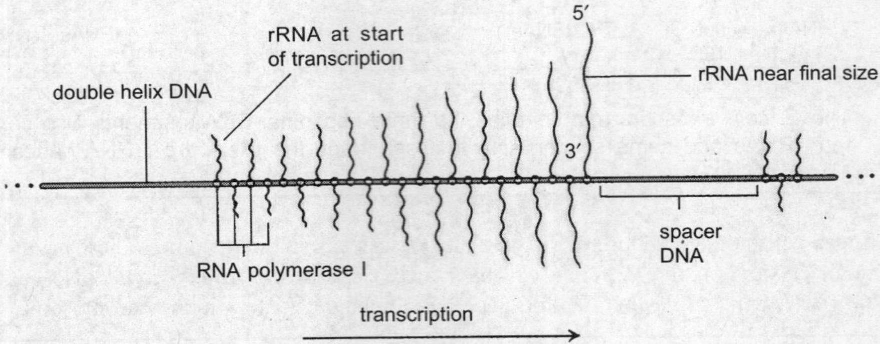

Fig. 9.7. Details of the transcription of large ribosomal RNA gene. Note the polarity of the process and the spacer DNA (after Tamarin, 2002).

Processing of Ribosomal RNAs

Three of four rRNAs (185, 5.8 and 28S) are made by chemically modifying and cleaving a single large (45S) precursor rRNA (Fig. 9.8). Extensive chemical modifications occur in the 13,00 – nucleotide – long precursor rRNA before the rRNAs are cleaved out of it and assembled into ribosomes. (Note. The 5SrRNA does not require chemical modification). Such modifications include about 10 methylations of the 2'- OH positions on nucleotide sugars and 100 isomerizations of uridine nucleotides to pseudouridine (Fig. 9.9A). The functions of these modifications are not

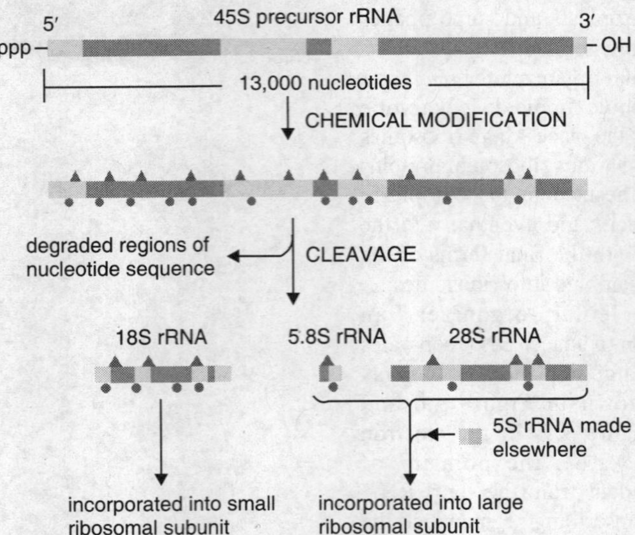

Fig. 9.8. The chemical modification and nucleolytic processing of a enkaryotic 45S precursor rRNA molecule into three separate ribosomal RNAs (i.e. 18S, 5.8S and 28SrRNAs) (Alberts *et al.*, 2002).

understood in detail, but they probably aid in the folding and assembly of the final rRNAs and may also delicately alter the function of ribosomes. Each modification is made at a specific position in the precursor RNA. These positions are specified by several hundred **"guide RNAs"**, which locate theselves through base pairing to the precursor rRNA and thereby brings on RNA modifying enzyme to the appropriate position (Fig. 9.9B). Other guide RNAs promote, probably by causing conformational changes in the precursor RNA, the cleavage of the precursor rRNA into the rRNAs. All of these guide RNAs are members of a large class of RNA, called **small nucleolar RNAs** (or **snoRNAs**), so named because these RNAs perform their functions in a sub-compartment of the nucleus, the nucleolus. Many snoRNAs are encoded in the introns of other genes especially those encoding ribosomal proteins. They are therefore synthesized by RNA polymerase II and processed from excised intron sequence.

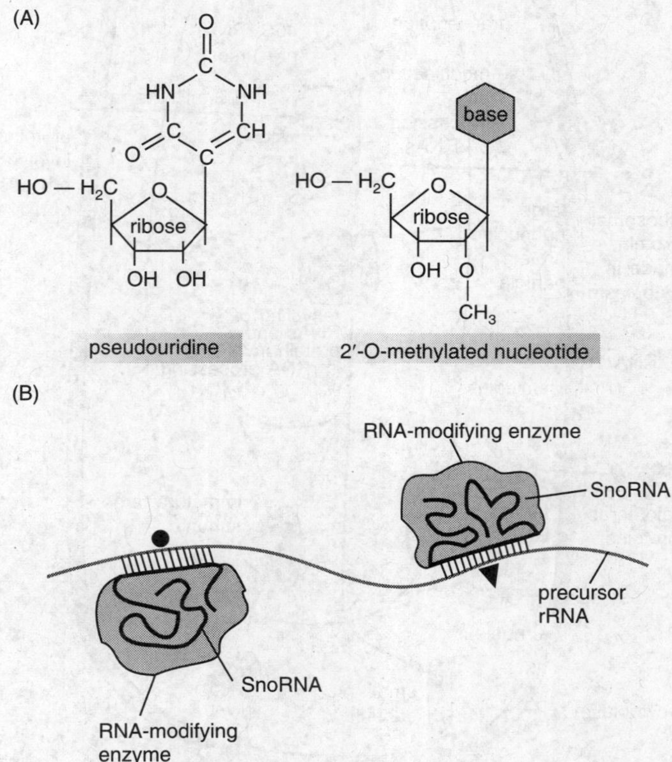

Fig. 9.9. Modifications of the precursor rRNA by guide RNAs. A – Two prominent covalent modifications occur after rRNA synthesis; the differences from the initially incorporated nucleotide are indicated by red atoms. B – The sno RNAs locate the sites of modification by base – pairing to complementary sequences on the precursor rRNA. The snoRNAs are bound to proteins and the complexes are called snoRNPs. The snoRNPs contain the RNA modification activities, presumably contributed by the proteins but possibly contributed by the snoRNAs themselves (after Alberts *et al.*, 2002).

Role of Nucleolus in Transcription

In addition to the important role in ribosome biogenesis (Fig. 9.10), the nucleolus is also the site where other RNAs are produced and other RNA protein complexes are assembled. For example, the U6 snRNP which functions in pre-mRNA splicing, is composed of one RNA molecule and at least seven proteins. The U6 snRNA is chemically modified by snoRNAs in the nucleolus before its final assembly there into the U6 snRNP. Other important RNA protein complexes including **telomerase** and the **signal recognition particle,** are also believed to be assembled at the nucleolus. Finally, the tRNAs are processed as well in the nucleolus. Thus the nucleolus can be thought of as a large factory at which many different noncoding RNAs are processed and assembled with proteins to form a large variety of ribonucleoprotein complexes.

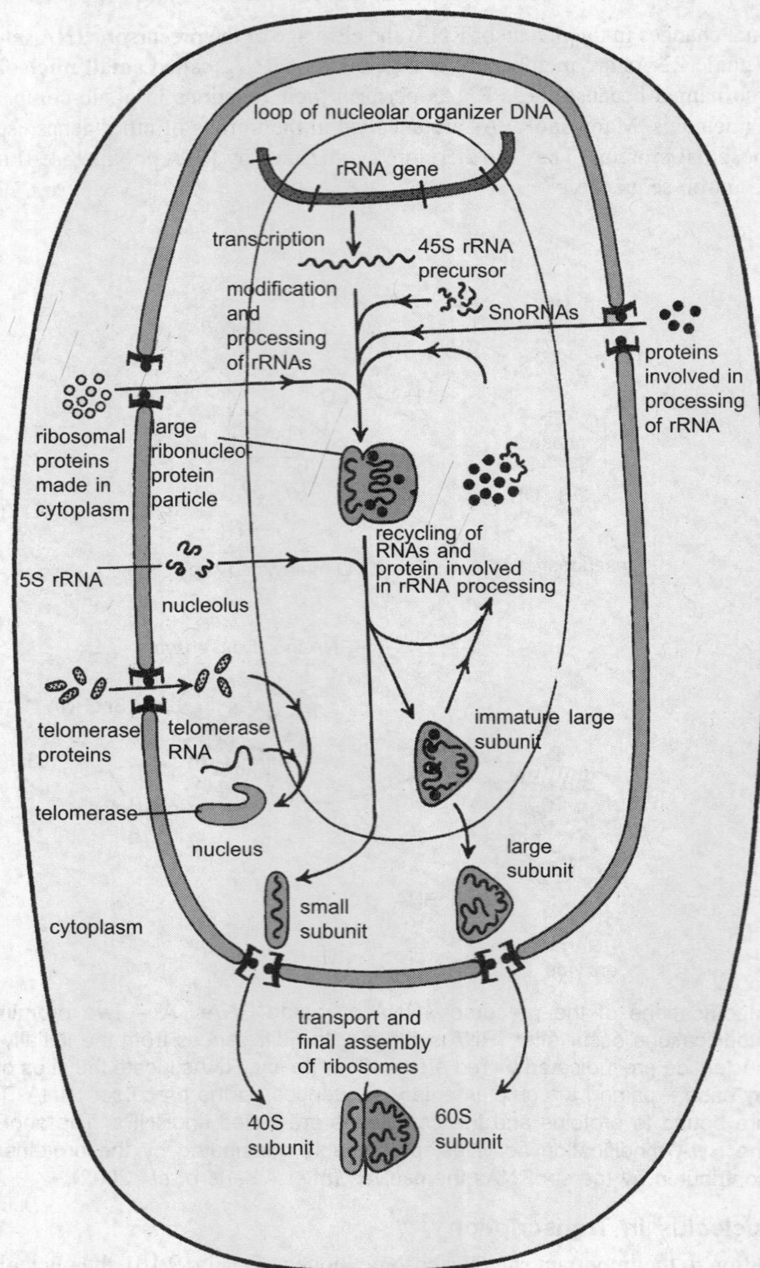

Fig. 9.10. The function of the nucleolus in ribosome and other ribonucleoprotein synthesis. The 45S precursor rRNA is packaged in a large ribonucleoprotein particle containing many ribosomal proteins imported from the cytoplasm while this particle remains in the nucleolus, selected pieces are added and others discarded as it is processed into immature large and small ribosomal subunits. The two ribosomal subunits are thought to attain their final functional form only as each is individually transported through the nuclear pores into the cytoplasm. Other ribonucleoprotein complexes (*e.g.,* telomerase) are also assembled in the nucleolus (after Alberts *et al.,* 2002).

9.2. TRANSFER RNA (tRNA)

The RNA which possesses the capacity to combine specifically with only one amino acid in a reaction mediated by a set of amino acid-specific enzymes, called **aminoacyl-tRNA synthetases;** transfers that amino acid from the "amino acid pool"to the site of protein synthesis and recognizes the codons of the mRNA is known as the **soluble RNA (sRNA)** or **transfer RNA (tRNA).** Thus, tRNA molecule has to perform several highly complex functions during protein synthesis–it interacts with a specific synthetase enzyme, possesses a site for binding an amino acid, possesses a second site for interacting with a ribosome, and contains an anticodon that must be exposed to the codons of mRNA.

Structure of tRNA. Robert Holley (1965) and his colleagues reported the complete nucleotide sequence of alanine tRNA of yeast (Holley received the Nobel Prize in 1968 for his work along with **Khorana** and **Nirenberg**). Nucleotide sequences are now known for more than 100 different "species" of tRNA.

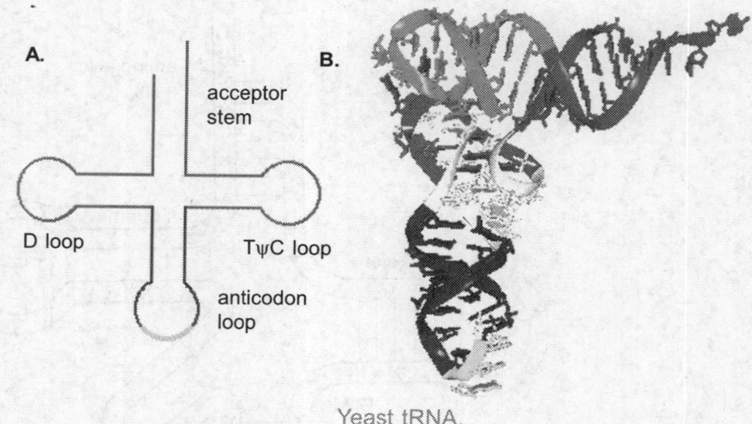

Transfer RNA has several unique characteristics; I. It is a relatively small molecule of 75 to 90 ribonucleotides and is, thus, smaller than either mRNA or any of the rRNAs, and has a sedimentation coefficient of 4S.

2. The ratios of A:U and G:C are near unity which suggests the formation of DNA-like double helical segments (secondary structure). In these double helical segments, G:C base pairs are more common than A:U as suggested by the ratio AU:GC = 0.7

3. All tRNA molecules have a tertiary structure, the details for which are now known and Mg^{2+}ion concentration is important for its stabilization.

4. A number of "unusual" nucleotides are found in tRNA (*e.g.*, pseudouridine ψ or psi), inosine (1), dihydroxyuridine (DHU), etc.). Many of "ünusual" nucleotides are methylated derivatives of common ones (*e.g.*, I – methylguanylic acid, I-methyladenylic acid, ribothymidylic acid and 5-methylcytosine).

The significance of these unusual bases of tRNA was understood well by molecular biologists during the construction of two-dimensional model from the primary-sequences of nucleotides of known tRNA. Thus, it was realized that most bases of tRNA pair according to Watson-Crick's pairing rule, but unusual bases fail to do so because they carry substitutions or alterations in those positions that usually participate in hydrogen bonding. Consequently, the presence of these bases forces the model builder to construct several non-base-paired loops in the tRNA molecule. By working on these lines, **R.Holley** (1965) first of all proposed a colover leaf model for yeast $tRNA^{ala}$. The clover leaf model of tRNA because accommodated several of the known functions of tRNA, so, it gained general acceptance. A typical clover-leaf model tRNA (Fig. 9.11) depicts the following structural peculiarities :

(*i*) All tRNA molecules have guanine residue G at the 5′ terminal end and unpaired (single stranded) C-C-A sequence at the 3′ end. This is called **amino acid attachment** site, because the amino acid becomes covalently attached to adenylic acid or A of CCA sequence during polypeptide synthesis.

(*ii*) The **amino acid stem** or **helix** consists of seven paired bases.

(*iii*) The **T-stem** is composed of five paired bases-the last (*i.e.,* nearest the **T-loop** or TψC loop is C-G. T-loop contains seven unpaired bases and is involved in the binding of rRNA molecules to the ribosomes.

(*iv*) The **anticodon stem** includes five paired bases. The **anticodon loop** consists of seven unpaired bases, the third, the fourth and the fifth of which (from the 3') end of the molecule) constitute the **anticodon**. The anticodon permits temporary complementary pairing with three bases (triplet codon) on mRNA.

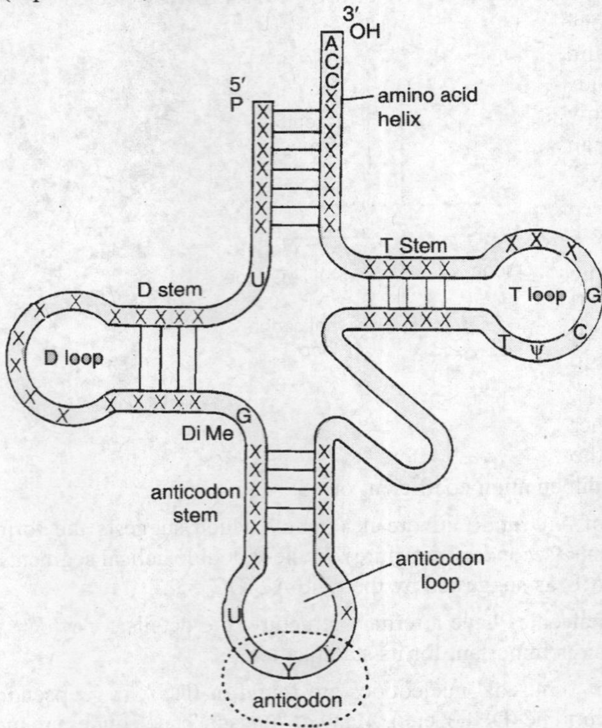

Fig. 9.11. Generalized two-dimensional clover-leaf model of tRNA, based on analyses of several yeast tRNA molecules by various investigators. Note the common 3' terminal –CCA, the TCG in the T-loop, the anticodon-U (here YYYU) in the anticodon loop, the DiMeG between the anticodon loop and D loop and the U as the first unpaired based on the 5' strand. As diagrammed here, the anticodon is read from right to left 3'?5' (A=adenosine, C=cytidine, G=guanosine, T=ribothymidine, U=uridine, ψ = pseudouridine, DiMeG=dimethylguanosine, Y=any base of the anticodon (after Burns and Bottino, 1989).

(*v*) The base on the 3′ side of anticodon is a purine.

(*vi*) Immediately adjacent to the 5′ side of the anticodon, uracil and another pyrimidine occurs.

(*vii*) A purine, often dimethylguanylic acid, is located in the "corner" between the anticodon stem and the D stem.

(*viii*) The **D-stem** is composed of three or four base pairs (depending on the "species" of tRNA). The **DHU-loop** or **D-loop** is also variable in size containing 8 to 12 unpaired bases. The D-loop helps in binding of amino-acyl synthetase.

(*ix*) The extra arm is variable in nucleotide composition and is lacking entirely in some tRNA.

Three-dimensional structure of tRNA. In order to understand the structure-function relationship of tRNA, its three dimensional structure (TDS) was worked out by the help of X-ray crystallography study. **A. Klug,** the Nobel laureate of 1982, has contributed much to the TDS of tRNAs. **S.H.Kim** (1973) proposed a most acceptable TDS model of tRNA of yeast cells). According to **Kim**, TDS of tRNA takes the shape of letter L with a thickness of 10A°. Each arm of the L doubled over by bonds holding complementary base together. Such an L-shape can also easily derived from two dimensional clover-leaf model (Fig. 9.12).

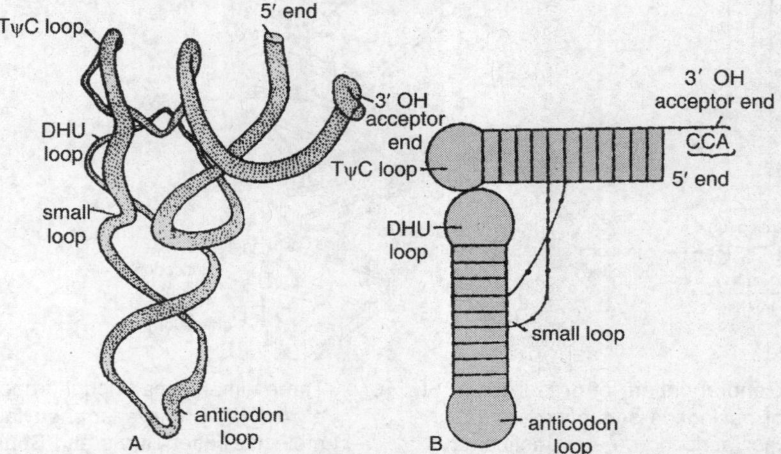

Fig. 9.12. Three dimensional structure of yeast phenylalanine tRNA. A-Actual appearance of the folding of the molecule. The polynucleotide chain is represented as a continuous coiled tube; B-The way that the clover leaf representation must be transformed in order to show the physical connections between various parts of the molecule. Two double-stranded helical regions are seen, each oriented at right angles to the other produces a L-shaped structure.

Extended anticodon hypothesis. Recently it is reported that the performance of anticodon, when isolated from tRNA, is weak and inaccurate. However, the performance of this anticodon triplet is enhanced if a matching sequence is present on the anticodon loop and on the stem on either side of anticodon triplet. As a result, an **extended anticodon hypothesis** has been proposed which suggests that the structures of anticodon loop and that of the proximal anticodon stem are related to the sequence of anticodon (**Michael Yarus,** 1982). Thus, the anticodon is extended into the nearby sequence and consists of in all 12 nucleotides arranged in the following way : (*i*) two nucleotides (*i.e.*, CU,ψU or UU) at the 5′ side of anticodon loop; (*ii*) three nucleotides of anticodon (3′ nucleotide of anticodon being very important and termed **cardinal nucleotide);** (*iii*) two nucleotides at the 3′ side of anticodon loop; and (*iv*) five pairs of nucleotides in the anticodon stem which can be conveniently written by giving only the bases on 3′ side of the stem (Fig. 9.13).

Processing of tRNA. Transfer RNA is transcribed from several particular sites on template DNA and comprised of about 15 per cent of the RNA present at any one time in an *E.coli* cell. Like the synthesis of rRNA and mRNA, synthesis of tRNA molecules is initiated at a promoter site and completed at a terminator sequence. Transfer RNAs are also processed from larger precursors. In prokaryotes, tRNA processing includes removal of nucleotides from the precursor or primary transcript and modification of some internal nucleotides. For example, in *E.coli,* the precursor for the amino acid tyrosine (pre $tRNA^{tyr}$) consists of 128 bases; in processing to $tRNA^{tyr}$ 41 nucleotides are removed from the 5′ end and two are removed from the 3′end (Fig. 9.15).

In addition to processing, methylation and inclusion of other "unusual" intercalary bases takes place after transcription, as does the addition of the 3′ terminal –C-C-A. The final functional tRNA molecule has 85 bases. Three-dimensional shape is achieved through hydrogen bonding. The processing of eukaryotic tRNAs resembles with that of prokaryotes. However, in them it is far more complex.

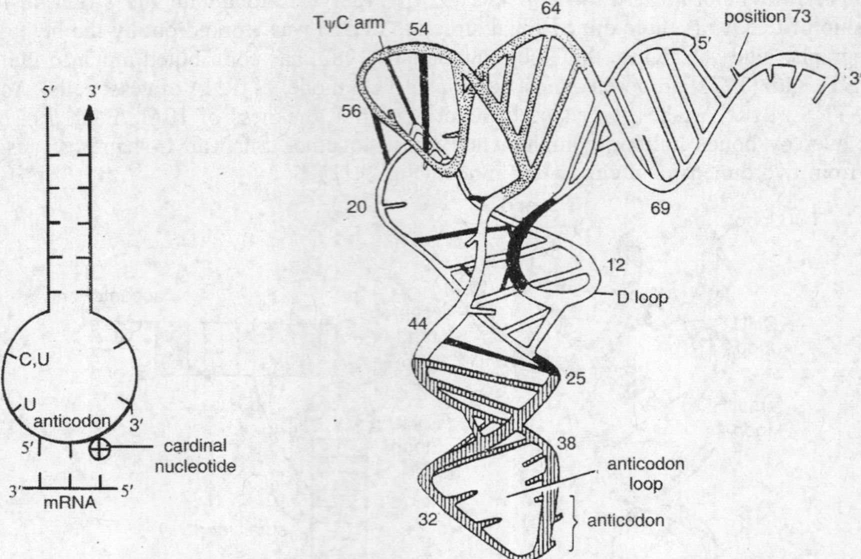

Fig 9.13. General pattern of organization of nucleotide sequence around the anticodon, *i.e.*, in anticodon and loop and stem of tRNA.

Fig. 9.14. Three dimensional model proposed by **Kim** *et al.*, (1974) for yeast phenylalanine tRNA molecule (after Burns and Bottino, 1989).

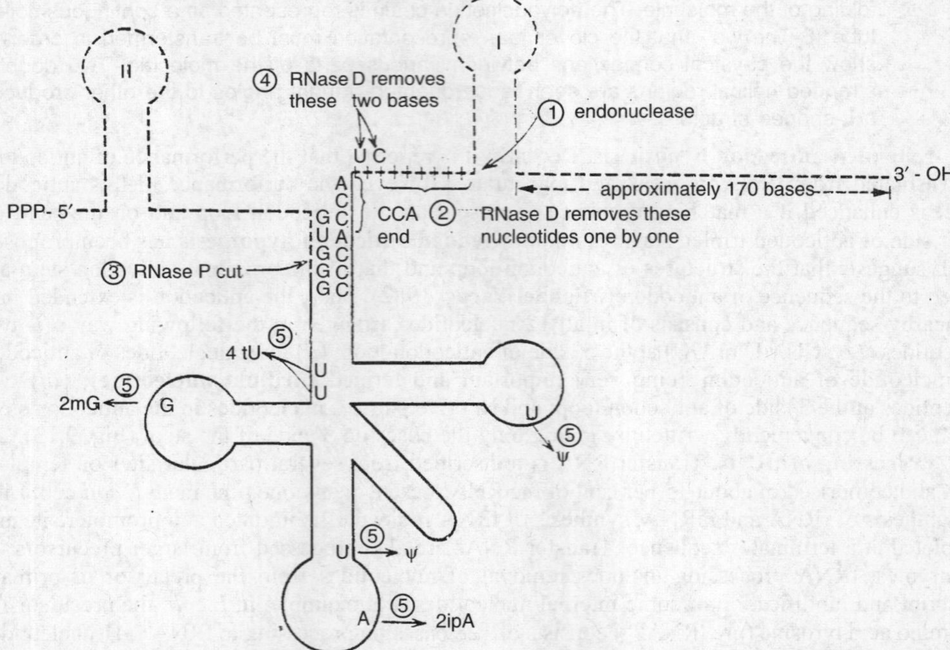

Fig. 9.15. Different stages in processing of the *E.coli* tRNA^{Tyr} gene transcript. The five stages have been given Arabic numbers. Step 3 generates the 5'-P end. Step 4 generates the 3'-OH end (the CCA end).In step 5 six bases, all in or near the loops of tRNA molecule, are modified to form pseudouridine (ψ,psi), 2-isopentenyladenosine (2 ip A), 2-O-methylguanosine (2mg), and 4-thiouridine (4tU). The continuous sequence that forms the final tRNA molecule is given in solid line (after Freifelder, 1985).

Splicing of tRNA Precursors

The tRNA precursor splicing reaction has been worked out in detail in the yeast *Saccharomyces cerevisiae.* Both *in vitro* splicing systems and temperature-sensitive splicing mutants have been used in dissecting the tRNA splicing mechanism in *S. cerevisiae.* The excision of introns from yeast tRNA precursors occurs in two stages (Fig.9.16).

1. In the first step, a nuclear membrane-bound splicing **endonuclease** enzyme makes two cuts precisely at the ends of the intron.

2. Then, in a fairly complex set of reactions, a splicing **ligase** enzyme joins the two halves of the tRNA to produce the mature form of the tRNA molecule.

The specificity of these reactions exists in the conserved three dimensional structural features of the tRNA precursors, not in the nucleotide sequences per se (i.e., by itself). Cleavage of the tRNA precurson yields 5′ – OH termini and 2′ – 3′ cyclic phosphate groups at the 3′ termini. The stage II ligation process actually involves four separate reactions :

1. The first reaction is the **addition** of a phosphate group to the 5′- OH terminus; this reaction requires **kinase** (enzyme) and a **phosphate donor** (ATP).

2. The second reaction, the 5′ phosphate group is activated by the transfer of an AMP group to the terminus from an AMP – ligase intermediate (the AMP originally having been derived from ATP also).

3. The 2′- 3′cyclic phosphate is opened by a cyclic phosphodiesterase activity that produces a 2'- phosphate and a free 3′ hydroxyl.

4. The final ligation reaction occurs via a nucleophilic attack of the free 3′- OH on the interior 5′phosphate with the release of AMP.

All four of these reactions are catalyzed by the splicing ligase. Finally, the 2′- phosphate group (remaining from the 2′- 3′cyclic phosphate produced by the original cleavage reaction) is removed by a phosphatase activity to yield the mature tRNA molecule. The overall two stage mode of tRNA intron excision appears to occur in other organisms as well, e.g., plants, mammals, etc. (**Gardner** *et al*., 2002).

Earlier we considered a simple definition of a gene as a length of DNA that codes for one protein. But we have just encountered a variance : genes code code for both tRNAs and rRNAs, yet neither is eventually translated into a protein. Their transcription function as final products without ever being translated. Thus, tRNA and rRNA are the major exceptions to the general rule that a gene codes for a protein.

Genes for tRNA. There are, probably, at least 30 to 40 different tRNA genes and tRNA molecules in *E.coli.* Higher organisms are found to contain 60 tRNA molecules and 60 tRNA genes. Since the cell uses only 20 amino acids in protein synthesis (and probably ony 20 synthetase enzymes), it follows that several tRNA will often have an affinity for the same amino acid. For example, *E.coli* cells contain five species of tRNA for leucine amino acid.

All the tRNA genes constitute far less than 1% of total genome in both *E.coli* and eukaryotic cells, yet some 10 to 15% of each cell's RNA may be in the form of tRNA. This discrepancy between the number of tRNA genes and gene transcripts occurs because of the following facts- (I) The tRNA molecules are relatively stable compared with many kinds of RNA. (2) The tRNA molecules are transcribed continuously and more quickly by tRNA genes than other RNAs because they are needed in plentiful amounts.

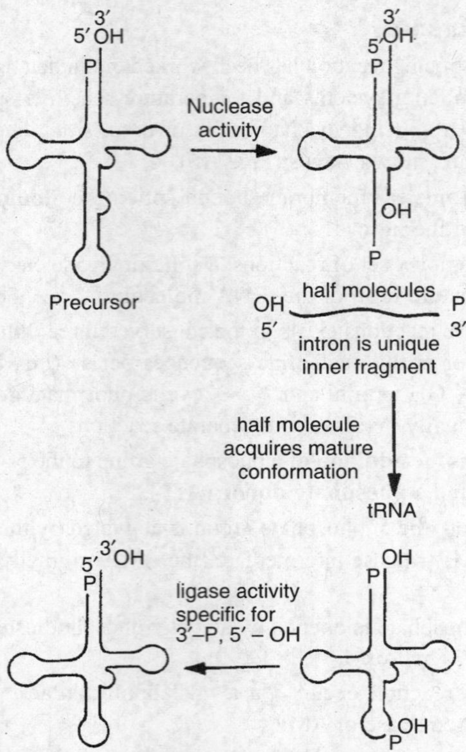

Fig. 9.16. Splicing mechanism of yeast tRNA, due to nuclease and ligase activities.

Questions And Problems

1. Describe the structure and composition of ribosomes. Describe the different sites and enzymes which are found in ribosomes and play a role in protein synthesis.
2. Describe the role of nucleolus in the processing of precursor of rRNAs and biogenesis of ribosomes.
3. Describe the primary, secondary and three dimensional structure (TDS) of transfer RNA structure and relate this with its role in protein synthesis.
4. Discuss the extended anticodon hypothesis of tRNA structure and describe how sequences other than anticodon determine the translation efficiency of tRNA.
5. Describe the process of splicing of tRNA precursor.
6. Write short notes on the following :
   (*i*) Protein modification; (*ii*) Clover–leaf model,
   (*iii*) Cardinal nucleotides.

Multiple Choice Questions

Choose the correct answer from the four alternatives given.

1. The function of nucleolus is the synthesis of
   (*a*) DNA (*b*) mRNA
   (*c*) rRNA (*d*) tRNA
2. Functional unit in synthesis of protein is
   (*a*) peroxisome
   (*b*) dictyosome
   (*c*) lysosome
   (*d*) polysome

3. An anticodon of tRNA represents
   (*a*) Wobble hypothesis
   (*b*) template hypothesis
   (*c*) gene flow hypothesis
   (*d*) Richmond and Long effect
4. Which of the following RNAs picks up specific amino acid from amino acid pool in the cytoplasm to ribosome during protein synthesis?
   (*a*) tRNA (*b*) mRNA
   (*c*) rRNA (*d*) all of them
5. Which site of the tRNA molecule hydrogen bonds to a mRNA molecule?
   (*a*) codon
   (*b*) anticodon
   (*c*) 5'end of tRNA
   (*d*) 3'end of tRNA
6. Clover leaf model belongs to
   (*a*) tRNA (*b*) DNA
   (*c*) centriole (*d*) flagella
7. An anticodon is complementary to the nucleotide triplet in
   (*a*) tRNA (*b*) rRNA
   (*c*) mRNA (*d*) cDNA
8. Which of the character is not applicable to tRNA?
   (*a*) it is the smallest of the RNAs
   (*b*) it acts as an adapter for amino acids
   (*c*) it has clover leaf structure
   (*d*) it is the largest of the RNAs
   (*e*) It bears anticodon
9. According to Wobble hypothesis
   (*a*) first base is unstable
   (*b*) second base is unstable
   (*c*) third base is unstable
   (*d*) the process of polypeptide chain elongation has been established
10. The codon for anticodon 3′UUA 5′ is
    (*a*) 5′AAU 3′ (*b*) 3′ AAU 5′
    (*c*) 5′AAT 3′ (*d*) 3′ AAC 5′
11. tRNA recognizes aminoacyl synthetase enzyme by
    (*a*) anticodon (*b*) DHU loop
    (*c*) T Ψ C loop (*d*) AA-site
12. Shape of 3 – D view of tRNA is
    (*a*) Z- shaped (*b*) X-shaped
    (*c*) Y-shaped (*d*) L-shaped
13. Amino acid binding site in tRNA is
    (*a*) 5′ end (*b*) anticodon loop
    (*c*) CCA 3′end (*d*) DHU loop
14. Genetic information for the synthesis of ribosomal RNA is coded in
    (*a*) DNA present in nucleus
    (*b*) nucleolar associated chromatin
    (*c*) granular zone of nucleolus
    (*d*) amorphous zone of nucleolus

ANSWERS

1. (*c*)	2. (*d*)	3. (*a*)	4. (*a*)	5. (*b*)
6. (*a*)	7. (*c*)	8. (*d*)	9. (*c*)	10. (*a*)
11. (*b*)	12. (*d*)	13. (*c*)	14. (*b*).	

CHAPTER

10

Genetic Code

Chapter Contents

As DNA is a genetic material, it carries genetic informations from cell to cell and from generation to generation. At this stage, an attempt will be made to determine that in what manner the genetic informations are existed in DNA molecule? Are they written in articulated or coded language on DNA molecule? If in the language of codes what is the nature of genetic code?

A DNA molecule is composed of three kinds of moieties: (*i*) phosphoric acid, (*ii*) deoxyribose sugar, and (*iii*) nitrogen bases. The genetic informations may be written in any one of the three moieties of DNA. But the poly-sugar-phosphate backbone is always the same, and it is, therefore, unlikely that these moiteies of DNA molecule carry the genetic informations. The nitrogen bases, however, vary from one segment of DNA to another, so the informations might well depend on their sequences. The sequences of nitrogen bases of a given segment of DNA molecule, actually, has been found to be identical to linear sequence of amino acids in a protein molecule. The proof of such a **colinearity** between DNA nitrogen base sequence and amino acid sequence in protein molecules has first obtained from an analysis of mutants of head protein of bacteriophage T_4

	U	C	A	G
U	UUU UUC Phe UUA UUG Lou	UCU UCC UCA UCG Ser	UAU UAC Tyr UAA UAG Stop	UGU UGC Cys UGA Stop UGG Trp
C	CUU CUC CUA CUG Leu	CCU CCC CCA CCG Pro	CAU CAC His CAA CAG Gin	CGU CGC CGA CGG Arg
A	AUU AUC AUA Ile AUG Met	ACU ACC ACA ACG Thr	AAU AAC Asn AAA AAG Lys	AGU AGC Ser AGA AGG Arg
G	GUU GUC GUA GUG Val	GCU GCC GCA GCG Ala	GAU GAC Asp GAA GAG Glu	GGU GGC GGA GGG Gly

The standard genetic code.

(**Sarabhai** *et al.,* 1964) and the A protein of tryptophan synthetase of *Escherichia coli* (**Yanofski** *et al*., 1964). The colinearity of protein molecules and DNA polynucleotides has given the clue that the specific arrangement of four nitrogen bases (*e.g.*, A, T, C and G) in DNA polynucleotides chains, somehow, determines the sequence of amino acids in protein molecules. Therefore, these four DNA bases can be considered as four alphabets of DNA molecule. All the genetic informations, therefore, should be written by these four alphabets of DNA. Now the question arises that whether the genetic informations are written in articulated language or coded language ? If genetic informations might have occurred in an articulated language, the DNA molecule might require various alphabets, a complex system of grammar and ample amount of space on it. All of which might be practically impossible and troublesome too for the DNA. Therefore, it was safe to conclude for molecular biologists that genetic informations were existed in DNA molecule in the form of certain special language of code words which might utilize the four nitrogen bases of DNA for its symbols. Any coded message is commonly called **cryptogram**.

10.1. BASIS OF CRYPTOANALYSIS

The basic problem of such a genetic code is to indicate how information written in a four letter language (four nucleotides or nitrogen bases of DNA) can be translated into a twenty letter language (twenty amino acids of proteins). The group of nucleotides that specifies one amino acid is a **code word** or **codon**. By the **genetic code** one means the collection of base sequences (codons) that correspond to each amino acid and to translation signals. We can consider here the classical but logical reasoning done by **George Gamov** (1954) about the possible size of a codon. The simplest possible code is a **singlet** code (a code of single letter) in which one nucleotide amino acid could be specified. A **doublet code** (a code of two letters) is also inadequate, because it could specify only sixteen (4 × 4) amino acids, whereas a **triplet code** (a code of three letters) could specify sixty four (4 × 4 × 4) amino acids. Therefore, it is likely that there may be 64 triplet codes for 20 amino acids. The possible singlet, doublet and triplet codes, which are customarily represented in terms of "**mRNA language**" [mRNA is a complementary molecule which copies the genetic informations (cryptogram of DNA) during its transcription], have been illustrated in Table 10.1.

Table 10.1. Possible singlet, doublet and triplet codes of mRNA.

Singlet code (4 words)	Doublet code (16 words)				Triplet code (64 words)			
					AAA	AAG	AAC	AAU
					AGA	AGG	AGC	AGU
					ACA	ACG	ACC	ACU
					AUA	AUG	AUC	AUU
					GAA	GAG	GAC	GAU
					GGA	GGG	GGC	GGU
A	AA	AG	AC	AU	GCA	GCG	GCC	GCU
G	GA	GG	GC	GU	GUA	GUG	GUC	GUU
C	CA	CG	CC	CU	CAA	CAG	CAC	CAU
U	UA	UG	UC	UU	CGA	CGG	CGC	CGU
					CCA	CCG	CCC	CCU
					CUA	CUG	CUC	CUU
					UAA	UAG	UAC	UAU
					UGA	UGG	UGC	UGU
					UCA	UCG	UCC	UCU
					UUA	UUG	UUC	UUU

The first experimental evidence in support to the concept of triplet code is provided by **Crick** and coworkers in 1961. During their experiment, when they added or deleted single or double base pairs in a particular region of DNA of T_4 bacteriophages of *E. coli*, they found that such bacteriophages ceased to perform their normal functions. However, bacteriophages with addition or deletion of three base pairs in DNA molecule, had performed normal functions. From this experiment, they concluded that a genetic code is in triplet form, because the addition of one or two nucleotides has put the reading of the code out of order, while the addition of third nucleotide resulted in a return to the proper reading of the message.

10.2. CODON ASSIGNMENT (Cracking the Code or Deciphering the Code)

The genetic code has been cracked or deciphered by the following kinds of approaches :

A. Theoretical Approach

The physicist **George Gamow** proposed the **diamond code** (1954) and the **triangle code** (1955) and suggested an exhaustive theoretical framework to the different aspect of the genetic code. **Gamow** suggested the following properties of the genetic code :

(*i*) **A triplet codon** corresponding to one amino acid of the polypeptide chain.
(*ii*) **Direct template translation** by codon-amino acid pairing.
(*iii*) Translation of the code in an **overlapping** manner.
(*iv*) **Degeneracy** of the code, *i.e.*, an amino acid being coded by more than one codon.
(*v*) **Colinearity** of nucleic acid and the primary protein synthesized.

	T	C	A	G
T	TTT Phe (F) TTC" TTA Leu (L) TTG"	TCT Ser (S) TCC" TCA" TCG"	TAT Tyr (Y) TAC TAA**Ter** TAG **Ter**	TGT Cys (C) TGC TGA **Ter** TGG Trp (W)
C	CTT Leu (L) CTC" CTA" CTG"	CCT Pro (P) CCC" CCA" CCG"	CAT His (H) CAC" CAA Gln (Q) CAG"	CGT Arg (R) CGC" CGA" CGG"
A	ATT Ile (I) ATC" ATA" ATG Met (M)	ACT Thr (T) ACC" ACA" ACG"	AAT Asn (N) AAC" AAA Lys (K) AAG"	AGT Ser (S) AGC" AGA Arg (R) AGG"
G	GTT Val (V) GTC" GTA" GTG"	GCT Ala (A) GCC" GCA" GCG"	GAT Asp (D) GAC" GAA Glu (E) GAT "	GGT Gly (G) GGC" GGA " GGG"

Degeneracy of genetic code.

(*vi*) **Universality** of the code, *i.e.*, the code being essentially the same for different organisms.

Some of these Gamow's proposals have been contradicted by the molecular biologists. For example, **Brenner** (1957) showed that the overlapping triplet code is an impossibility, and subsequent work has shown that the code is a **non-overlapping** one. Similarly, Gamow's idea of direct template relationship between nucleic acid and polypeptide chain was challenged when **Crick** proposed his **adopter hypothesis**. According to this hypothesis, **adaptor molecules** intervene between nucleic acid and amino acids during translation. In fact, it is now known that tRNA molecules act as adaptors between codons of mRNA and amino acids of the resulting polypeptide chain.

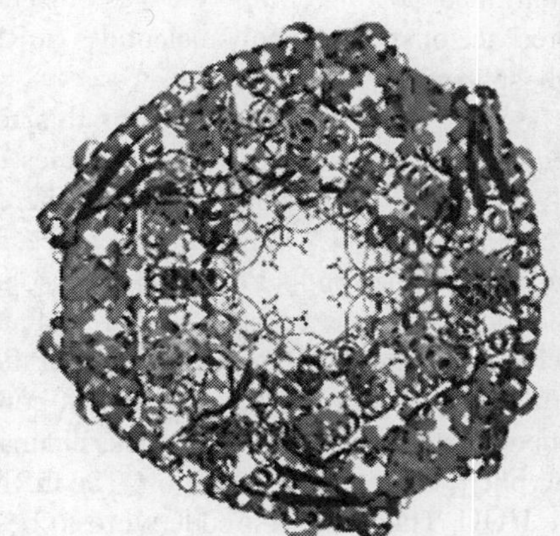

Polynucleotide phosphorylase.

B. The *in vitro* Codon Assignment

1. Discovery and use of polynucleotide phosphorylase enzyme. Marianne Grunberg-Manago and **Severo Ochoa** isolated an enzyme from the bacteria (e.g., *Azobacter vinelandii* or *Micrococcus lysodeikticus)* that catalyzes the breakdown of RNA in bacterial cells. This enzyme is called **polynucleotide phosphorylase. Manago** and **Ochoa** found that outside of the cell *(in vitro),* with high concentrations of ribonucleotides, the reaction could be driven in reverse and an **RNA** molecule could be made (see **Burns** and **Bottino**, 1989). Incorporation of bases into the molecule is random and does not require a DNA template. Thus, in 1955 **Manago** and **Ochoa** made possible the artificial synthesis of polynucleotides (= mRNA) containing only a single type of nucleotides (U, A, C, or G respectively) repeated many times.

Polynucleotide	**Configuration**
1. Polyuridylic acid or poly (U)	UUUUUU
2. Polyadenylic acid or poly (A)	AAAAAA
3. Polycytilic acid or poly (C)	CCCCCC
4. Polyguanidylic acid or poly (G)	GGGGGG

Thus, the action of polynucleotide phosphorylase can be represented in the following way :

$$(RNA)_n + \text{Ribonucleoside diphosphate} \xrightleftharpoons{\text{polynucleotide phosphorylase}} (RNA)_{n+1} + Pi$$

The polynucleotide phosphorylase enzyme differs from RNA polymerase used to transcribe mRNA from DNA in that : (*i*) it does not require a template or primer ; (*ii*) the activated substrates are ribonucleoside diphosphates (*e.g.*, UDP, ADP, CDP and GDP) and not triphosphates; and (*iii*) orthophosphate (*Pi*) is produced instead of pyrophosphates (*PPi*).

The deciphering of the genetic code was made possible by the use of synthetic (or artificial) polynucleotides and trinucleotides. The different types of techniques used include the use of polymers containing a single type of nucleotide (called **homopolymers**), the use of mixed polymers (**copolymers**) containing more than one type of nucleotides (**heteropolymers**) in random or defined sequences and the use of trinucleotides (or "**minimessengers**") in ribosome-binding or filter-binding.

2. Codon assignment with unknown sequence. ***(i)* Codon assignment by homopolymer.** The first clue to codon assignment was provided by **Marshall Nirenberg** and **Heinrich Matthaei** (1961) when they used *in vitro* system for the synthesis of a polypeptide using an artificially synthesized mRNA molecule containing only one type of nucleotide *(i.e.*, homopolymer). Prior to performing the actual experiments, they tested the ability of a cell-free protein synthesizing system to incorporate radioactive amino acids into newly synthesized proteins. Their cell-free extracts of *E.coli* contained ribosomes, tRNAs, aminoacyl-tRNA synthetase enzymes, DNA and mRNA. The DNA of this extract was eradicated by the help of **deoxyribonuclease** enzyme, thus, the template which might synthesize new mRNA was destroyed. When twenty amino acids were added to this mixture along with ATP, GTP, K^+ and MG^{2+}, they were incorporated into proteins. This incorporation continued so long as mRNA was present in such a cell-free suspension. It also continued in the presence of synthetic polynucleotides (mRNAs) which could be made with the help of polynucleotide phosphorylase enzyme.

The first successful use of this technique was made by **Nirenberg** and **Matthaei** who synthesized a chain of uracil molecules (poly U) as their synthetic mRNA (homopolymer). Poly (U) seemed a good choice, because there could be no ambiguity in a message consisting of only one base. Poly (U) was a good choice for other reasons : it binds well to ribosomes and, as it turned out, the product protein was insoluble and easy to isolate. When poly (U) was presented as the message to the cell free system containing all the amino acids, one amino acid was exclusively selected from the mixture for incorporation into the polypeptide, called **polyphenylalanine**. This amino acid was phenylalanine and it could be concluded that some sequence of UUU coded for phenylalanine. Other homogeneous chains of nucleotides (Poly A, Poly C and Poly G) were inactive for phenylalanine incorporation. The mRNA code word for phenylalanine was, therefore, shown to be UUU. Thus, the first code word to be deciphered was UUU.

This discovery was extended in the laboratories of **Nirenberg** and **Ochoa**. The experiment was repeated using synthetic **poly (A)** and **poly (C)** chains, which gave **polylysine** and **polyproline** respectively. Thus, AAA was identified as the code for lysine and CCC as the code for proline. A poly (G) message was found non-functional *in vitro*, since it attains secondary structure and, thus, could not attach the ribosomes. In this way three of 64 codons were easily accounted for.

(*ii*) **Codon assignment by heteropolymers (Copolymers with random sequences)**. Further exposition of the genetic code took place by using synthetic messenger RNAs containing two kinds of bases. This technique was used in the laboratories of **Ochoa** and **Nirenberg** and led the deduction of the composition of codons for the 20 amino acids. The synthetic messengers contained the bases at random (called **random copolymers**). For example, in a random copolymer using U and A nucleotides eight triplets are possible, such as UUU, UUA, UAA, UAU, AAA, AAU, AUU and AUA. Theoretically, eight amino acids could be coded by these eight codons. Actual experiments, however, yielded only six phenylalanine, leucine, tyrosine, lysine, asparagine and isoleucine. By varying the relative compositions of U and A in the random copolymer and determining the percentage of the different amino acids in the proteins formed, it was possible to deduce the composition of the code for different amino acids.

3. Assignment of codons with known sequences. **(*i*) Use of trinucleotides or minimessengers in filter binding (Ribosome-binding technique)**. Ribosome technique of **Nirenberg** and **Leder** (1964) made use of the finding that aminoacyl-tRNA molecules specificially bind to ribosome-mRNA complex. The binding does not require the presence of a long mRNA molecule; in fact, the association of a **trinucleotide** or **minimessenger** with the ribosome is sufficient to cause aminoacyl-tRNA binding. When a mixture of such small mRNA molecules-ribosomes and amino acid-tRNA complexes are incubated for a short time and then filtered through a **nitrocellulose membrane**, then the mRNA-ribosome-tRNA-amino acid complex is retained back and rest of the mixture passes through the filter. By using a series of 20 different amino acid mixtures, each containing one radioactive amino acid at a time, it is possible to find out the amino acid corresponding to each triplet by analysing the radioactivity absorbed by the membrane, *e.g.*, the triplet GCC and GUU retain only alanyl-tRNA and valyl-tRNA respectively. All 64 possible triplets have been synthesized and tested in this way. Forty five of them have given clear-cut results. Later on, with the help of longer synthetic messages it has been possible to decipher 61 out of the possible 64 codons.

C. The *in vivo* Codon Assignment

The cell free protein synthetic systems, though have proved of great significance in decipherment of the genetic code, but they could not tell us whether the genetic code so deciphered is used in the living systems of all organisms also. Three kinds of techniques are used by different molecular biologists to determine whether the same code is also used *in vivo* (*a*) amino acid replacement studies (*e.g.*, tryptophan synthetase synthesis in *E.coli* (**Yanofsky** *et al.* 1963) and haemoglobin synthesis in humans), (*b*) frameshift mutations (*e.g.*, investigations of **Terzaghi** *et al.* 1966, on lysozyme enzyme of T_4 bacteriophages, and (*c*) comparison of a DNA or mRNA polynucleotide cryptogram with its corresponding polypeptide clear text (*e.g., comparison* of amino acid sequence of the R_{17} bacteriophage coat protein with the nucleotide sequence of the R_{17} mRNA in the region of the molecule that dictates coat-protein synthesis by **S. Cory** *et al.,* 1970*)*.

Thus, *in vitro* and *in vivo* studies, so far described, gave the way to formulate a code table for twenty amino acids (see Table 10.2 and Table 10.3).

Table 10.2. The genetic dictionary. The trinucleotide codons are written in the 5′→3′ direction.

First base	Second base U		Second base C		Second base A		Second base G		Third base
U	UUU	Phe	UCU	Ser	UAU	Tyr	UGU	Cys	U
	UUC		UCC		UAC		UGC		C
	UUA	Leu	UCA		UAA	Non-sense codon†	UGA	Non-sense codon†	A
	UUG		UCG		UAG		UGG	Trp	G
C	CUU	Leu	CCU	Pro	CAU	His	CGU	Arg	U
	CUC		CCC		CAC		CGC		C
	CUA		CCA		CAA	Gln	CGA		A
	CUG		CCG		CAG		CGG		G
A	AUU	Ileu	ACU	Thr	AAU	Asn	AGU	Ser	U
	AUC		ACC		AAC		AGC		C
	AUA		ACA		AAA	Lys	AGA	Arg	A
	AUG*	Met	ACG		AAG		AGG		G
G	GUU	Val	GCU	Ala	GAU	Asp	GGU	Gly	U
	GUC		GCC		GAC		GGC		C
	GUA		GCA		GAA	Glu	GGA		A
	GUG		GCG		GAG		GGG		G

*AUG—Met or chain initiation codon.; †UAA, UAG, UGA—Stop Codon.

Table 10.3. Amino acids and their messenger RNA codons.

Amino acid	Codons
Alanine (Ala)	GCA, GCG, GCC, GCU
Arginine (Arg)	AGA, AGG, CGA, CGG, CGC, CGU
Aspartic acid (Asp)	GAC, GAU
Asparagine (Asn; Aspn)	AAC, AAU
Cysteine (Cys)	UGC, UGU
Glycine (Gly)	GGA, GGG, GGC, GGU
Glutamine (Gln ; Glun)	CAA, CAG
Glutamic acid (Glu)	GAA, GAG
Histidine (His)	CAC, CAU
Isoleucine (Ile)	AUC, AUU, AUA
Lysine (Lys)	AAA, AAG
Leucine (Lcu)	CUA, CUG, CUC, CUU, UUA, UUG
Methionine (Met) (Starting codon)	AUG
Phenylalanine (Phe)	UUC, UUU
Proline (Pro)	CCA, CCG, CCC, CCU
Serine (Ser)	AGC, AGU, UCA, UCG, UCC, UCU
Threonine (Thr)	ACA, ACG, ACC, ACU
Tryptophan (Trp)	UGG
Tyrosine (Tyr)	UAC, UAU
Valine (Val)	GUA, GUG*, GUC, GUU
Terminator (Nonsense codons)	UAA (ocher), UAG (amber), UGA (opal)

* GUG is also used as a start codon for some proteins.

10.3. CHARACTERISTICS OF GENETIC CODE

The genetic code has the following general properties :

1. The code is a triplet codon. The nucleotides of mRNA are arranged as a linear sequence of codons, each codon consisting of three successive nitrogenous bases, *i.e.*, the code is a triplet codon. The concept of triplet codon has been supported by two types of point mutations: frameshift mutations and base substitution.

(*i*) Frameshift mutations. Evidently, the genetic message once initiated at a fixed point is read in a definite frame in a series of three letter words. The framework would be disturbed as soon as there is a deletion or addition of one or more bases. When such frameshift mutations were intercrossed, then in certain combinations they produce wild type normal gene. It was concluded that one of them was deletion and the other an addition, so that the disturbed order of the frame due to mutation will be restored by the other (Fig.10.1).

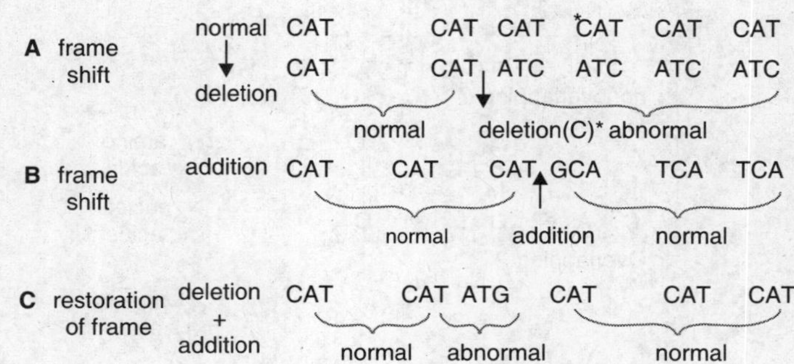

Fig. 10.1. Frame-shift mutations: A—Deletion; B—Addition; C—Restoration of frame.

(*ii*) Base substitution. If in a mRNA molecule at a particular point, one base pair is replaced by another without any deletion or addition, the meaning of one codon containing such an altered base, will be changed. In consequence, in place of a particular amino acid at a particular position in a polypeptide, another amino acid will be incorporated. For example, due to substitution mutation, in the gene for tryptophan synthetase enzyme in *E.coli*, the GGA codon for glycine becomes a missence codon AGA which codes for arginine. **Missence codon** is a codon which undergoes an alteration to specify another amino acid.

A more direct evidence for a triplet code came from the finding that a piece of mRNA containing 90 nucleotides, corresponded to a polypeptide chain of 30 amino acids of a growing haemoglobin molecule. Similarly, 1200 nucleotides of "satellite" tobacco necrosis virus direct the synthesis of coat protein molecules which have 372 amino acids.

2. The code is non-overlapping. In translating mRNA molecules the codons do not *overlap* but are "read' sequentially (Fig.10.2). Thus, **a non-overlapping code** means that a base in a mRNA is not used for different codons. In Figure 10.3., it has been shown that an overlapping code can mean coding for four amino acids from six bases. However, in actual practice six bases code for not more than two amino acids. For example, in case of an overlapping code, a single change (of substitution type) in the base sequence will be reflected in substitutions of more than one amino acid in corresponding protein. Many examples have accumulated since 1956 in which a single base substitution results into a single amino acid change in insulin, tryptophan synthetase, TMV coat protein, alkaline phosphatase, haemoglobin, etc.

Fig. 10.2. Bases in mRNA are read sequentially in the 5' to 3' direction, in group of three bases.

Recently, however, it has been shown that in the bacteriophage $\phi \times 174$ there is a possibility of overlapping of genes and codons (**Barrel** and coworkers, 1976; **Sanger**, *et al*., 1977).

3. The code is commaless. The genetic code is commaless, which means that no codon is reserved punctuations. It means that after one amino acid is coded, the second amino acid will be automatically coded by the next three letters and that no letters are wasted as the punctuation marks (Fig.10.4)

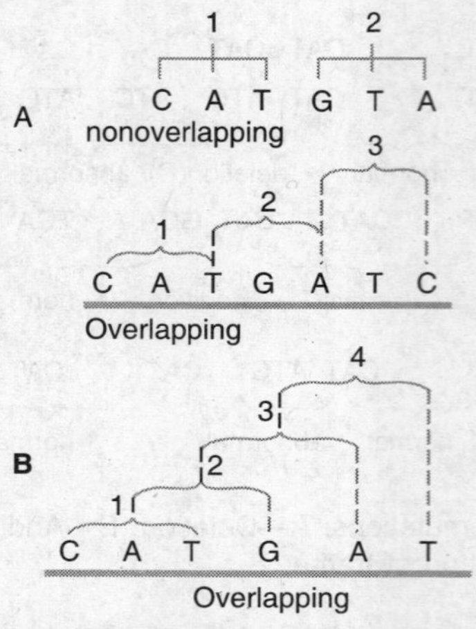

Fig.10.3. A—Non-overlapping codons; B—Overlapping of codon due to one base; C—Overlapping of codon due to two bases.

amino acids
nucleic acids
1 2
C A T G A T
non-overlapping comma free

amino acids
nucleic acids
1 2
C A T X G A T
non-overlapping with comma

Fig. 10.4. Two forms of genetic code. A—Genetic code without comma; B—Genetic code with comma.

4. The code is non-ambiguous. Non-ambiguous code means that a particular codon will always code for the same amino acid. In case of ambiguous code, the same codon could have different meanings or in other words, the same codon could code two or more than two amino acids. Generally, as a rule, the same codon shall never code for two different amino acids. However, there are some reported exceptions to this rule: the codons AUG and GUG both may code for *methionine* as initiating or starting codon, although GUG is meant for *valine*. Likewise, GGA codon codes for two amino acids *glycine* and *glutamic acid* (Box 10.1).

5. The code has polarity. The code is always read in a fixed direction, *i.e.*, in the $5' \rightarrow 3'$ direction. In other words, the codon has a **polarity**. It is apparent that if the code is read in opposite directions, it would specify two different proteins, since the codon would have reversed base sequence :

Codon :	UUG	AUC	GUC	UCG	CCA	ACA	AGG
Polypeptide : →	Leu	Ile	Val	Ser	Pro	Thr	Arg
	Val	Leu	Leu	Ala	Thr	Thr	Gly ←

6. The code is degenerate. More than one codon may specify the same amino acid; this is called **degeneracy** of the code. For example, except for *tryptophan* and *methionine*, which have a single codon each, all other 18 amino acids have more than one codon. Thus, nine amino acids, namely *phenylalanine, tyrosine, histidine, glutamine, asparagine, lysine, aspartic acid, glutamic acid* and *cysteine*, have two codons each. Isoleucine has three codons. Five amino acids, namely *valine, proline, threonine, alanine* and *glycine*, have four codons each. Three amino acids, namely *leucine, arginine* and *serine*, have six codons each (see Table 10.3).

The code degeneracy is basically of two types : partial and complete. **Partial degeneracy** occurs when first two nucleotides are identical but the third (*i.e.*, 3′ base) nucleotide of the degenerate codons differs, *e.g.*, CUU and CUC code for leucine. **Complete degeneracy** occurs when any of the four bases can take third position and still code for the same amino acid (*e.g.*, UCU, UCC, UCA and UCG code for serine).

Degeneracy of genetic code has certain biological advantages. For example, it permits essentially the same complement of enzymes and other proteins to be specified by microorganisms varying widely in their DNA base composition. Degeneracy also provides a mechanism of minimizing mutational lethality.

7. Some codes act as start codons. In most organisms, AUG codon is the **start** or **initiation** codon, *i.e.*, the polypeptide chain starts either with **methionine** (eukaryotes) or **N-formylmethionine** (prokaryotes). methionyl or N-formylmethionyl-tRNA specifically binds to the **initiation site** of mRNA containing the AUG initiation codon. In rare cases, GUG also serves as the initiation codon, *e.g.*, bacterial protein synthesis. Normally GUG codes for valine, but when normal AUG codon is lost by deletion, only then GUG is used as initiation codon.

Box 10.1.

Genetic code is degenerate not ambiguous

After the start and stop codons, the remaining 60 codons are far more than enough to code for the other 19 amino acids–and indeed there are repeats. Thus we say that the genetic code is **degenerate,** that is, an amino acid may be represented by more than one codon. The degeneracy is not evenly divided among the amino acids. For example, methionine and tryptophan are represented by only one codon each, whereas leucine is represented by six different codons.

The term "degeneracy" should not be confused with "ambiguity". To say that the code was ambiguous would mean that a *single codon could specify either of two (or more) different amino acids;* there would be doubt whether to put in, say leucine or something else. The genetic code is not ambiguous. Degeneracy in the code means that *there is more than one clear way to say "put leucine here".* In other words, a given amino acid may be encoded by more than one codon, but a codon can code for only one amino acid. But just as people in different places prefer different ways of saying the same thing —" Good-bye" ! "See you" | "Ciao"! and "So long"! have the same meaning — different organisms prefer one or others of the degenerate codons (Ciao means informal hello or good bye). These preferences are important in genetic engineering.

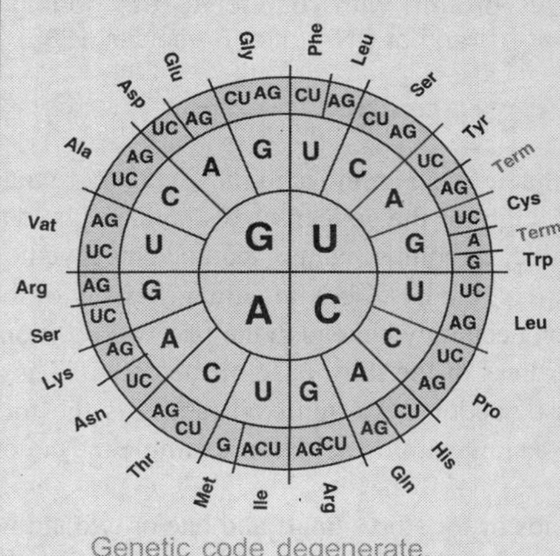

Genetic code degenerate.

8. Some codes act as stop codons. Three codons UAG, UAA, UGA are the chain **stop** or **termination** codons. They do not code for any amino acids. These codon are not read by any tRNA molecules (via their anticodons), but are read by specific proteins, called **release factors** (e.g., RF-1, RF-2, RF-3 in prokaryotes and RF in eukaryotes). These codons are also called as **nonsense codons**, since they do not specify any amino acid.

The UAG was the first termination codon to be discovered by **Sidney Brenner** (1965). It was named **amber** after a graduate student named **Bernstein** (= the German word for 'amber' and amber means brownish yellow) who helped in the discovery of a class of mutations. Apparently, to give uniformity the other two termination codons were also named after colours such as **ochre** for UAA and **opal** or **umber** for UGA. (*Ochre* means yellow red or pale yellow; *opal* means milky white and *umber* means brown). The existence of more than one stop codon might be a safety measure, in case the first codon fails to function.

9. The code is universal. Same genetic code is found valid for all organisms ranging from bacteria to man. Such universality of the code was demonstrated by **Marshall, Caskey** and **Nirenberg** (1967) who found that *E.coli* (bacterium), *Xenopus laevis* (amphibian) and guinea pig (mammal) amino acyl-tRNA use almost the same code. **Nirenberg** has also stated that the genetic code may have developed 3 billion years ago with the first bacteria, and it has changed very little throughout the evolution of living organisms.

Codon and Anticodon

The codon words of DNA would be complementary to the mRNA code words (i.e., DNA codes run in $3' \rightarrow 5'$ direction and mRNA code words run in $5' \rightarrow 3'$ direction) and so thereby the three bases forming the **anticodon** of tRNA (i.e., bases of anticodons run in $3' \rightarrow 5'$ direction). Three bases of anticodon pair with the mRNA on the ribosomes at time of aligning the amino acids during protein synthesis (translation of mRNA into proteins which proceeds in N_2 — COOH direction). For example, one of two mRNA and DNA code words for the amino acid phenylalanine is UUC and AAG respectively, and the corresponding anticodon of tRNA is CAA. This indicates that codon and anticodon pairing is antiparallel. In this case, C pairs with G and U pairs with A.

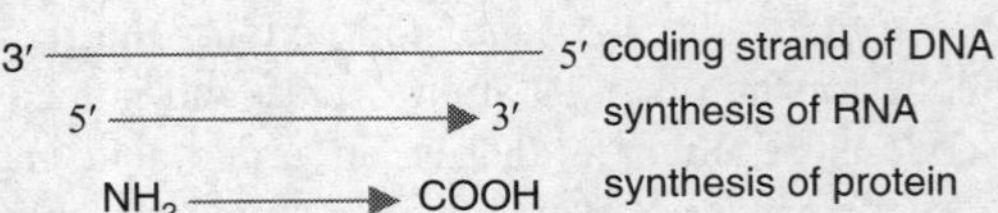

Fig. 10.5. Directions of synthesis of RNA and protein with respect to the coding strand of DNA (after Freifelder, 1985).

10.4. WOBBLE HYPOTHESIS

The genetic code is a degenerate code, meaning that a given amino acid may have more than one codon (Box 10.1). As you can see from table 10.2, eight of the sixteen boxes contain just one amino acid per box (*e.g.,* serine, leucine, proline, arginine, threonine, valine, alanine and glycine). (A **box** is determined by the first and second positions, *e.g.,* the UUX box in which X is any of the four bases). Therefore, these eight amino acids, the codon need only be read in the first two positions because the same amino acid will be represented regardless of the third base of the codon. These eight groups of codons are termed **unmixed families** of codons. An unmixed family is the four codons beginning with the same two bases that specify a single amino acid. For example the codon family GUX codes for valine.

Mixed families of codons code for two amino acids or for stop signals and one or two amino acids. Six of the mixed-family boxes are split in half so that the codons are differentiated by the presence of a purine or a pyrimidine in the third base. For example, CAU and CAC both code for histidine : in both, the third base, U (uracil) or C (cytosine) is a pyrimidine. Only two of the families of codons are split differently.

The lesser importance of the third position is the genetic code is linked with two facts about transfer RNAs. First, although there would seem to be a need for sixty-two transfer RNAs— since there are sixty-one (*i.e.,* one codons specifying amino acids and an additional codon for initiation) there are actually only about fifty different transfer RNAs in an *E.coli* cell. Second, a rare base such as inosine (I) can appear in the anticodon, usually in the position that is complementary to the third position of the codon. These two facts researchers believe that some kind of conservation of transfer RNAs is occurring and that rare base may be involved.

At this stage, one point should be very clear that both messenger RNA and transfer RNA bases are usually numbered from the $5'$ side. Thus, the number one base of the codon is complementary to the number-three base of the anticodon (Fig. 10.6). Thus, the codon base of lesser importance is the number-three base, whereas its complement in the anticodon is the number-one base.

Since the first position of the anticodon (5′) is not as constrained as the other two positions, a given base at that position may be able to pair with any of several bases in the third position of the codon. **Crick** (1966) characterized this ability as **wobble** (wobble means to sway or move unsteadily) (Fig. 10.6). Table 10.4 shows the possible pairings that would produce a transfer RNA system compatible with the known code. For example, if an isoleucine transfer RNA has the anticodons 3′ – UAI – 5′, it is compatible with the three codons for that amino acid (see Table 10.4 and Table 10.5) : 5′ – AUU' – 3′, 5′ – AUC – 3′ and 5′ – AUA – 3′. That is inosine is the first (5′) position of the anticodon can recognise U, C, or A in the third (3′) position of the codon, and thus one transfer RNA complements with all three codon for isoleucine.

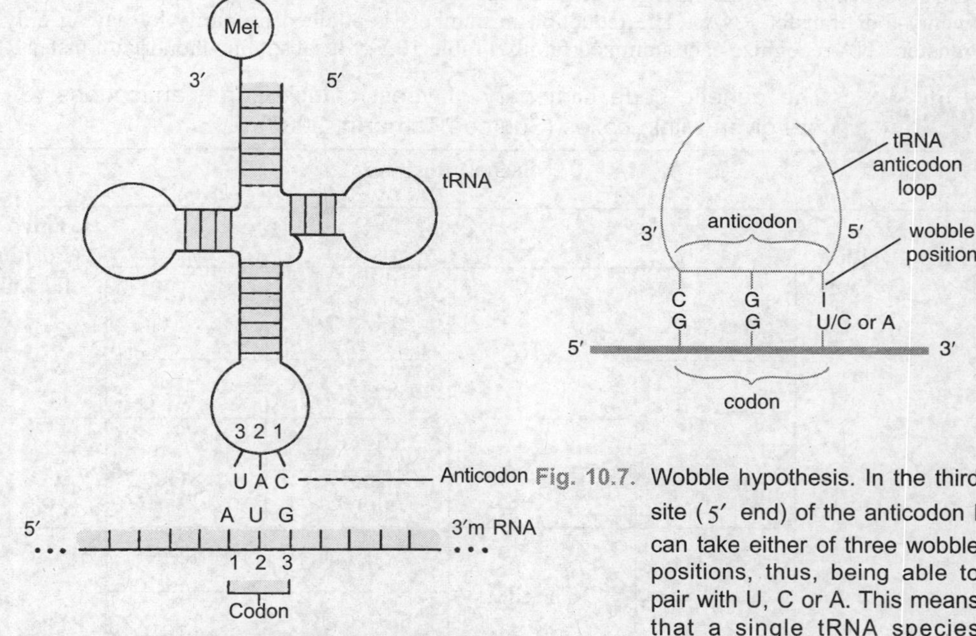

Fig. 10.6. Codon and anticodon base positions are numbered from the 5′ end. The 3′ position in the codon (5′ in the anticodon) is the wobble base (after Tamarin, 2002).

Fig. 10.7. Wobble hypothesis. In the third site (5′ end) of the anticodon I can take either of three wobble positions, thus, being able to pair with U, C or A. This means that a single tRNA species carrying an amino acid can recognize three codons in mRNA: GCU, GCC and GCA (After Suzuki *et al.*, 1986).

Table 10.4. Pairing combinations at the third codon position (Source : **Tamarin,** 2002).

Number-one base in tRNA (5′ end)	Number-three base in *mRNA* (3′ end)
G	U *or* C
C	G
A	U
U	A *or* G
I	A, U, *or* C

10.5. DEVIATIONS FROM UNIVERSALITY OF GENETIC CODE

Molecular biologists concluded by 1979 that the genetic code was universal. That is the code dictionary (Table 10.2) was the same for *E. coli,* human beings and oak trees, as well as other species investigated up to that time. The universality of the genetic code was demonstrated, for example, by taking the ribosomes and messenger RNA from rabbit reticulocytes and mixing them with the aminoacyl-tRNA and other translational components of *E. coli.* Rabbit haemoglobin was synthesized.

Around 1980, however researchers noted divergence when sequencing mitochondrial genes for structural proteins. It was discovered that there were *two* kinds of deviations from universality in the way mitochondrial transfer RNAs read the code. First, fewer transfer RNAs were needed to read the genetic code, second, there were several cases in which the mitochondrial and cellular systems interpreted a codon differently.

According to Crick's wobble rules (Table 10.4), thirty-two transfer RNAs (including one for initiation) can complement all sixty-one non-terminating codons. For example, unmixed families of codons require two transfer RNAs, and mixed families of codons require one, two or three transfer RNAs, depending on the family. The yeast mitochondrial coding system apparently needs only twenty-four transfer RNAs. The reduction in numbers is attained primarily by having only one transfer RNA recognize each unmixed family (Table 10.5). Because mitochondrial transfer RNAs

Table 10.5. The genetic code dictionary of yeast mitochondria; anticodons (3′→5′) are given within boxes (Source : **Tamarin,** 2002).

First Position (5′ End)	Second Position: U	C	A	G	Third Position (3′ End)
U	Phe AAG	Ser AGU	Try AUG	Cys AGG	U C
U	Leu AAU		*stop*	Trp ACU	A G
C	Thr GAU	Pro GGU	His GUG	Arg GCA	U C
C			Glon GUU		A G
A	IIe UAG	Thr UGU	Asn UUG	Ser UCG	U C
A	Met UAC		Lys UUU	Arg UCU	A G
G	Val CAU	Ala CGU	Asp CUG	Gly CCU	UC
G			Glu CUU		A G

for unmixed families of codons have a U in the first (*i.e.,* wobble) position of the anticodon, apparently, given the structure of the mitochondrial transfer RNAs, the U can pair with U, C, A or G. Most probably, evolutionary pressure has minimized the number of transfer RNA genes in the DNA of the mitochondrion, in keeping with its small size. Reduction from thirty-two to twenty-four is a 25% savings. Recent evidence suggests that mammalian mitochondria may need only twenty-two transfer RNAs.

It has also been found that yeast mitochondria read the CUX family (*i.e.,* CUU, CUC, CUA, CUG) as threonine rather than as leucine (Table 10.2 and Table 10.5) and the terminator UCA (opal) as tryptophan rather than as termination. However, there have been found differences among mitochondria of different groups of organisms reading the CUX family. Human and *Neurospora* mitochondria appear to read the CUX codons as leucine just as cellular systems do. Of the groups so far analyzed, only yeast reads the CUX family as threonine. Similarly, human and *Drosophila* mitochondria read AGA and AGG as stop signals rather than as arginine (Table 10.6).

Table 10.6. Common and alternative meanings of codons (Source: **Tamarin**, 2002).

Codon	General meaning	Alternative meaning
1. CUX	Leu	Thr in yeast mitochondria
2. AUA	Ile	Met in mitochondria of yeast, *Drosophila* and Vertebrates
3. UGA	Stop	Trp in mycoplasmas and mitochondria other than higher plants.
4. AGA/AGG	Arg	*Stop* in mitochondria of yeast and vertebrates in mitochondria of *Drosophila*
5. CGG	Arg	Trp in mitochondria of higher plants
6. UAA/UAG	*Stop*	Gln in ciliated protozoa
7. UAG	*Stop*	Ala or Leu in mitochondria of some higher plants

The ciliated protozoan *Paramecium* species read the UAA and UAG stop codons as glutamine within the cell. In addition, a prokaryote (*Mycoplasma capricolum*) reads UGA codon as tryptophan. We do not yet know how general this finding is : scientists have analysed the genetic code of very few species. We can thus conclude that *the genetic code seems to have universal tendencies among prokaryotes, eukaryotes and viruses.* Mitochondria however, read the code slightly differently : different, woble rules apply, and mitochondria and cells read at least one terminator and one unmixed family of codons differently. Also the mitochondrial deviations are not universal among all types of mitochondria. Further work, involving the sequencing of more mitochondrial DNAs, should explain the pattern of deviations among the mitochondria of diverse species. Further, it has also become clear that not every organism reads all codons in the same way. For example, ciliated protozoa and a mycoplasma read some stop signals as coding for amino acids. Likewise, nuclear variants are known in the following codons : CUG, AUA, UAA, UAG, UGA, CGG and AGA. Mitochondrial variants are known in CUX, AUA, UAA, UAG, AAA, UGA, CGX, AGA and AGG.

Site specific variation. In this type of variation in codon reading, the interpretation of a codon depends on its specific location. For example, codons GUG and, rarely, UUG can serve as prokaryotic initiation codons. This means that they are recognized by $tRNA_f^{Met}$. However, they are not recognised by $tRNA_m^{Met}$. (*i.e.,* GUG and UUG are not misread internally in messenger RNAs). In some cases, two of the termination codons (UGA and UAG, but not UAA) are misinterpreted as codons for amino acids. That is, termination will not occur at the normal place, resulting in a longer-than usual protein. In some cases, these "read-through" proteins are vital — the organism depends on their existence. For example, in the phage $\theta\beta$, the coat-protein gene is read through about 2% of the time. Without this small number of read through proteins, the phage coat (capsid) cannot be constructed properly.

Selenocysteine codon. Site-specific variation also involves the amino acid selenocysteine (*i.e.,* cysteine with a selenium atom replacing the sulphur). Although many proteins have unusual amino acids, almost all are due to post-translational modifications of normal amino acids. However, the amino acid selenocysteine is inserted directly into some proteins, such as formate dehydrogenase in *E. coli*, which has selenium in its active site. Selenocysteine is inserted into the protein by a new transfer RNA that recognises the termination codon, UGA, if that codon involved in a particular stem-loop secondary structure in the messenger RNA. The selenocysteine transfer RNA is originally charged with a serine that is then modified to a selenocysteine. In addition to the stem-loop structure 3′ (downstream) from the amber codon (UAG), a selenocysteine elongation factor (SELB) is also needed at the ribosome. This same mechanism may occur in eukaryotes, but not all of the components have yet been recognized.

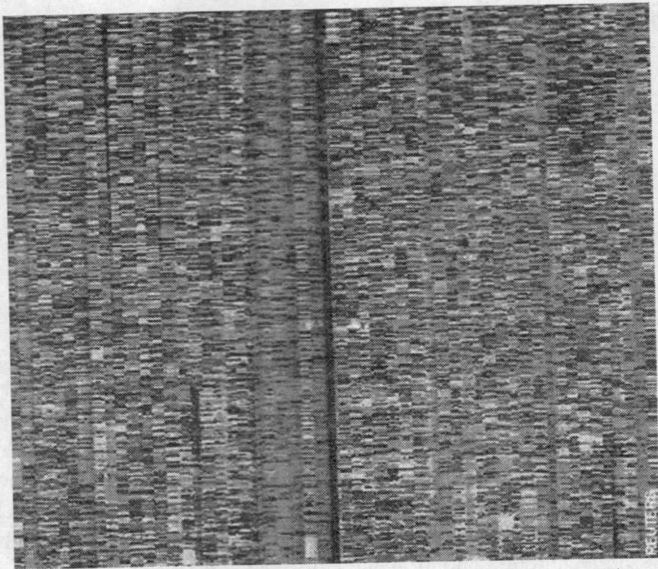

A part of the 3 billion letter map of human genetic code.

10.6. EVOLUTION OF GENETIC CODE

Certain modern theories have suggested that the genetic code has the wobble in it because it originally arose from a code in which only the first two bases were needed for the small number of amino acids in use several billion years ago. As new amino acids with useful properties became available, they were incorporated into proteins by a code modified by the third base, albeit with less specificity. This view has been supported by the fact that codons starting with the same nucleotide come from the same biosynthetic pathway. This indicates that in early evolution, as biosynthetic pathways were extended to new amino acids, the newcomers were incorporated by use of the second and the third bases of the code.

Let us see whether there is a relationship between the codons and the amino acids they code for, or is the code just one of many random/chance possibilities. That is, whether the genetic code is highly evolved or just a "frozen accident". Recent computer simulations of random codes indicate that the current genetic code is far outside the range of random in its ability to protect the organism from mutation. This suggests that the genetic code is not a frozen accident, but rather is highly evolved. Numerous examples in the current code support this view.

For example, in the unmixed family $5' - \text{CUX} - 3'$, any mutation in the third position produces another codon for the same amino acid. Wobble in the third position and codon arrangement ensures that less than half of the mutations in the third codon position result in the specification of a different amino acid.

Some patterns also have been observed in the genetic code. For example, the mutation of one codon to another results in an amino acid of similar properties. A high probability exists that such a mutation will produce a functional protein. All the codons with U as the middle base, for example, are for amino acids that are hydrophobic (*e.g.,* phenylalanine, leucine, isoleucine, methionine and valine). Mutation in the first or third positions for any of these codons still codes a hydrophobic amino acids. Both of the two negatively charged amino acids, aspartic acid and glutamic acid, have codons that start with GA. All of the aromatic amino acids—phenylalanine, tyrosine and tryptophan–have codons that begin with uracil. Such patterns minimize the negative effects of mutation (see **Tamarin,** 2002).

QUESTIONS AND PROBLEMS

1. What do you understand by 'genetic code'? How will you show that the minimum size of a code word should be triplet?
2. A basic concept of molecular biology is the collinearity of gene, DNA, RNA and protein. Discuss the various lines of evidence supporting or providing this concept.
3. Gamow (1954) pointed out that since the genetic language contains only four letters, A, U = (T), C and G, if all code words are of the same size, codons must be at least three bases long.
   (*a*) Show why codons cannot consist of one or of two bases.
   (*b*) Discuss two lines of research that indicate that codons are three base long.
   (*c*) Since codons are three base long, 64 different triplets can exist. Illustrate these using the branching method.
   (*d*) How many of the 64 triplets will contain (1) no adenine, (2) at least one adenine?
4. (*a*) Explain what is meant by a degenerate code and illustrate your answer to show degeneracy in translation.
   (*b*) Which are the nonsense triplets and why are they so termed?
   (*c*) Do codons have the same meaning *in vitro as in vivo* ?
5. Describe the general properties of the genetic code.
6. Give a brief account of the experiments which helped in deciphering the genetic code.
7. Discuss the experiments conducted by H.G. Khorana and his coworkers for cracking the genetic code.
8. The size of haemoglobin gene in humans is estimated to consist of approximately 450 nucleotide pairs. The protein product of the gene is estimated to consist of about 150 amino acid residues. Calculate how many nucleotides are present in a genetic codon for each amino acid.
9. How many triplet codons can be made from the four ribonucleotides A, U, G and C containing (a) no uracils; (b) one or more uracils?
10. Assume a series of different one-base changes in the codon GGA, which produces following several new codons: (a) UGA; (b) GAA; (c) GGC, (d) CGA. Which of these represent (s) degeneracy, which missense, and which nonsense?
11. If single-base changes occur in DNA (and, therefore, in mRNA), which amino acid, tryptophan or arginine, is most likely to be replaced by another in protein synthesis?
12. Write short notes on the following :
    (1) Homopolymers; (2) Copolymers of repetitive sequences; (3) Poly U; (4) Frame shift mutation; (5) Overlapping genes; (6) Wobble hypothesis; (7) Split genes.
13. (a) How was the genetic code first decoded ? What refinements have since been incorporated in the technique.
14. In what sense and to what extent is the genetic code (a) degenerate and (b) universal?
15. What evidence supports the hypothesis of colinearity between the nucleotide sequence in a gene and the amino acid sequence in a polypeptide?
16. Why is colinearity between codons and polypeptides significant?
17. **A. Garen** (1968) extensively studied a particular nonsense (Chain-termination) mutation in the alkaline phosphatase gene of *E. coli.* This mutation resulted in the termination of the alkaline phosphatase polypeptide chain at a position where the amino acid tryptophan occurred in the wild-type polypeptide. Garen induced revertants (in this case, mutations altering the same codon) of this mutant with chemical mutagens that induced single base-pair substitutions and sequenced the mutant polypeptides. Seven different types of revertants were found, each with a different amino acid at the tryptophan position of the wild-type polypeptide (termination

position of the mutant polypeptide fragment). The amino acids present at this position in the various revertants included tryptophan, serine, tyrosine, leucine, glutamic acid, glutamine and lysine. Was the nonsense mutation studied by Garen an *amber* (UAG), an *ochre* (UAA), or an *opal* (UGA) nonsense mutation? Explain the basis of your deduction.

Answers to Problems

8. Approximately 3 nucleotides code for each amino acid.
9. (*a*) Since uracil represents 1 among 4 nucleotides; the probability that uracil will be the first letter of the codon is 1/4 and the probability that U will not be the first letter is 3/4. The same reasoning holds true for the second and third letters of the codon. The probability that none of the three letters of the codon are uracils is $(3/4)^3 = 27/64$.
   (*b*) The number of codons containing at least one uracil is $1 - 27/64 = 37/64$.
10. (*a*) Nonsense; UGA is a stop signal and does not translate.
    (*b*) Missense; GGA codes for glycine, GAA for glutamic acid.
    (*c*) Degeneracy; both GGA and GGC code for glycine.
    (*d*) Missense; GGA codes for glycine, CGA for arginine.
11. Tryptophan, which has only one codon. Arginine has six codons.
13. (*a*) Synthetic RNA molecules (polyuridylic acid molecules) containing only the base uracil were prepared when these synthetic molecules were used to activate *in vitro* protein synthesis systems, small polypeptides containing only the amino acid phenylalanine (polyphenyl alanine molecules) were synthesized. Codons composed only of uracil were therefore shown to specify phenylalanine. Similar experiments were carried out using synthetic RNA molecules with different base compositions. (b) Better *in vitro* systems activated with synthetic RNA molecules with known repeated base sequences were developed ultimately, *in vitro* systems in which specific aminoacyl- t RNAs were shown to bind to ribosomes activated with specific mini-mRNAs, which were trinucleotides of known base sequence, were developed and used in codon identification.
14. (*a*) The genetic code is degenerated in that all but 2 of the 20 amino acids are specified by two or more codons. Some amino acids are specified by six different codons. The degeneracy occurs largely at the third or 3′ base of the codons. "*Partial degeneracy*" occurs where the third base of the codon may be either of the two purines or either of the two pyrimidines and the codon still specifies the same amino acid. "*Complete degeneracy*" occurs where the third base of the codon may be any one of the four bases and the codon still specifies the same amino acid. (b) The code appears to be almost completely universal. Known exceptions to universality include strains carrying suppressor mutations that alter the reading of certain codons (with low efficiencies in most cases) and the use of UGA as a tryptophan codon in yeast and human mitochondria.
15. Colinearity received strong support from studies that showed a direct correlation between the linear sequence of mutational sites in a gene (established by genetic mapping experiments) and the linear sequence of mutational defects (amino acid substitutions or chain terminations) in the polypeptide gene-product (established by purification and amino acid sequencing, etc., of mutant polypeptides).
16. Colinearity is an important prediction of our present concepts of noncoding sequences ("introns" or "intervening sequences") within eukaryotic genes does not violate the concept of colinearity; their presence simply means that the colinear structures frequently contain interruptions.
17. *Amber* (UAG). This is the only nonsense codon that is related to tryptophan, serine, tyrosine, leucine, glutamic acid, glutamine and lysine codons by a single base pair substitution in each case.

MULTIPLE CHOICE QUESTIONS

Choose the correct answer from the four alternatives given.

1. Genetic code is sequence of nitrogenous bases on
   (*a*) mRNA (*b*) tRNA
   (*c*) rRNA (*d*) DNA
2. The terminator codons are
   (*a*) UAA, UAG, UGA
   (*b*) AUG, UAG, UGA
   (*c*) UAC, AUG, UAG
   (*d*) AUG, ACG, GAG
3. Who discovered the entire genetic code ?
   (*a*) Nirenberg (*b*) Ochoa
   (*c*) Khorana (*d*) Crick
4. A codon consists of how many nitrogen bases in it
   (*a*) 1 (*b*) 2
   (*c*) 3 (*d*) 4
5. The arrangement of three bases in DNA represents
   (*a*) protein (*b*) amino acid
   (*c*) plasmid (*d*) nucleic acid
6. Degeneration of a genetic code is attributed to the
   (*a*) first member of codon
   (*b*) second member of codon
   (*c*) entire codon
   (*d*) third member of codon
7. In the genetic code dictionary, how many codons are used to code for all the 20 essential amino acids ?
   (*a*) 20 (*b*) 64
   (*c*) 61 (*d*) 60
8. What would happen if in a gene encoding a polypeptide of 50 amino acids, 25th codon (UAU) is mutated to UAA ?
   (*a*) a polypeptide of 24 amino acids will be formed
   (*b*) two polypeptide of 24 and 25 amino acids will be formed
   (*c*) a polypeptide of 49 amino acids will be formed
   (*d*) a polypeptide of 25 amino acids will be formed
9. Which of the chemical characteristics is not common to all living beings ?
   (*a*) ribosomes are sites of protein synthesis
   (*b*) types of protein present in the body
   (*c*) energy is stored by high phosphate bond,
   (*d*) similar triplet codes for amino acid

ANSWERS

1. (*a*) 2. (*a*) 3. (*a*) 4. (*c*) 5. (*b*)
6. (*d*) 7. (*c*) 8. (*c*) 9. (*d*).

CHAPTER

11

Phenotypic Expression of Gene

Chapter Contents

11.1. GENE EXPRESSION

The process by which the biological information contained in the gene is made available to the cell, is called **gene expression**.

The Central Dogma

The process that we know as gene expression was put forward by **Francis Crick** in 1958 in a lecture called "On protein synthesis" presented to the Society For Experimental Biology. Crick postulated that the biological information contained in the DNA of the gene is transferred first to RNA and then to protein (Fig. 11.1). Crick also stated that the information flow is unidirectional, that proteins cannot themselves direct synthesis of RNA and that RNA cannot direct synthesis of DNA (In other words, once information has passed to protein it cannot get out again). This part of central dogma was shaken in 1970 when **Howard Temin** and **David Baltimore** independently discovered that certain viruses do transfer biological information from RNA to DNA. However, the central dogma still remains today as one of the underlying concepts of molecular genetics.

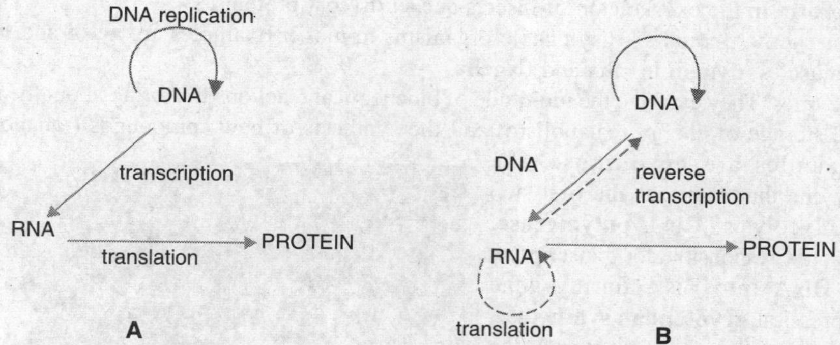

Fig. 11.1. Crick's central dogma. A-Transfer of biological information that occur in all cells. B-Transfer of biological information that occurs in a few organisms (Viruses).

Gene expression occurs in the following steps :

1. Transcription (The first stage of gene expression). All genes undergo the first stage of gene expression which is called **transcription** (Fig. 11.2). During transcription the template strand of the gene directs synthesis of an RNA molecule, the sequence of this RNA transcript being determined by **complementary base pairing**.

2. Translation (The second stage of gene expression). For some genes the RNA transcript is itself the end- product of gene expression. For others the transcript undergoes the second stage of gene expression, called **translation.** During translation the RNA molecule (called **messenger RNA or mRNA**) directs synthesis of a polypeptide, the amino acid sequence of which is determined by the nucleotide sequence of the mRNA (which is of course itself derived from the nucleotide sequence of the gene). Each **triplet** of adjacent ribonucleotides specifies a single amino acid of the polypeptide, the identity of the amino acid corresponding to each triplet being set by the **genetic code.**

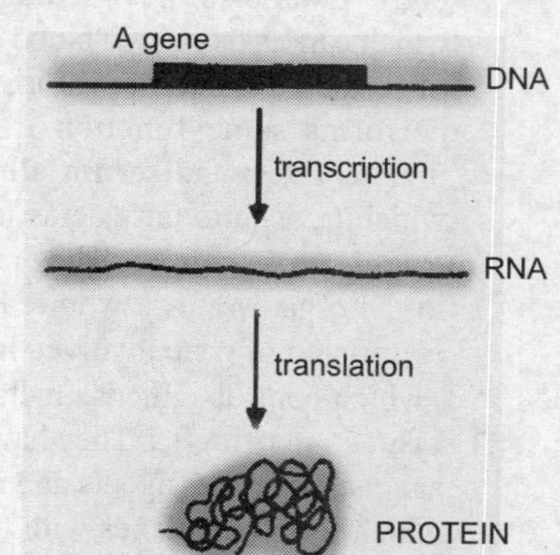

Fig. 11.2. The two stages of gene expression.

3. Protein synthesis (Key to expression of biological information). For all genes the end- point of gene expression is synthesis of either an RNA molecule or a polypeptide. How can this simple information bring about the utilization of the biological information contained in genes and thereby enable construction (building) of a living organism ?

The answer lies in the functional flexibility of protein molecules. This is because polypeptides of different amino acid sequences can have quite different chemical properties and can therefore play quite different roles in the cell as follows :

***(i)* Structural proteins.** They form parts of the framework of organisms. Examples are **collagen** which is associated with tendons, bone and cartilage in vertebrates,

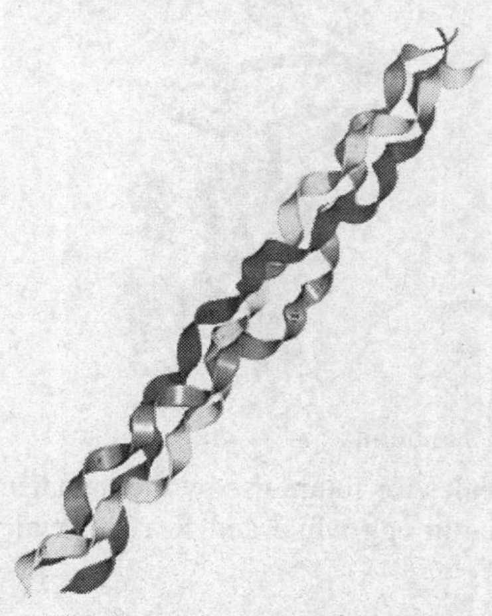

Collagen : A structural protein.

sclerotin in the exoskeleton of insects and virus-coat proteins.

(*ii*) **Contractile proteins.** They enable organisms to move. Examples are **actin** and **myosin** in muscles, **dynein** in cilia and flagella.

(*iii*) **Enzymes.** They catalyze the multitude of biochemical reactions that bring about the release and storage of energy (catabolism) and the synthesis of new compounds (anabolism) . Examples are **hexokinase**, the enzyme that catalyses the first step in glycolysis; **RNA polymerase**, which is responsible for transcription of DNA into RNA during gene expression; **tryptophan synthetase**, which catalyzes the last step in the lengthy biosynthetic pathway that converts phosphoenol- pyruvate into the amino acid tryptophan.

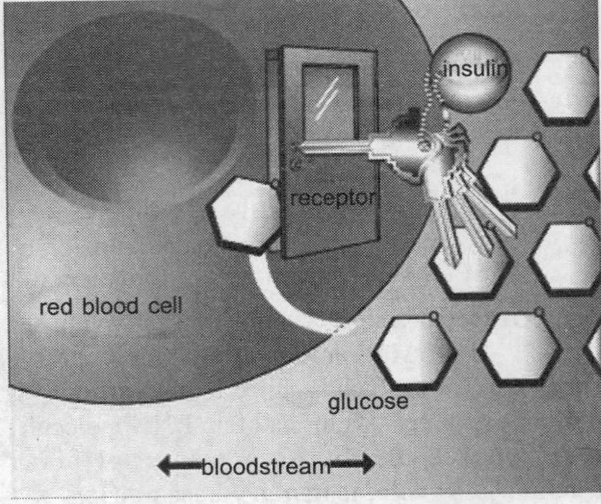

Insulin : A Regulatory Protein.

(*iv*) **Transport proteins.** They carry important molecules around the body. Examples are **haemoglobin**, which carries oxygen in blood stream of vertebrates, **haemocyanin**, which performs same function in some invertebrates; and **serum albumin**, which transports fatty acids in the blood.

(*v*) **Regulatory proteins.** They control and coordinates biochemical reactions in the cell and in the organism as a whole. Examples are **catabolic activator protein**, which regulates expression of gene involved in sugar metabolism in *E.coli* , and **hormones** such as insulin , which controls glucose metabolism in vertebrates.

(*vi*) **Protective proteins.** They have evolved to greatest sophistication in vertebrates and protect against infectious agents and injury. Examples are **immunoglobulins** and other antibodies which form complexes with foreign proteins, and **thrombin** and other components of the blood clotting mechanism.

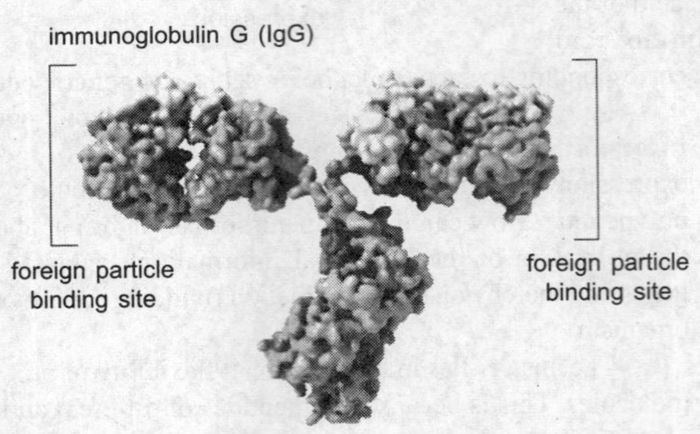

Immunoglobulin : A protective protein.

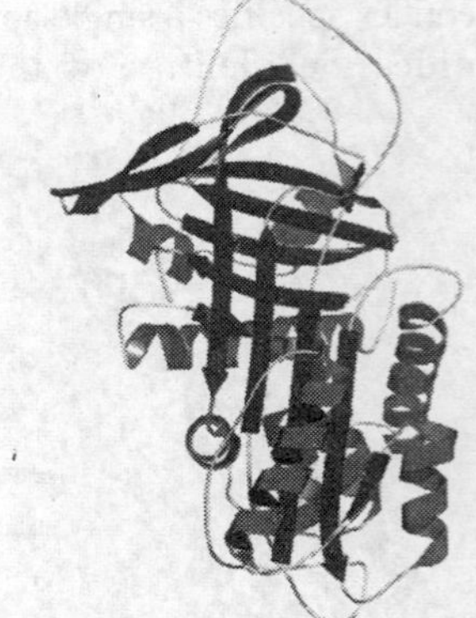

Ovalbumin : A storage protein.

(*vii*) **Storage proteins.** They store compounds and molecules for future use by the organism. Examples are **ovalbumin**, which stores amino acids in egg white and **ferritin**, which stores iron in the liver.

The development and functioning of a living organism can be looked on as the coordinated activity of a vast range of different protein molecules. The biological information obtained in the genes, **the blueprint for life itself** , is simply the instructions for sythesizing these proteins at the correct time and in the correct place.

11.2. PHENOGENESIS

DNA of genes has two essential functions : (i) replication or self-reproduction and (ii) intervention in **phenogenesis**. The phenogenesis is a mechanism by which the phenotype of an organism is realized or produced under the control of DNA in a given environment which includes not only external factors such as temperature and the amount or quality of light, but also internal factors such as hormones and enzymes. The phenotype of an organism is not the direct outcome of action of its DNA, but is the result of various embryological and biochemical activities of its cells from zygotic to multicellular stage. All these cellular activities basically involve a variety of structural and functional or enzymatic proteins. Enzymatic proteins or **enzymes** perform catalytic functions, causing the splitting or union of various cellular molecules. All of the biochemical reactions of cell constitute the subject of **intermediary metabolism**. Each reaction of intermediary metabolism occurs as stepwise conversions of one substance into another. All the steps which transform a precursor substance to its end product which is ultimately expressed in to a structural or functional phenotypic trait, constitute a **biosynthetic pathway**. Each step of a biosynthetic pathway is catalyzed by a specific enzyme which in its turn being produced by a specific gene. It has become clear now that (1) DNA of genes itself, does not have an enzymatic character, therefore, does not involve directly in a biosynthetic pathway of cell ; (2) between the gene and final product or between the gene and enzyme there is a long, tortuous path; (3) the immediate or primary gene product is a kind of RNA called **mRNA** (messenger ribonucleic acid) which is complementary to the DNA of the genes and which controls the assemblage of the amino acids to form enzymes, at the surface of cytoplasmic ribosomes: Thus, to produce a particular phenotypic trait DNA transcribes mRNA which translates into an enzymatic or structural protein which ultimately produces a phenotypic trait by the process of phenogenesis. The relationship between genes and enzymes had been suspected very early in the history of biochemical or physiological genetics.

Box 11.1.

All enzymes are not proteins.

For several decades, the dogma in biology has been that molecular reactions occurring in living cells are catalyzed by enzymes which are composed of polypeptide. We how know that the introns* of some precursor RNA molecules such as the rRNA precursors in *Tetrahymena thermophilia* are removed autocatalytically ("self-spliced") with no involvement of any protein (**Kruger** *et al.*, 1982). Some primary transcripts of genes in mitochondria and chloroplasts are also excised autocatalytically. An RNA molecule that has catalytic activity is called **ribozymes** (see **Brown**, 2002).

**Introns* are intervening sequence, of DNA bases within eukaryotic genes that are not represented in the mature RNA transcripts because they spliced out of the primary RNA transcript.

11.3. GENE-ENZYME RELATION

The understanding of correlationship between genes and phenotype of an organism requires a consideration of the consequences of gene malfunction, since the expression of abnormal gene effect is a major source of the information gained concerning normal gene function. The groundwork for a functional relationship between genes and enzymes was laid in 1902 when **Bateson** reported that a rare human defect alcaptonuria was inherited as a recessive trait. Then in 1909, **Garrod**, 'in his book, "**Inborn errors of metabolism**" suggested relationship between genes and specific chemical reactions. Alcaptonuria was among the heritable diseases with which he dealt at length. But his remarkable piece of work remain unnoticed until 1940 when **Beadle, Tatum, Ephrussi** and other geneticists had got modern understanding of gene action by performing experiments on *Neurospora* and *Drosophila.* They found that mutational changes of genes can be related to losses of specific enzymes. This concept was widely known as "**one gene one enzyme hypothesis**." This hypothesis can be understood well by knowing the phenomenon of **metabolic block**.

Metabolic Block

A biosynthetic pathway, often, requires more than two enzymes and each enzyme is specified by a single gene. If, theré occurs any mutation in any gene, then either the production of normal enzyme will be ceased or a defective enzyme will be produced. This defective enzyme instead of determining the normal phenotype, will determine an abnormal, defective or mutant phenotype. Such mutant genes, which alter the enzymes of a biosynthetic pathway in such a manner that a defective phenotype is resulted are said to cause a "**metabolic blocks**." The metabolic blocks have been found to cause various abnormal phenotypes or diseases in variety of organisms and provide best opportunities to understand the relation of genes with enzymes and phenotypes of an organism more properly.

Thus, the mutations in genes that control sequential biological steps have proved to be invaluable for "dissecting" the number and order of reactions in given biosynthetic pathway. In particular, a field of study known as **biochemical genetics** has combined genetics and biochemistry to elucidate the nature of metabolic pathways, most notably in haploid organisms whose growth requirements are known and whose gene expression is not complicated by allelic interactions.

Approach of Biochemical Genetics

The approach followed in biochemical genetics is to assemble a collection of mutant strains that cannot synthesize a particular component and thus require it for growth. These strains are subjected to complementation tests to estimate how many separate genes are involved, and the genes are mapped to determine their relationships. Strains in a given complementation group are then tested for their ability to grow when supplied with known metabolic precursors of the final component. It then can grow in the presence of a certain precursor, they must suffer from a **genetic lesion** affecting a step before the synthesis of that precursor. If instead they cannot grow when a certain precursor is supplied, the genetic lesion must affect a step that follows the synthesis of that precursor.

11.4. EXAMPLES OF BIOCHEMICAL GENETICS

1. Phenylalanine Metabolism in Human beings

The metabolism of the aromatic amino acids—phenylalanine and tyrosine in human beings provides a best example of a gene-controlled, enzyme catalyzed biochemical reaction. In human beings, phenylalanine is an essential amino acid which must be supplied in the dietary proteins. Once in the body, phenylalanine may follow any of three paths. It may be (1) incorporated into cellular proteins, (2) converted to phenylpyruvic acid, or (3) converted to tyrosine. The conversion of phenylalanine into tyrosine takes place in the presence of phenylalanine hydroxylase enzyme, in the liver cells. Tyrosine is converted in turn to 3-4-dihydroxy phenylalanine (nick-named DOPA) by another enzyme and DOPA serves as a precursor for the hormones adrenaline and noradrenaline and for the black pigment, melanin. Tyrosine itself serves as a precursor of the hormones thyroxine and triiodothyronine. Excess tyrosine is degraded to carbon dioxide and water by a series of steps which involves the formation of *p*-hydroxyphenyl pyruvate, 2-5 dihydroxy phenyl pyruvate, homogentisic acid, maleylacetoacetic acid, fumaryl acetoacetic acid and fumaric and acetoacetic acid. Excess phenylalanine is degraded by a series of steps to compounds which include phenylpyruvic acid and phenyl lactic acid.

Genetic disorders of the phenylalanine metabolism and resulted diseases. Five rare diseases in human beings result from improper functioning of five enzyme systems (*i.e.*, metabolic blocks) involved in the metabolism of phenylalanine, tyrosine and their derivatives. All of these diseases have been found to be due to mutant, autosomal recessive genes in homozygous conditions. These diseases are following :

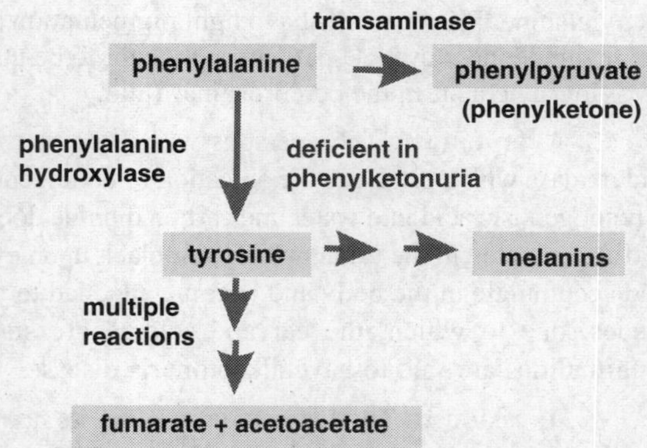

Phenylketonuria : Genetic Disorder of Phenylalanine Metabolism.

1. Phenylketonuria. Persons with genotype *pp* fail to produce enzyme phenylalanine hydroxylase (parahydroxylase) with the result that phenylalanine fails to convert into tyrosine and consequently, the concentration of phenylalanine rises in the blood plasma, cerebrospinal fluid and urine. The urine of phenylketonuric (**PKU**) patient contains (in addition to phenylalanine) elevated amounts of phenylpyruvic acid, phenyl lactic acid and other derivatives of

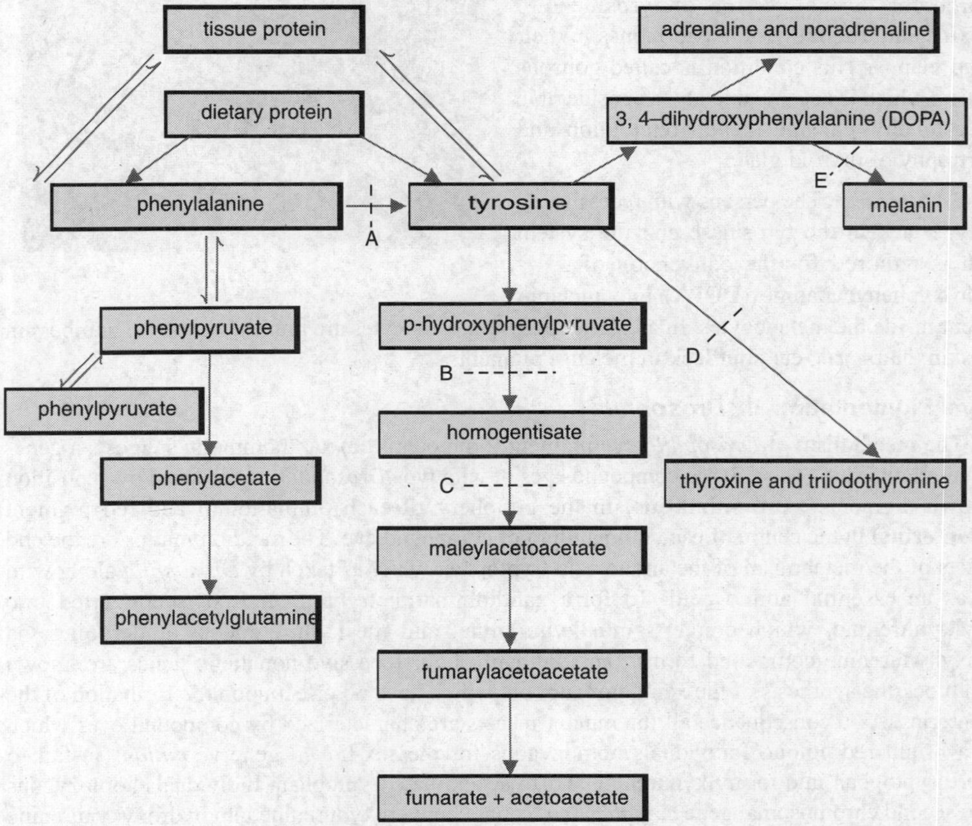

Fig. 11.3. Possible genetic blocks in the normal metabolism of the amino acids phenylalanine and tyrosine in human beings. These blocks lead to the production of the following genetic syndromes. A—phenylketonuria; B—tyrosinosis ; C—alkaptonuria ; D—goitrous cretinism; E—albinism (after DeRoberits, *et.al.*, 1970).

phenylalanine. PKU patients have light pigmentation and are physically and mentally retarded. The feeble mindedness in PKU patients is thought to be due to an impairment of the brain tissues by the phenylpyruvic acid in the cerebrospinal fluid.

2. Alkaptonuria. The persons with genotype *hh* fail to produce the enzyme homogentisic acid oxidase which catalyzes the oxidation of homogentisic acid. Therefore, in them, normal oxidation of homogentisic acid into water and carbon dioxide does not occur and large amounts of homogentisic acid are excreted in the urine, which turn black upon exposure to the air. Moreover, the homogentisic acid accumulate in the body and become attached to the collagen of cartilage and other connective tissues, due to which, the ear and sclerae are stained black. Persons with such phenotypic abnormalities are said to have alkaptonuria disease.

3. Tyrosinosis. The recessive gene, *t* in its homozygous condition, blocks the conversion of *p*-hydroxyphenylpyruvate into 2, 5-dihydroxyphenyl pyruvate. This leads to the accumulation of tyrosine, excesses of which are excreted via urine. This condition is called tyrosinosis. It is reported in only one human and cause no harmful effect.

4. Goitrous cretinism. The persons with *cc* genotype fail to produce the enzyme which is required for the conversion of tyrosine into thyroxine and triiodothyronine hormones in their thyroid glands. This condition is called goitrous cretinism which is accompanied by a considerable degree of physical and mental retardation and hypertrophy of thyroid gland.

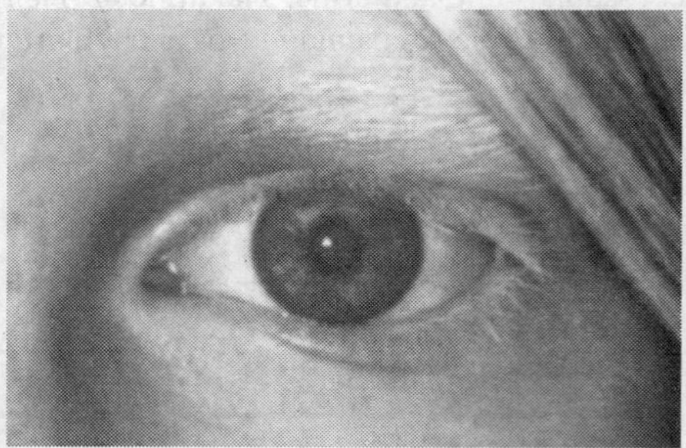

Albinism.

5. Albinism. The persons with recessive *aa* genotype lack in the tyrosinase enzyme system which is required for the conversion of 3, 4-dihydroxyphenyl alanine (DOPA) into melanin pigment inside the melanocytes. In an albino patient melanocytes are present in normal numbers in their skin, hairs, iris, etc., but lack in melanin pigment.

2. Eye Pigmentation in *Drosophila*

The metabolism of *Drosophila* eye pigment (ommochrome) xanthommatin suggests a gene-enzyme-relationship. The dull red compound eyes of wild type *Drosophila* is produced by deposition of brown pigment (**xanthommatins**) in the periphery of each ommatidium and red pigment (**drosopterins**) in the centre of each ommatidum of compound eye. The xanthommatins are the end products of the metabolism of the amino acid tryptophan which is taken by *Drosophila* along with food as an essential amino acid. To form xanthommatin, tryptophan first is converted into formylkynurenine, which converts into kynurenine, and the kynurenine by undergoing via hydroxykynurenine compound forms xanthommatin. Four recessive nonallelic genes are known which block the synthesis of the xanthommatin but which have no effect upon the formation of the drosopterins. As a consequence all the mutant insects are characterized by compound eyes which have a bright red colour. Individuals homozygous for the sex-linked gene *vermilion* (*v*) fail to convert tryptophan into formylkynurenine and so accumulate tryptophan. Individuals homozygous for the second chromosomal gene *cinnabar* (*cn*) cannot convert kynurenine into hydroxykynurenine and as a consequence, they accumulate kynurenine. Individuals homozygous for the third chromosomal mutants *scarlet* (*st*) or *cardinal* (*cd*) produce the water soluble compounds, formylkynurenine, kynurenine and hydroxykynurenine.

Table 11.1. Gene control of tryptophan metabolism in *Drosophila*.

Biochemical reactions	Mutants	Accumulation
Biochemical reaction ↓ Formylkynurenine	Vermilion (*vv*)	Tryptophan
↓ Kynurenine ↓	Cinnabar (*cn*)	Kynurenine
Hydroxykynurenine ↓ Xanthommatine	Scarlet (*st*) or Cardinal (*cd*)	Formylkynurenine Kynurenine and hydroxylkynurenine.

3. The Arginine Pathway in *Neurospora*

The concept of one gene-one enzyme-one phenotypic effect relationship, is best exemplified by *Neurospora crassa*. The wild type *Neurospora crassa* can live on a simple artificial medium containing inorganic salts, a source of organic carbon and the vitamin biotin (such simple medium is called **minimal medium**). Evidently, it has the inborn capacity to synthesize all the other vitamins, amino acids and nitrogenous bases essential to normal development. On the assumption that this capacity was genetically conferred, **Beadle** and **Tatum** argued that mutation could lead to nutritional requirements over and above those satisfied by a minimal medium and that the nature of the biosynthetic (metabolic) block could be determined from the nature of the growth supplement which restored normal development.

In fact, they found that the treatment of conidia of *N. crassa* by X-rays or ultra-violet rays, increased the frequency of such mutant strains which could not survive on minimal medium but developed normally when one or other supplement was added to it. In each instance, genetic tests show that the mutant strain produced by irradiation differs from the normal wild type by a single

Table 11.2. The arginine pathway In *Neurospora*.

Wild type Pathway	Mutant number	Requirement	Accumulated metaboilte	Enzyme absent
+ ↓	arg 4-9	Ornithine, Citrulline or Arginine	—	—
Ornithine + ↓	arg-12	Citrulline or Arginine	—	Ornithine transcarb-amylase
Citrulline + ↓	arg-1	Arginine only	Citrulline	Arginino-succinate synthetase
Arginino succinic acid + ↓ Arginine	arg-10	Arginine only	Arginino-succinic acid	Arginino-succinase

gene and chemical test shows that the addition of a single substance to the minimal medium will enable the mutant strain to grow normally. This indicates that each normal gene of wild strain produces a single enzyme which regulates one step in the biosynthesis of this particular chemical ; the mutant gene does not produce this enzyme.

For example, wild type *N. crassa* can synthesize arginine amino acid such as follows:-

Sugar + ammonia (in minimal medium) —→ glutamic acid —→ glutamic semialdehyde —→ ornithine —→ citrulline —→ arginine.

But there are about 12 mutant strains of *N. crassa* which loss the capacity to synthesize arginine. These mutant strains are called **arginine mutant strains** and numbered as arg-1, arg-2...... arg-12. It has been found that out of these 12 arginine mutants, some-strains (arg-1 and arg-10) have specific requirement for arginine only ; other strains (*e.g.*, arg-12) grow normally if either arginine or citrulline is added to the environment ; still different strains (*e.g.*, arg-4 to arg-9) are able to grow on a supplement of arginine, citrulline or ornithine.

The existence of these three types of mutants indicates a sequence of reactions involved in the synthesis of arginine. Since, some mutants can utilize citrulline but not ornithine, the latter must precede citrulline in the arginine synthetic pathway. Similarly, \citrulline must precede arginine. Some of the steps of arginine pathway as exemplified by these three kinds of mutants of *N. crassa* have been summarized in Table 11.2.

4. Arginine Biosynthesis Pathway of *E. coli*

Certain mutant strains of *E. coli* require arginine for growth. Biochemical genetics of these strains of *E. coli* have shown that arginine biosynthesis pathway of *E. coli* involved eight genes and strains having mutation in any one gene from these eight contain specific growth requirements. For example, arg **H** strains, can grow only if they are given arginine ; no other precursor will do. The arg **G** strains, on the other hand, will grow either in the presence of arginine or of arginosuccinic acid. The arg **F** strains can grow if supplied with arginine, arginosuccinic acid or citrulline, and so on. The arginine biosynthetic pathway of *E. coli* can be represented as follows :

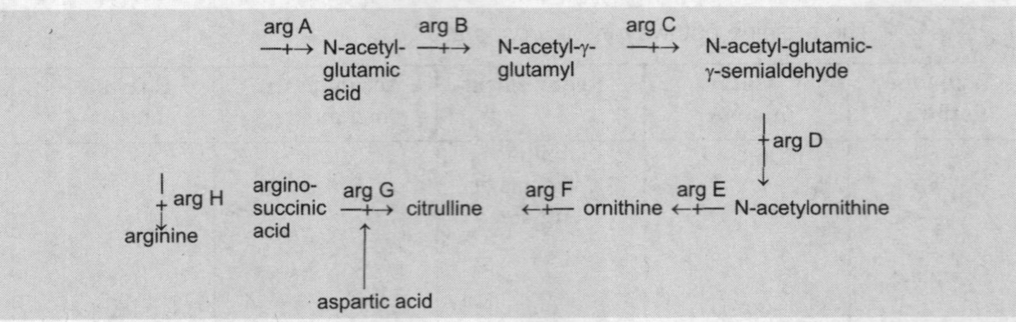

Fig. 11.4. Pathways of arginine biosynthesis and a spectrum of mutant strains of *E. coli* blocked at the indicated positions.

Besides, *Neurospora*, *E. coli*, Lactobacillus (**Volcani** and **Snell**, 1948) and *Aspergillus* (**Pontecorvo** 1953) have similar arginine pathway and so have confirmed the concept of one gene-one enzyme-one phenotype concept more strongly.

11.5. GENETIC DISSECTIONS OF COMPLEX BIOCHEMICAL PATHWAYS

The approach of biochemical genetics has later on used to dissect quite complex metabolic pathways of certain microbes (bacteriophages and *Chlamydomonas*). Perhaps the most elegant of these studies, initiated by **R. Edgar** and **W. Wood**, concerns morphogenesis of the phage T_4, a process that involves the elaborate and sequential self assembly of numerous different kinds of proteins.

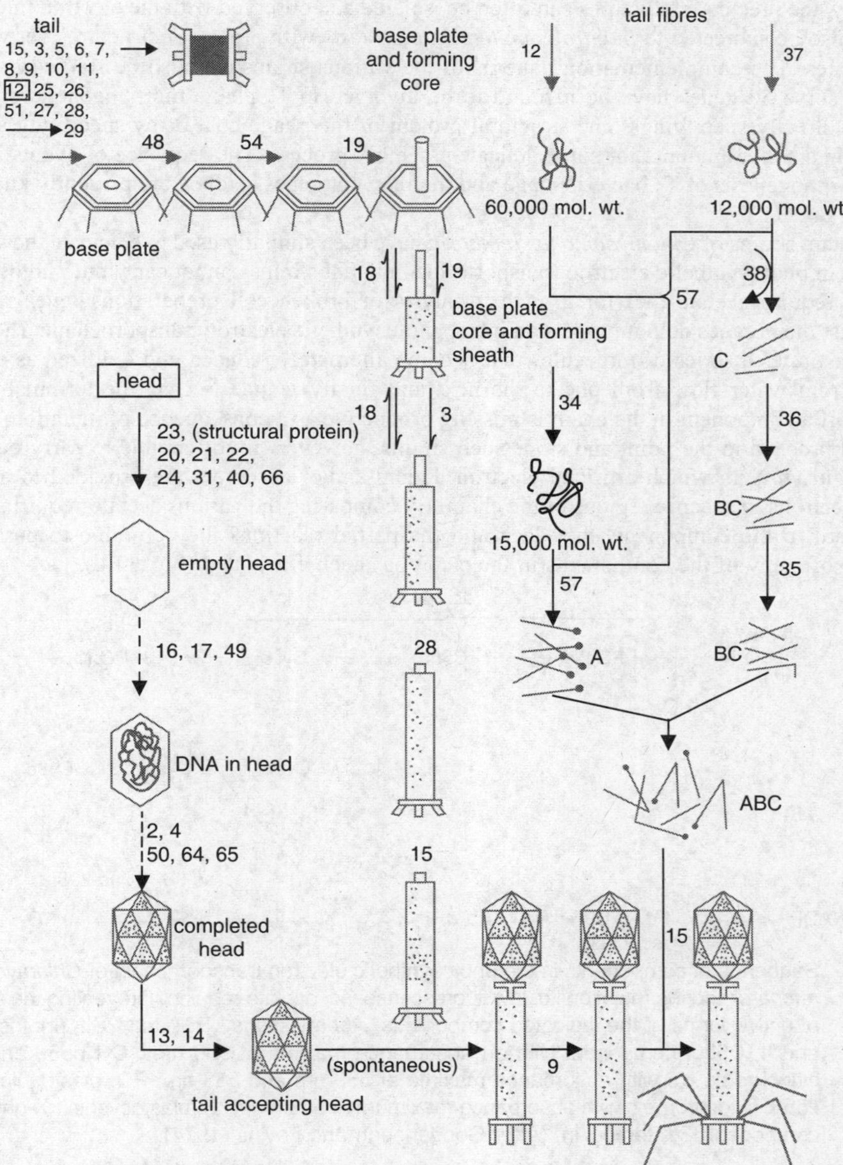

Fig. 11.5. Morphogenetic pathway of T_4 bacteriophage showing individual pathways for head, tail, and tail-fibre assembly. Dashed arrows indicate steps that have not yet been demonstrated *in vitro*. Number (s) associated with each arrow indicates the gene controlling that step or protein in the assembly process. Gene 12 participates both in the tail and in the tail fibre sequence (After Goodenough and Levine, 1974).

The analysis begins by isolating conditional lethal strains of T_4 bacteriophage that carry *amber* or temperature-sensitive mutations in genes essential to the assembly of mature phages. Each strain is allowed to infect *E. coli* cells under non-permissive conditions, (*Note* : Permissive means supporting genetic replication as of a virus). Whereupon phage assembly proceeds up to the point at which an essential protein is missing or inactive. Depending on the protein that is absent, different precursor phage structures will accumulate in the infected cells. If the cells are now lysed artificially, the precursor structures can often be isolated and observed with the electron microscope and can also be subjected to *in vitro complementation tests* with precursor structures derived from other lysates. The complementation patterns of the various strains can also be studied *in vivo*. In this way some 50 genes have been found to be involved in T_4 phage morphogenetic processes, either by directly specifying some structural protein of the phage coat or by specifying a protein required in smaller amounts for some delicate assembly process. The sequence of events involved in the morphogenesis of T_4 bacteriophage and the relevent genes, as they are presently known, are shown in Fig. 11.5.

Mutant strains of *Chlamydomonas reinhardi* have been similarly used to elucidate the sequence of events in photosynthetic electron transport. Such mutant strains cannot carry out photosynthesis and thus require acetate (ac) for growth. Analysis of broken cell preparations reveals that the chloroplast membranes do not possess an intact photosynthetic electron transport chain. The normal chain constitutes a series of molecules that become alternately reduced and oxidized as electrons derived from water flow from one to the next and finally reduce NADP, the terminal electron acceptor. If a component in the chain is missing or inactive as a consequence of mutation, electron flow will proceed to the point and stop. Such strains, however, are often able to carry out partial reactions *in vitro*, in which artificial electron donors and/or acceptors are provided to accept or donate electrons in specific regions of the chain. By comparing the various acetate-requiring strains with regard to the components they lack and the partial reactions they are able to perform the sequence of many of the components in the chain has been elucidated (Fig. 11.6).

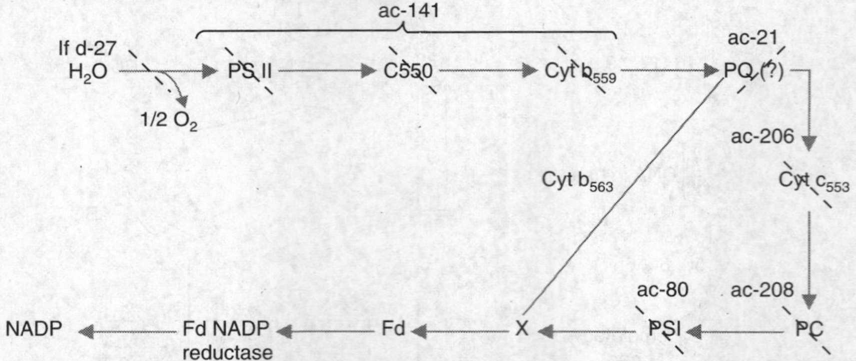

Fig. 11.6. Sequence of components in the photosynthetic electron transport chain of *Chlamydomonas reinhardi* and the mutation lfd, low fluorescence; ac, acetate-requiring preventing the synthesis of active forms of the indicated components. *Abbreviations* : PSI and PSII, photosystems I and II; C550, component with an absorbance maxima at 550 nm ; Cyt b559 and b 563. cytochrome b's with absorbance maxima at 559 nm and 563 nm; PQ. plastoquinone; Cyt c553, cytochrome c with absorbance maximum at 553 nm ; PC, plastocyanin ; X-unidentified component; Fd, ferridoxin, (After Goodenough and Levine, 1974).

When it has become confirmed that genes have relation with the phenotype of the organism via structural and enzymatic proteins, the synthesis of which depends on them (genes), then it becomes necessary for us to have some basic knowledge about the structure of proteins.

11.6. STRUCTURE OF PROTEINS

Proteins, like nucleic acids, are polymers and they consist of a sequence of conjoined monomeric α-amino acids (**AA**) units of the general type $H_2N—CH(R)—COOH$. In biological systems about 40 different amino acids are usually found, but most biological proteins basically contain only 20 kinds of amino acids. These twenty kinds of naturally occurring amino acids have been classified into following four categories by **Lehninger** (1970)—(a) **Amino acids with non-polar R groups**—Alanine (ala), Valine (val), Leucine (leu), Isoleucine (ile), Proline (pro), Phenylalanine (phe), Tryptophan (trp), Methionine (met); (b) **Amino acids with uncharged polar R groups**—Glycine (gly), Serine (ser), Threonine (thr), Cysteine (cys), Tyrosine (tyr), Asparagine (asn), Glutamine (glu); (c) **Amino acids with charged polar groups at pH 6.0 to 7.0**—(i) Basic amino-acids (postively charged at pH 6.0)—Lysine (lys), Arginine (arg), Histidine (his); (ii) Acidic amino acids (negatively charged at pH 6.0)—Aspartic acid (asp), Glutamic acid (glu). Here R is stand for radical.

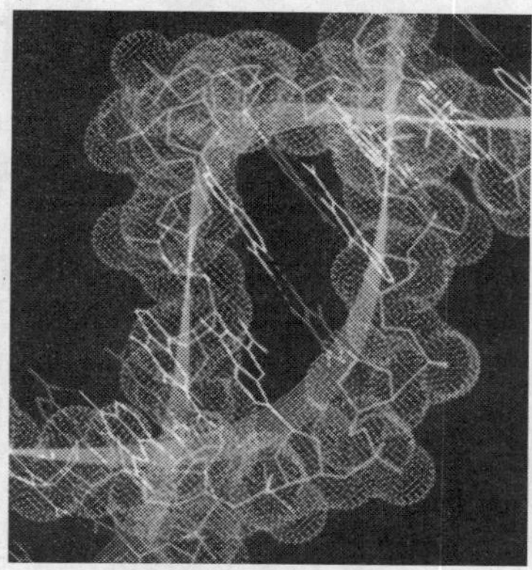

Structure of Protein.

Peptide linkages and polypeptides. α -amino acids are capable of forming **polypeptide chains** through the association of the amino (—NH_2) group of one amino acid and the hydroxyl (—OH) group of the carboxyl (COOH) radical of another amino acid. The resultant amide (—CONH) bonds or linkages between amino acids of a polypeptide chain are called **peptide bonds** or **linkages**. The respective number of peptide bonds in a peptide or amino acidic chain determines the name of that peptide such as, a **dipeptide** has two such bonds, a **tripeptide** has three such peptide bonds and a **polypeptide** has more than two such

$$NH_2-\underset{R'}{\overset{H}{C}}-\overset{O}{\overset{\|}{C}}-OH + H-\underset{H}{N}-\underset{R'}{\overset{H}{C}}-\overset{O}{\overset{\|}{C}}-OH \rightarrow NH_2-\underset{R'}{\overset{H}{C}}-\overset{O}{\overset{\|}{C}}-\underset{H}{N}-\underset{R'}{\overset{H}{C}}-\overset{O}{\overset{\|}{C}}-OH + H_2O$$

amino acid — amino acid — dipeptide (peptide bond)

Fig. 11.7. Formation of a dipeptide.

peptide bonds. A polypeptide chain may has variable number of amino acids according to the specific protein of a species. Further, protein molecules may have different numbers of polypeptide chains (see table 11.3).

Table 11.3. Composition and size of some important proteins.

	S.N.	Protein	Source	Number of polypep-tide chain	Molecular's weight of chains	Number of aminoacids
1. Eukaryotes	1.	Cytochrome—C	Human heart	1	12,100	104
	2.	Ribonuclease	Bovine pancreas	1	15,800	124
	3.	Myoglobin	Whale sperm	1	17,816	153
	4.	Chymotrypsinogen	Bovine pancreas	1	2500	246
	5.	Haemoglobin	Human blood	2α, 2β	Single chain = 1600 (total M.W = 64,500)	141 146
2. Prokaiyotes	6.	Lactate dehydrogenase	Human blood	4	3400	—
	7.	TMV		2300	17.420	157
	8.	TMYV		120 or 180	21,300	200
	9.	Bacteriophage T_2		2000	80.000	—

Each polypeptide chain can be considered as starting from an amino acid with a free —NH_2 group (N-terminal) and ending with one free carboxyl (COOH) group (C-terminal). In the formation of polypeptide chains each amino acid contributes an identical peptide group to the back bone of the molecule (—NH. CH.CO—) together with distinguishing (R) group. Each kind of protein, thus, has a unique number and sequence of such R groups which determine the functional possibilities of the protein molecule.

STRUCTURAL LEVELS OF PROTEINS

Biochemists have recognised following structural levels to proteins :

1. Primary structure. The linear sequence of amino acid components in a polypeptide chain constitute the primary molecular structure of a protein. The newly synthesized protein molecule has this structure.

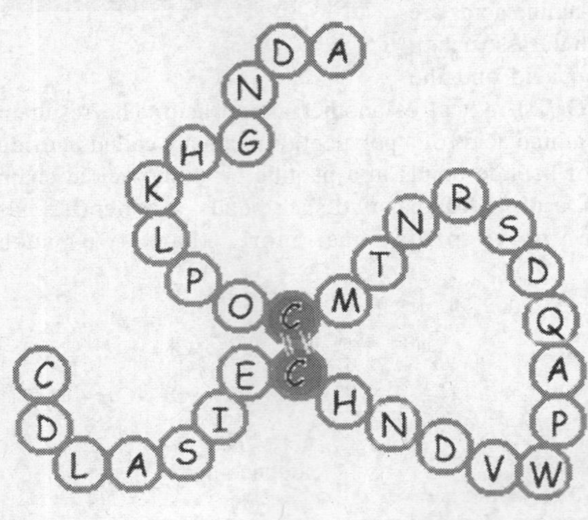

Primary structure of protein.

Secondary structure of protein.

2. Secondary structure. The functional groups of the polypeptide chain have a tendency to form H-bonds between them. Thus, when the sections of primary polypeptide chain are twisted or coiled into a helix, hydrogen bonds are formed between —CO group and an —OH group of amino acids facing each other due to coiling of a polypeptide chain and thus, a **secondary structure** of protein is resulted. In biological proteins, the polypeptide chain remains coiled either in alpha (α) like shape or in beta (β) like shape and consequently are called α**-helix** and β**-helix** respectively.

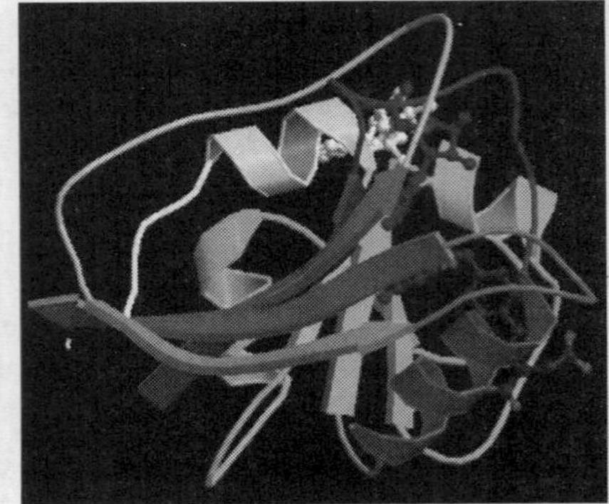

Protein : Tertiary structure.

3. Tertiary structures. In order to compress the very long spiral chain in a globular form, there occur extensive folding over and bending of the helices, which result into tertiary structural features. The tertiary configuration is stabilized by certain intra-chain bonds, especially disulphide (—S— S—) bridges between cysteine residues of polypeptide chain. When, two cysteine amino acids confront each other, they are linked by disulphide bond to form a double amino acid chain. For example, **insulin** enzyme contains two polypeptide chains held together by two —S—S— bridges. **Ribonuclease** enzyme, on the other hand, contains four intra-chain disulphide bridges when gives a three dimensional arrangement of the molecule.

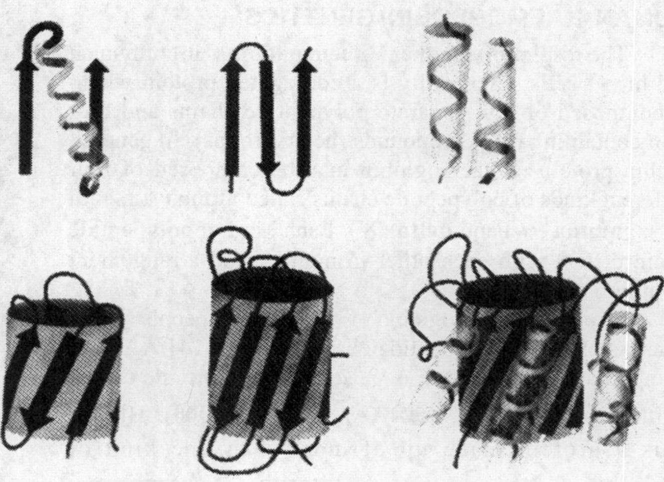

Tertiary structure of protein.

Because, tertiary structure permits interaction over a vast surface and creates interstices between the polypeptide chains into which particular reaction pattern can fit with optimal accuracy, so most tertiary proteins act like catalysts. Enzymes have tertiary structures.

4. Quaternary structure. Sometimes, equivalent inter-chain bonds or bridges link together two or more otherwise independent polypeptide chains and so join subunits into a complex quaternary or multimeric edifice. Most proteins with a molecular weight higher than, 20,000 possess such a quaternary structure and are composed of more than one polypeptide structures. The best example of quaternary molecular structure is haemoglobin (Hb) molecule which has two α-helix and two β-helix, each with molecular weight of about 16,000 and each associated with one haem group. The four chains are disposed in tetrahedral arrangement but barely touch one another, being held together mainly by H-bonds and by electrostatic attractions.

The molecular weight, number of total amino acids and number of polypeptide chains of some biologically important proteins have been tabulated in Table 11.3.

Species Specificity of Proteins

Each cell contains hundreds of different proteins and each kind of cell contains same proteins which are unique to it. There is evidence that every species of plant, animal and microorganisms have certain proteins which differ from those of every other species. The degree of difference in the proteins of two species depends upon the evolutionary relationship of the form involved. Organisms less closely related by evolution have proteins which differ more markedly than those of closely related forms. These facts formulate the **theory of species specificity** which states that the constituent proteins are characteristic of each species differing at least slightly from those of more distantly related species.

11.7. ONE GENE-ONE POLYPEPTIDE CHAIN HYPOTHESIS

The hypothesis of one gene-one protein-one phenotype relationship has not been found correct by later findings of physiological genetics. There are various evidences in human beings and in other organisms which suggest that the synthesis and structure of a single protein molecule can be controlled by two or more genes. Recent works have confirmed that single gene controls the production of a single polypeptide chain of protein molecules, therefore, the concept of one gene-one enzyme, now is termed as the concept of **one-gene one-polypeptide** and it is exemplified by following examples :

HUMAN HAEMOGLOBIN GENETICS

Haemoglobin.

The respiratory pigment, **haemoglobin** (Hb) of human red blood cells, chemically is a conjugated protein which is composed of four separate polypeptide chains and four iron containing ring compounds (heme groups). In general, globin protein of haemoglobin may be composed of four different kinds of polypeptide chains called **alpha** (α), **beta** (β), **gamma** (γ) and **delta** (δ). Each kind of polypeptide chain differs with each other in number and arrangement of amino acids.

98 per cent haemoglobin of normal adult people (HbA) consists of two α-polypeptide chains (each of 141 AA) and two β-chains (each of 146 AA), so that the molecule can be written as $\alpha^A\alpha^A\beta^A\beta^A$ (574AA, M. Wt. 67,000). HbA is thus a protein made up of more than one kind of

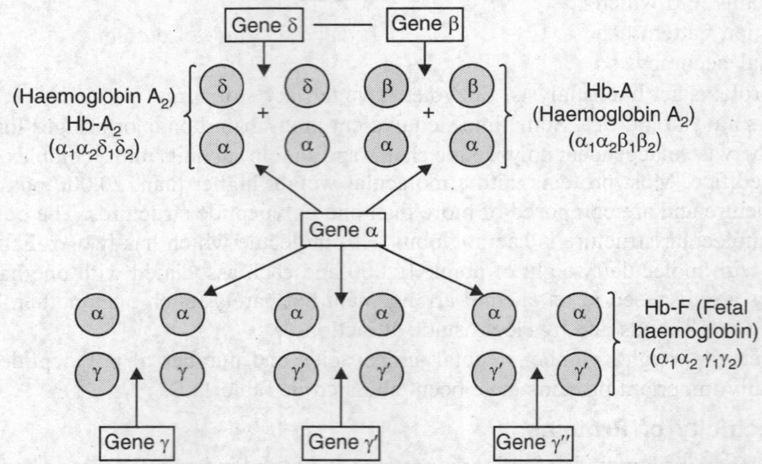

Fig. 11.8. Haemoglobins in man, each complete Haemoglobin (Hb) molecule possessing four polypeptides, two α and two either β, δ, or γ. Genes specifying haemoglobin polypeptides are shown as circles. Connected boxes represent closely linked genes. Alleles are known for genes α and β (after Goodenough and Levine, 1974),

polypeptide and is therefore called a **heteropolymeric** protein. The rest 2 per cent haemoglobin of normal adult human blood contains two delta (δ) polypeptide chains instead of beta chains, such haemoglobin (**HbA$_2$**) can be symbolized $\alpha^A\alpha^A\delta^A\delta^A$. The haemoglobin of foetal blood has two gamma (γ) polypeptide chains instead of β -polypeptide chains, such haemoglobin (**HbF**) can be symbolized as $\alpha^A\alpha^A\gamma^F\gamma^F$. From these variants of haemoglobin [*viz.*, $\alpha^A\alpha^A\beta^A\beta^A$ (HbA), $\alpha^A\alpha^A\delta^A\delta^A$ (HbA$_2$), and $\alpha^A\alpha^A\gamma^F\gamma^F$ (Hb^F)] it becomes evident that α^A polypeptide chains are controlled by a single gene, while, β, γ, δ polypeptide chains are controlled by different forms (alleles) of another single gene or some closely linked genes.

In 1949 **Pauling** and his co-workers reported that the formation of an abnormal haemoglobin (**haemoglobin S, Hb^S**) which differed in its electrophoretic mobility from normal haemoglobin (Hb^A) was the cause of hereditary disease the **sickle-cell anemia** in humans. The red blood cells of individuals suffering from this fatal form of hemolytic anemia undergo a reversible alteration in shape when the oxygen tension of the plasma falls slightly, and they assume elongate, filamentous,

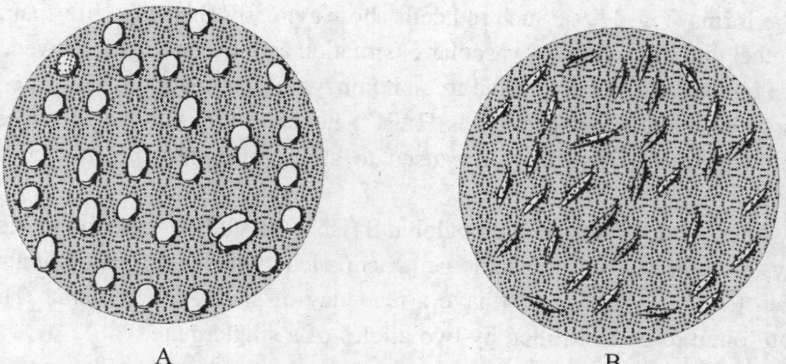

Fig. 11.9. A—Normal erythrocytes; B—Erythrocytes from venous blood of a patient with sickle-cell anemia (after DeRobertis, *et al.*, 1970).

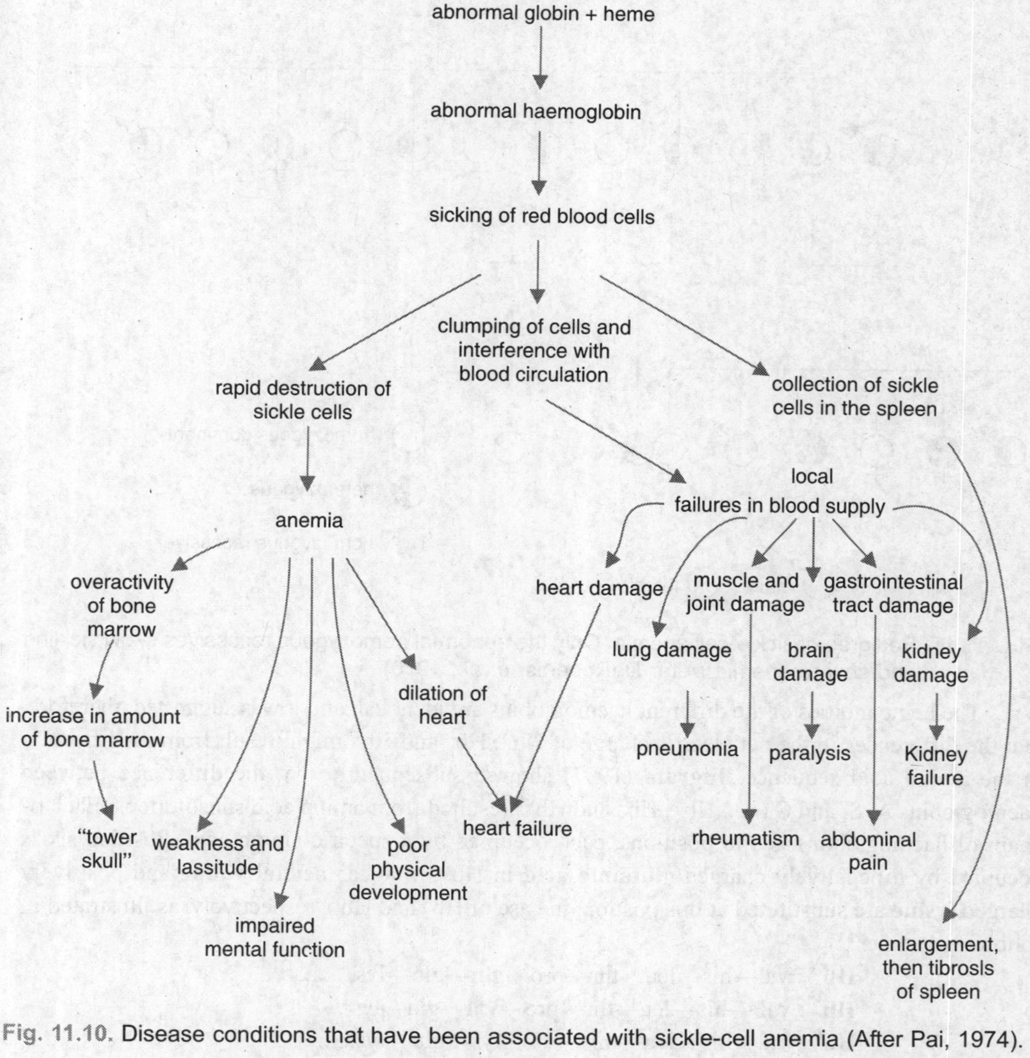

abnormal globin + heme

abnormal haemoglobin

sicking of red blood cells

rapid destruction of sickle cells

clumping of cells and interference with blood circulation

collection of sickle cells in the spleen

anemia

local failures in blood supply

overactivity of bone marrow

heart damage

muscle and joint damage

gastrointestinal tract damage

lung damage

brain damage

kidney damage

increase in amount of bone marrow

dilation of heart

pneumonia

paralysis

Kidney failure

heart failure

"tower skull"

weakness and lassitude

poor physical development

rheumatism

abdominal pain

impaired mental function

enlargement, then tibrosls of spleen

Fig. 11.10. Disease conditions that have been associated with sickle-cell anemia (After Pai, 1974).

and sickle-like forms (Fig. 11.9). Such red cells show a greatly shortened life span, since they tend to clump together and often causing vascular obstruction and are rapidly destroyed. The individuals with sickle-cell anemia have been found to be homozygous for an abnormal gene (H^S). The blood cells of normal individuals (of genotypes H^A/H^A) never sickle, but the blood cells of perfectly healthy heterozygotes (H^A/H^S) can be caused to sickle provided the oxygen concentration is drastically reduced.

Shortly after the discovery of haemoglobin S (Hb^S) the study of certain families which showed as aberrant system of inheritance of sickle-cell anemia led to the identification of a second abnormal haemologlobin C (Hb^C). Hb^C is also have a tendency of sickle-cell anaemia. The Hb^S and Hb^C traits, later on found to be controlled by two alleles of a single gene.

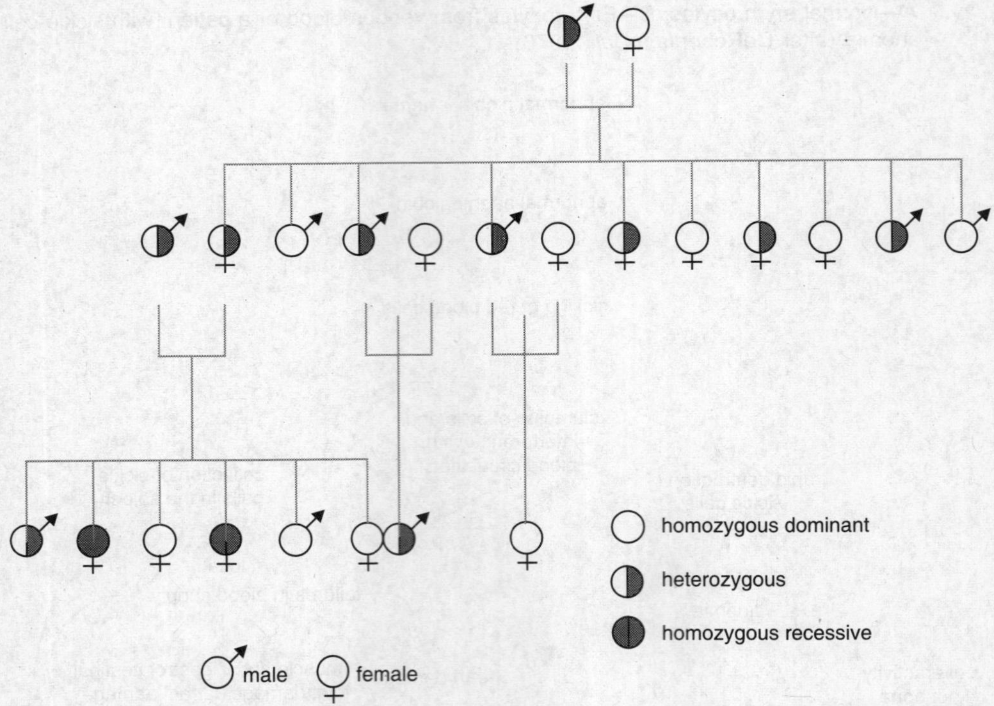

Fig. 11.11. Pedigree of sickle-cell anemia. Only the abnormal homozygous recessives show sickling and severe anemia (After DeRobertis, *et al.*, 1970).

The heme moities of the different haemoglobins are identical, and it was suggested, therefore, that the differences in the net electric charge of Hb^A, Hb^S and Hb^C might result from modification in the amino acid sequence. **Ingram** (1957) showed subsequently that the difference between haemoglobins A, S, and C (*i.e.*, Hb^A, Hb^S and Hb^C) resulted from amino acid substitutions. Each β-chain of haemoglobin has 146 positions, each occupied by a specific amino acid. Position six is occupied by a negatively charged **glutamic acid** in Hb^A; whereas neutral **valine** and positively charged **lysine** are substituted at this postion, in case of Hb^S and Hb^C, respectively, as illustrated in following manner :

Hb^A : val—his—leu—thr—pro—**glu**—glu—lys.........
Hb^S : val —his—leu—thr—pro—**val**—glu—lys.....
Hb^C : val—his—leu—thr—pro—**Iys—**glu—lys......

These amino acids substitutions result in Hb^S and Hb^C molecules bearing a net charge relative to Hb^A of 2+ and 4+. Such charge differences account for the observed differences in the electrophoretic points of the three molecules (Fig. 11.12).

The available informations, therefore, indicates that in the case of haemoglobin at least two chromosomal genes are concerned with the formation of a single protein and that at each locus there occurs a variety of alleles. The genetics of haemoglobin also gives the evidence in support of gene's relation with molecular structure of proteins (*i.e.*, arrangement of amino acids in the polypeptide chains of a protein).

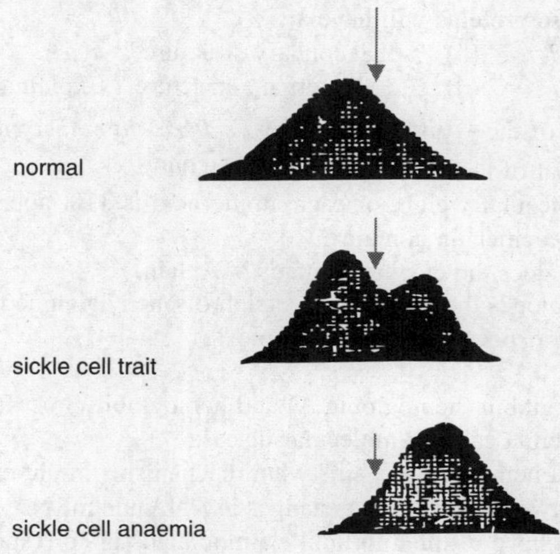

Fig. 11.12. Electrophoretic behaviour of various human haemoglobins. Normal homozygous dominant, sickle-cell trait heterozygous, sickle-cell anemia. The arrows indicate the reference point of origin of the electrophoretic pattern (after DeRobertis, *et. al.*, 1970).

Box 11.2.

Sickle cell disease is caused by replacing a polar glutamic acid residue with a nonpolar valine on the outer surface of the β-globin molecule. This causes increased intermolecular adhesion, leading to aggregation of deoxyhemoglobin and distortion of the red cell. Sickled red blood cells have decreased survival time (leading to anaemia) and tend to occlude capillaries, leading to ischemia and infraction of organs down stream of the blockage (**Strachan** and **Reid**, 1996).

Example of Sickle cell anemia shows drastic influence of a change in single amino acid in a protein having 560 amino acids (2 chains of α and 2 chains of β, each 140 amino acids). There may however be other cases, where changes in one or two amino acids do not affect the phenotype as in insulin.

QUESTIONS AND PROBLEMS

1. (a) In any chain of biochemical reactions how could you tell whether a mutation produced a block before or after a given step ?
   (b) Is a genetic block always complete ? Why ?
   (c) "A mutant gene can change the rate of development by changing the efficiency of catalysis of a single enzyme". Explain.

2. Present evidence for the one-gene-one enzyme hypothesis and explain why it has now been modified to the form one gene, one polypeptide. Illustrate your discussion with examples.
3. (a) What are the primary consequences to the biochemical constitution of cells or tissues of a metabolic block ?
   (b) In what respect can gene-caused disease such as *diabetes mellitus* and *phenylketoneuria* be "cured" (alleviated) ?
4. Ribonuclease contains one polypeptide chain only (124 amino acids); insulin contains two : A, with 21 amino acids and B, with 30 amino acids. Alkaline phosphates has two chains of same kind, whereas human adult haemoglobin contains four polypeptide chains of two kinds, α and β. Which of these proteins will have a :
   (a) Primary structure ? (b) Secondary structure ?
   (c) Tertiary structure ? (d) Quaternary structure ? Explain.
5. The molecular weight of the α and β chains of *Escherichia coli* tryptophan synthetase are 29,500 and 49,500. That of the entire enzyme is 159,000.
   (a) If the average molecular weight of each amino acid is 110, approximately how many amino acids does each chain contain ?
   (b) How many chains does the enzyme contain ? Explain.
6. A single gene change blocks the normal conversion of phenylalanine to tyrosine
   (a) Is the mutant gene expected to be pleiotropic ?
   (b) Explain.
7. How can normal hemoglobin (hemoglobin A) and hemoglobin S be distinguished ?
8. Why is sickle-cell anaemia called a molecular disease ?
9. Assuming that the beta hemoglobin chain? what mechanisms might explain the differences that now exist in these two chains ? What changes in DNA and mRNA codons would account for the differences that have resulted in unlike amino acids in corresponding positions ?

ANSWERS TO PROBLEMS

6. (a) Yes.
   (b) A block would result in the accumulation of phenylalanine and a decrease in the amount of tyrosine, which would be expected to result in several phenotypic expressions.
7. These hemoglobins can be distinguished by mobility of molecules in an electric field (electrophoretic mobility) and by determining the amino acids sequences of their β-polypeptides.
8. The label 'molecular disease' became common in speaking of sickle-cell anaemia because its molecular basis (the substitution of a valine residue at amino acid position number 6 in β-chain) was recognized quite early during the emergence of the science of molecular biology. Actually, most if not all inherited diseases probably have very similar molecular bases. We just don't know what the molecular defects are in the most instances.
9. Mutations : transition, transversions and frameshifts.

MULTIPLE CHOICE QUESTIONS

Choose the correct answer from the four alternatives given.

1. In which of the following disorders, blood has a defective hemoglobin?
   (*a*) hemophilia
   (*b*) hematuria
   (*c*) hematoma
   (*d*) sickle cell anaemia
2. Sickle cell anaemia is caused due to substitution of
   (*a*) valine at 6th position in β chain of Hb by valine

(b) glutamic acid at 6th position in β chain of Hb by valine
(c) glutamic acid at 6th position in α chain of Hb
(d) valine at 6th position in α chain of Hb by glutamic acid

3. The first protein to have its primary structure determined was
(a) urease (b) insulin
(c) glucagon (d) histone

4. The enormous diversity of protein molecules is mainly due to the diversity of
(a) peptide bonds
(b) amino groups on the amino acids
(c) *R* group on the amino acids
(d) amino acid sequences within the protein molecule

5. Albinism is a congenital human disorder resulting from the lack of the enzyme
(a) catalase
(b) fructokinase
(c) tyrosinase
(d) xanthine oxidase

6. Inability to convert dihydroxyphenyl alanine into melanin lead, to which of the conditions given below
(a) alkaptonuria (b) albinism
(c) cretinism (d) glycosauria

7. Who proposed the central dogma of molecular biology ?
(a) Beadle and Tatum
(b) Watson and Crick
(c) Klug
(d) Crick

8. One of these can not be prepared directly from DNA
(a) another DNA
(b) mRNA
(c) protein
(d) tRNA

9. Which one of the following is not a fibrillar protein ?
(a) elastin (b) collagen
(c) myosin (d) albumin

10. The ribozyme is a RNA
(a) without sugar
(b) with extra phosphate
(c) with enzymatic activity
(d) without phosphate

Answers

1. (d) 2. (b) 3. (b) 4. (d) 5. (c) 6. (b)
7. (d) 8. (c) 9. (d) 10. (c).

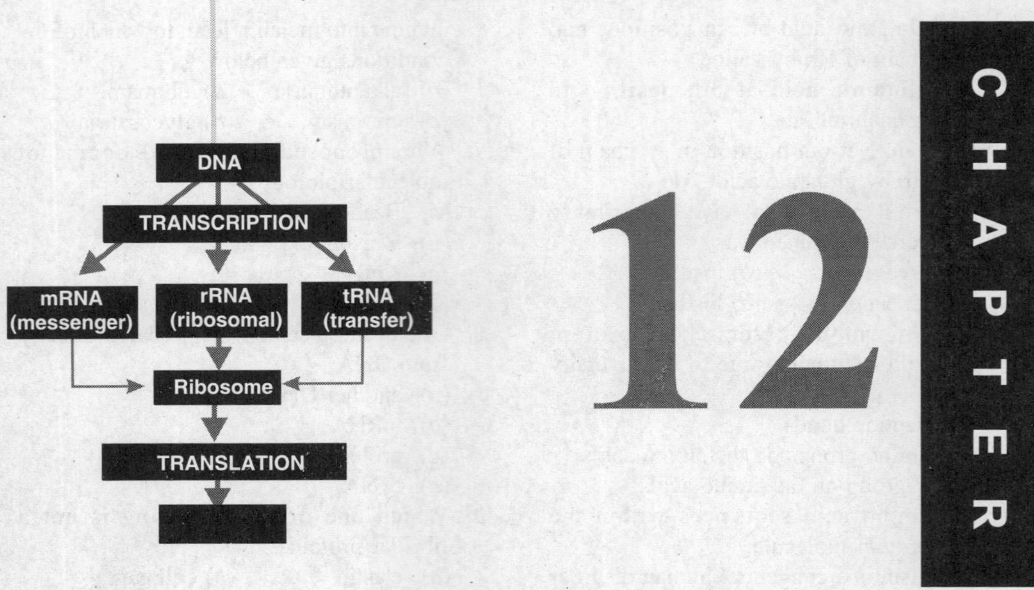

12

Translation : Protein Synthesis

Chapter Contents

DNA, with its correct mechanism of replication, serves to carry genetic information from cell to cell and from generation to generation. This information is translated into proteins that determine the phenotype. Virtually, all the phenotypes examined so far are the result of biochemical reactions that occur in the cell. All of these reactions require enzymes and enzymes are proteins. In fact, more than 2000 types of enzymes have been identified in the living organisms. Each enzyme is a unique molecule catalyzing a specific chemical reaction. Other phenotypes are due primarily to the kinds and amounts of non-enzymatic proteins (including the structural proteins) present, for example, haemoglobin, myoglobin, gamma-globulin (e.g., immunoglobulins or Igs), insulin, cytochrome C, fibroin (silk protein), or collagen. Proteins are composed of one or more, long linear polymers of amino acid residues (polypeptide chains) that are synthesized almost exclusively in the cytoplasm. The topic of this chapter is how the information present in the sequence of bases (=triplet codons) of the mRNA is translated into a sequence of amino acids in proteins. However, at the outset, one point should be clear that the genetic information of mRNA is undirectionally read beginning at the 5′–hydroxyl end by one or more ribosomes (polysomes) and 3′ – end signify the C-terminal amino acid. The 5′ – end of the mRNA correspond to the N-terminal amino acid and ending at 3′-end in the completed protein.

12.1. CENTRAL DOGMA AND CENTRAL DOGMA REVERSE

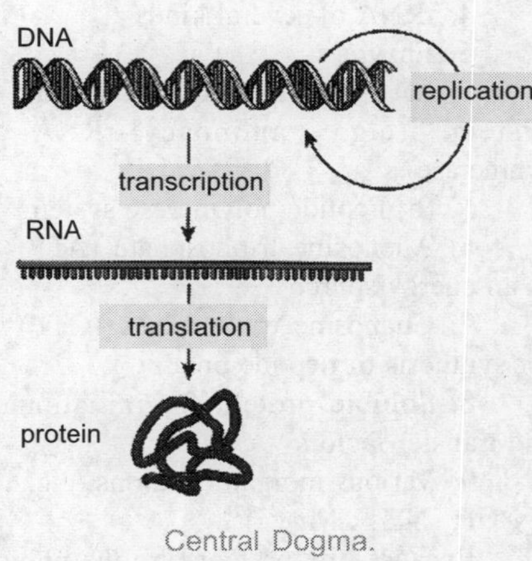

Central Dogma.

The process of synthesis of protein involves one of the central dogma of molecular biology, which postulates that genetic information flows from nucleic acids to protein. (It was first forwarded by **Crick** in 1958). The first step of this central dogma is known as transcription and does not involve a change of code since DNA and mRNA are complementary. The second step involves a change of code from nucleotide sequences to amino acid sequences and is called translation. It can be illustrated as follows:

$$\text{DNA} \xrightarrow{\text{Replication}} \text{DNA} \xrightarrow{\text{transcription}} \text{RNA} \xrightarrow{\text{translation}} \text{Proteins}$$

Thus, according to this central dogma, the flow of information is one way, *i.e.*, from DNA the information is transferred to RNA (mRNA) and from RNA to proteins. In 1968, **Barry Commoner** suggested a circular flow of information, *i.e.*, DNA transcribes RNA, RNA translates into proteins, proteins synthesize RNA and RNA synthesizes DNA, as has been illustrated in following figure:

$$\begin{array}{ccc} \text{DNA} & \longrightarrow & \text{RNA} \\ \uparrow & & \downarrow \\ \text{RNA} & \longleftarrow & \text{Proteins} \end{array}$$

Later on, **Temin** (1970) reported the existence of an enzyme **"RNA-dependent DNA polymerase" (reverse transcriptase)** which could synthesize DNA from a single-stranded RNA template. **D. Baltimore** (1970) also reported the activity of this enzyme in certain tumour viruses. This exciting finding in molecular biology gave rise to the concept of "**central dogma reverse**" or **teminism**, suggesting that the sequence of information flow is not necessarily from DNA to RNA to protein but can also take place from RNA to DNA. The central dogma reverse can be illustrated as follows:

$$\text{DNA} \underset{\text{inverse transcription}}{\overset{\text{transcription}}{\rightleftarrows}} \text{RNA} \xrightarrow{\text{translation}} \text{Proteins}$$

In fact, the step of translation of this central dogma is the time in the flow of information between genes and proteins to change the language being used. In going from DNA to RNA, the language (nucleotide sequences) remained the same. However, in going from RNA to protein the language is changed from a nucleotide sequence to amino acid sequence. Just as in the process of translation from one language to another, this process of using information in RNA to make protein is called **translation**.

12.2. MINIMUM NECESSARY MATERIALS FOR TRANSLATION

Success in polypeptide synthesis in *in vitro* cell free systems shows that the minimum necessary materials for translation are the following:

1. Amino acids (*i.e.*, 20 amino acids forming the pool of amino acids in the cytoplasm)
2. Ribosomes (each of which comprises two subunits which exist as separate subunits prior to the translation of mRNA and contain two tRNA binding sites: the **P site** or **peptidyl site** and an **A** or **aminoacyl site**. One more site called **E** or **exit site** has been recognized in the ribosomes by some workers but this three site model (*i.e.,* A, P and E) still is not popular.

3. mRNA
4. tRNA of several kinds
5. Enzymes
   (a) Amino-acid activating system (e.g., aminoacyl-tRNA-synthetase).
   (b) Peptide polymerase system
6. Adenosine triphosphate (ATP) as an energy source
7. Guanosine triphosphate (GTP) for synthesis of peptide bonds
8. Soluble protein for initiation and transfer factors
9. Various inorganic cations (e.g., K^+, NH^+ NH_4^+, Mg^{2+}).

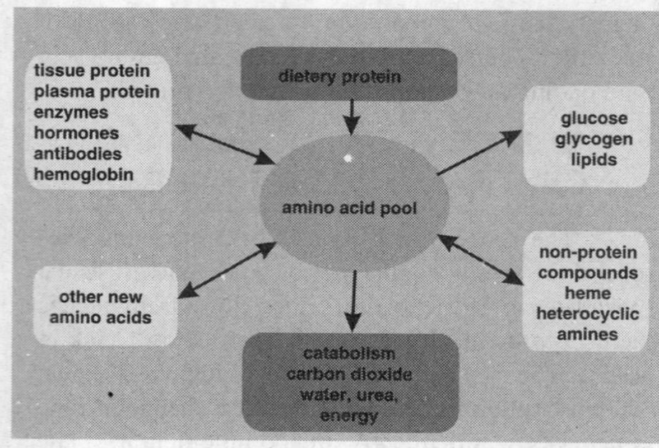

Amino acid pool.

Because, the proteins are the polymers of amino acids therefore, protein synthetic process requires the amino acids as the raw material. All the naturally occurring proteins of living organisms fundamentally are the polymers of about 20 amino acids. These amino acids occur in the matrix forming an **"amino acid pool"** and are readily available for the process of protein synthesis.

All amino acids present in amino acid pool are called *α-amino acids* because one carbon, the α-carbon has four specific groups attached to it: an amino group, a carboxyl (acidic) group, a hydrogen and one of the twenty different R groups (side chains), imparting the specific properties of that amino acid. (Technically, proline is called an **imino acid** because of its structure). Having these four groups attached to C imparts a property known as **chirality** on the amino acid: such as left and right-handed gloves, the mirror images cannot be superimposed. (*Chirality* means handedness or the property by which a form is distinguishable from its mirror image). Because of optical properties, the two forms of each amino acids are referred to as D and L, in which D comes from *d*extrorotatory (right turning) and L comes from *l*evorotatory (left turning). All biologically active amino acids are of L form, and hence we need not refer to this designation. Proteins (polypeptides) are synthesized when peptide bonds form between any two amino acids- called **residues** when

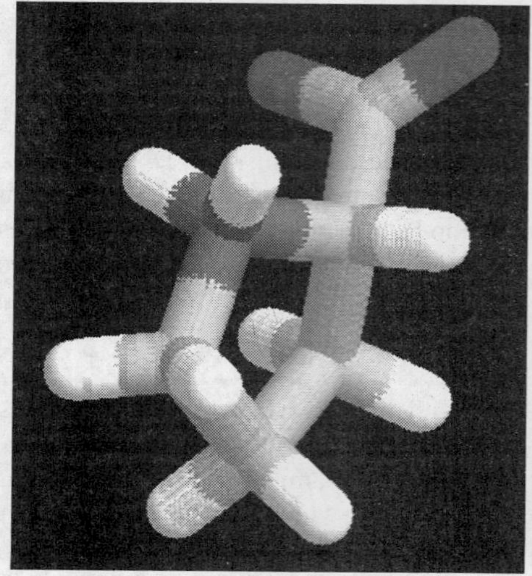

L-proline.

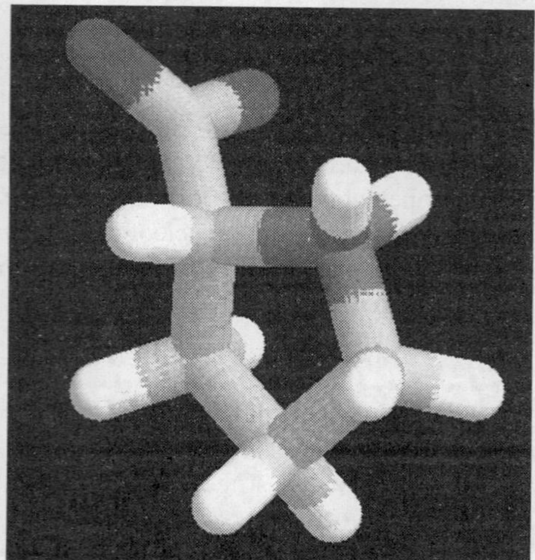

D-proline.

Chirality of proline.

incorporated into a protein – can join and all chains will have an amino (N – terminal) end and a carboxyl (C–terminal) end.

Proteins (polypeptides) are synthesized by the formation of peptide bonds between two amino acids. The polypeptide chain synthesized in such a way will have an amino end (N-terminal) and a carboxyl (C-terminal) end.

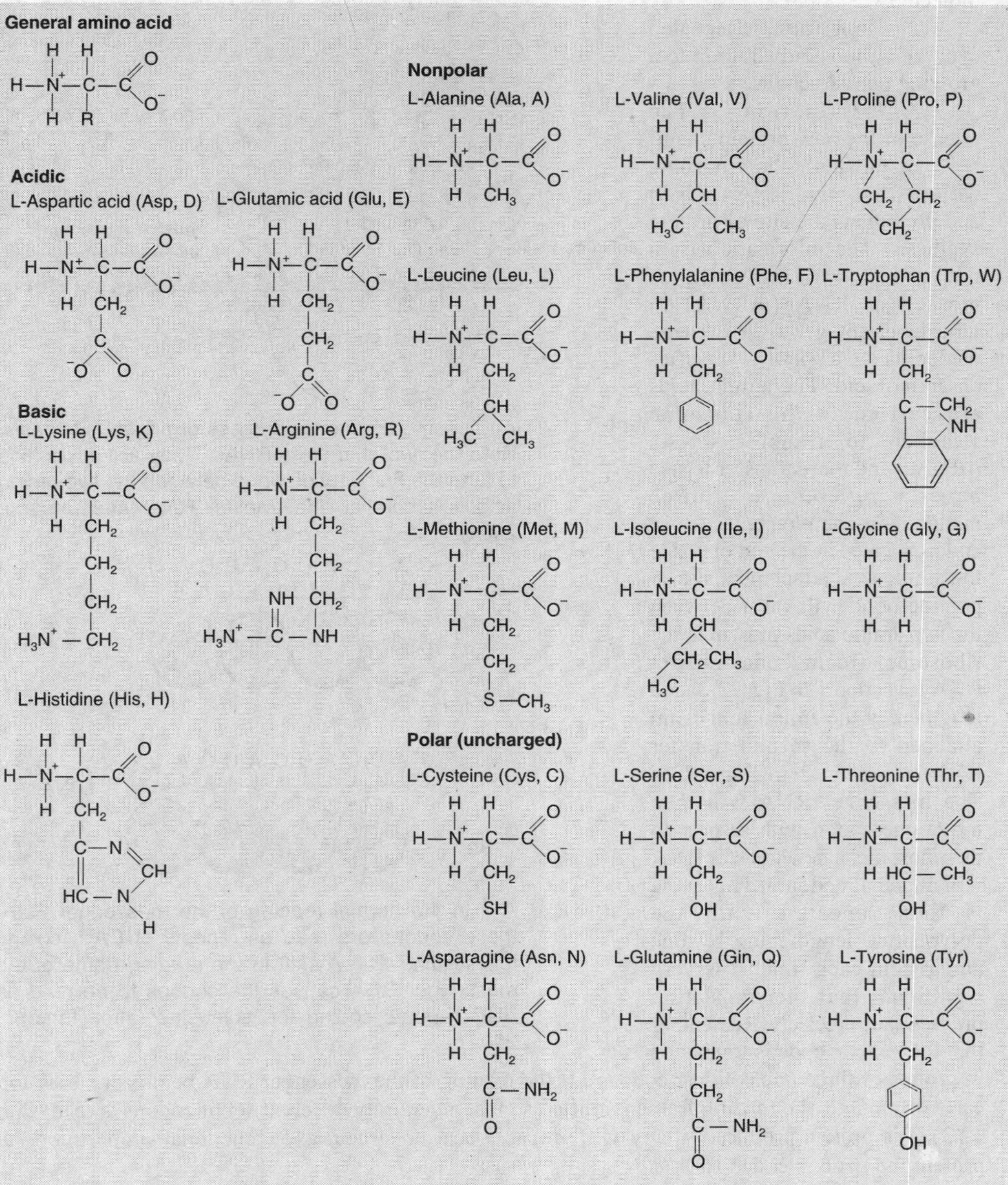

Fig. 12.1. The twenty amino acids found in proteins and their three– and one – letter abbreviations. At physiological pH, the amino acids usually exist as ions. Note the classification of the various R groups (after Tamarin, 2002).

Overview of Mechanism of Protein Synthesis

The RNA-directed assembly of a protein is called *translation* and *protein synthesis.* The actual mechanism of protein synthesis can be divided into three stages:

1. **Initiation.** The assembly of a ribosome on a mRNA molecule.

2. **Elongation.** Repeated cycle of amino acid addition to a growing peptide chain.

3. **Termination.** The release of the new protein chain.

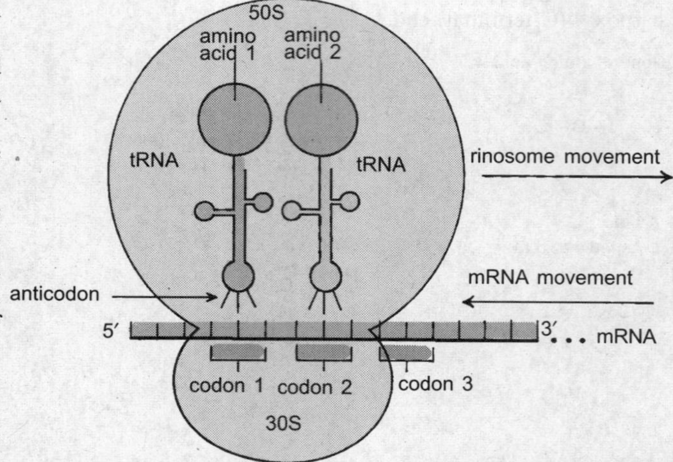

Fig. 12.2. A ribosome during the process of protein synthesis. Note the two charged mRNAs. They are in position to form the first peptide bond between the two amino acids attached to the transfer RNA (afterTamarin, 2002).

In nutshell, the ribosome with its ribosomal RNA (rRNA) and proteins is the site of protein synthesis. The information from the gene is in the form of messenger RNA (mRNA) in which each group of three nucleotides – a *codon* – specifies an amino acid. The amino acids are carried to the ribosome attached to transfer RNAs (tRNAs) and these transfer RNAs have *anticodons,* three nucleotides complementary to a codon, located at the end opposite the amino acid attachment site. A peptide bond will form between the two amino acids present at the ribosome, freeing one transfer RNA (at codon 1 in Fig. 12.2) and lengthening the amino acid chain attached to the second transfer RNA (at codon 2 in Fig. 12.2). The messenger RNA will then move one codon with respect to the ribosome, a new transfer RNA will attach at codon 3. This cycle is then repeated, with the polypeptide lengthening by one amino acid each time. It is very significant that the translation process start precisely. Remember that the genetic code is translated in groups of three nucleotides (codons). If the reading of the messenger RNA begins one base too early or too late, the reading frame is shifted so that an entirely different set of codons is read (Fig. 12.3). The protein produced, if any, will probably bear no structural or functional similarity to the protein the gene is coded for.

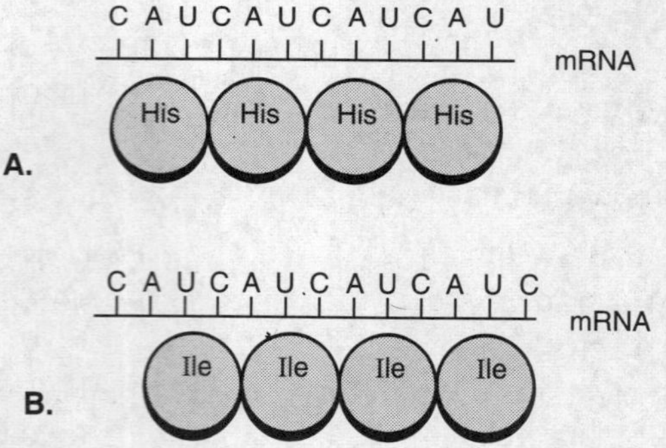

Fig. 12.3. A – In the normal reading of the messenger RNA, these codons are read as repeats of CAU, coding for histidine. B – A shift in the reading frame of the messenger RNA causes the codons to be read as AUC repeats coding for isoleucine (after Tamarin, 2002).

12.3. MECHANISM OF TRANSLATION OR PROTEIN SYNTHESIS

I. Aminoacylation of tRNA (Attachment of Amino Acid to Transfer RNA)

The function of tRNA (transfer RNA) is to ensure that each amino acid incorporated into a protein corresponds to a particular codon in the messenger RNA. The transfer RNA serves this function through its peculiar structure: it has an anticodon at one end and an amino acid attachment site at the other end. The "correct" amino acid, the amino acid corresponding to the anticodon, is attached to the transfer RNA by enzymes known as **aminoacyl tRNA synthetases** (e.g., arginyl-tRNA synthetase, leucyl-tRNA synthetase). A transfer RNA with an amino acid attached is said to be "charged". The transfer RNAs for each amino acid are designated by the convention tRNA^{Leu} (for leucine), tRNA^{His} (for histidine) and so on.

An aminoacyl – tRNA synthetase joins a specific amino acid to its transfer RNA in a two stage reaction that takes place on the surface of the enzyme (Box 12.1): In the first stage, the amino acid is activated with ATP. In the second stage of the reaction, the amino acid is attached with a

> **Box 12.1.**
>
> Each activating enzyme (*i.e.*, aminoacyl – tRNA synthetase) is specific for one amino acid and for a tRNA. The enzyme has a three part active site that recognizes three smaller molecules:
>
> 1. a specific amino acid (AA);
> 2. ATP; and
> 3. a specific transfer RNA.

high energy bond to the 2′ or 3′ carbon of the ribose sugar at the 3′ end of the transfer RNA: The first step in aminoacyl – tRNA synthesis involves the ***activation*** of the amino acid using energy from ATP:

$$\text{Amino acid + ATP} \xrightarrow{\text{Aminoacyl synthetase}} \text{amino acid} \sim \text{AMP + P + P}$$

The aminoacyl ~ AMP intermediate is not normally released from the enzyme before undergoing the second step in aminoacyl tRNA synthesis, namely the reaction with appropriate tRNA:

$$\text{Amino acid} \sim \text{AMP + tRNA} \xrightarrow{\text{Aminoacyl -tRNA synthetase}} \text{amino acid} \sim \text{tRNA + AMP}$$

The resultant aminoacyl ~ tRNAs (amino acid ~ tRNAs) are immediate precursors of protein (= polypeptide) synthesis.

In the Figure 12.4, we denote high energy bonds, bonds that liberate a lot of free energy when hydrolysed, as "~". Thus, during the process of protein synthesis, the energy for the formation of the peptide bond will be present when it is needed, at the point of peptide bond formation.

Few Characteristics of Aminoacyl-tRNA Synthetases

In bacteria, there are twenty aminoacyl-tRNA synthetases, one for each amino acid. A particular enzyme recognizes a particular amino acid, as well as, all the transfer RNAs that code for that amino acid. In eukaryotes, there are separate sets of twenty cytoplasmic and twenty mitochondrial aminoacyl-tRNA synthetases, all coded in the aminoacyl-tRNA synthetases, all coded in the nucleus.

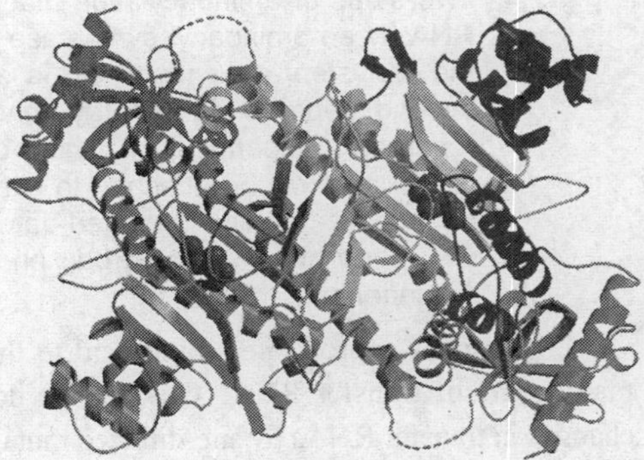

Aminoacyl t-RNA synthetase.

Aminoacyl-tRNA synthetases are a heterogeneous group of enzymes. In *E.coli*, these enzymes vary from monomeric proteins (one subunit) to tetrameric proteins, made up of two copies each of two subunits. The enzymes fall into two categories based on sequence similarity, structural features, and whether the amino acid is attached at the 2′-OH (in class I enzymes) or 3′-OH (in class II enzymes) of the 3′-terminal adenosine of the transfer RNA (Table 12.1; **Ibba** *et al.,* 2000).

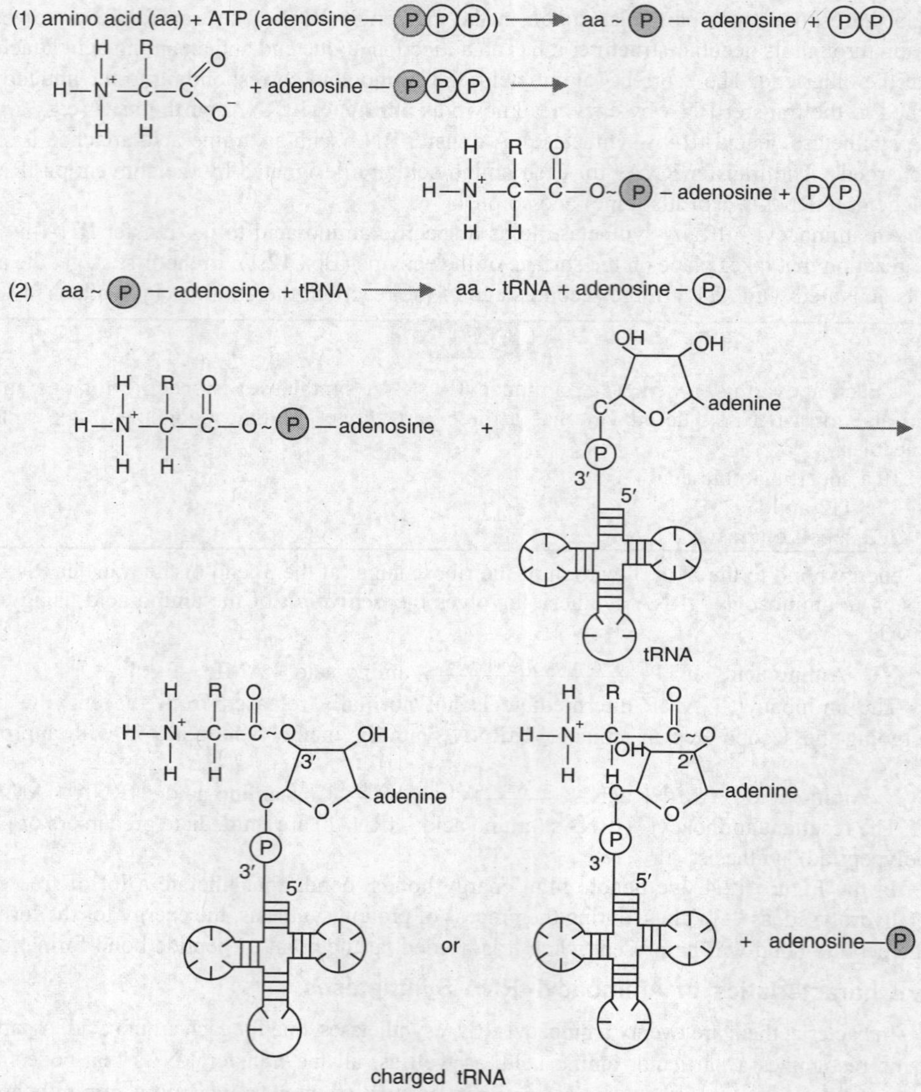

(1) amino acid (aa) + ATP (adenosine — ⓅⓅⓅ) ⟶ aa ~ Ⓟ — adenosine + ⓅⓅ

(2) aa ~ Ⓟ — adenosine + tRNA ⟶ aa ~ tRNA + adenosine – Ⓟ

tRNA

or

+ adenosine—Ⓟ

charged tRNA

Fig. 12.4. Two steps of aminoacylation that is attachment of a specific amino acid to its transfer RNA by an aminoacyl synthetase enzyme. High – energy bonds are indicated by ~. In the first step, an amino acid is attached to AMP with a high – energy bond. In the second step, the high-energy bond is transferred to the tRNA, which is then referred to "charged". Depending on which class of aminoacyl-tRNA synthetase is involved, the amino acid will be attached to either the 2′ or 3′ carbon of the sugar of the 3′ terminal adenosine. The link is formed between the —COOH group of the amino acid and the —OH group attached to carbon on the sugar of the last nucleotide, which is always an A(= adenine).

To add its appropriate amino acid to the appropriate transfer RNA, a synthetase recognizes many parts of transfer RNA. This can be demonstrated by experiments that introduces specific changes in transfer RNAs by site-directed mutagenesis. In seventeen of the twenty synthetase enzymes of *E. coli*, recognition involves part of the anticodon itself.

Table 12.1. Features of aminoacyl-tRNA synthetases (Source : **Brown**, 2002).

Feature	Class I enzymes	Class II enzymes
1. Structure of the enzyme's active sites	Parallel β - sheet	Antiparallel β - sheet
2. Interaction with tRNA	Minor groove of the acceptor stem	Major groove of the acceptor stem
3. Orientation of the bound tRNA	V loop faces away from the enzyme	V loop faces the enzyme
4. Amino acid attachment	To the 2′– OH of the terminal nucleotide of the tRNA	To the 3′– OH of the terminal nucleotide of the tRNA
5. Enzymes for *	Arg, Cys, Glu, Gln, Ile, Leu, LysI, Met, Trp, Tyr, Val	Ala, Asn, Asp, Gly, His, LysI, Phe, Pro, Thr, Ser

* The aminoacyl-tRNA synthetase for lysine in a class I enzyme in some archaea and bacteria and a class II enzyme in all other organisms.

Further, an aminoacyl-tRNA synthetase enzyme can initially make errors and attach the "wrong" amino acid to a tRNA. For example, isoleucine (Ile) synthetase will attach valine about one in 225 times. This type of error occurs because a similar, but smaller, amino acid can sometimes occupy the active site of the enzyme (compare chemical formulae of isoleucine and valine in Fig. 12.1). However, because of a proof-reading step, only 1 in 270 to 1 in 800 of the errors are released intact from the enzyme. The amino acids on the rest of the incorrectly charged transfer RNAs are hydrolyzed before the transfer RNAs are released. Thus, only about one incorrectly charged transfer RNA occurs per 60,000 to 80,000 formed.

In various bacteria, the number of amino acyl-tRNA synthetases in a particular organism is below twenty. For example, in some archaea, there is no cysteinyl-tRNA synthetase (enzyme). However, the prolyl-tRNA synthetase activates the tRNAs for both cysteine and proline with their appropriate amino acids. Similarly, in some eubacteria, there is no glutaminyl-tRNA synthetase; the glutaminyl-tRNA is charged with glutamic acid, rather than glutamine. An amido transferase (enzyme) then converts the glutamic acid to glutamine (Fig.12.1). This process is called **transamidation**. Some archaea also use process of transamidation

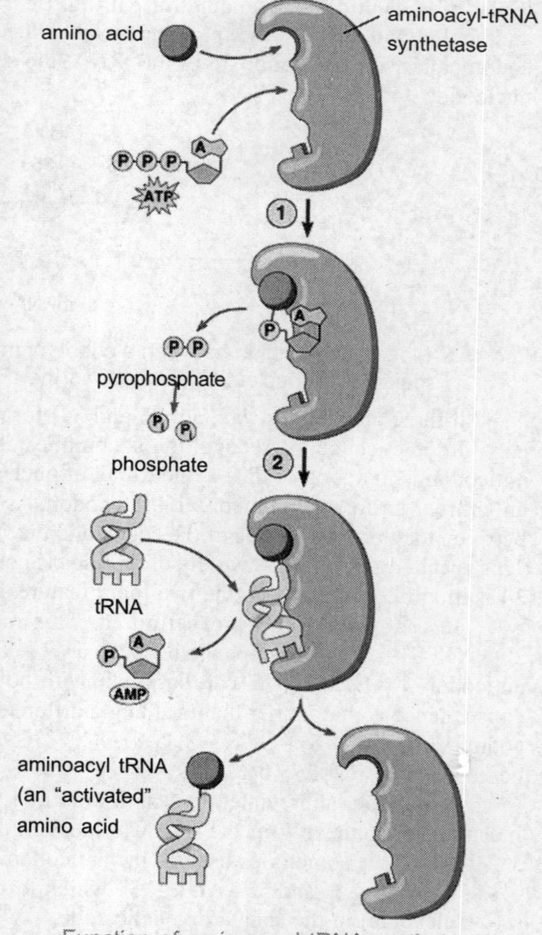

Function of amino acyl tRNA synthetase enzyme.

to synthesise asparagine-$tRNA^{Asn}$ from aspartic acid-$tRNA^{Asn}$ (**Ibba** *et al.*, 2000).

Lastly, three of sixty-four codons are used to terminate translation. Thus, sixty-one transfer RNAs are needed because there are sixty-one different *nonterminator codons.* About fifty transfer RNAs are known in *E. coli.* The number fifty can be explained by **Wobble phenomenon** (see chapter 10, Genetic Code), which occurs in the third position of the codon.

Box 12.2.

Wobble helps in the reduction of number of tRNAs

Aminoacylation represents the first level of specificity demonstrated by a tRNA. The second level is the specificity of the interaction between the anticodon of the tRNA and codon of the mRNA being translated. This specificity ensures that protein synthesis follows the rules of the genetic code.

In principle, codon-anticodon recognition is an easy process involving base-pairing between the anticodon of the tRNA and a codon in the mRNA (Figure 12.5). The specificity of aminoacylation ensures that the tRNA carries the amino acid denoted by the codon that it pairs with, and the ribosome controls the topology of the interaction in such a way that only a single triplet of nucleotides is available for pairing. Because base-paired polynucleotides are always antiparallel and because the mRNA is read in the 5′→3′ direction, the first nucleotide of the codon pairs with nucleotide 36 of the tRNA, the second with nucleotide 35 and the third with nucleotide 34.

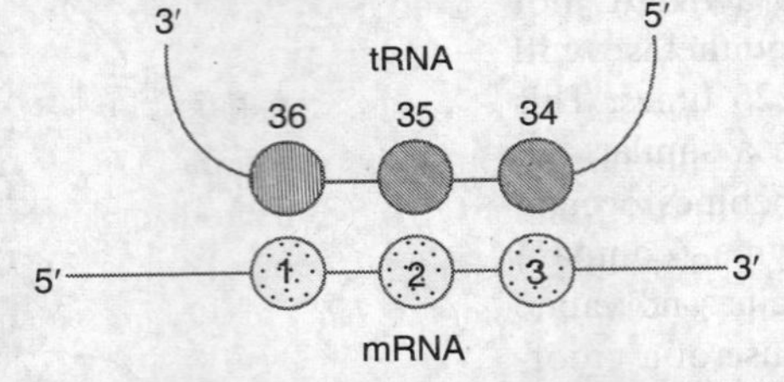

Fig. 12.5. Mode of interaction between a codon of mRNA and an anticodon of tRNA. The numbers indicate the nucleotide positions in the tRNA.

All three nucleotides (*i.e.*, 36, 35 and 34) form an anticodon.

In practice, codon recognition is complicated by the possibility of **Wobble.** Because the anticodon is in a loop of tRNA, the triplet of nucleotides is slightly curved and so cannot make an entirely uniform alignment with the codon. As a result, a non-standard base pair can form between the third nucleotide of the codon and the first nucleotide (number 34) of the anticodon. This is called "wobble". A variety of pairings is possible, especially if the nucleotide at position 34 is modified. In bacteria, the two main features of wobble are (**Ikemura,** 1981).

1. G-U base pairs are permitted. This means that an anticodon with the sequence 3′ – XX G – 5′ can base-pair with both 5′ – XX C – 3′ and 5′ – XX U – 3′. Similarly, the anticodon 3′ – XX U – 5′ can base-pair with both 5′ – XX A – 3′ and 5′ – XXG – 3′. The consequence is that, rather than needing a different tRNA for each codon, the four members of codon family [*e.g.*, 5′ – GCN – 3′ (*i.e.*, GCC, GCA, GCU and GCG) all coding for alanine] can be decoded by just two tRNA.

2. Inosine, abbreviated to I, is a modified purine that can base pair with A, C and U. Inosine can occur only in the tRNA because mRNA is not modified in this way. The triplet 3′ – UAI – 5′ is sometimes used as the anticodon in a $tRNA^{Ile}$ because it pairs with 5′ – AUA – 3′, 5′ – AUC – 3′ and 5′ – AUU – 3′, which form the three codon family for this amino acid (*i.e.* Isoleucine) in the standard genetic code.

Thus, *wobble tends to reduce the number of tRNAs in a cell by enabling one tRNA to read two or possibly three codons.* Hence, bacteria can decode their mRNAs with as few as 30 tRNAs. Eukaryotes also make use of wobble but in a restricted way. For example, human genome has 48 tRNAs. Of these, 16 tRNAs are predicted to use wobble to decode two codons each, with the remaining 32 tRNAs being specific for just a single triplet (see **Brown,** 2002).

In eukaryotes, the bacterial G – U wobble is used with eight tRNAs but in every case the wobble involving an anticodon with the sequence 3′ – XXG – 5′. The alternative version of G – U wobble, where the anticodon sequence is 3′ – XXU – 5′, appears not be used in eukaryotes, possibly because this could result in a $tRNA^{Ile}$ with the anticodon 3′ – UAU – 5′ reading the methionine codon 5′ –AUG – 3′ (Fig. 12.6). Eukaryotes may therefore have a means of preventing this type of wobble from occurring (**Percudani,** 2001).

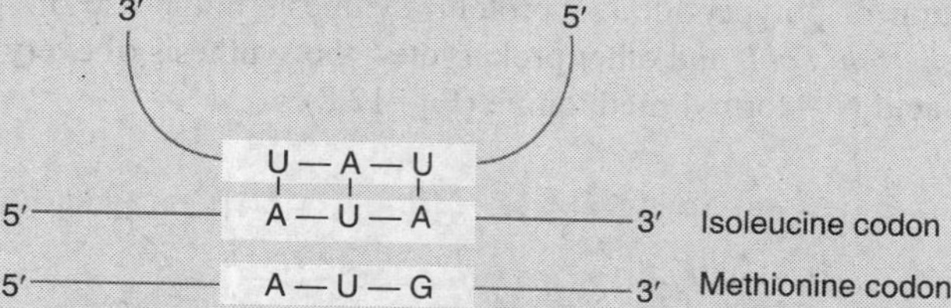

Fig. 12.6. A tRNA with the anticodon 3′ – UAU – 5′ could read the isoleucine codon 5′ – AUA – 3′ as well as the methionine codon.

Eight other human tRNAs have anticodons containing inosine (3′ – XXI – 5′) but, these decode only 5′ – XXC- 3′ and 5′ – XXU – 3′ codons. The base pairing between I and A is weak, which means that 5′ – XXA – 3′ codons are only inefficiently recognized by an 3′ – XXI – 5′ anticodon. To avoid this inefficiency, in every example of wobble involving inosine in the human tRNA set, the 5′ – XXA – 3′ codon is recognized by a separate tRNA. Note, however that recognition by a separate tRNA does not preclude the 5′ – XXA – 3′ codon from also being decoded by the tRNA 3′ – XXI – 5′, albeit inefficiently. This does not compromise the specificity of the genetic code, because wobble involving inosine is limited to those codon families in which all four triplets specify the same amino acid.

Superwobble. Human mitochondria use only 22 tRNAs, with some of these tRNAs the nucleotide in the wobble position of the anticodon is virtually redundant because it can base-pair with any nucleotide, enabling all four codons of a family to be recognized by the same tRNA. This phenomenon has been called **superwobble.**

Experiment to show that the transfer RNA, not the amino acid, is recognized during protein synthesis.

F. Chapeville and colleagues, in 1962, isolated transfer RNA with cysteine attached. They chemically converted the cysteine to alanine by using Raney nickle, a catalytic form of nickle that

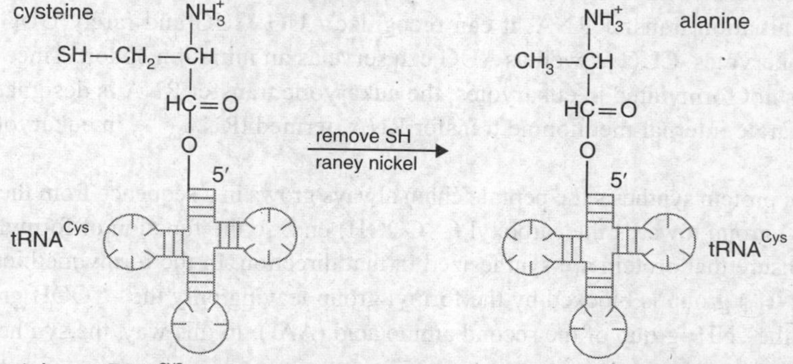

Fig. 12.7. Cysteine – $tRNA^{cys}$ treated with Raney nickel becomes alanine – tRNAcys by the removal of the SH group of cysteine.

removes the SH group of cysteine (Fig. 12.7). When such modified tRNAs were used in protein synthesis, alanine was incorporated where cysteine should have been. This result demonstrated that the transfer RNA, not the amino acid, was recognized during protein synthesis. Thus, the synthetase puts a specific amino acid on a specific tRNA; then, during protein synthesis, the anticodon on the tRNA – not the amino acid itself – determines which amino acid is incorporated.

A. Initiation of Translation

I. Initiation in prokaryotes

1. Interaction between initiation codon and formylmethionine – tRNA. An important feature of initiation of polypeptide (= protein) synthesis is the use of specific initiating tRNA molecule. In *Escherichia coli* and other prokaryotes, the synthesis of every protein begins with the modified amino acid N – formyl methionine (Fig. 12.8).

Methionine

N-formyl methionine

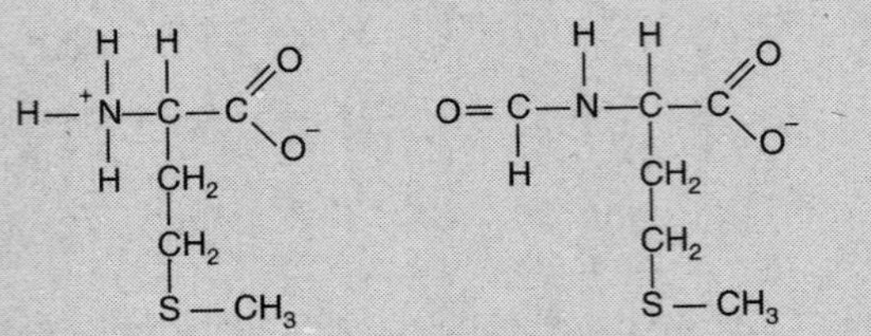

Fig. 12.8. The structure of the amino acid methionine and N-formyl methionine.

However, none of the completed protein in *E.coli* contains N-formyl methionine. Many of these proteins do not even have methionine as their first amino acid. Evidently, before a protein becomes functional, the initial amino acid is modified or removed (called **processing**). In eukaryotes, the initial amino acid is also methionine but it does not have a N-formyl group.

The codon for methionine is 5′ – AUG – 3′ and is called **initiation codon**. Methionine has two tRNAs with the same anticodon (3′ – UAC – 5′) but different structures (Fig. 12.7). One of these tRNAs ($tRNA_f^{met}$) serves as a part of the initiation complex. Before the initiation of translation, this tRNA will have its methionine chemically modified to N-formyl methionine (fMet). The other transfer RNA will not have its methionine modified ($tRNA_m^{met}$). The translation process will use it to insert methionine into proteins, where it is required, in all but the first position. The cell thus has a device to make use of methionine in the normal way as well as to use a modified form of it (*i.e.,* formyl methionine) to initiate protein synthesis. Because of the structure of the prokaryotic initiation transfer RNA, it can recognize AUG, GUG and rarely UUG as initiation codons. In eukaryotes, CUG as well as AUG can serve as an initiation codon. Since the initiation methionine is not formylated in eukaryotes, the eukaryotic transfer RNA is designated $tRNA_i^{met}$; there is a separate internal methionine transfer RNA, termed $tRNA_m^{met}$, in eukaryotes as well as prokaryotes.

Since in protein synthesis the peptide chain always grows in a sequence from the free terminal amino (–NH_2) group towards the carboxyl (—COOH) end, so the function of formylmethionine – tRNA is to ensure that protein are synthesized in that direction. In the formylmethionine – tRNA, the amino (–NH_2) group is blocked by the formyl group leaving only the –COOH group available to react with the –NH_2 group of the second amino acid (AA_2). In this way, the synthesis of protein chain follows in the correct sequence.

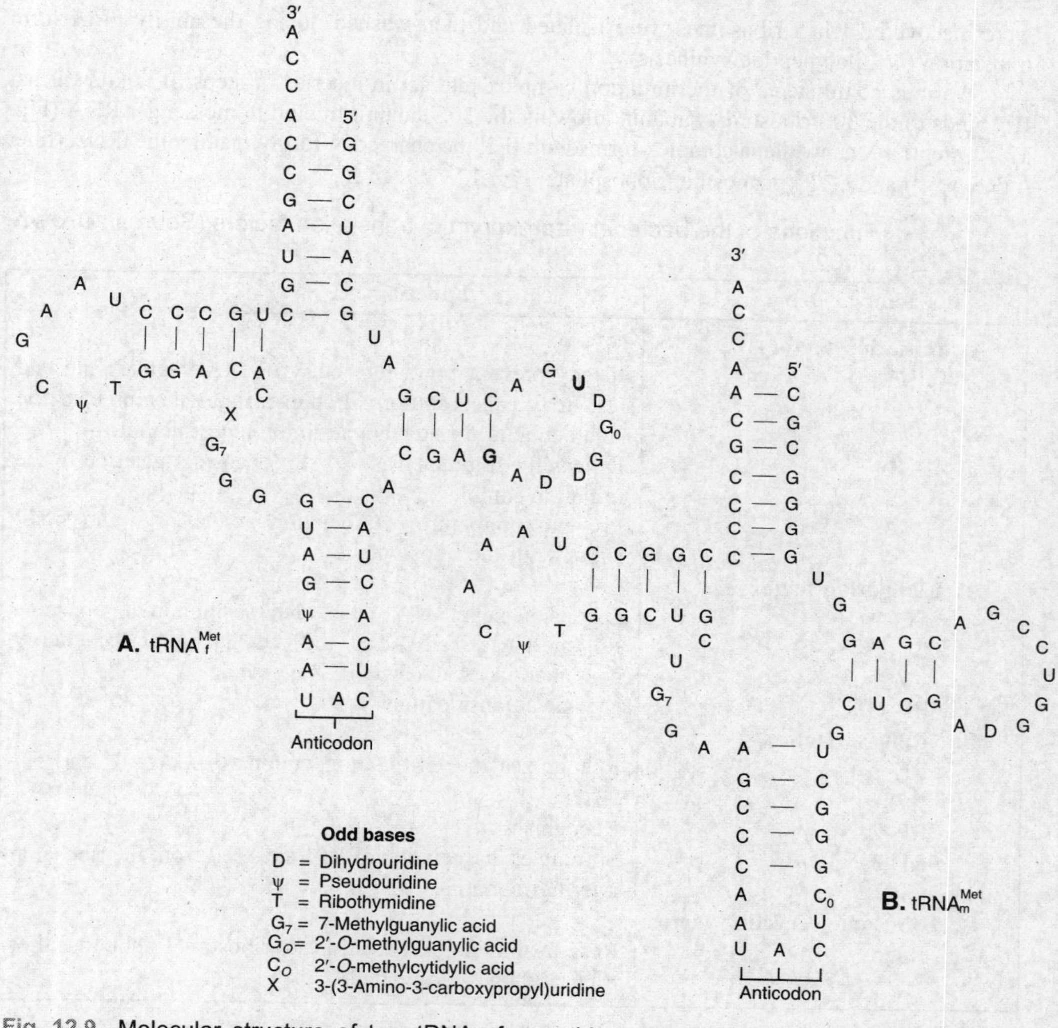

A. tRNA$_f^{Met}$
Anticodon
B. tRNA$_m^{Met}$
Anticodon
Odd bases
D = Dihydrouridine
ψ = Pseudouridine
T = Ribothymidine
G_7 = 7-Methylguanylic acid
G_O = 2′-O-methylguanylic acid
C_O 2′-O-methylcytidylic acid
X 3-(3-Amino-3-carboxypropyl)uridine

Fig. 12.9. Molecular structure of two tRNAs for methionine in *E.coli*. A – The initiator tRNA; B – The interior or internal tRNA (after Tamarin, 2002).

2. Formation of initiation complex

The 30S and 50S ribosomal subunits of bacteria usually dissociate from each other when not involved in translation. To begin translation, an **initiation complex** forms, consisting of following components in prokaryotes (*i.e., E.coli*): the 30S subunit of the ribosome, a messenger RNA, the charged N-formyl methionine tRNA (fmet – tRNA$_f^{Met}$), and three **initiation factors (IF1, IF2, IF3)**. Initiation factors (as well as elongation and termination factors) are proteins loosely associated with the ribosome (Table 12.2). They

met
IF2
UAC
5′-CUAUCCGCGCAGGAGGCCUAUGCCAUCGAACUUUUAA-3′
IF3
30 S subunit

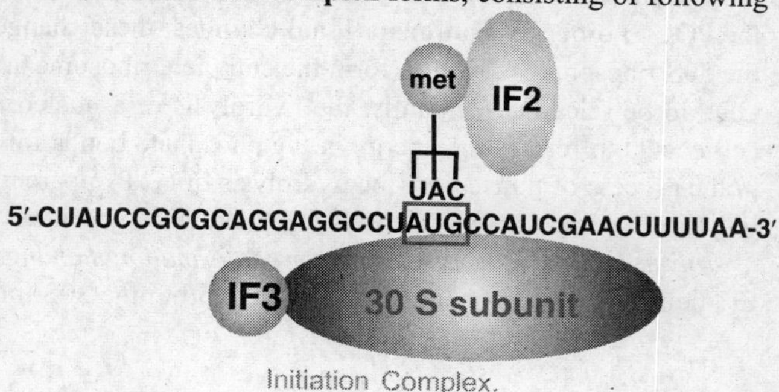

Initiation Complex.

were discovered when ribosomes were isolated and then washed, losing the ability to perform translation (*i.e.,* polypeptide synthesis).

Various components of the initiation complex interact in a series of steps. It is known that IF3 binds to the 30S ribosomal subunit, allowing the 30S subunit to bind to messenger RNA (Fig. 12.10, step 1). Meanwhile, a complex forms with IF2, the charged N-formylmethionine tRNA (fmet – $tRNA_f^{Met}$) and GTP (guanosine triphosphate; Fig. 12.10, step 2).

Table 12.2. Functions of the bacterial (=prokaryotic) translation factors (Source : **Brown,** 2002).

Factor	Function
A. Initiation factors	
IF–1	It may prevent premature entry of tRNA into A site and also may cause conformational changes that prepare the small subunit for attachment to the large subunit.
IF–2	Directs the initiator $tRNA_f^{Met}$ to its correct position in the initiation complex.
IF–3	Prevents premature reassociation of the large and small subunits of the ribosome.
B. Elongation factors	
EF–Tu	Directs the next tRNA to its correct position in the ribosome
EF–Ts	Regenerates EF-Tu after the latter yielded the energy contained in its attached GTP molecule
EF–G	Mediates translocation
C. Release factors	
RF–1	Recognizes the termination codons 5′–UAA–3′ and 5′–UAG–3′.
RF–2	Recognises 5′–UAA–3′ and 5′–UGA–3′
RF–3	Stimulates dissociation of RF1 and RF2 from the ribosome after termination
D. Ribosome recycling factor	
RRF	Responsible for dissociation of ribosomal subunits after translation has terminated

It is IF_2 that brings the initiator transfer RNA to the ribosome. IF2 binds only to the charged initiator transfer RNA, and without IF2, the initiator tRNA cannot bind to the ribosome. The final step in initiation-complex formation is bringing together the first two components (Fig. 12.10; step 3).

Here, GTP has unusual function. The hydrolysis of GTP to GDP + Pi (inorganic phosphate, the PO_4^{-3}) produces **conformational changes**; these changes allow the initiation complex to join the 50S ribosomal subunit to form the complete ribosome and then allow the initiation factors and GDP to be released. Generally, the hydrolysis of a nucleoside triphosphate (*i.e.,* ATP, GTP) in a cell occurs to release the energy in the phosphate bonds for use in a metabolic process. However, in the process of translation, the hydrolysis of GTP apparently changes the shape so that it and the initiation factors can be released from the ribosome after 70S particle has been formed. Thus, *hydrolysis of GTP in translation is for conformational change rather than covalent bond formation.* IF1 helps the other two initiation factors bind to the 30S initiation complex.

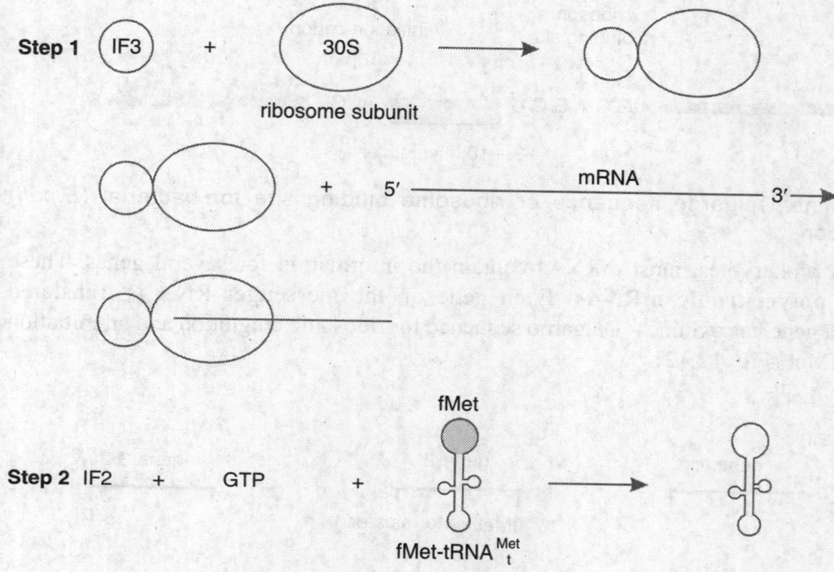

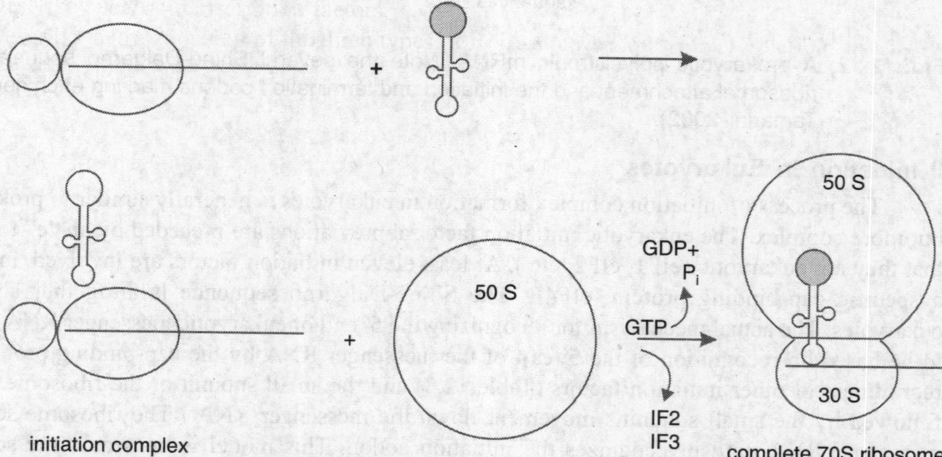

Fig. 12.10. The prokaryotic 70S ribosome forms in a three step process. In the first step, the 30S ribosomal subunit and the mRNA combine. In the second step, the initiator tRNA combines with IF2. In the final step, the components from steps 1 and step 2 combine to form the initiation complex, followed by the formation of the 70S ribosome (after Tamarin, 2002).

How does 70S ribosome recognizes mRNA? The ribosome recognizes the prokaryotic mRNA through complementarity of a region at the 3′ end of the 16S ribosomal RNA and a region slightly upstream from the initiation sequence (AUG) on the mRNA (Fig. 12.13A, 12.11). This consensus sequence is 5′ – AGGAGGU – 3′ and is called **Shine-Dalgarno sequence** (Fig.12.11). It is located between 3 to 10 nucleotides upstream of the initiation codon.

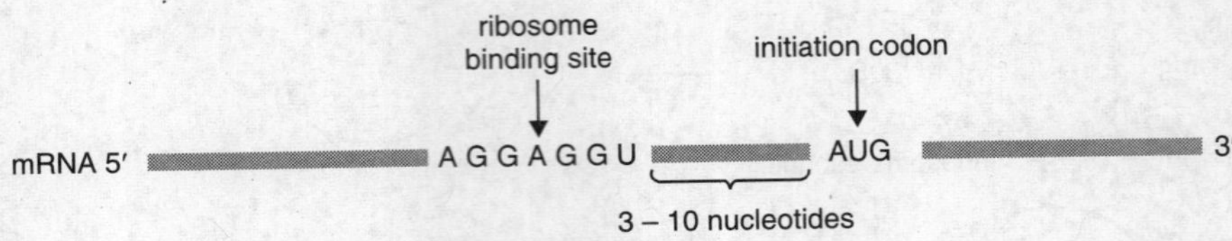

Fig. 12.11. The Shine–Dalgarno sequence or ribosome binding site for bacterial (*E.coli*) translation.

Moreover, in prokaryotes, most mRNAs contain the information for several genes. These RNAs are called **polycistronic mRNAs**. Each gene on the messenger RNA is translated independently: each gene has a Shine – Dalgarno sequence for ribosome reognition and an initiation codon (AUG) for fMet (Fig. 12.12).

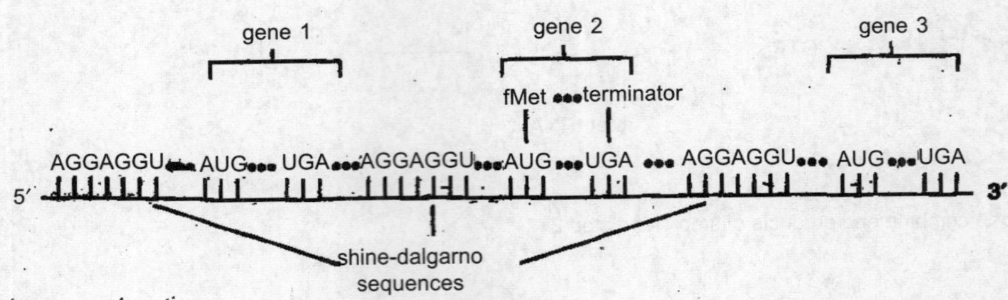

Fig. 12.12. A prokaryotic polycistronic mRNA. Note the several Shine-Dalgarno sequences for ribosomal attachment and the initiation and termination codons marking each gene (after Tamarin, 2002).

II Initiation in Eukaryotes

The process of initiation complex formation in eukaryotes is generally similar to prokaryotes but more complex. The eukaryotic initiation factor abbreviations are preceded by an "e" to denote that they are eukaryotic (eIF1, eIF2, etc.). At least eleven initiation factors are involved, including a specific cap-binding protein, eIF4E. The Shine-Dalgarno sequence is altogether absent in eukaryotes. The actual mechanism for recognizing the 5′ end of eukaryotic messenger RNA appears to be based on recognition of the **5′ cap** of the messenger RNA by the cap-binding protein with recruitment of other initiation factors (Table 12.3) and the small subunit of the ribosome. This is followed by the small subunit's movement down the messenger RNA . The ribosome scans the messenger RNA until it recognizes the initiation codon. This model is known as the **scanning hypothesis.**

In many eukaryotes, a process called **shunting** occurs, in which the first AUG does not serve as the initiation codon; rather, scanning begins, but it bypasses a region of the messenger RNA upstream of the initiation codon, called the **leader** or **5′ untranslated region** (5′UTR), in favour of an AUG further down the messenger RNA. The cause of shunting seems to be secondary structure in the messenger RNA, upstream from the AUG codon that actually serves as the initiation codon.

In some eukaryotes, very small genes called **open reading frames (ORFs)** are present in this region of the messenger RNA and play some role in shunting. In the genes of some plant and animal viruses, ORFs are translated, and then the main gene is translated by the same ribosome in a process called **reinitiation** (Fig.12.13 B).

In some eukaryotes, ribosomes can initiate protein synthesis within the mRNA if that mRNA contains a sequence called an **IRES** (*i.e.*, internal ribosome entry site). These sequences were discovered in RNA genome of the human poliovirus and rhinovirus (for cold) and in several cellular messenger RNAs. They are at least four hundred nucleotides long.

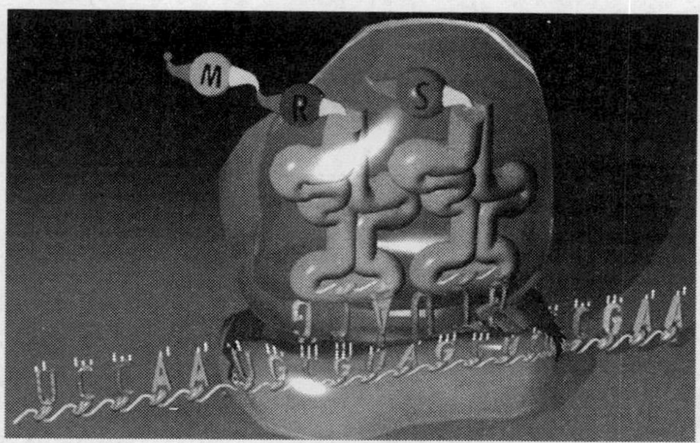

An open reading frame.

Table 12.3. Eukaryotic translation factors (Source: **Brown**, 2002).

A. Initiation factors	
eIF-1	Component of pre-initiation complex
eIF-1A	Component of pre-initiation complex
eIF-2	Binds to initiator tRNAMet within the ternary complex component of the pre-initiation complex; phosphorylation of eIF-2 results in a global repression of translation
eIF-3	Component of the pre-initiation complex; makes direct contact with eIF-4G and forms the link with the cap binding complex
eIF-4A	Component of the cap binding complex, a helicase that aids scanning by breaking intramolecular base pairs in the mRNA
eIF-4B	Aids scanning possible by acting as a helicase that breaks intramolecular base pairs in the mRNA
eIF-4E	Component of the cap-binding complex, possibly the component that makes direct contact with the cap structure at the 5′ end of mRNA
eIF-4F	The cap-binding complex, comprising eIF-4A, eIF-4E and eIF-4G, which makes the primary contact with the cap structure at the 5′ end of the mRNA
eIF-4G	Component of the cap-binding complex; forms a bridge between the cap binding complex and eIF-3 in the preinitiation complex; in at least some organisms,eIF-4G also forms an association with the poly (A) tail, via the polyadenylate-binding protein.
eIF-5	Aids release of the other initiation factors at the completion of initiation
eIF-6	Associated with the large subunit of the ribosome, prevents large subunits from attaching to small subunits in the cytoplasm
B. Elongation factors	
eEF-1	Complex of four subunits (eEF-1a, eEF-1b, eEF-1d and eEF-1g); directs the next tRNA to its correct position in the ribosome
eEF-2	Mediates translocation
C. Release factors	
eRF-1	Recognizes the termination codon
eRF-2	Possibly stimulates dissociation of eRF-1 from the ribosome after termination; possibly causes the ribosome subunits to dissociate after termination of translation

(*i*) Initiation of eukaryotic translation with scanning. The first step involves assembly of the *pre-initiation complex*. This structure comprises of the 40S ribosomal subunit, a 'ternary complex' made up of the initiation factor eIF-2 bound to the initiator $tRNA^{Met}$ and a molecule of GTP, and three additional initiation factors, eIF-1, eIF-1A and eIF-3. As in bacteria, the initiator tRNA is distinct from the normal tRNA ^{Met} that recognizes

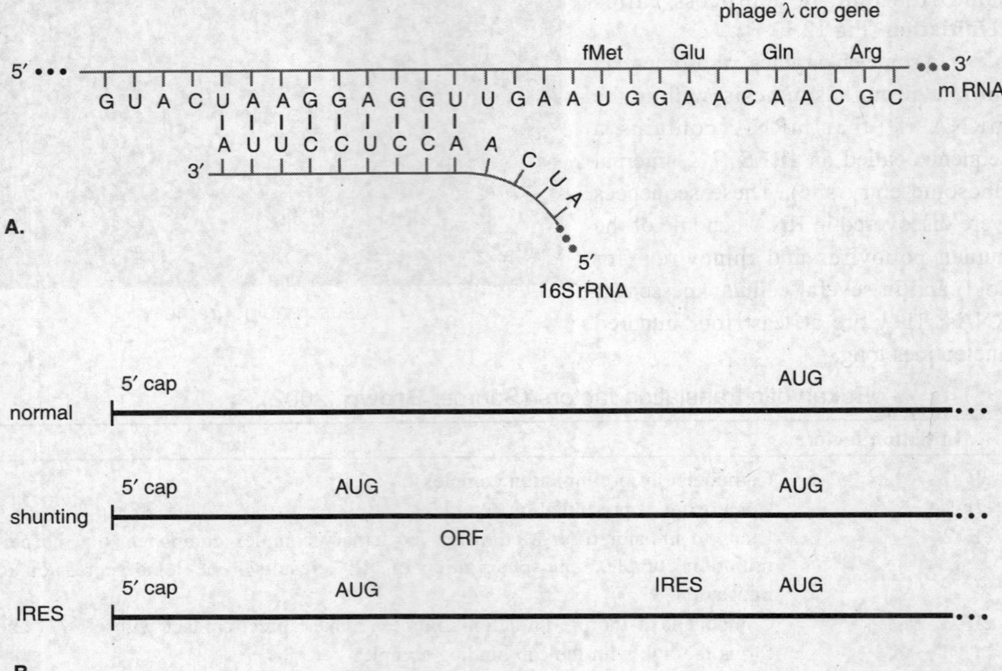

Fig. 12.13. Initiation signals for translation. A-The Shine-Dalgarno hypothesis for prokaryotic translation. The Shine-Dalgarno sequence (AGGAGGU) is on the prokaryotic messenger RNA just upstream from the initiation codon AUG. Complementarity exists between this sequence and a complementary sequence (UCCUCCA) on the 3′ end of the 16S ribosomal RNA. B-Scanning, shunting and internal ribosome entry in eukaryotic messenger RNA. In the scanning model, the initiation codon of the gene is the first AUG encountered. In shunting, an open reading frame (ORF) may or may not be present to provide secondary structure in the messenger RNA to shunt scanning to the main gene. If the open reading frame (ORF) is translated, reinitiation of translation at the same ribosome may occur at the main gene. Finally, an internal ribosome entry site (IRES) allows translation to begin with the messenger RNA without scanning (after Tamarin, 2002).

internal 5′–AUG-3′ codons but, unlike bacteria, it is aminoacylated with normal methionine, not the formylated version.

After assembly, the pre-initiation complex associates with the 5′ end of the mRNA. This step requires the **cap binding complex** (sometimes called eIF-4F) which comprises the initiation factors eIF-4A, eIF-4E and eIF-4G. The contact with the cap might be made by eIF-4E alone (Fig. 12.14) or involve a more general interaction with the cap binding complex (**Pestova** and **Hellen**, 1999). The factor eIF-4G serves as a bridge between eIF-4E, bound to the cap, and eIF-3, attached to pre-initiation complex gets attached to the 5′ region of the mRNA. Attachment of the pre-initiation complex to the mRNA is also influenced by the poly (A) tail, at the distant 3′ end of the mRNA. This interaction is thought to be mediated by the polyadenylate-binding protein (PADP) which is attached to the poly (A) tail. In yeast and plants, PADP is found to form an association with eIF-4G and for such an association mRNA has to bend back on itself.

With its attachment to the 5′ end of the mRNA, the **initiation complex** has to scan along the mRNA molecule and find the initiation codon. The leader regions of eukaryotic mRNAs can be several tens or even hundreds of nucleotides in length and often contain regions that form hairpins and other base paired structures. These are probably removed by a combination of eIF- 4A and eIF-4B. The eIF-4A, and possibly also eIF-4B, has a helicase activity and is able to break intra-molecular base pairs in the mRNA, freeing the passage for the initiation complex (Fig. 12.14B). The initiation codon (AUG) is recognizable because it is contained in a short consensus sequence; 5′-ACCAUGG-3′, which is called **Kozak consensus** (see **Brown**, 2002).

As soon as, the initiation complex is positioned over the initiation codon, the large subunit (60S) of the ribosome attaches. As in bacteria, this requires hydrolysis of GTP and leads to release of the initiation factors. Two final initiation factors are involved at this stage: eIF-5 which aids release of the other factors and eIF-6, which is associated with the unbound large subunit and prevents it from attaching to a small subunit in the cytoplasm.

A. attachment of the pre-initiation complex to the mRNA

initiator tRNA
eIF-2
M
2
3
pre-initiation complex
small subunit
cap binding complex
eIF-3
cap
AUG
3′
E
3
G
A
cap binding complex

B. scanning

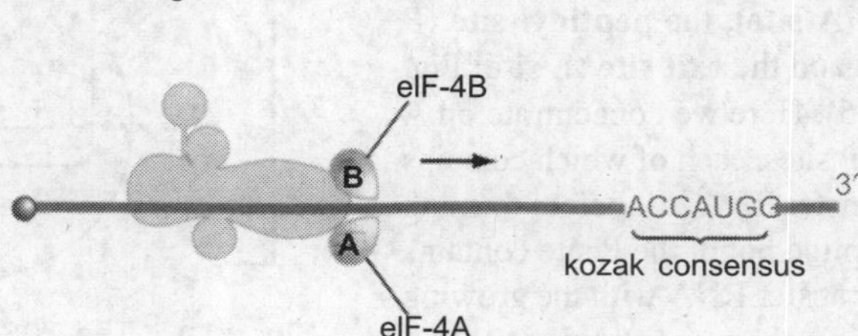

Fig. 12.14. Initiation of eukaryotic translation. A-Assembly of the pre-initiation complex and its attachment to the mRNAs (according to Hentze 1997). B-The pre-initiation complex scans along the mRNA until it reaches a initiation codon, which is recognisable because it is located within Kozak consensus sequence. Scanning is aided by eIF-4A, eIF-4B, which are thought to have helicase activity. Most probably, eIF-3 remains attached to the pre-initiation complex during scanning, as shown here. Scanning is an energy- dependent process that requires hydrolysis of ATP. M=methionine (after Brown, 2002).

(*ii*) **Initiation of eukaryotic translation without scanning.** Transcripts (mRNA) of picornaviruses (*i.e.*, RNA viruses such as poliovirus and rhinovirus) are not capped but instead have an internal ribosome entry site (IRES). The presence of IRES on their transcripts means that picorna viruses can block protein synthesis in the host cell by inactivating the cap binding complex, without affecting translation of their own transcripts.

Quite interestingly, no virus proteins are required for the recognition of an IRES by a host ribosome. In other words, the normal eukaryotic cell possesses proteins and/or other factors that enable it to initiate translation by the IRES method (**Holick** *et al.,* 2000). Some nuclear genes are reported to have IRES. For example, IRES are present on mRNAs for the mammalian immunoglobulin heavy chain binding protein and the *Drosophila* antennapedia protein. IRES are also found on several mRNAs whose protein products are translated when the cell is put under stress, for example by exposure to heat, irradiation, on low oxygen conditions. Under these circumstances, cap dependent translation is globally suppressed. The presence of IRESs on the 'survival' mRNAs therefore enables these to undergo preferential translation at the time when their products are needed (see **Brown,** 2002).

Three Sites of Ribosomes

When the initiator transfer RNA joins the 30S ribosomal subunit of prokaryotes with its mRNA attached, it fits into one of three sites in the ribosome. These sites or cavities in the ribosome, are referred to as the **aminoacyl site (A site),** the **peptidyl site (P site)** and the **exit site (E site;** Fig. 12.15). Here we concentrate on A and P sites, each of which contains a transfer RNA just before forming a peptide bond: the P site contains the transfer RNA with the growing peptide chain (peptidyl – tRNA); the A site contains a new transfer RNA with its single amino acid (aminoacyl-tRNA). The E site helps eject the depleted transfer RNAs after a peptide bond forms. When a complete 70S ribosome of Figure 12.10 has formed, the initiation fMet – $tRNA_f^{Met}$ is placed directly into the P site (Fig. 12.15), *the only charged tRNA that can be placed directly there.* The association of tRNA and ribosome is aided by a G-C base pairing between the 3′- CCA terminus of all transfer RNAs and a guanine in the 23S ribosomal RNA.

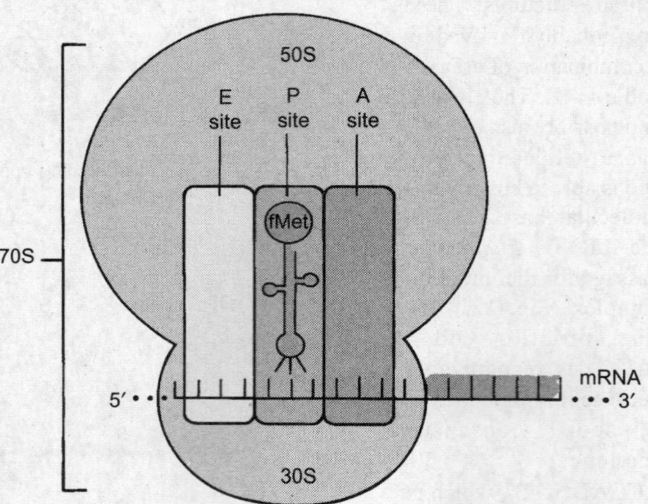

Fig. 12.15. The 70S ribosome contains an A site, a P site and an E site that can receive tRNAs. The mRNA runs through the bottom of the sites (after Tamarin, 2002).

B. Elongation

The main differences between translation in bacteria and eukaryotes occur during the initiation phase; the events after the large subunit of the ribosome becomes associated with the initiation complex are similar in both types of organisms.

1. Positioning a second transfer RNA

The second charged tRNA binds (due to a codon-directed binding) to the first ribosome at the latter's 'A' site with the help of the proteins, called elongation factors (e.g. EF-Tu). EF-Tu carries a molecule of GTP. Correct hydrogen bonding with the mRNA template dictates the selection of a new tRNA, and the activity of the EF-Tu ensures the proper positions of the tRNA in the A site. Such a placement activity needs energy that is provided by the hydrolysis of the molecule of GTP to GDP and phosphate. After performing its function, the **EF-Tu** protein dissociates from the ribosome, and in the cytoplasm is subsequently regenerated to its active form by another elongation factor, the **EF-Ts**. At this point, both sites of the ribosome are occupied by tRNA's each of which carries an amino acid, and each of which is hydrogen bonded to the template mRNA (Fig. 12.16).

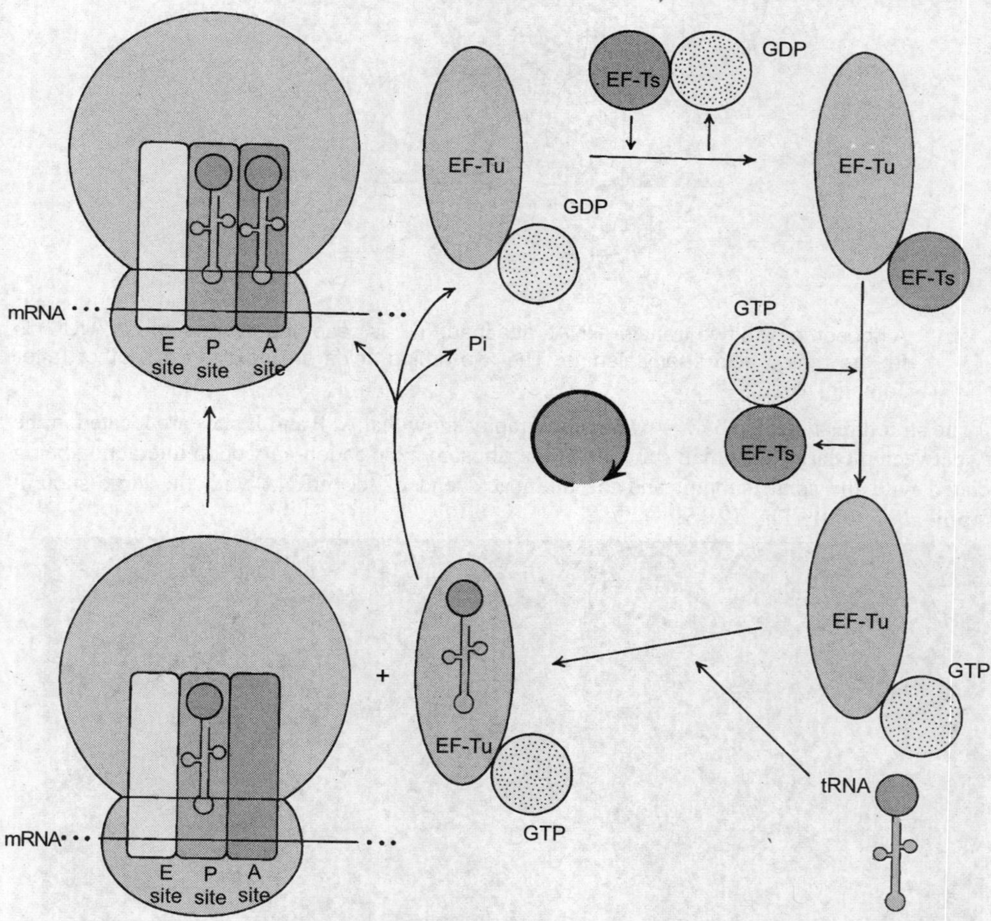

Fig. 12.16. The EF – Ts / EF – Tu cycle. EF – Ts and EF – Tu are required for a transfer RNA to attach to the A site of the ribosome. At *top centre,* there is EF – Tu attached to a GDP. The GDP is then displaced by EF – Ts, which in turn is displaced by GTP. A transfer RNA attaches and is brought to the ribosome. If the codon-anticodon fit is correct, the transfer RNA attaches at the A site with the help of the hydrolysis of GTP to GDP + Pi, allowing EF – Tu to release. The EF – Tu is now back where we started. Since EF – Tu has a strong affinity for GDP, the role of EF – Ts is to displace the GDP, and later to be replaced by GTP (after Tamarin, 2002)

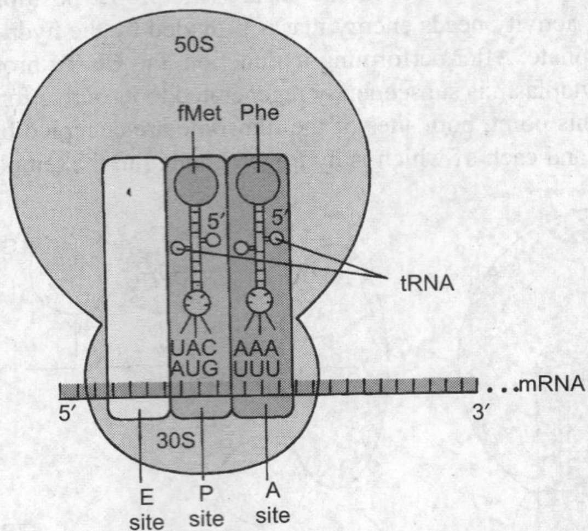

Fig. 12.17. A ribosome with two transfer RNAs attached. In this case, the second codon (UUU) is for the amino acid phenylalanine. The two amino acids are next to each other (after Tamarin, 2002).

The structures revealed by X-ray crystallography show that A, P and E sites are located in the cavity between the large and small subunits of the ribosome, the codon-anticodon interaction being associated with the small subunit and the aminoacyl end of the mRNA with the large subunit (**Yusupov** *et al.* 2001; Fig. 12.16).

Fig. 12.18. X-ray crystallographic structure of bacterial ribosome. The structure on the left is the large subunit of the *Thermus thermophilus* ribosome; that on the right is the small subunit.

Further, it takes several milliseconds for the GTP to be hydrolysed, and another few milliseconds for the EF-Tu/GDP to actually leave the ribosome. During those two intervals of time, the codon-anticodon fit of the transfer RNA is scrutinized. If the correct tRNA is in place, a peptide bond forms. If not, the charged tRNA is released and a new cycle of EF-Tu/GTP-mediated testing of tRNAs begins. The error rate is only about one mistake in ten thousand amino acids incorporated into protein. The speed of amino acid incorporation is about fifteen amino acids per second in prokaryotes and about two to five per second in eukaryotes.

(ii) **Peptide bond formation.** The next step is the formation of a peptide bond between the two amino acids. To accomplish this job, the first amino acid (N-formylmethionine) is removed from its attachment to its tRNA and transferred to the free-NH_2 terminus of the second amino acid. The first amino acid is, thus, placed "on top of" the second. The ensuing peptide bond, thus, joins the carboxyl group of the first amino acid with the amino group of the second amino acid (Fig. 12.19). The resulting compound is a dipeptide whose carboxyl end is still bonded to the second tRNA, but whose amino end is free. The reaction is catalyzed by an enzyme associated with 50S subunit and called **peptidyl transferase.** This enzymatic centre, an integral part of the 50S subunit, was originally believed to be composed of parts of several of the 50S proteins. Now, however, it is believed to have ribozymic activity (enzymatic activity) of 23S rRNA of 50S ribosomal subunit.

The enzymatic activity involve a bond transfer from the carboxyl end of N-formyl methionine to the amino end of the second amino acid (phenylalanine in Fig. 12.17). Every subsequent peptide bond is identical, regardless of the amino acids involved. The energy used is contained in the high-energy ester bond between the transfer RNA in the P site and its amino acid (Fig. 12.19). Immediately after the formation of the peptide bond, the transfer RNA with the dipeptide is in the A site, and a depleted tRNA is in the P site.

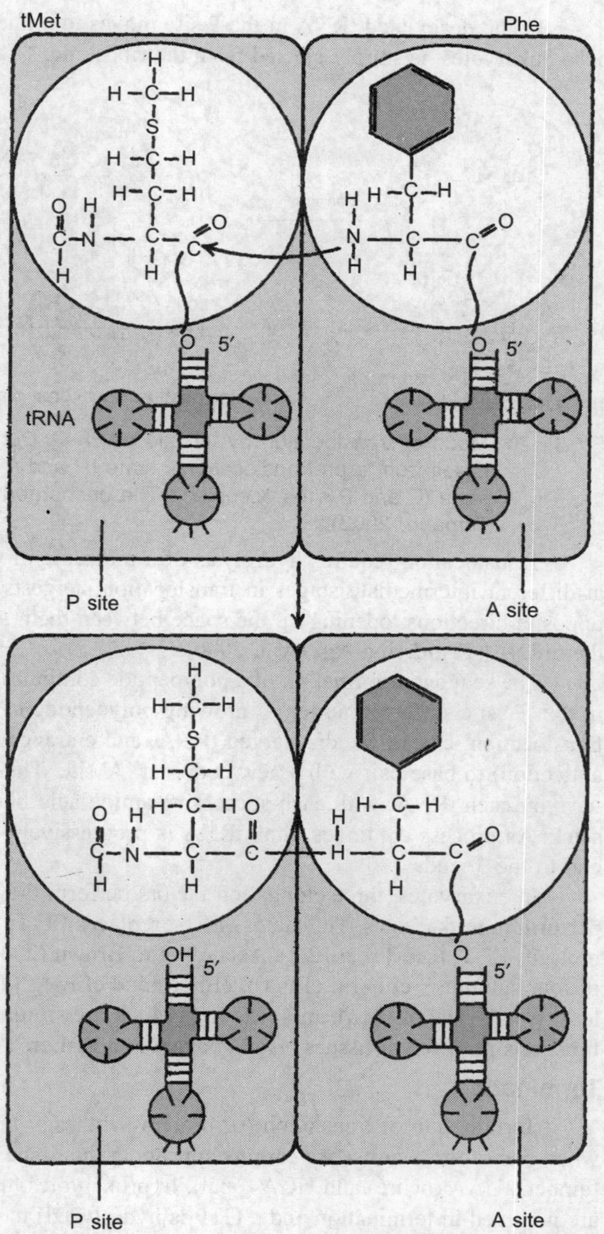

Fig. 12.19. Peptide bond formation on the ribosome (after Tamarin, 2002).

(iii) **Translocation**

The next stage in elongation is *translocation* of the ribosome in relation to the tRNA and the mRNA. Elongation factor **EF-G**, earlier called ***translocase***, catalyses the translocation process. The ribosome must be converted from *pretranslocational state* to the *post-translocational* state by the action of EF-G, which physically moves the mRNA and its associated transfer RNA (Fig. 12.20).

Thus, during translocation following three things happen at once:

1. The ribosome moves along the three nucleotides, so that the next codon enters the A site.
2. The dipeptide-tRNA in the A site moves to the P site.

3. The deacylated tRNA in the P site moves to a third position, the E or exit site in bacteria or in eukaryotes, is simply ejected from the ribosome.

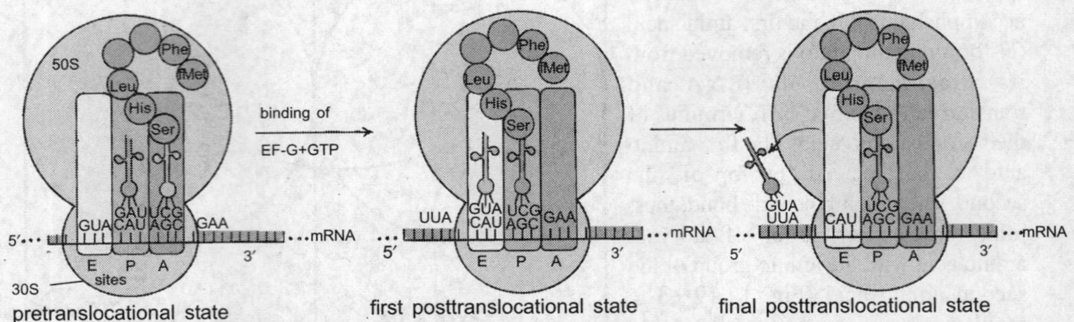

Fig.12.20. Mode of translocation by the aid of EF-G. During translocation the ribosome converts istself from a pretranslocational state (P and A sites occupied) to a post-translocational state (E and P sites occupied). The uncharged tRNA in the E site is then ejected (after Tamarin, 2002).

Translocation requires hydrolysis of a molecule of GTP. Electron microscopy of ribosomes at different intermediate stages in translocation suggests that the two subunits rotate slightly in opposite directions, opening up the space between them and enabling the ribosome to slide along the mRNA (**Frank** and **Agarwal,** 2000).

The sequential formation of a polypeptide continues in the manner described above. A tRNA in the 'P' site shifts its burden of growing polypeptide to the next succeeding tRNA, followed by translocation, exit of the discharged tRNA, and entrance of a new charged tRNA (having correct anticodon) to base pair with a new codon at 'A' site. Thus, the growing polypeptide is adopted in turn, by each tRNA, with each successive amino acid being added in effect, to the bottom of the stack. As process continues, the mRNA is progressively translated, codon by codon, from the 5′ end to the 3′ end.

In eukaryotes, three elongation factors perform the same tasks that EF-Tu, EF-Ts and EF-G perform in prokaryotes. The factor eEF1α replaces EF-Tu, factor eEF1ßγ replaces EF-Ts and eEF2 replaces EF-G (see Table 12.2). According to Brown (2002), elongation factor eEF-1 is a complex of four subunits : eEF-1a, eEF-1b, eEF-1d and eEF-1g (Table 12.3). The first of these exists in at least two forms, eEF-1a1 and eEF-1a2, which are similar proteins that probably have equivalent functions in different tissues (**Hafezparast** and **Fisher,** 1998).

Termination

Termination of translation (or protein synthesis) in both prokaryotes and eukaryotes occur when one of three **nonsense codons** appears in the A site of the ribosome. These codons are UAG (amber), UAA (ochre) and UGA (opal). In prokaryotes, three proteins called **release factors (RF)** are involved in termination and a GTP is hydrolyzed) to GDP + Pi.

When a nonsense codon enters the A site on the ribosome, a release factor recognizes it. **RF1** and **RF2** are class 1 release factors: they recognize termination/stop codon and then promote hydrolysis of the bond between the terminal amino acid and its tRNA in the P site. Class 2 release factors (*e.g.,* **RF3**) do not recognize stop codons, but they stimulate class1 release factors to act. RF1 recognizes the termination or stop codon UAA and UAG and RF2 recognizes UAA and UGA (Fig.12.20). Both do so because they have tripeptides that mimic anticodons to recognize the stop codon: proline-alanine-threonine in RF1 and serine-proline-phenylalanine in RF2. In this **molecular mimicry,** a protein mimics the shape of a nucleic acid in order to function properly.

The next base in mRNA past the stop codon is usually an adenine, required for efficient termination. After the release factors act, with the hydrolysis of a GTP, the ribosome has completed its task of translating mRNA into a polypeptide. Final release of all factors and dissociation of the two subunits of the ribosome take place with the help of IF3, which rebinds to the 30S subunit, and a **ribosome recycling factor (RRF).** RRF has a tRNA-like structure like the eRF-1 (**Selmer** *et al.,* 1999). RRF probably enters the P or A site and "unlocks" the ribosome. Dissociation requires energy, which is released from GTP by EF-G and also requires the initiation factor IF-3 to prevent the subunits from attaching together again. The disassociated ribosome subunits enter the cytoplasmic pool, where they remain until used again in another round of translation.

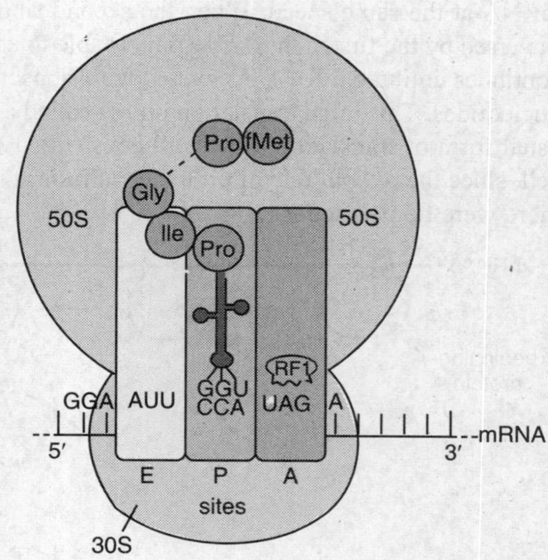

Fig. 12.21. Chain termination at the ribosome. One of two release factors recognizes a termination or stop codon in the A site in ribosome. In this case, RF1 recognizes a UAG. The complex then falls apart releasing the peptide (after Tamarin, 2002).

Eukaryotes have just two release factors: eRF-1, which recognizes the termination codon, and eRF-3, which might play the same role as RF-3 although this has not proven (see **Brown,** 2002). The structure of eRF-1 has been solved by X-ray crystallography, showing that the shape of this protein is very similar to that of a tRNA (**Kisselev** and **Buckingham,** 2000).

Further, in contrast to the often polycistronic mRNA of bacteria, mRNA of eukaryotes are monocistronic, containing the coding sequence only for one polypeptide. The initiation codon for this sequence is located near the 5′ end of the message. Thus, the association of mRNA takes place on 5′ end and not at the initiation codon AUG as in prokaryotes.

Comparison of Prokaryotic and Eukaryotic Translation

Table 12.4. Comparison between prokaryotic and eukaryotic translation (Source : **Tamarin,** 2000).

	Prokaryotes	Eukaryotes
1. Initiation codon	AUG, occasionally GUG, UUG	AUG, occasionally GUG, CUG
2. Initiation amino acid	N-formyl methionine	Methionine
3. Initiation tRNA	$tRNA_f^{met}$	$tRNA_i^{Met}$
4. Internal methionine tRNA	$tRNA_{in}^{met}$	$tRNA_m^{Met}$
5. Initiation factors	IF1, IF2, IF3	eIF factors
6. Elongation factor	EF-Tu	eIF1α
7. Elongation factor	EF-Ts	eIF1 βγ
8. Translocation factor	EF-G	eIF 2
9. Release factors	RF1, RF2, RF3, RRF	eRF1, eRF3

Polysomes and Coupled Transcription – Translation

The unit of translation is never simply a ribosome traversing a mRNA molecule, but is a more complex structure. After about 25 amino acids have been joined in a polypeptide chain, the AUG initiation site of the encoding mRNA molecule is completely free of the ribosome. A second initiation complex then forms. The overall configuration is of two 70S ribosomes moving along the mRNA at the same speed. When the second ribosome has moved along a distance similar to that traversed by the first, a third ribosome is able to attach. This process, *i.e.,* movement and reinitiation, continues until the mRNA is covered with ribosomes at a density of about one 70S ribosome per 80 nucleotides. This large translation unit is called a **polyribosome** or simply a **polysome.** This is the usual form of translation unit in all cells (Fig 12.22). The use of polysomes is advantageous to a cell, since the overall rate of protein synthesis is increased compared to the rate that would occur if there were no polysomes.

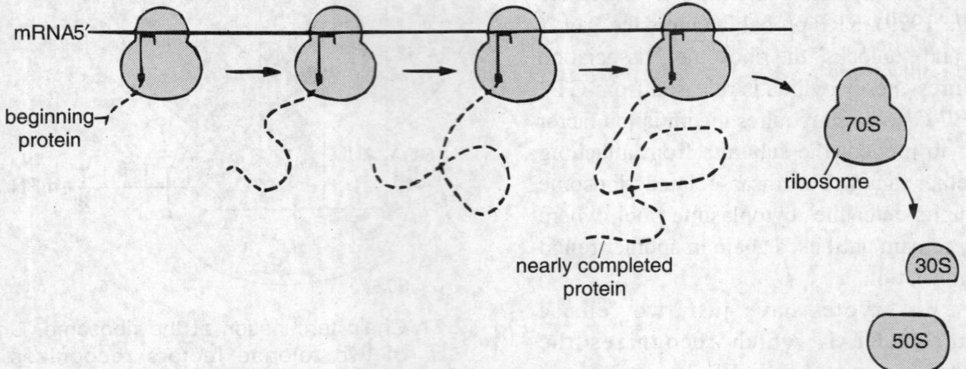

Fig. 12.22. Polysome formation: diagram shows the relative movement of the 70S ribosome and the mRNA, and growth of the protein chain (after Freifelder, 1985).

Further, a mRNA molecule being synthesized has a free 5′ terminus and translation also occurs in the 5′ → 3′ direction, so each cistron contained in the mRNA immediately starts its translation. As a result, the ribosome binding site is transcribed first, followed in order by the AUG codon, the region encoding the amino acid sequence, and finally the stop codon. Thus, in bacteria in which no nuclear membrane separates the DNA and the ribosomes, there is no obvious reason why the 70S initiation complex should not form before the mRNA is released from the DNA. With

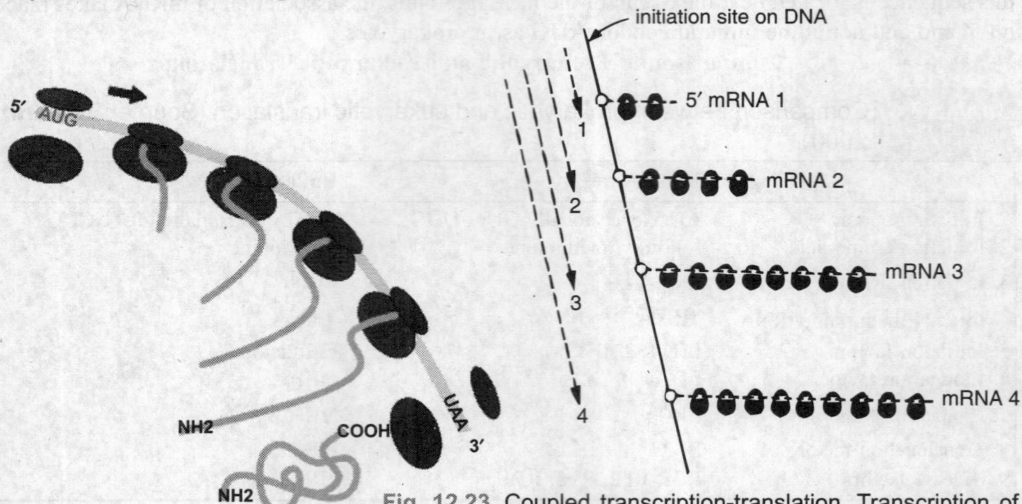

Fig. 12.23. Coupled transcription-translation. Transcription of a section of the DNA of *E.coli* and translation of the nascent mRNA. The dashed arrows show the distances of each RNA polymerase from the transcription initiation site, mRNA 4 probably becomes shorter due to partial digestion off its 5′ end by a RNase (after Freifelder, 1985).

prokaryotes (*E.coli)* this does indeed occur; this process is called **coupled transcription-translation.** This coupled activity does not occur in eukaryotes, because the mRNA is synthesized and processed in the nucleus and later on transported through the nuclear membrane to the cytoplasm where the ribosomes are located. Coupled transcription-translation, too, speeds up protein synthesis in the sense that translation does not have to await release of mRNA from the DNA.

12.4. UNSUAL EVENTS DURING TRANSLATION ELONGATION

The simple codon-by-codon translation of an mRNA is viewed as the standard way in which proteins are synthesized. But following three unusual translation elongation events have been reported:

1. Frameshifting;
2. Translation slippage; and
3. Translation bypassing

1. Frameshifting

This event occurs when a ribosome pauses in the middle of an mRNA, moves back one nucleotide or, less frequently, forward one nucleotide, then continues translation (**Farabaugh**, 1996). The result is that the codons that are read after the pause are not contiguous with the preceding set of codons, *i.e.*, they lie in a different reading frame (Fig.12.24A). Such frame shifts may be spontaneous or programmed.

(*i*) **Spontaneous frameshifts.** They occur randomly and are deleterious because the polypeptide synthesized after the frameshift has the incorrect amino acid sequence.

(*ii*) **Programmed frameshifts.** They occur only in some mRNAs. A few mRNAs utilize programmed frameshifting to induce the ribosome to change frame at a specific point within transcript. Programmed frameshifting occurs in all types of organisms, from bacteria through to humans, as well as during expression of a number of viral genomes. An example occurs during synthesis of **DNA polymerase III** (enzyme) in *E.coli*, the main enzyme involved in replication of DNA. Two of the subunits of DNA polymerase III, γ (gamma) and τ (tau), are coded by a single gene, *dnaX*. Subunit τ is the full-length translation product of the *dnaX* mRNA, and subunit γ is a shortened version. Synthesis of γ subunit invoves a frameshift in the middle of *dnaX* mRNA, the ribosome encountering a termination codon immediately after frameshift and so producing the truncated γ version of the translation product. It is thought that the frameshift is induced by *three* features of the *dnaX* mRNA:

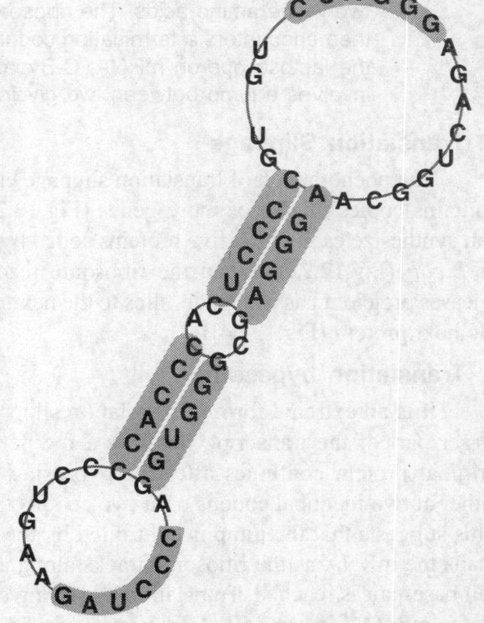
Antizyme mRNA frameshifting.

1. A hairpin loop, located immediately after the frameshift position, which stalls the ribosome.

2. A sequence similar to a ribosome binding site immediately upstream of the frameshift position, which is thought to base-pair with the 16S rRNA, again causing the ribosome to stall.

3. The codon 5′ –AAG–3′ at the frameshift position. The presence of a modified nucleotide at the wobble position of the $tRNA^{lys}$ that decodes 5′ –AAG–3′ means that the codon-anticodon interactions is relatively weak at this position, enabling the frameshift to occur.

(A) Programmed frameshifting in the dnaX mRNA

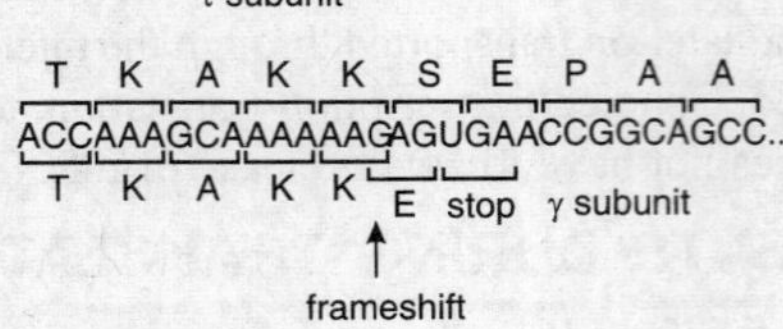

(B) Translational slippage in the lactose operon mRNA

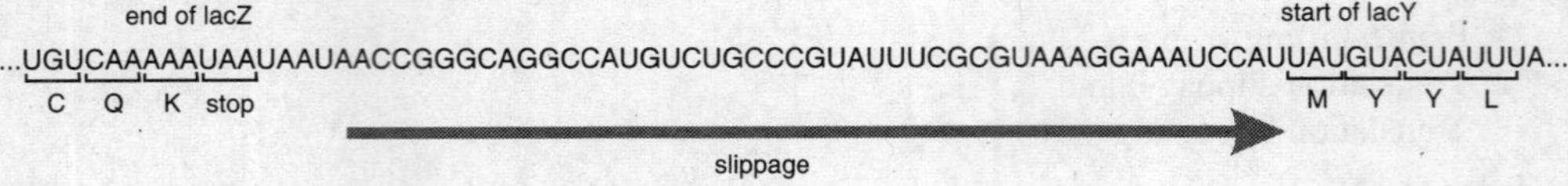

(C) Translational bypassing in the T4 gene 60 mRNA

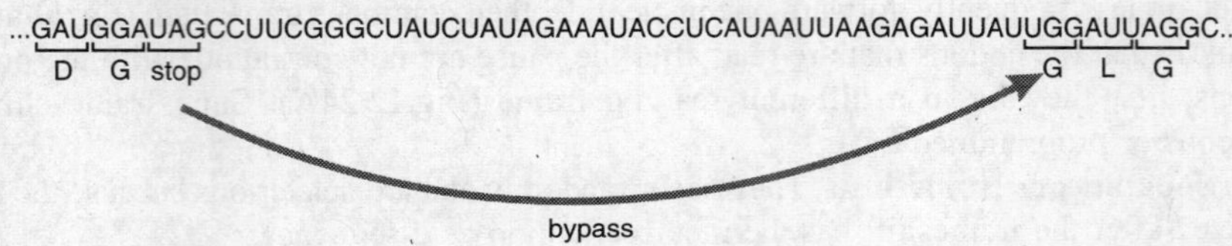

Fig. 12.24. Three unusual translation elongation events occurring in *Escherichia coli*. A-Programmed frameshifting during the translation of the *dnaX* mRNA. During the synthesis of γ subunit the ribosome shifts back one nucleotide, immediately after a series of As coding for two lysine amino acids. The ribosome inserts a glutamic acid into the polypeptide and then encounters a termination codon. B-Slippage between the *lac Z* and *lac y* genes of the lactose operon mRNA. C-Bypassing during translation of the T4 gene 60 mRNA involves a jump between two glycine codons (after Brown, 2002).

2. Translation Slippage

The phenomenon of translation slippage enables a single ribosome to translate an mRNA that contains copies of two or more genes (Fig.12.24B). Due to this phenomenon, a single ribosome can synthesize each of the five proteins coded by the mRNA transcribed from the tryptophan operon of *E.coli* (Fig.12.25). When the ribosome reaches end of one series of codons (*e.g.*, UAA), it releases protein it has just made, slips to the next initiation codon (*e.g.*, AUG), and begins synthesizing the next protein (Fig.12.24B).

3. Translation bypassing

It is an extreme form of translation slippage (**Herr** *et al.*, 2000). In translation bypassing, a larger part of the transcript, possibly a few tens of base pairs, is skipped and elongation of the original protein continues after the bypassing event (Fig. 12.24C). The bypass starts and ends either at two identical codons or at two codons that can be translated by the same tRNA by wobble. This suggests that the jump is controlled by the tRNA attached to the growing polypeptide, which scans the mRNA as the ribosome tracks along and halts the bypass when a new codon to which it can base-pair is reached. Translation bypassing of 44 nucleotides occurs in *E.coli* during translation of the mRNA for *gene 60* of T4 bacteriophage, which codes for the DNA topoisomerase subunit. Similar events have also been identified in a variety of other bacteria. Bypassing could result in two different proteins being synthesized from one mRNA- one protein from normal translation and one from bypassing, but whether this is its general function is not yet known (see **Brown**, 2002).

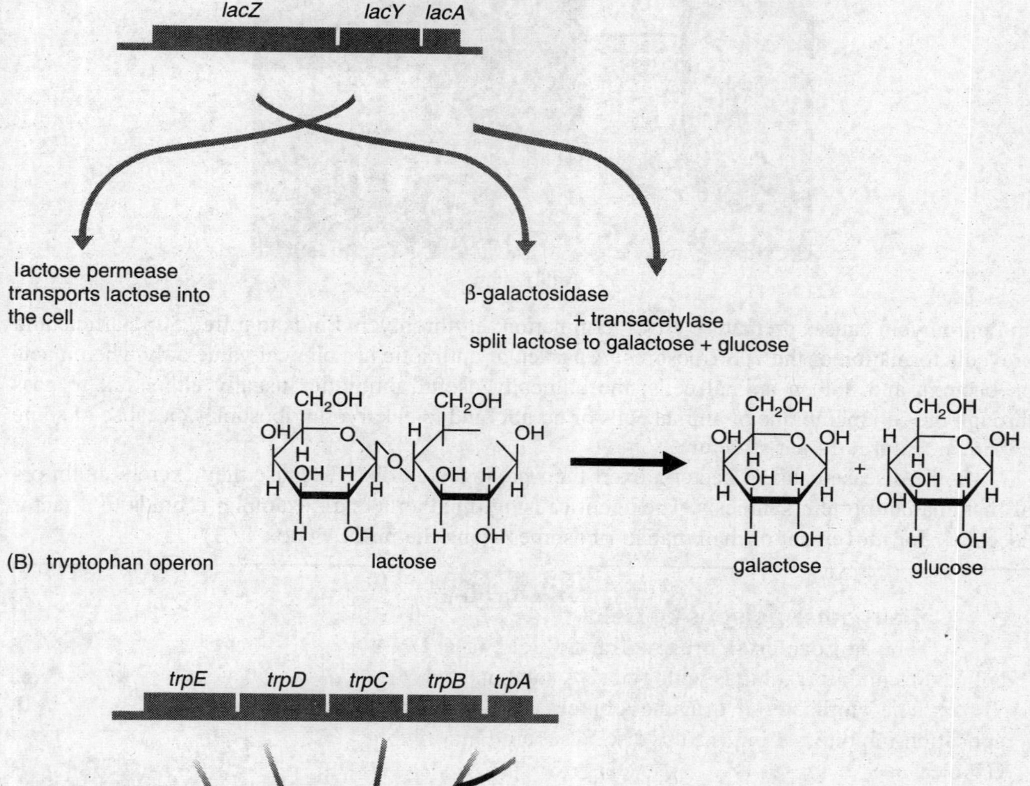

Fig. 12.25. Two operons of *E.coli*. *A*-The lactose operon. The three genes are called *lacZ, lacY* and *lacA*, the first two genes are separated by 52 bp and the second two genes by 64 bp. All three genes are expressed together, *lacY* coding for the lactose permease which transports lactose into the cell, and *lacZ* and lacA coding for enzymes that split lactose into its component sugars—galactose and glucose. *B*- The tryptophan operon, which contains five genes coding for enzymes involved in the multistep biochemical pathway that converts chorismic acid into the amino acid tryptophan. The genes in the tryptophan operon are closer together than those in the lactose operon : *trpE* and *trpD* overlap by 1bp, as do *trpB* and *trpA; trpD* and *trpC* are separated by 4bp and *trpC* and *trpB* by 12bp (after Brown, 2002).

12.5. ANTIBIOTICS AND PROTEIN SYNTHESIS

Many antibacterial agents (called antibiotics) have been isolated from fungi. Most of these are inhibitors of protein synthesis. For example, streptomycin and neomycin bind to a particular protein in the 30S particle and thereby prevent binding of $tRNA^{fmet}$ to the 'P' site; the tetracycline's inhibit binding of charged tRNA; lincomycin and chloramphenicol inhibit the peptidyl transferase:

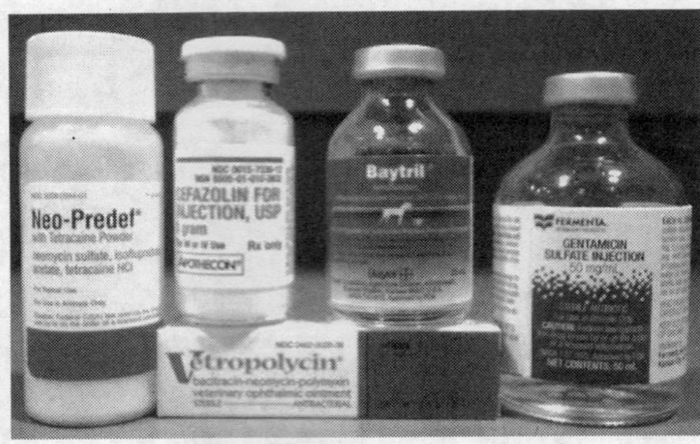

Antibiotics.

and puromycin causes premature chain termination; erythromycin binds to a free 50S particle and prevents formation of the 70S ribosome. A particular antibiotic has clinical value only when it acts on bacteria and not on animal cells; the clinically useful antibiotics usually either fail to pass through the cell membrane of animal cells or do not bind to eukaryotic ribosomes, because of some unknown feature of their structure.

Some disease-causing bacteria exert their pathogenic effect, because they excrete inhibitors of mammalian protein synthesis. The agent causing diphtheria is an example; it binds to a factor necessary for movement of mammalian ribosomes along the mRNA (Box 12.3)

Box 12.3.

Certain other poisons of Cells

1. The antibacterial drug nalidixic acid is a DNA-replication-inhibitor; it binds with gyrase A subunit.

2. The antibacterial drug novobiocin is also a DNA-replication-inhibitor; it binds to gyrase B subunit and inhibits ATP cleavase.

3. The centrosome duplication is arrested in the mammalian cells by antibiotic amphidicolin. DNA polymerases L, S and E are sensitive to inhibition by amphidicolin (see **Mathews,** 2000).

4. Microtubular poisons are benomyl and nocodazole.

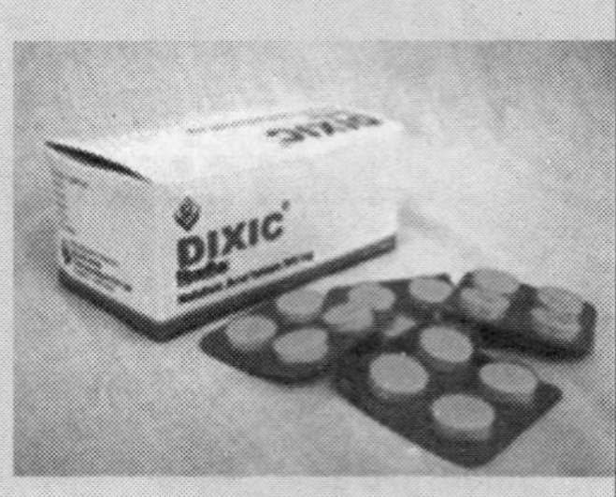

Nalidixic acid.

Questions And Problems

1. How do the function of rRNA, mRNA and tRNA differ during the process of protein synthesis?
2. Give an account of the mechanism of protein synthesis; compare it with prokaryotes and eukaryotes.
3. (*a*) Explain how an amino acid is activated and then attached to its specific tRNA.
   (*b*) What is the evidence that mRNA may be long enough to code for several proteins?
4. The first step in protein synthesis is the formation of an initiation complex (**Nomura** *et. al.,* 1967). What is the initiation complex, and what events follow to begin polypeptide synthesis?
5. Differentiate between transcription and translation.
6. How many different mRNAs could specify the amino acid sequence met-phe-ser-pro ?
7. Using the information in Table 10.2 convert the following mRNA segments into their polypeptide equivalents:
   (*a*) .... 5′GAAAUGGCAGUUUAC3′....; (*b*) .....3ÚUUCGAGAUGUCAA 5′......; and
   (*c*) ..... 5′AAA ACC UAG AAC CCA3′.....

8. A certain codon is determined to be AUG.
   (*a*) Of what nucleic acid molecule is this codon a part ?
   (*b*) What is the corresponding anticodon ?
   (*c*) Of what nucleic acid molecule is this anticodon a part ?
   (*d*) What is the deoxyribonucleotide responsible for this codon ?
9. Alpha (α) polypeptide chain of human haemoglobin consists of 141 amino acid residues. Would you expect the DNA segment responsible for the ultimate synthesis of this chain to be shorter than, longer than, or about the same length as the functional mRNA molecule for this chain ?
10. An antibiotic is a microbial product of low molecular weight that specifically interferes with the growth of microorganisms when it is present in exceedingly small amounts. Specify some of the physiological activities that might be interrupted by an appropriate antibiotic and the reason why human cells are not harmed.
11. Which processes in protein synthesis require hydrolysis of GTP ?
12. Write short notes on the following:
   (*i*) Central dogma; (*ii*) Teminism; (*iii*) Reverse transcription; (*iv*) Kozak's scanning hypothesis; (*v*) Shine-Dalgarno sequence; (*vi*) rho factor; (*vii*) Aminoacyl RNA synthetase; (*viii*) Elongation factors; (*ix*) Chaperones; (*x*) Peptidyl transferase.
13. Write a short essay on aminoacyl-tRNA synthetases. Make sure that you cover the following points: the two classes of synthetase; fidelity of aminoacylation ; modification of attached amino acids after aminoacylation.
14. Give a detailed description of initiation of translation in (a) *E. coli,* and (b) eukaryotes.
15. Describe various unusual events which occur during translocation elongation.

Answers to Problems

6. 1 × 2 × 6 × 2 = 48
7. (*a*)- glu-met-ala-val-tyr-.
   (*b*)- phe-ala-arg-cys-asn-(since genetic code is always read in 5′ → 3′ direction).
   (*c*) -lys-thr- (nonsense), chain terminates prematurely.
8. (*a*) By definition, codon occurs only in mRNA.
   (*b*) UAC
   (c) By definition, anticodons occur only in tRNA
   (*d*) TAC
9. Longer.
10. Cell wall formation is interfered with by penicillins and cephalosporins . DNA replication is prevented by the bleomycins and anthracyclines. Rifamycins interrupt the transcription of DNA into RNA. Translation is disrupted by erythromycin, tetracyclines, chloramphenicol, neomycin, lincomycin, puromycin, and streptomycin. Antibiotics are toxic for microorganisms but safe for humans because all of these metabolic processes are subtly different in bacteria and humans.
11. Formation of the 70S initiation complex and translation.

Multiple Choice Questions

Choose the correct answer from the four alternatives given.

1. The 'central dogma' states that biological information flaws in which of these patterns ?
   (*a*) DNA → Protein → RNA
   (*b*) RNA → DNA → Protein
   (*c*) Protein → RNA → DNA
   (*d*) DNA → RNA → Protein
2. One of these cannot be prepared directly from DNA

(*a*) another DNA (*b*) mRNA
(*c*) Protein (*d*) tRNA

3. Amino acid sequence, in protein synthesis is decided by the sequence of
(*a*) *r*RNA (*b*) *t*RNA
(*c*) mRNA (*d*) *c*RNA

4. One of the following is not involved in translation
(*a*) mRNA (*b*) tRNA
(*c*) DNA (*d*) rRNA

5. Codes of mRNA and amino acid residues of proteins are
(*a*) colinear
(*b*) coplanar
(*c*) zig-zag
(*d*) none of these

6. The name of Temin and Baltimore is associated with
(*a*) photorespiration
(*b*) RNA synthesis
(*c*) reverse transcription
(*d*) all of these

7. Enzyme involved in reverse transcription
(*a*) primase
(*b*) reverse transcriptase
(*c*) DNA polymerase
(*d*) terminase

8. RNA retrovirus has a special enzyme that
(*a*) polymerizes host DNA
(*b*) disintegrates host DNA
(*c*) translates host DNA
(*d*) transcribes viral RNA to DNA

9. The starting tRNA of prokaryotes is loaded with
(*a*) valine
(*b*) methionine
(*c*) formylated methionine
(*d*) tryptophan

10. Initiation of polypeptide chain formation is always brought about at the site of codon coding for an amino acid called
(*a*) isoleucine (*b*) cysteine
(*c*) phenylalanine (*d*) methionine

11. Which of the following is associated with protein synthesis ?
(*a*) aminoacyl tRNA
(*b*) initiation factors IF3, IF2, IF1
(*c*) peptidyl transferase
(*d*) transposons

12. The site of protein synthesis in a cell is
(*a*) mitochondria
(*b*) chloroplasts
(*c*) ribosomes
(*d*) plasma membrane

13. During translation, initiation in prokaryotes, a GTP molecule is needed in
(*a*) formation of formyl-met-tRNA
(*b*) binding of 30S subunit of ribosome with mRNA
(*c*) association of 50S subunit of ribosome with initiation complex
(*d*) association of 30S-mRNA with formyl-met-tRNA

14. Which step of translation does not consume high energy phosphate bond?
(*a*) translocation
(*b*) amino acid activation
(*c*) peptidyl transferase reaction
(*d*) aminoacyl tRNA binding to A site

15. Translocase is an enzyme required for
(*a*) DNA replication
(*b*) RNA synthesis
(*c*) initiation of protein synthesis
(*d*) chain elongation in protein synthesis

Answers

1. (*d*)	2. (*c*)	3. (*c*)	4. (*c*)
5. (*a*)	6. (*c*)	7. (*b*)	8. (*d*)
9. (*c*)	10. (*d*)	11. (*a*)	12. (*c*)
13. (*d*)	14. (*d*)	15. (*d*).	

CHAPTER 13

Post-Translational Processing of Proteins

Chapter Contents

The process of gene expression is not over when the genetic code has been used to create the sequence of amino acids that constitutes a protein. To be useful to the cell, this new polypeptide chain must fold up into its unique three-dimensional conformation, bind any small molecule cofactors required for its activity, be appropriately modified by protein kinases or other protein-modifying enzymes, and assemble correctly with the other protein subunits with which it functions. Thus, the polypeptide that emerges from the ribosome is inactive, and before taking on its functional role in the cell, it must undergo at least the first of following four types of post-translational processing.

1. Protein folding
2. Proteolytic cleavage
3. Chemical modification
4. Inteine splicing

13.1. PROTEIN FOLDING

The proteome is the final product of genome expression and comprises all the proteins present in a cell at a particular time. A 'typical' mammalian cell, for example a hepatocyte, is thought to contain 10,000 – 20,000 different proteins, about 8×10^9 individual molecules in all, representing approximately 0.5ng of protein or 18 – 20% of the total cell weight (**Brown,** 2002). A protein, like a DNA molecule is a linear unbranched polymer. The monomeric subunits of a protein are called **amino acids** and the resulting polymers, or polypeptides, are rarely more than 2000 units in length.

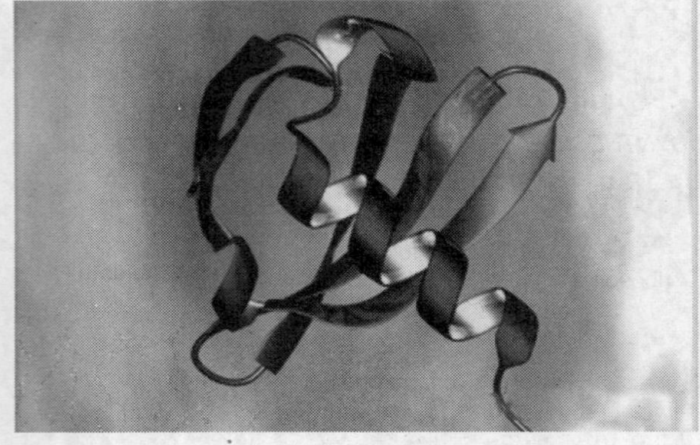

Protein folding.

Four levels of Protein Structure

Proteins are found to have four distinct levels of structure. These levels are hierarichal, the protein being built up stage by stage, with each level of structure depending on the one below it.

1. Primary structure. The primary structure of protein is formed by joining amino acids into a polypeptide. The amino acids are linked by

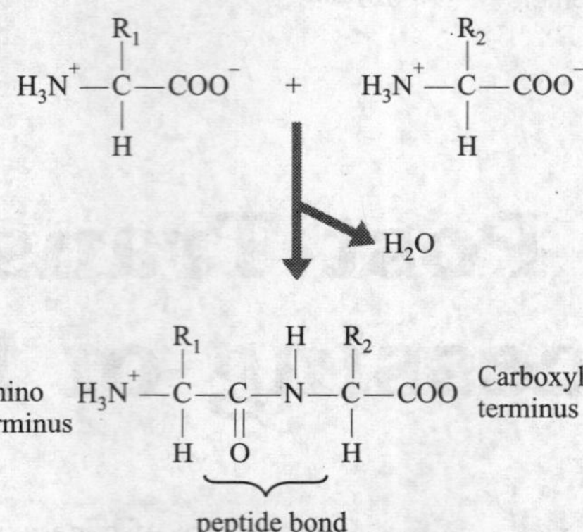

Fig. 13.1. In polypeptides, amino acids are linked by peptide bonds.

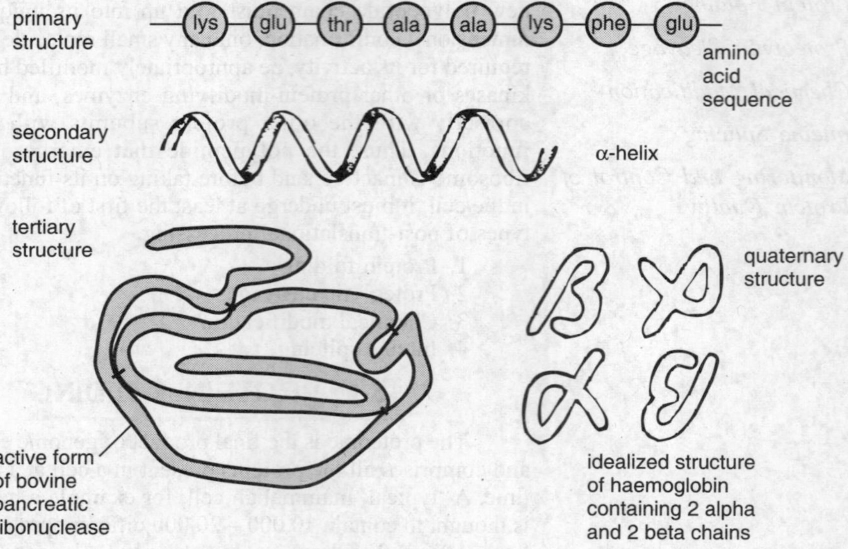

Fig. 13.2. Stages in the development of a functional protein (after Stansfield, 1969).

peptide bonds which are formed by a condensation reaction between the carboxyl group of one amino acid and the amino group of a second amino acid (Fig. 13.1). The two ends of the polypeptide are chemically distinct: one has a free amino group and is called the **amino, NH_2, - or N – terminus**; the other has a free carboxyl group and is called the **carboxyl, COOH** – or **C – terminus**. The direction of the polypeptide can therefore be expressed as either $N \rightarrow C$ (left to right) or $C \rightarrow N$ (right to left).

2. Secondary structure: The secondary structure of protein refers to the different conformations that can be taken up by the polypeptide. The two types of secondary structure are the **β–helix** and **α–sheet** (Fig. 13.3), both of which are stabilized by hydrogen bonds that form between different amino acids in the polypeptide. Most polypeptides are long enough to be folded into a series of secondary structures one after another along the molecule.

3. Tertiary structure: The tertiary structure of protein results from folding the secondary structural components of the polypeptide into a three-dimensional configuration (Fig. 13.4). The tertiary structure is stabilized by various chemical forces, notably hydrogen bonding between individual amino acids, and hydrophobic forces, which dictate that amino acids with non-polar (*i.e.*, water hating or hydrophobic) side groups must be shielded from water by embedding within the internal regions of the protein. There may also be covalent linkages called **disulphide bridges** between cysteine amino acids at various places in the polypeptide.

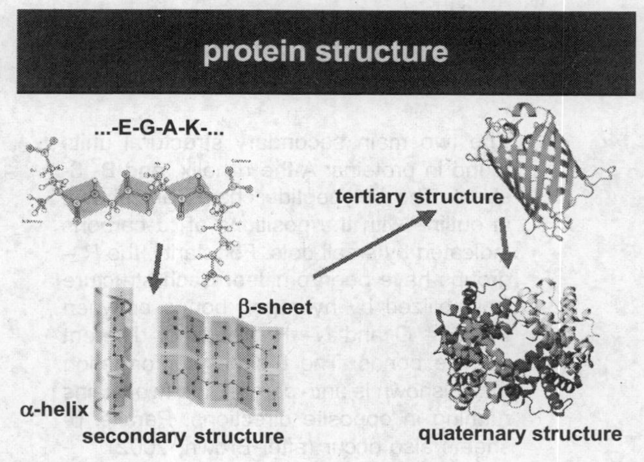

Different structures of proteins.

4. Quaternary structure: The quarternary structure of proteins involves the association of two or more polypeptides, each folded into its tertiary structure, into a multisubunit protein. Not all proteins form quaternary structures, but it is a feature of many proteins with complex functions, including several involved in genome expression. Some quaternary structures held together by disulphide bridge between different polypeptides, but many proteins comprise looser associations of subunits stabilized by hydrogen bonding and hydrophobic effects. However, according to the functional requirements, quaternary proteins can revert to their component polypeptides, or change their subunit composition.

All the information that a polypeptide needs in order to adopt its correct three-dimensional structure is contained within its amino acid sequence. This is one of the central principles of molecular biology. The notion that the amino acid sequence contains all the information needed to fold the polypeptide into its correct tertiary structure derives from experiments carried out on ribonuclease in the 1960s (**Anfinsen**, 1973). The enzyme ribonuclease (Fig. 13.2) is a small protein, just 124 amino acids in length, containing four disulphide bridges and with a tertiary structure that is made up predominantly of α–sheet, with very little β–helix.

Experiment showing spontaneous folding pathways for proteins: Studies of folding of proteins were carried out with ribonuclease that had been purified from cow pancreas and resuspended in buffer. Addition of urea, a compound that disrupts hydrogen bonding, resulted in a decrease in the activity of the enzyme (measured by testing its ability to cut RNA) and an increase

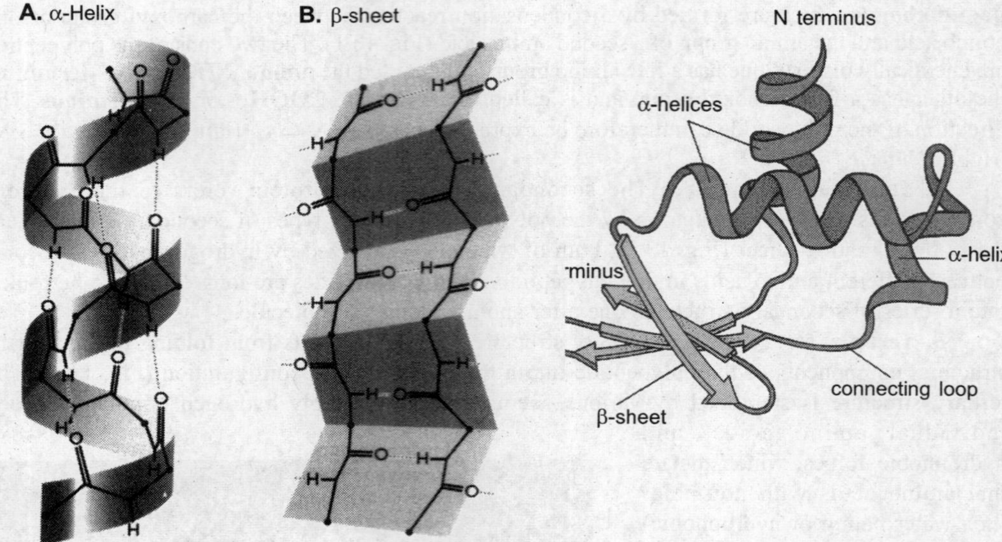

Fig. 13.3. The two main secondary structural units found in proteins: A-the α-helix, and B- β-sheet. The polypeptide chains are shown in outline with the positions of α-carbons indicated by small dots. For clarity, the R – groups have been omitted. Each structure is stabilized by hydrogen bonds between the C = O and N– H groups of different peptide bonds. The β-sheet conformation that is shown is anti- parallel, the two chains running in opposite directions. Parallel β-sheets also occur (after Brown, 2002).

Fig. 13.4. The tertiary structure of a protein. This imaginary protein structure comprises three α-helices, shown as coils, and a four-stranded β-sheet, indicated by the arrows (after Brown, 2002).

in the viscosity of the solution (Fig. 13.5), indicating that the protein was being **denatured** by unfolding into an unstructured polypeptide chain. Critical observation was that when the urea was removed from the experimental solution, the viscosity decreased and the enzyme activity reappeared. The conclusion is that the protein refolds spontaneously when the denaturant (*i.e.,* urea) is removed. In these initial experiments the four disulphide bonds remained intact because they were not disrupted by urea but same result occurred when the urea treatment was combined with addition of a reducing agent to break the disulphide bonds: the activity was still regained on renaturation. This shows that the disulphide bonds are not critical to the protein's ability to refold, they merely stabilise the tertiary structure once it has been adopted.

More detailed study of the spontaneous folding pathways for ribonuclease and other small proteins has led to the following two sets of the process:

1. The secondary structural motifs along the polypeptide chain form within a few milliseconds of the denaturant being removed.

2. During the next few seconds or minutes, the secondary structural motifs interact with one another and the tertiary structure gradually takes shape, often *via* a series of intermediate conformations (Box 13.1). In other words, the protein follows a **folding pathway.** However, there may be more than one possible pathway that a protein can follow to reach its correctly folded structure (**Radford**, 2000). Such *in vitro* folding pathways were found true for only the small proteins; large proteins fail to refold.

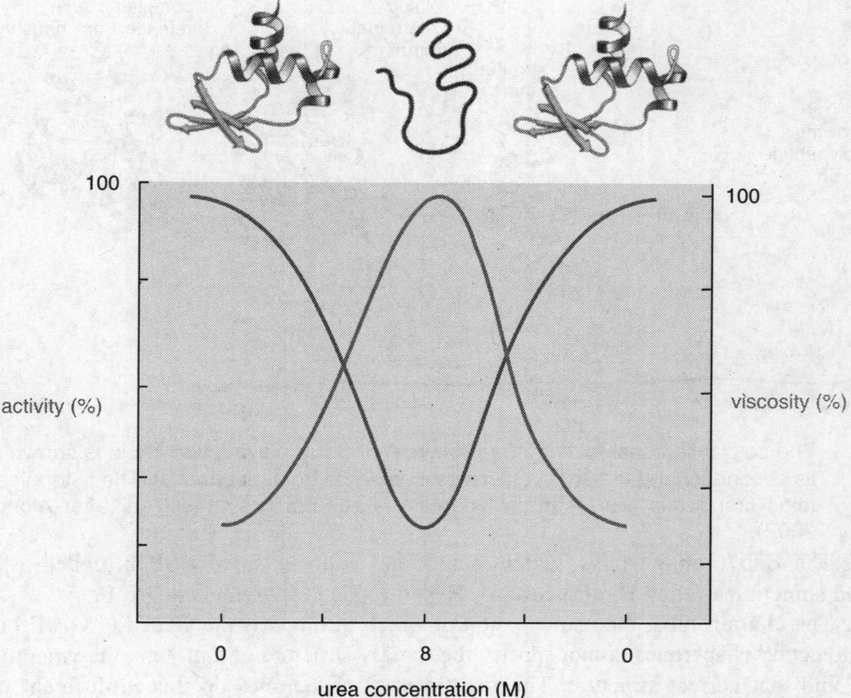

Fig. 13.5. Denaturation and spontaneous renaturation of a small protein. As the urea concentration increases to 8M, the protein becomes denatured by unfolding: its activity decreases and viscosity of the solution increases. When the urea is removed from solution by dialysis, this small protein re-adopts its folded conformation. The activity of the protein increases back to the original level and the viscosity of the solution decreases (after Brown, 2002).

Box 13.1.

Experiments have demonstrated that once a protein domain in a multidomain protein emerges from the ribosome, it forms a compact structure within a few seconds that contains most of the final secondary structure (β-helices and α-sheets) aligned in roughly the right way. For many protein domains, this unusually open and flexible structure, which is called a **molten globule**, is the starting point for a relatively slow process in which many side chain adjustments occur that eventually form the correct tertiary structure. Nevertheless, because it takes several minutes to synthesize a protein of average size, a great deal of folding process is completed by the time the ribosome releases the C – terminal end of a protein (**Federov** and **Baldwin,** 1997).

Role of Molecular Chaperones in Folding of Proteins

Most of our current understanding of protein folding in the cell is based on the discovery of proteins that help other proteins to fold. These are called **molecular chaperones** and have been studied most vividly in *E.coli*. It is clear that both eukaryotes and archaea possess equivalent proteins (chaperones), although some of the details of the way they work are different (**Hartl**, 1996; **Slavotinek** and **Biesecker**, 2001).

Molecular chaperones in *E.coli*. The molecular chaperones in *E.coli* can be divided into two groups: Hsp 70 chaperones (Hsp = heat shock proteins) and chaperonins.

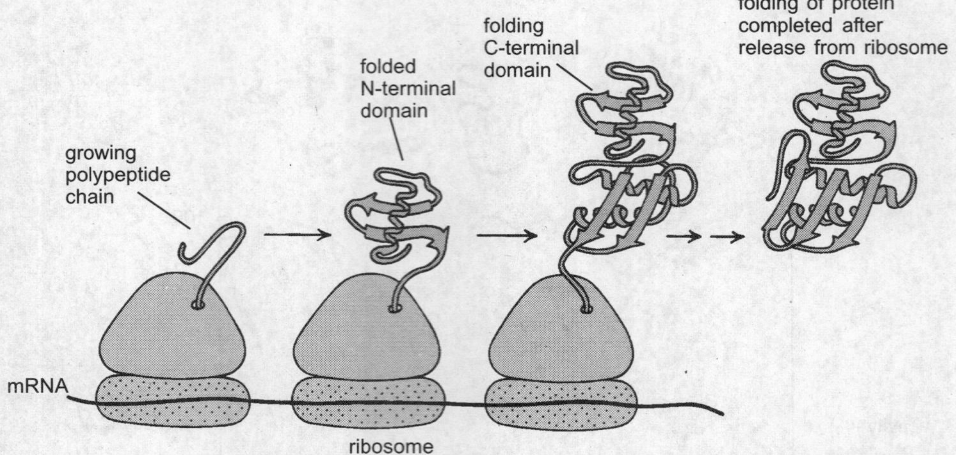

Fig. 13.6. The co-translational folding of a protein. A growing polypeptide chain is shown acquiring its secondary and tertiary structure as it emerges from a ribosome. The N-terminal domain folds first, while the C-terminal domain is still being synthesized (after Alberts *et al.*, 2002).

1. The **Hsp70 chaperones**, which include the proteins called **HSP70** (coded by the *dnak* gene and sometimes called **Dnak protein**), **Hsp40** (coded by *dnaJ*) and Grp E.

2. The **chaperonins**, the main version of which in *E.coli* is the **GroEL** – **GroES** complex.

Molecular chaperones do not specify the tertiary structure of a protein, they merely help the protein find that correct structure. The two types of chaperones do this in different ways. For example, the Hsp70 family (of chaperones) bind to hydrophobic regions of proteins, including proteins that are still being translated (Fig. 13.7A). They prevent protein aggregation by holding the protein in an open conformation until it is completely synthesized and ready to fold. The Hsp70 chaperones are also involved in other processes that require shielding of hydrophobic regions in proteins, such as transport through membranes and disaggregation of proteins that have been damaged by heat stress.

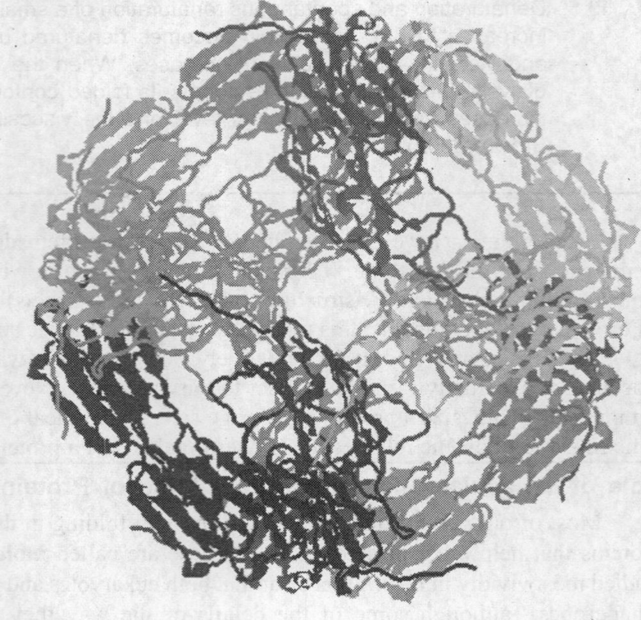

Molecular chaperones facilitate correct folding of protein subunits.

The chaperonins work in quite a different way. GroEL and GroES form a multi-subunit structure that looks like a hollowed out bullet with a central cavity (Fig. 13.7B; **Xu** *et al.,* 1997). A single unfolded protein enters the cavity and emerges folded. The mechanism for this is not known but it is postulated that GroEL/GroES acts as a cage that prevents the unfolded protein from aggregating with other proteins, and that the inside surface of the cavity changes from hydrophobic to hydrophilic in such a way as to promote the burial of hydrophobic amino acids within the protein. It is later found that the cavity unfolds proteins that have folded incorrectly, passing these unfolded proteins back to the cytoplasm so that they can have a second attempt at adopting their correct tertiary structure (**Shtilerman** *et al.*, 1999).

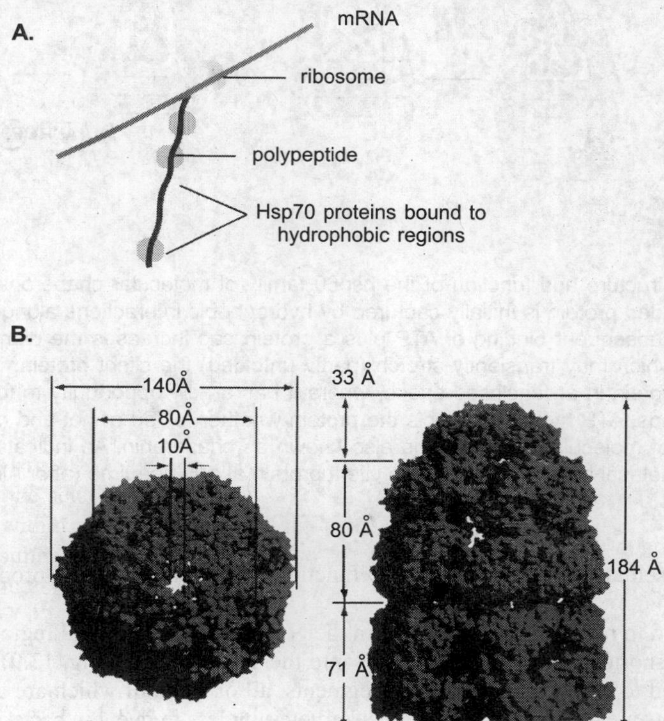

Fig. 13.7. Molecular chaperones of *E.coli*. A – Hsp70 chaperones bind to hydrophobic regions in unfolded polypeptides, including those that are still being translated and hold the protein in an open conformation until it is ready to be folded. B – The structure of the GroEL/GroES chaperonin (*i.e.,* GroEL bound to its GroES cap). On the left is a view from the side. 1Å is equal to 0.1nm. The GroES part of the structure is made up of seven identical protein subunits and is shown in gold. The GroEL components consist of 14 identical proteins arranged into two rings (shown in red and green), each containing seven subunits. The main entrance into the central cavity is thought at the bottom of the structure shown on the right. The images were creations of **Dr. Zhaohni Xu** and coworkers in 1977 (after Brown, 2002).

Molecular chaperones and chaperonins in eukaryotes. Eukaryotic cells have at least two families of molecular chaperones, known as the **hsp60** and **hsp70** proteins. Different family members function in different organelles. For example, mitochondria contain their own hsp60 and hsp70 molecules that are distinct from those that function in the cytosol (where hsp60 is called **TCP – 1** in vertebrate cells), and a special hsp70 (called **BIP**) helps to fold proteins in the endoplasmic reticulum. The hsp70 machinery act early in the life of many proteins, binding to a string of about seven hydrophobic amino acids before protein leaves the ribosome.

The hsp60 – like proteins form a large barrel-shaped structure that acts later in proteins life, after it has been fully synthesized. This type of chaperone forms an "*isolation chamber*" into which misfolded proteins are fed, preventing their aggregation and providing them with a favourable environment in which they attempt to refold (Fig. 13.8).

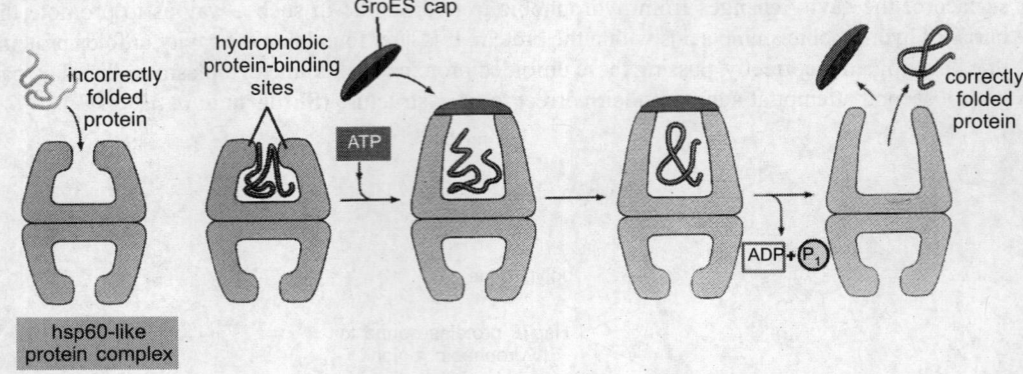

Fig. 13.8. The structure and function of the hsp60 family of molecular chaperones. As indicated, a misfolded protein is initially captured by hydrophobic interactions along one rim of barrel. The subsequent binding of ATP plus a protein cap increases the diameter of the barrel rim, which may transiently stretch (partly unfolded) the client protein. This also confines the protein in an enclosed space, where it has a new opportunity to fold. After about 15 seconds, ATP hydrolysis ejects the protein whether folded or not and cycle repeats. This type of molecular chaperone is also known as chaperonin. As indicated only half of the symmetrical barrel operates on a client protein at any one time (after Alberts *et al.,* 2002).

13.2. PROTEOLYTIC CLEAVAGE

Proteolytic cleavage has following *two* functions in post-translational processing of proteins (Fig. 13.9).

1. It is used to remove short pieces from the N – and/or C – terminal regions of polypetides, leaving a single shortened molecule that folds into the active protein (Fig. 13.9).

2. It is used to cut **polyproteins** into segments, all or some of which are active proteins.

These events are relatively common in eukaryotes but less frequent in bacteria. Not all proteins undergo proteolytic cleavage.

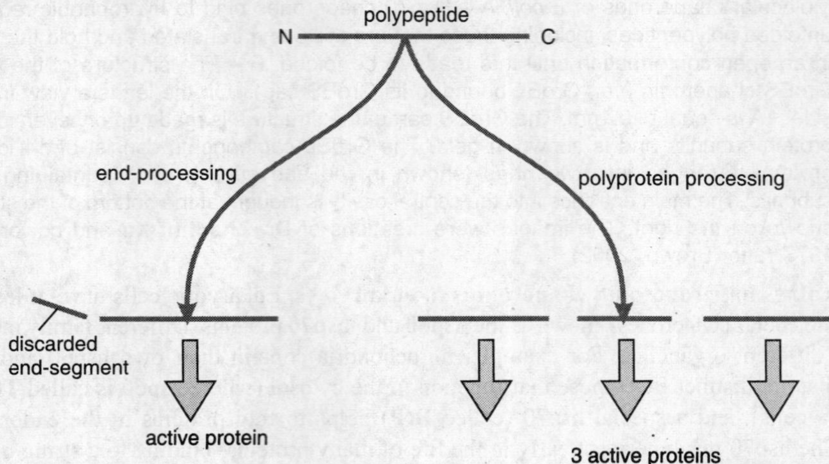

Fig. 13.9. Processing of protein by proteolytic cleavage. On the left, the protein is processed by removal of the N – terminal segment. C – terminal processing also occurs with some proteins. On the right, a polyprotein is processed to give three different proteins (after Brown, 2002).

1. Cleavage of the Ends of Polypeptides

Processing by proteolytic cleavage is common with secreted polypetides whose biochemical activities might be injurious to the cell producing the protein. An example is provided by **melittin**, the most abundant protein in bee venom and the one responsible for causing cell lysis after injection of the bee sting into the person or animal being stung. Melttin (protein) lyses cells in bees as well as animals so have to be initially synthesized as an inactive precursor. This precursor, called **promelittin**, has 22 amino acids at its N – terminus. The pre-sequence is removed by an extracellular protease that cuts it at all eleven positions (Fig. 13.10A), releasing the active venom protein. The protease does not cleave within the active sequence because its mode of action is to release dipeptides with the sequence X – Y, where X is alanine, aspartic acid or glutamic acid, and Y is alanine or proline, these motifs are not found in the active sequence (Fig. 13.10A).

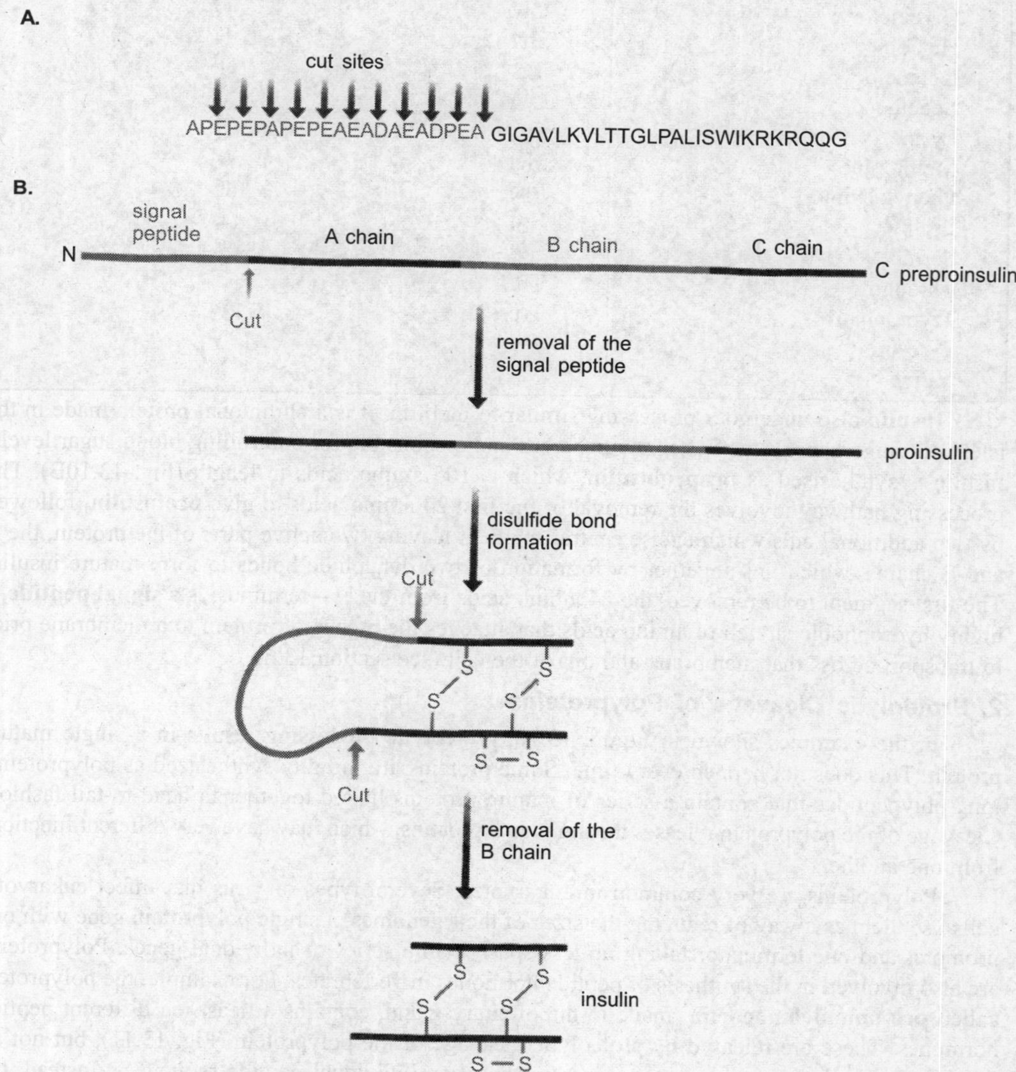

Fig. 13.10. Post-translational processing by proteolytic cleavage. A-processing of promelittin, the bee-sting venom. Arrows indicate the cut sites. B-processing of preproinsulin (after Brown, 2002).

Table 13.1. Amino acid abbreviations.

Amino acid	Abbreviation Three-letter	One-letter
Alanine	Ala	A
Arginine	Arg	R
Asparagine	Asn	N
Aspartic acid	Asp	D
Cysteine	Cys	C
Glutamic acid	Glu	E
Glutamine	Gln	Q
Glycine	Gly	G
Histidine	His	H
Isoleucine	Ile	I
Leucine	Leu	L
Lysine	Lys	K
Methionine	Met	M
Phenylalanine	Phe	F
Proline	Pro	P
Serine	Ser	S
Threonine	Thr	T
Tryptophan	Trp	W
Tyrosine	Tyr	Y
Valine	Val	V

Insulin also undergoes processing similar to melittin. It is a chromonal protein made in the islets of Langerhans in the vertebrate pancreas and responsible for controlling blood sugar levels. Insulin is synthesised as **proproinsulin**, which is 105 amino acids in length (Fig. 13.10B). The processing pathway involves the removal of the first 24 amino acids to give **proinsulin,** followed by two additional cuts which excise central segment leaving two active parts of the protein, the A and B chains, which link together by formation of two disulphide bonds to form mature insulin. The first segment to be removed, the 24 amino acids from the N – terminus, is a **signal peptide**, a highly hydrophobic stretch of amino acids that attaches the precursor protein to a membrane prior to transport across that membrane and out of the cell (see section 13.6).

2. Proteolytic Cleavage of Polyproteins

In the examples shown in Figure 13.10, proteolytic processing results in a single mature protein. This does not happen every time. Some proteins are initially synthesized as polyproteins, long polypeptides that contain a series of mature proteins linked together in head-to-tail fashion. Cleavage of the polyprotein releases the individual proteins, which may have very different functions from one another.

Polyproteins are very common in eukaryotes. Several types of virus that infect eukaryotic cells use them as a way of reducing the sizes of their genomes, a single polyprotein gene with one promoter and one terminator taking up less space than a series of individual genes. Polyproteins are also involved in the synthesis of peptide hormones in vertebrates. For example, the polyprotein called **pro-opiomelanocortin**, made in the pituitary gland, contains at least ten different peptide hormones. These are released by proteolytic cleavage of the polyprotein (Fig. 13.11), but not all can be produced at once because of overlaps between individual peptide sequences. Instead, the exact cleavage pattern is different in different cells.

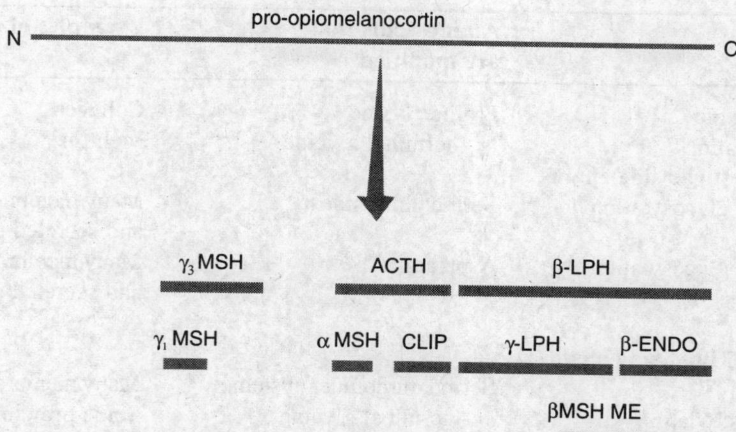

Fig. 13.11. Processing of pro-opiomelanocortin polyprotein. Abbreviations: ACTH, adrenocorticotropic hormones; CLIP, corticotropin-like intermediate lobe protein; ENDO, endorphin; LPH, lipoprotein; ME, met-encephalin; MSH, melanotropin (after Brown, 2002).

13.3. CHEMICAL MODIFICATION

The genome has the capacity to code for 21 different amino acids: the 20 specified by the standard genetic code, and selenocysteine, which is inserted into polypeptides by the context-dependent reading of a 5′ – UGA – 3′ codon. This range is increased dramatically by post-translational chemical modification of proteins, which results in an immense supply of different types of amino acids.

The most simple sort of chemical modifications involve addition of a small chemical group (*e.g.*, an acetyl, methyl or phosphate group; Table 13.2) to an amino acid side chain, or to the amino or carboxyl groups of the terminal amino acids in a polypeptide (**Bradshaw** *et al.*, 1998). More than 150 different modified amino acids, have been reported in different proteins; each modification is found to be carried out in a highly specific manner, *i.e.,* each amino acid being modified in a highly specific manner and the same amino acids being modified in same way in every copy of the proteins, *e.g.*, histone H3 (Fig. 13.12).

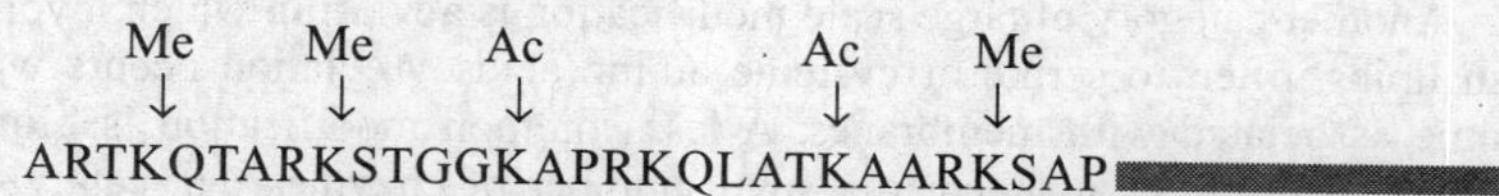

Fig. 13.12. Post-translational modification of calf histone H3. The first 30 amino acids of this 135 – amino acid protein are listed using the one – letter abbreviations (see Table 13.1). Five modifications occur-three methylations and two acetylations.

Table 13.2. Examples of post-translational chemical modifications (Source: **Brown,** 2002)

Modification	Amino acids that are modified	Examples of proteins
I. Addition of small groups		
1. Acetylation	Lysine	Histones
2. Methylation	Lysine	Histones
3. Phosphorylation	Serine, threonine, tyrosine	Some proteins involved in signal transduction

Modification	Amino acids that are modified	Examples of proteins
4. Hydroxylation	Proline, lysine	Collagen
5. N - formylation	N – terminal glycine	Melittin
II. Addition of sugar side chains		
6. O – linked glycosylation	Serine, threonine	Many membrane proteins and secreted proteins
7. N – linked glycosylation	Asparagine	Many membrane proteins and secreted proteins
III. Addition of lipid side chains		
8. Acylation	Serine, threonine, cysteine	Many membrane proteins
9. N - myristoylation	N – terminal glycine	Some protein kinases involved in signal transduction
IV. Addition of biotin		
10. Biotinylation	Lysine	Various carboxylase enzymes

We have observed how acetylation and methylation of histone H3 and other histones have an important influence on chromatin structure and hence on genome expression. Other types of chemical modification have important regulatory roles, *e.g.*, phosphorylation is used in transduction.

Glycosylation is a complex type of modification of amino acids. It involves the attachment of large carbohydrate side chains to polypeptides (**Drickamer** and **Taylor,** 1998). There are following two general types of glycosylation (Fig. 13.12).

1. O – linked glycosylation. It is the attachment of a sugar side chain *via* the hydroxyl group of a serine or threonine amino acid.

2. N – linked glycosylation. It involves attachment through the amino group on the side chain of asparagine.

Glycosylation results in attachment to the proteins of grand structures comprising branched networks of 10 to 20 sugar units of various types. These side chains help to target proteins to particular sites in the cells and determine the stability of proteins circulating in the bloodstream.

Another category of large scale modification is **acylation** which involves attachment of long chain lipids, often to serine or cysteine amino acids. Acylation occurs with many proteins that become associated with membranes. A less common modification is **biotinylation,** in which a molecule of biotin is attached to a small number of enzymes that catalyze the carboxylation of organic acids such as acetate and propionate (**Chapman**, **Smith** and **Cronan,** 1999).

13.4. INTEINE SPLICING

The final type of post-translational processing is **inteine splicing**. It is a protein version of the more extensive intron splicing that occurs in context to pre-RNAs. Inteins are internal segments of proteins that are removed soon after translation, the two external segments or **exteins** becoming linked together (Fig. 13.13). The first intein was discovered in 1990 in *S. cerevisiae* and more than 100 inteins have been reported till today. Most inteins are known in bacteria, archaea and lower eukaryotes. In a few cases, there is more than one intein in a single protein.

Most inteins are approximately 150 amino acids in length and like pre-mRNA introns, the sequences at the splice junctions of inteins have some similarity in most of the known examples. The first amino acid of the downstream extein is cysteine, serine or threonine. A few other amino acids within the intein sequence are also conserved. These conserved amino acids are involved in the splicing process, which is catalyzed by the intein itself (**Paulus,** 2000).

A. O-linked glycosylation

Sia
Gal
GalNAc-Sia
O
O CH$_2$
—C—C—N—
H H

B. N-linked glylcosylaton

Sia Sia Sia
Gal Gal Gal
GlcNAc GlcNAc GlcNAc
Man Man
Man
GlcNAc
GlcNAc—Fuc
NH
C=O
CH$_2$
—C—C—N—
H H

Fig. 13.13. Glycosylation of proteins. A – O – linked glycosylation. The structure shown is found in a number of glycoproteins. It is drawn here attached to a serine amino acid but it can also be linked to a threonine. B – N – linked glycosylation usually results in larger sugar structures than are seen with O – linked glycosylation. The figure shows a typical example of a complex glycan attached to an asparagine amino acid. Abbreviations: Fuc, fucose; Gal, galactose; GalNAc, N – acetylgalactosamine; GlnNAc, N – acetyl glucosamine; Man, mannose; Sia, sialic acid (after Brown, 2002).

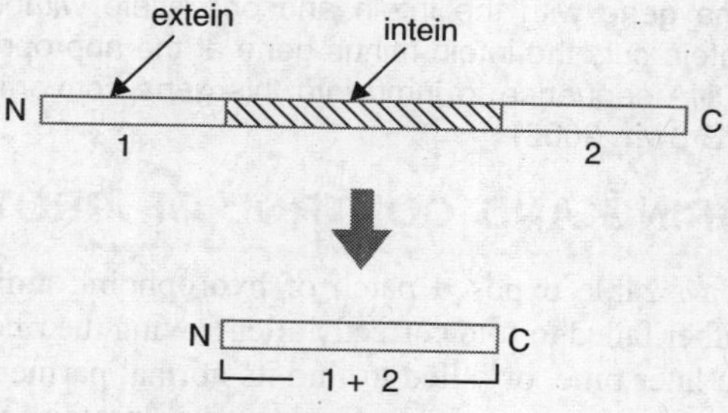

Fig. 13.14. Intein splicing.

Characteristics of Inteins

1. Self – catalysis. The first interesting feature of inteins was discovered when the structure of two inteins were determined by X-ray crystallography (**Duan** *et al.*, 1997; **Klabunde** *et al.*, 1998). These structures are similar in some respects to that of a *Drosophila* protein called **Hedgehog**, which is involved in the development of the segmentation pattern of the embryo of fruit fly. Hedgehog is an autoprocessing protein that cuts itself into two. The structural similarity with inteins lies in the part of the Hedgehog protein that catalyzes its self-cleavage. Molecular biologists believe that possibly the same protein structure has evolved twice, or possibly inteins and Hedgehog protein shared a common link at some stage in the evolutionary past.

2. Intein homing. The second interesting feature of intein is that with some inteins the excised segment is a sequence-specific endonuclease. The intein cuts DNA at the sequence corresponding to its insertion site in a gene coding for intein-free version of the protein from which it is derived

(Fig. 13.15). If the cell also contains a gene coding for the intein containing protein, then the DNA sequence for the intein is able to jump into the cut site converting the intein-minus gene into an intein-plus version, a process called **intein homing (Pietrokovski,** 2001).

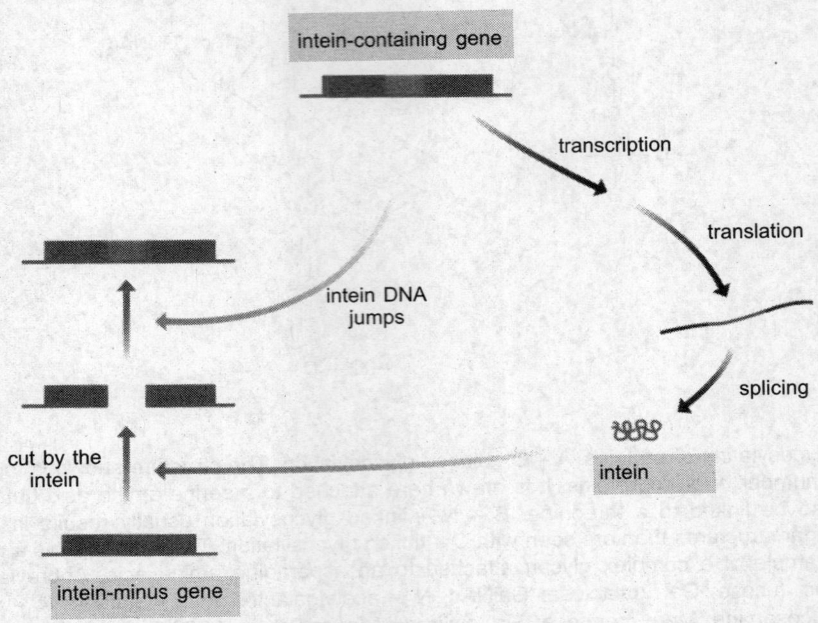

Fig. 13.15. Process of intein homing. The cell is heterozygous for the intein-containing gene, possessing one gene with the intein and one allele without the intein. After protein splicing, the intein cuts the intein minus gene at the appropriate place, allowing a copy of the intein DNA sequence to jump into this gene converting it into the intein – plus version (after Brown, 2002).

13.5. MONITORING AND CONTROL OF PROTEIN QUALITY

A protein that has a sizable exposed patch of hydrophobic amino acids on its surface is usually abnormal: it has either failed to fold correctly after leaving the ribosome, suffered an accident that partly unfolded it at a later time, or failed to find its normal partner subunit in a larger protein complex. Such a protein is useless to the cell. Further, such protein complex can be dangerous. Many proteins with an abnormally exposed hydrophobic region can form large aggregates, precipitating out of solution. We shall see that in rare cases, such aggregates to form and cause severe human diseases. But in the vast majority of cells, powerful protein quality control mechanisms prevent such disasters.

Cells have evolved elaborate mechanisms that recognize and remove the hydrophobic patches on proteins. Two of mechanisms depend on the molecular chaperones just discussed, which bind to the patch and attempt to repair the defective protein by giving it another chance to fold. At the same time, by covering the hydrophobic patches, these chaperones transiently prevent protein aggregation. Proteins that very rapidly fold correctly on their own do not display such patches and are therefore bypassed by chaperones.

All of the options of quality control that a cell makes for a difficult-to-fold newly synthesized protein have been outlined in Figure 13.16. As indicated, when attempts to refold a protein fail, a third option is called into play that completely destroys the protein by proteolysis. The proteolytic pathway begins with the recognition of an abnormal hydrophobic patch on a proteins surface and it ends with the delivery of the entire protein to a protein destruction machine, a complex protease called **proteosome**. As described next, this process depends on an elaborate protein marking system that also carries out other central functions in the cell by destroying selected normal proteins.

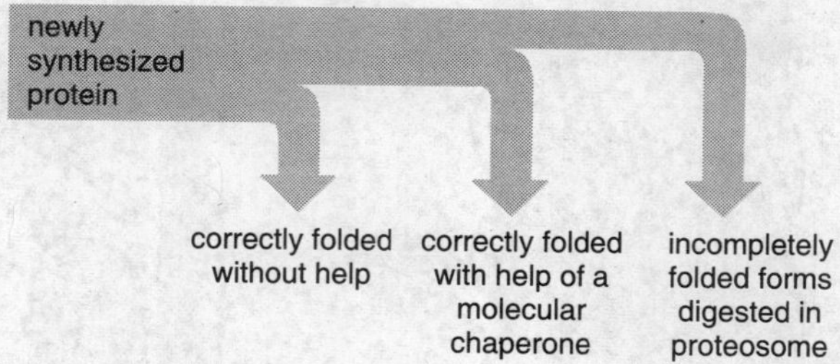

Fig. 13.16. The cellular mechanisms that monitor protein quality after protein synthesis. A newly synthesized protein sometimes folds correctly and assembles with its partners on its own, in which case it is left alone. Incompletely folded proteins are helped to refold by molecular chaperons: first by a family of hsp70 proteins, and if this fails, then by hsp70-like proteins. In both cases the client proteins are recognized by an abnormally exposed patch of hydrophobic amino acids on their surface. These processes complete with a different system that recognizes an abnormally exposed patch and transfers proteosome for complete destruction. The combination of all of these processes is needed to prevent massive protein aggregation in a cell, which can occur when hydrophobic regions on protein clump together and precipitate the entire mass out of solution (after Alberts *et al.*, 2002).

1. Protein Degradation

The protein synthesis and processing events that we have studied so far result in new proteins that take up their place in the cells proteome. These proteins either replace existing one that have reached the end of their working lives or provide new protein functions in response to the changing requirements of the cell. The concept that the proteome of a cell can change over time requires not only *de novo* protein synthesis but also the removal of proteins whose functions are no longer required. This removal must be highly selective so that only the correct proteins are degraded, and must also be rapid in order to account for the abrupt changes that occur under certain conditions, for example during key transitions in the cell cycle (**Hunt,** 1997).

In eukaryotes, most breakdown involves a single system, involving **ubiquitin** and the proteasome (Fig. 13.17). A link between ubiquitin and protein degradation uses first established in 1975 when it was shown that this abundant 76-amino acid protein is involved in energy dependent proteolysis reactions in rabbit cells (**Varshavsky,** 1997). Subsequent research identified a series of three enzymes that attach ubiquitin molecules, singly or in chain, to lysine amino acids in proteins that are targeted for breakdown. Whether or not a protein becomes ubiquitinated depends on the presence or absence within it of amino acid motifs that act as **degradation susceptible signals**. These signals have not been completely characterized but yeast is known to contain about 10 diffierent such signals. These signals include the following two categories.

(i) The N–degron, a sequence element present at the N-terminus of a protein;

(ii) PEST sequence, internal sequences that are rich in proline (P), glutamic acid (E), serine (S) and threonine (T).

These sequences are permanent features of the proteins that contain them and so cannot be straight forward 'degradation signal' – if they were so then these proteins would be broken down as soon as they are synthesized. Instead, they must determine susceptibility to degradation and hence the general stability of a protein in the cell. How this might be linked to the controlled breakdown of selected proteins at specific times, for instance during the cell cycle, is little understood.

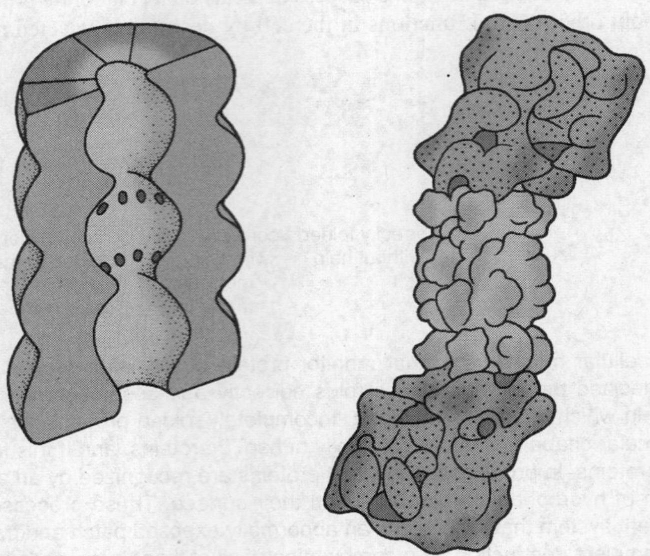

Fig. 13.17. The proteasome. A – A cut-away view of the structure of the central 20S cylinder, as determined by X-ray crystallography, with the active sites of the proteases indicated by *red dots.* B – The structure of the entire proteasome, in which the central cylinder (*yellow*) is supplemented by a 19S cap (*blue*) at each end. The complex cap structure selectively binds these proteins that have been marked for destruction (after Alberts *et al.*, 2002).

The second component of the ubiquitin-dependent degradation pathway is the **proteosome** the structure within which ubiquitinated protein are broken down. In eukaryotes, the proteosome is a large, multi-subunit structure with a sedimentation coefficient of 26S, comprising a hollow cylinder of 20S and two "caps" of 19S (**Groll** *et al.*, 1997; **Ferrell** *et al.,* 2000). Archaea also have proteasomes of about the same size but these are less complex, being composed of multiple copies of just two proteins; eukaryotic proteasomes contain 14 different types of protein subunits. The entrance into the cavity within the proteasome is narrow, and a protein must be unfolded before it can enter. This unfolding probably occurs through an energy dependent process and may involve structure similar to chaperonins (see **Brown,** 2002). After unfolding, the protein can enter the proteosome within which it is cleaved into short peptides 4 – 10 amino acids in length. These are released back into the cytoplasm where they are broken down into individual amino acids which can be reutilized in protein synthesis.

2. Abnormal Protein Folding Causes Destructive Human Diseases

When all of a cell's protein quality controls fail, large protein aggregates tend to accumulate in the affected cell. Some of these aggregates, by adsorbing critical macromolecules to them, can severly damage cell and even cause cell death. The protein aggregates released from dead cells tend to accumulate in the extracellular matrix that surrounds the cells in a tissue, and in extreme cases they can also damage tissues. Since the brain is composed of a highly organised collection of nerve cells, it is especially vulnerable. These protein aggregates primarily tend to causes diseases of neurodegeneration. Prominent among these are **Huntington's disease** and **Alzheimer's disease** – the latter causing age-related dementia in more than 20 million people in today's world (**Alberts** *et al.*, 2002).

A protein aggregate, which have to survive, grow and damage an organism, must be highly resistant to proteolysis both inside and outside the cell. Many of the protein aggregates that cause problems from fibrils built from a series of polypeptide chains that are layered one over the other as a continuous stack of β-sheets. This is so-called **cross beta filaments** (Fig. 13.18), tends to be highly resistant to proteolysis. This resistance presumably explains why this structure is observed in so many of the neurological disorders caused by protein staining deposits, called **amyloid**.

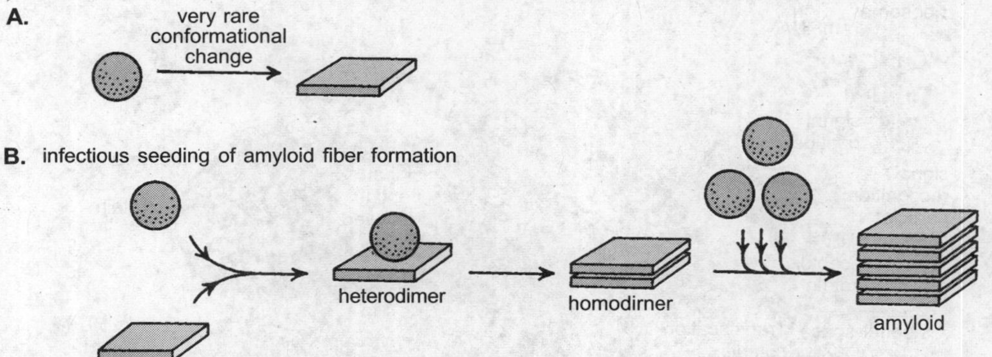

Fig. 13.18. Protein aggregates that cause human disease. A – Schematic illustration of the type of conformational change in a protein that produces material for a cross – beta filament. B– Diagram illustrating the self infectious nature of protein aggreagation that is central to prion disease. Prp is highly unusual because the misfolded version of the protein, called Prp*, induces the normal Prp protein it contacts to change conformation, as shown. Most of the human diseases caused by protein aggregation are caused by the over- production of a variant protein that is especially prone to aggregation, but because this structure is not infectious in this way, it cannot spread from one animal to another. (Drawing of a cross beta filament, a common type of protease-resistant protein aggregate found in a variety of human neurological diseases. Because the hydrogen-bond interactions in a β-sheet form between polypeptide backbone atoms, a number of different abnormally folded proteins can produce this structure (after Alberts *et al.*, 2002).

One particular variety of these neurological diseases is quite notorious. These are **prion diseases**. Unlike Huntington's or Alzheimer's disease, a prion disease can spread from one organism to another, providing that the second organism eats a tissue containing the protein aggregate. A set of diseases – called **scrapie** in sheep, **Creutzfeldt – Jacob disease** (**CJD**) in humans, and **bovine spongiform encephalopathy** (**BSE**) in cattle – are caused by a misfolded, aggregated form of protein called **PrP** (for prion protein). The PrP is normally located on the outer surface of the plasma membrane, most prominently in neurones. Its normal function is not known. However, PrP has the unfortunate property of being convertible to a very special abnormal conformation (Fig. 13.18A). This conformation not only forms protease-resistant, cross-beta filaments; it also is "infectious" because it converts normally folded molecules of PrP to the same form. This property creates a positive feedback loop that produces the abnormal form of prp, called **PrP*** (Fig. 13.18B) and thereby allows PrP to spread rapidly from cell to cell in the brain, causing death of both animals and humans. It is dangerous to eat the tissues of animals that contain PrP*, as witnessed most recently by spread of BSE (commonly referred to as the "mad cow disease") from cattle to humans in great Britain.

13.6. SIGNAL HYPOTHESIS (Protein Targeting)

Ribosomes are either free in the cytoplasm or associated with membranes, depending on the type of protein being synthesised. Membrane-bound ribosomes, which are indistinguishable from free ribosomes, synthesise proteins that enter membranes. These proteins either become a part of the membrane or in eukaryotes, either pass into membrane bound organelles (*e.g.*, the Golgi apparatus, mitochondria, chloroplasts, vacuoles) or transported outside the plasma membrane. The **signal hypothesis** of **Gunter Blobel**, a 1999 Nobel laureate, and his colleagues, explains the mechanism for membrane attachment. The mechanism applies to both prokaryotes and eukaryotes. Here, signal hypothesis has been described for mammals.

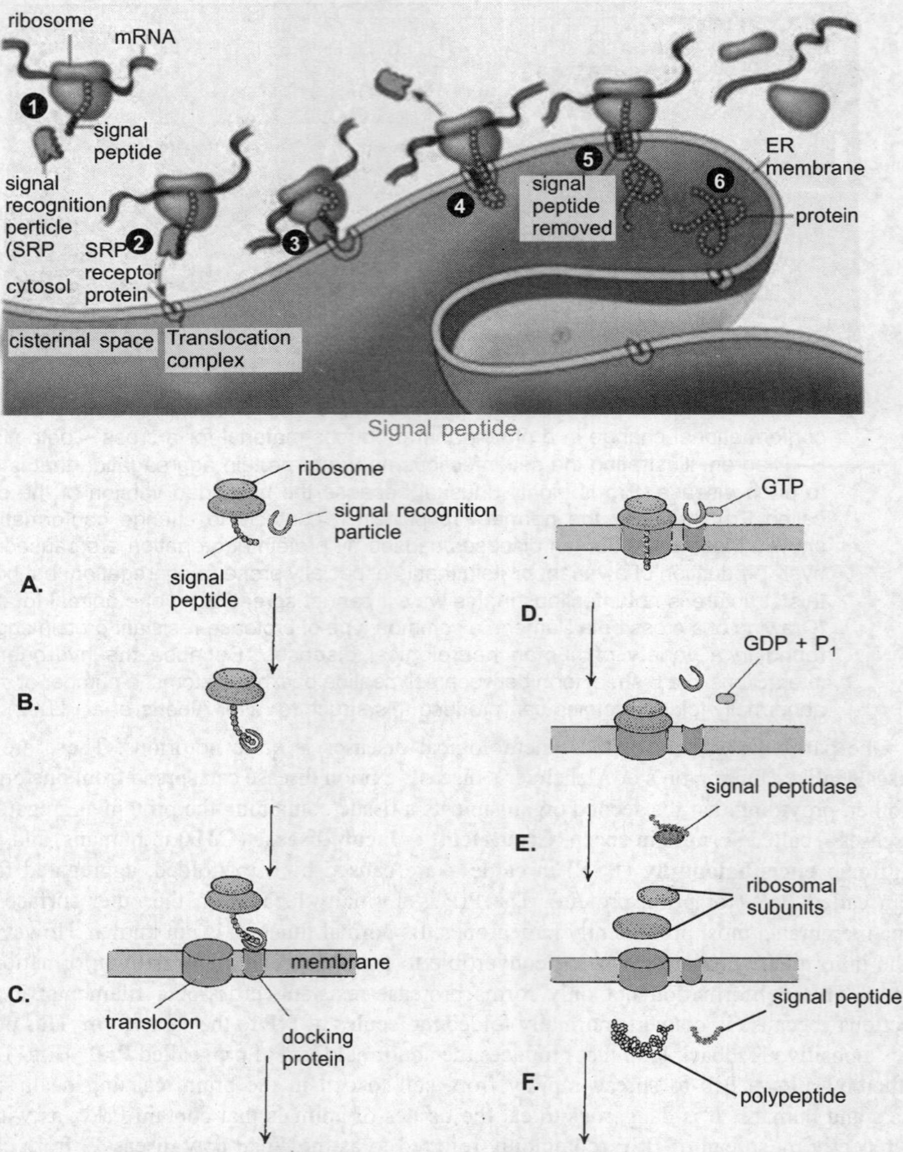

Fig. 13.19. Bobel's signal hypothesis for protein transportation. A signal recognition particle recognizes a ribosome with a signal peptide (A), then draws the ribosome to docking protein located near a translocon in the membrane. With the addition of GTP, the signal recognition particle releases the signal peptide; the hydrolysis of GTP to GDP + Pi causes the signal recognition particle to leave the docking protein(B). Peptide synthesis then resumes, with the newly synthesized peptide passing through the translocon in the membrane. A signal peptide on the other side of the membrane removes the signal peptidase (enzyme). When translation is completed, the ribosome dissociate and drops free of the translocon (after Tamarin, 2002).

The signal for membrane insertion is in the form of polypeptide. It is coded into the first one to three dozen amino acids of membrane-bound proteins. This **signal peptide** takes part in a chain of events that leads the ribosome to attach to the membrane and to insertion of the protein. The *first step* occurs when the signal peptide becomes exposed on outside of the ribosome. A ribonucleoprotein particle, called **signal recognition particle (SRP)**, which consists of six different proteins and a 7S RNA (about 300 nucleotide long), recognizes the signal peptide. In *second step*, the complex of signal recognition particle, ribosome, and signal peptide then passes to a membrane, where the SRP binds to a receptor called a **docking protein (DP) or signal recognition particle receptor** (Fig. 13.19). During this time, protein synthesis or translation halts. The ribosome is brought into direct contact with the membrane, and other proteins which help anchor the ribosome. In *third step*, the protein synthesis resumes, with the nascent protein usually passing directly into a **translocation channel** (called **translocon**). Once through the membrane, the signal peptide is cleaved from the protein by an enzyme called **signal peptidase** (Box 13.2).

Box 13.2.

Evidence for signal hypothesis came about through *recombinant DNA techniques*. A signal sequence was placed in front of the α-globin gene, whose protein product is normally not transported through a membrane. When this gene was translated, the ribosome became membrane bound, and the protein passed through the membrane.

Since different proteins enter different membrane-bound compartments (*e.g.*, the Golgi apparatus), some mechanism must direct a nascent protein to its proper membrane. This specificity seems to depend on the exact signal sequence and membrane – bound glycoproteins called **signal sequence receptor**. Apparently, after the ribosome binds to the docking protein, the signal peptide interacts with a signal sequence receptor, which presumably determines whether that protein is specific for that membrane. If it is, the remaining processes continue. If not, the ribosome may be released from the membrane.

The signal peptide does not seem to have a consensus sequence similar to transcription or translation recognition boxes. Rather similarities (at least for the endoplasmic reticulum and bacterial membrane bound proteins) include a positively charged (basic) amino acid (commonly lysine or arginine) near the beginning (N – terminal end), followed by about a dozen hydrophobic (nonpolar) amino acids, commonly alanine, isoleucine, leucine, phenylalanine and valine (**Sasavage** *et al.*, 1982; Table 13.3).

Table 13.3. The signal peptide of the bovine prolactin protein (Source: **Tamarin,** 2002).

NH_2 – Met Asp Ser Lys Gly Ser Ser Gln Lys Gly Ser Arg Leu Leu Leu Leu Leu Val Ser Asn Leu Leu Leu Cys Gln Gly Val Val Ser/Thr Pro Val ....Asn Asn Cys - COOH

*The slash line separates the signal peptide from the rest of the protein which consists of 199 residues.

Questions And Problems

1. Describe four levels of protein structure.
2. Describe the experiments which showed that a small protein such as ribonuclease can fold spontaneously *in vitro*. Why are larger proteins unable to fold spontaneously?
3. Distinguish between the activities of Hsp70 chaperones and chaperonins in protein folding.
4. Give examples of proteins that are processed by proteolytic cleavage.

5. What is an intein? How are inteins spliced?
6. Describe the process thought to be responsible for protein degradation in eukaryotes.
7. Are the similarities between inteins and introns purely coincidental?
8. What is signal peptide? What role does it play in eukaryotes? What is its fate?

ANSWERS TO PROBLEMS

8. A signal peptide is a sequence of amino acids at the amino terminal end of a protein that signals that the protein should enter a membrane. Although the concept is the same, the situation in eukaryotes is somewhat more complex because there are many different membrane-bound organelles, each having their own membrane specific requirements. Signal peptides are usually cleaved off the protein after the protein enters or passes through the membrane.

MULTIPLE CHOICE QUESTIONS

Choose the correct answer from the four alternatives given.

1. Peptide linkage is
   (*a*) —CO—NH—
   (*b*) —CO—NH_2
   (*c*) —COOH—NH_2
   (*d*) —CH—N
2. Tertiary structure of proteins having amino acids cystein is achieved through
   (*a*) ionic bonds
   (*b*) covalent bonds
   (*c*) disulphide bonds
   (*d*) hydrogen bonds

ANSWERS

1. (*a*) 2. (*c*).

CHAPTER

14

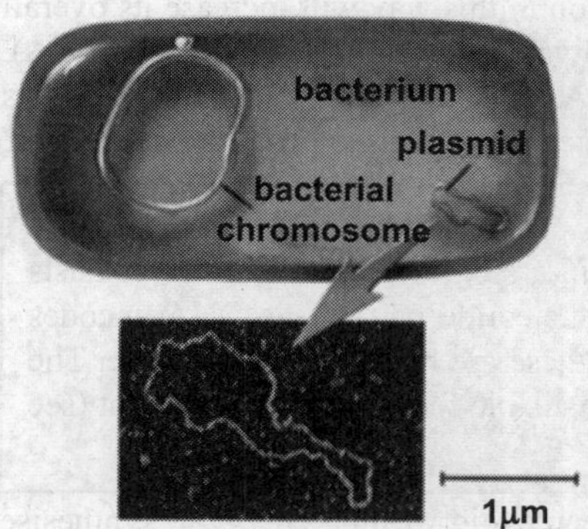

Regulation of Gene Action in Bacteria and Viruses

Chapter Contents

Bacteria are exposed to a wide range of environmental conditions. For example, *E. coli* may encounter rapidly changing growth conditions as they pass from mammalian intestinal tract to sewer systems to polluted rivers, lakes, ponds and so on. Each of these ecological niches will provide different organic molecules for use as energy sources. From this fact, it can be concluded that natural selection will have preserved those organisms that have evolved ways of adapting to the wide range of environmental conditions encountered during their evolution. Indeed, the available information indicates that most prokaryotes such as *E. coli* exhibit remarkable capacities to adapt to diverse environmental conditions.

To a great extent, the adaptability of bacteria and other prokaryotes depends on their ability to "turn on" and "turn off" the expression of specific sets of genes in response to the specific demands of the environmental milieu (*i.e.,* surroundings). In other words, prokaryotes exhibit a outstanding ability to **regulate the expression of specific genes in response to environmental signals** (Box 14.1). The expression of particular genes is "turned on" when the products of these genes are needed for growth in a given environment. Their expression is "turned off" when their products are no longer needed for growth in the existing environmental surroundings. Clearly, the ability of an organism to regulate gene expression in this way will increase its overall "fitness" (*i.e.,* its ability to grow and leave progeny under a variety of environmental conditions). The synthesis of gene transcripts (*i.e.,* various types of RNA molecules) and translation products (*i.e.,* proteins) require the expenditure of great amount of energy. By "turning off" the expression

Box 14.1.

The chromosome of the bacterium *Escherischia coli*, a single-celled organism, consists of a single circular DNA molecule of about 4.6×10^6 nucleotide pairs. This DNA encodes approximately 4300 proteins, although only a fraction of these are made at any one time. The expression of many of them is regulated according to the available food in the environment (see **Alberts** *et al.*, 2002).

of genes when their products are not needed, an organism can avoid wasting energy to synthesise products that maximize the growth rate in the existing environmental surroundings.

Here, one can ask three sort of questions:

1. What are the mechanisms by which these organisms regulate gene expression in response to changes in the environment?

2. Is there a single mechanism by which the expression of different genes or sets of genes are regulated?

3. Or are different genes controlled by different mechanisms?

Some genes, for example, the genes specifying ribosomal RNAs, ribosomal proteins and transfer RNAs, are certainly expressed at some time in virtually all cells regardless of the environment conditions. The products of these genes are required for growth of all cells in all environments. However, the products of many other genes are required for growth only in certain environments, and *the expression of these genes is regulated such that the products are synthesized only when they are needed*. As a result, the expression of these genes is continually being "turned on" and "turned off" in response to changes in the environment. From this discussion, one can conclude that gene expression can be (and is) regulated at several different levels: for example, **transcription, mRNA processing, mRNA turnover, translation** and **enzyme function**. However, extensive data indicate that the *regulation of transcription is the most important mode of control of gene expression, at least in prokaryotes*.

The regulatory mechanisms of transcription in both prokaryotes and eukaryotes are of two basic types. The *first*, and best understood, category includes mechanisms involved in rapid turn-on and turn- off of gene expression in response to environmental changes. Such sort of mechanisms are very significant in microorganisms (bacteria) because these organisms are exposed to sudden changes in environment. They provide microorganisms with a great deal of "plasticity", an ability to rapidly adjust their metabolic processes in order to achieve maximal growth and reproduction under highly variable environmental conditions. These quick responding on – off switches seem to be less important in higher eukaryotes. This might be expected since the circulatory systems of higher eukaryotes buffer their cells against many sudden environmental changes.

The second major category of regulatory mechanisms includes the so called **preprogrammed circuits of gene expression**. In these cases, some event (*e.g.,* infection by a virus) triggers the expression of one set of genes. The product (or products) of one (or more) of these genes function by turning off the transcription of the first set of genes and/or turning on the transcription of the second set of genes. In turn, one or more of the products of the second set acts by turning on a third set, and so on. In these cases, *the sequential expression of genes is genetically preprogrammed,* and the genes usually cannot be turned on out of sequence. Such preprogrammed sequences of gene expression in viral infections are well worked out. In most of these preprogrammed sequences, it seems the circuit is cyclical. For example, in viral infections some event associated with the packaging of the viral DNA or RNA inside the protein coat somehow seems to reset the program so that the first set of genes will again be expressed when a progeny virus subsequently infects another host cell.

Discovery of Prokaryotic Gene Regulatory Proteins

Genetic analyses in bacteria carried out in the 1950s provided the first evidence for the existence of **gene regulatory proteins** that turn specific sets of genes on or off. One of these regulators, the **lambda repressor**, is encoded by a bacterial virus, the bacteriophage lambda. The repressor shuts off the viral genes that code for the protein components of new virus particles and thereby enables the viral genome to remain a silent passenger in the bacterial chromosome, multiplying with the bacterium when conditions are favourable for bacterial growth. The lambda repressor was among the first gene regulatory proteins to be characterised. Other bacterial regulators respond to nutritional conditions by shutting off genes encoding specific sets of metabolic enzymes when they are not needed. For example, **lac repressor**, the first of these bacterial proteins to be recognised, turns off the production of the proteins responsible for lactose metabolism when this sugar is absent from the medium.

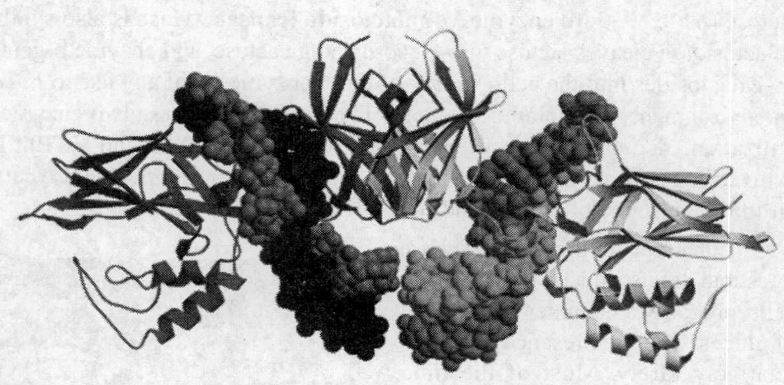

Gene regulatory proteins.

The first step toward understanding gene regulation was the isolation of mutant strains of bacteria and bacteriophage lambda that were unable to shut off specific sets of genes. It was proposed at the time, and later proven, that most of these mutants were deficient in proteins acting as specific repressors for these sets of genes. Because, these proteins, like most gene regulator proteins, are present in small quantities, it was difficult and time consuming to isolate them. They were eventually purified by fractionating cell extracts. Once isolated, the proteins were shown to bind to specific DNA sequences close to the genes that they regulate. The precise DNA sequences that they recognized were then determined by a combination of classical genetics, DNA sequencing and DNA footprinting experiments.

14.1. CONSTITUTIVE GENES AND INDUCIBLE GENES

Certain gene products, such as tRNA molecules, rRNA molecules, ribosomal proteins, RNA polymerase components (polypeptides), and other enzymes catalyzing metabolic processes that are commonly known for cellular “house keeping” functions, are essential components of almost all living cells. Genes that specify products of this type are continually being expressed in most cells. Such genes are said to be expressed **constitutively** and are called **constitutive genes**.

Rest of the gene products are needed for cell growth only under certain environmental conditions. Constitutive synthesis of such gene products would clearly be wasteful, using energy that could otherwise be utilized for more rapid growth and reproduction under the existing environmental conditions. The *evolution* of regulatory mechanisms that would provide for the synthesis of such gene products only when and where they were needed would clearly provide organisms with a selective advantage over organisms lacking these mechanisms. This undoubtedly explains why presently existing organisms, including the "primitive" bacteria and viruses, exhibit highly developed and very efficient mechanisms for the control of gene expression.

E.coli and most other bacteria are capable of growth using any one of several carbohydrates (*e.g.,* glucose, sucrose, galactose, arabinose, lactose) as an energy source. If glucose is present in the environment, it will be preferentially metabolished by *E. coli* cells. In the absence of glucose, however, *E. coli* cells can grow very well on other carbohydrates. Cells of *E. coli* growing in medium containing the sugar lactose, for example, as the sole carbon source synthesise two enzymes, **β – glalactosidase** and **β–galactoside permease**, that they uniquely required for the catabolism of lactose (Note: A third enzyme, **α-galactoside transacetylase** is also synthesized). The enzyme β-galactosidase cleaves lactose into glucose and galactose, and enzyme β-galactoside permease pumps α -galactosides into the cell. Neither of these enzymes is of any use to *E. coli* cells when present in an environment not containing lactose. The synthesis of these two enzymes, ofcourse, requires the utilization of great amount of energy (in the form of ATP and GTP). Thus, *E. coli* cells have evolved a regulatory mechanism by which the synthesis of these lactose catabolic enzymes is turned on in the presence of lactose and turned off in its absence.

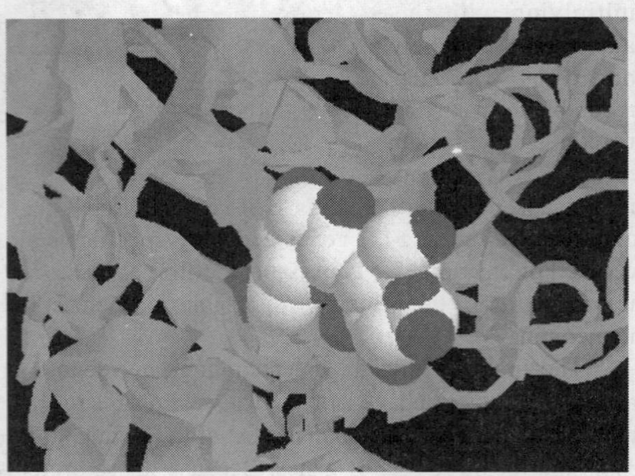

Beta galactosidase enzyme in *E.coli*.

In natural environments (such as intestinal tracts and sewers), *E. coli* cells probably encounter an absence of glucose and the presence of lactose relatively rarely. Most of the time, therefore, *E. coli* cells growing on a carbohydrate other than lactose are transferred to medium containing lactose as the only carbon source, they rapidly begin sythesising the enzymes required for lactose utilization. This process, by which the expression of genes is turned on in response to a substance in the environment, is called **induction**. Genes whose expression are so regulated are called **inducible genes**. The substances or molecules responsible for induction are known as **inducers** (*e.g.,* in this case lactose is inducer).

Enzymes that are involved in **catabolic** (*i.e.,* degradative) **pathway**, such as in lactose, galactose or arabinose utilization, are characteristically inducible. Researches made it evident that *induction occurs at the level of transcription*. Induction alters the rate of synthesis of enzymes, not the activity of existing enzyme molecules. Induction should not be confused with **enzyme activation**, in which the binding of a small molecule to an enzyme increases the **activity** of the enzyme (but does not affect its rate of synthesis).

Further, *E. coli* and other bacteria possess the metabolic capacity to synthesise most of the organic molecules (such as amino acids, purines and vitamins) required for their growth. For example, *E. coli* has five genes coding for enzymes that are required in the synthesis of tryptophan (an amino acid). These five genes must be expressed in *E. coli* cells growing in an environment devoid of tryptophan in order to provide adequate amounts of this amino acid for ongoing protein synthesis. When *E. coli* cells are present in an environment containing concentrations of tryptophan sufficient to support optimal growth, the continued synthesis of the tryptophan biosynthetic enzymes would be a waste of energy, because these bacteria have the capacity to take in external tryptophan. Thus, a regulatory mechanism has evolved in *E. coli* by which the synthesis of the tryptophan is present in the external milieu (surrounding environment). This process of "turning off" the expression of sets of genes is called **repression**. A gene whose expression has been turned off in this way is said to be repressed; when its expression is turned on, a gene of this type is said to be **depressed**.

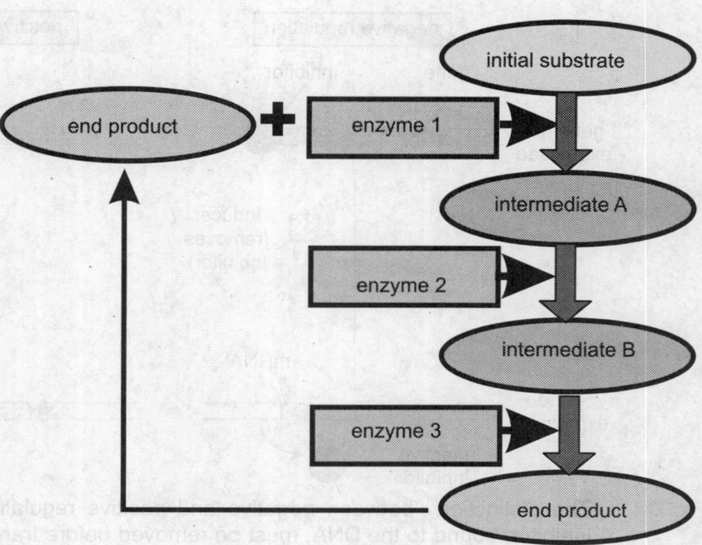

Feedback inhibition.

Enzymes that are components of **anabolic** (biosynthetic) **pathways** are frequently subject to **repression** (are **repressible**). *Repression, like induction occurs at the level of transcription.* However, repression should not be confused with **feedback inhibition**, in which the binding of an end product to the first enzyme in a biosynthetic pathway **inhibits the activity** of the enzyme (but does not affect its synthesis).

14.2. TRANSCRIPTIONAL CONTROL MECHANISMS

In bacteria, there occur several mechanisms of gene regulation at the level of transcription. A notable method depends on whether the enzymes being regulated act in catabolic (degradative) or anabolic (synthetic) metabolic pathways. For example, in a multistep catabolic system, the availability of the molecule to be degraded commonly determines whether the enzymes in the pathway will be synthesised. In contrast, in a biosynthetic pathway the final product is often the regulatory molecule. Even when a single protein molecule is translated from a monocistronic mRNA molecule, the protein may be **autoregulated**, *i.e.,* the protein itself may inhibit initiation of transcription and high concentrations of the protein may cause less transcription of the mRNA that encodes the protein. The molecular mechanisms for each of the regulatory patterns differ greatly and are of the following two types – negative regulation and positive regulation (Fig. 14.1). In a **negative regulated system**, an inhibitor is present in the cell and prevents transcription. An antagonist of the inhibitor, called an **inducer**, is needed to allow initiation of transcription. In a **positively regulated system**, an **effector** molecule (which may be protein, a small molecule, or a molecular complex) activates a promoter; no inhibitor must be abolished. Negative and positive regulation are not mutually exclusive, and some systems are both positively and negatively regulated, utilizing two regulators to respond to different conditions in the cell. Thus, a catabolic system may be regulated positively or negatively. In a biosynthetic (anabolic) pathway, the final product usually regulates negatively its own synthesis; in the simplest type of negative regulation, absence of the product increases its synthesis and presence of the product decreases its synthesis.

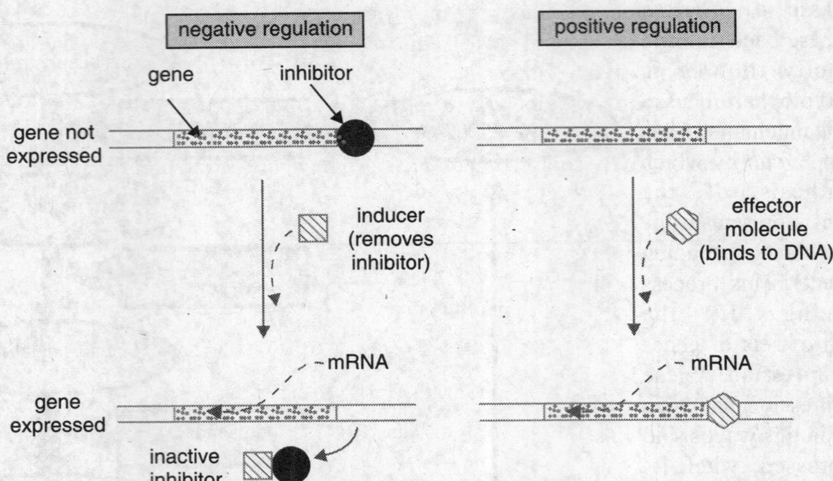

Fig. 14.1. The distinction between negative and positive regulation. In negative regulation an inhibitor, bound to the DNA, must be removed before transcription can occur. In positive regulation, an effector molecule must bind to the DNA (after Freifelder, 1985).

The Operon Model

Induction and repression of gene expression can be accomplished by essentially the same mechanism. This mechanism was first accurately described in 1961 when **Francois Jacob** and **Jacques Monod**, both 1965 Nobel Prize recipients, proposed the **operon model** to explain the regulation of genes encoding the enzymes required for lactose utilization in *E. coli*. **Jacob** and **Monod** proposed that the transcription of one or a set of contiguous structural genes (*i.e.,* genes coding for polypeptides) is regulated by two controlling elements (Fig 14.2). One of these controlling elements, called the **regulator gene** (or **repressor gene**), codes for a protein called the **repressor**; under the appropriate conditions, the repressor binds to the second element, the **operator**. The operator is always located contiguous to the structural gene or genes whose expression it regulates.

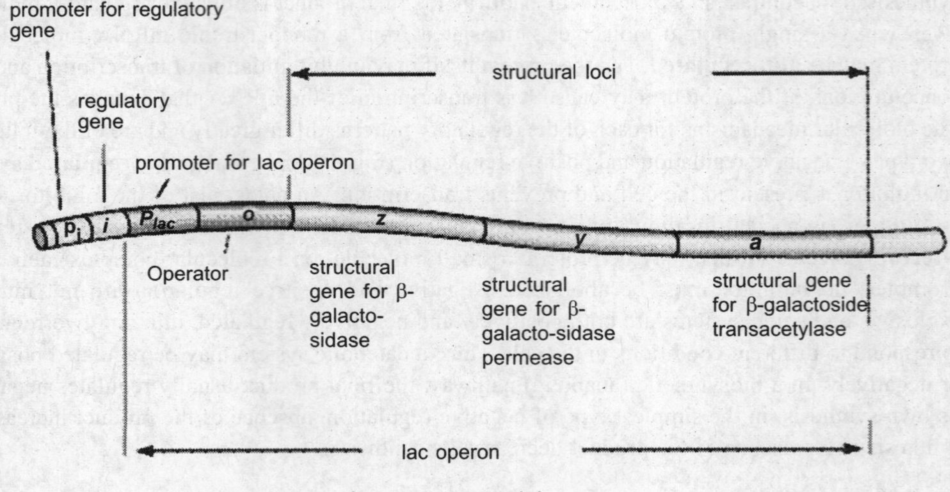

Lac operon model.

When the repressor is bound to the operator, transcription of the structural genes cannot occur. We now know that this results because the binding of the repressor to the operator stearically prevents RNA polymerase (enzyme) from binding at the **promoter site** (the RNA polymerase binding site), which is always located contiguous with the operator sequence. The **operator** is usually located between the promoter and structural genes. (Note: The promoter was not recognized at the time of Jacob and Monod's proposal, but has since been shown to be as essential component of an operon). Thus, the complete contiguous unit, including the structural genes, the operator and the promoter is called the operon (**Gardner** *et al.*, 2002). In other words, an operon is a unit of prokaryotic gene expression which includes co-ordinately regulated (structural) genes and control elements (promoter operator) which are recognized by regulatory gene products (**Turner** *et al.*, 2000).

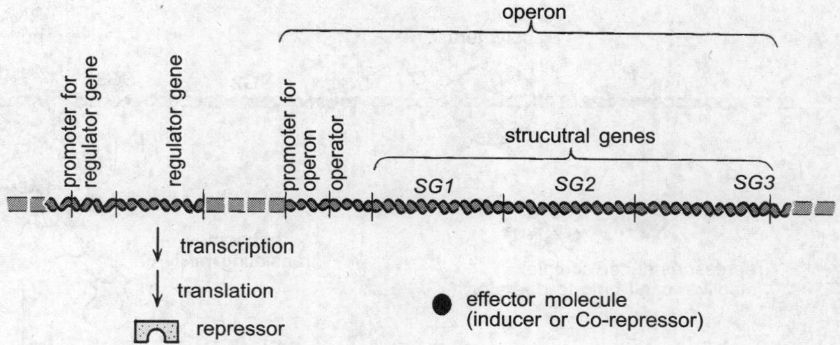

Fig. 14.2. The operon model for regulation of gene expression. The operon consists of one or more structural genes (three – *SG1*, *SG2* and *SG3* – are arbitrarily shown) and the adjoining operator and promoter sequences. The promoter for the operon (*PO*) is the site at which RNA polymerase enzyme must bind to initiate transcription of structural genes. The operator (*O*) is the site at which protein repressor (the products of the regulator gene or repressor gene) binds. The regulator gene need not be closely linked to the operon; in fact, it can be located at any position in the genome. The transcription of the regulator gene is initiated by RNA polymerase, which binds to its promoter (labeled *PR*, for promoter for regulator gene). When the repressor is bound to the operator, it stearically prevents RNA polymerase from binding to adjoining promoter (*PO*) and from initiating transcription of the structural genes. Whether the repressor binds to the operator or not, depends on the presence or absence of a metabolite called an effector molecule (after Gardner *et al.,* 2002).

Whether the repressor will bind to the operator and turn off the transcription of the structural genes in an operon is determined by the presence or absence of **effector molecules** (small molecules such as amino acids and sugars) in the environment. In the case of **inducible operons** (*e.g.,* lac operon), these effector molecules are called **inducers**. Those active on repressible operons (*e.g., trp* operon) are called **co–repressors** act by binding to (or forming a complex with) the repressors.

The only essential difference between inducible operons and repressible operons is whether the naked repressor or the repressor effector molecule complex is active in binding to the operator. 1. In case of an inducible operon, the free repressor binds to the operator, turning off (Fig. 14.3). When the effector molecule or the inducer (*e.g.,* allolactose in *lac* operon) is present, it binds to the repressor, that is, the **repressor-inducer complex cannot bind to the operator**. Thus, the addition of inducer turns on (or induces) the transcription of the structural genes in the operon (Fig. 14.3). 2. In case of a repressible operon (*e.g., trp* operon), the situation is just reversed. The free repressor-cannot bind to the operator. Only the repressor-effector molecule (co-repressor) complex is active

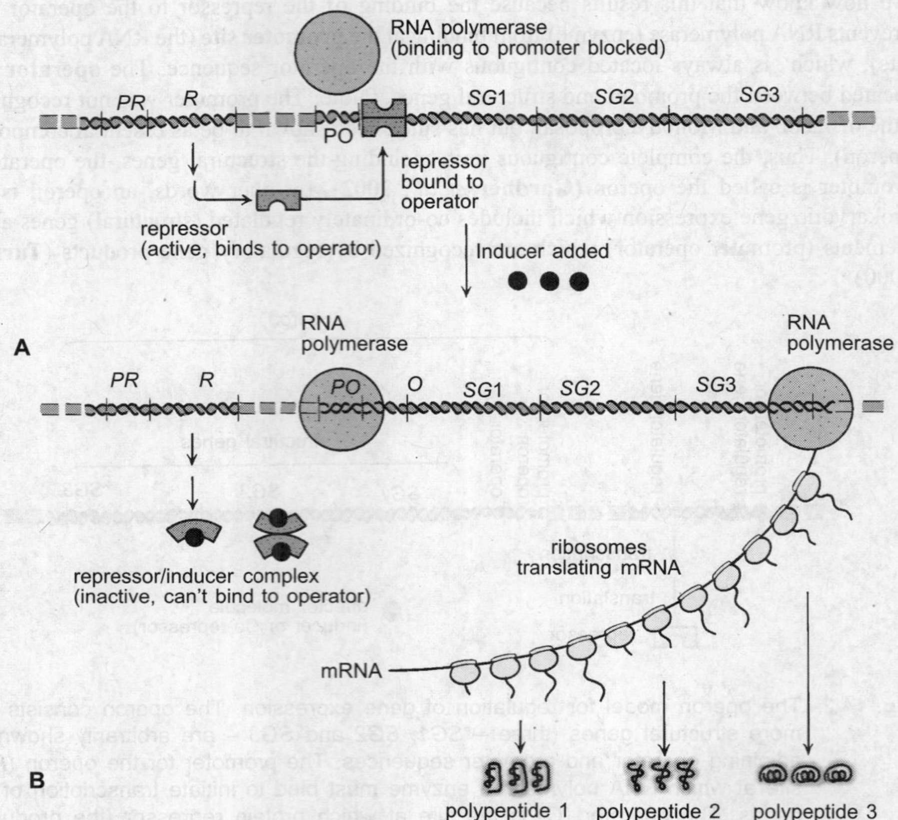

Fig. 14.3. Mode of regulation of gene expression for an inducible operon (*e.g., lac* operon). A – The product of the regulator gene (*R*), the repressor, in the absence of the effector molecule (called an inducer, for inducible operon, *e.g.,* allolactose in *lac* operon), binds to the operator, preventing RNA polymerase from binding to the promoter for the operon (*PO*). Thus, transcription of the structural genes cannot occur. B – When inducer is added, it binds to the repressor, causing it to be released from the operator (*O*). This, in turn, allows RNA polymerase to bind to the promoter (*PO*) and initiate transcription of the structural genes. The resulting multigenic (= polycistronic) mRNA is rapidly translated by ribosomes, producing the three polypeptide products of the structural genes (after Gardner *et al.,* 2002).

in binding to the operator (Fig. 14.4). Thus, transcription of the structural genes in repressible operon is turned on in the absence of and turned off in the presence of the effector molecule (co-repressor). Except for this difference in the operator repressible operons are comparable.

A single mRNA transcript carries the coding information of an entire operon. Thus, the mRNAs of operons consisting of more than one structural gene are polygenic or polycistronic. For example, the tryptophan operon mRNA of *E.coli* is a huge macromolecule carrying the coding sequences that specify five different polypeptides because of their cotranscription, all the structural genes in an operon are coordinately expressed.

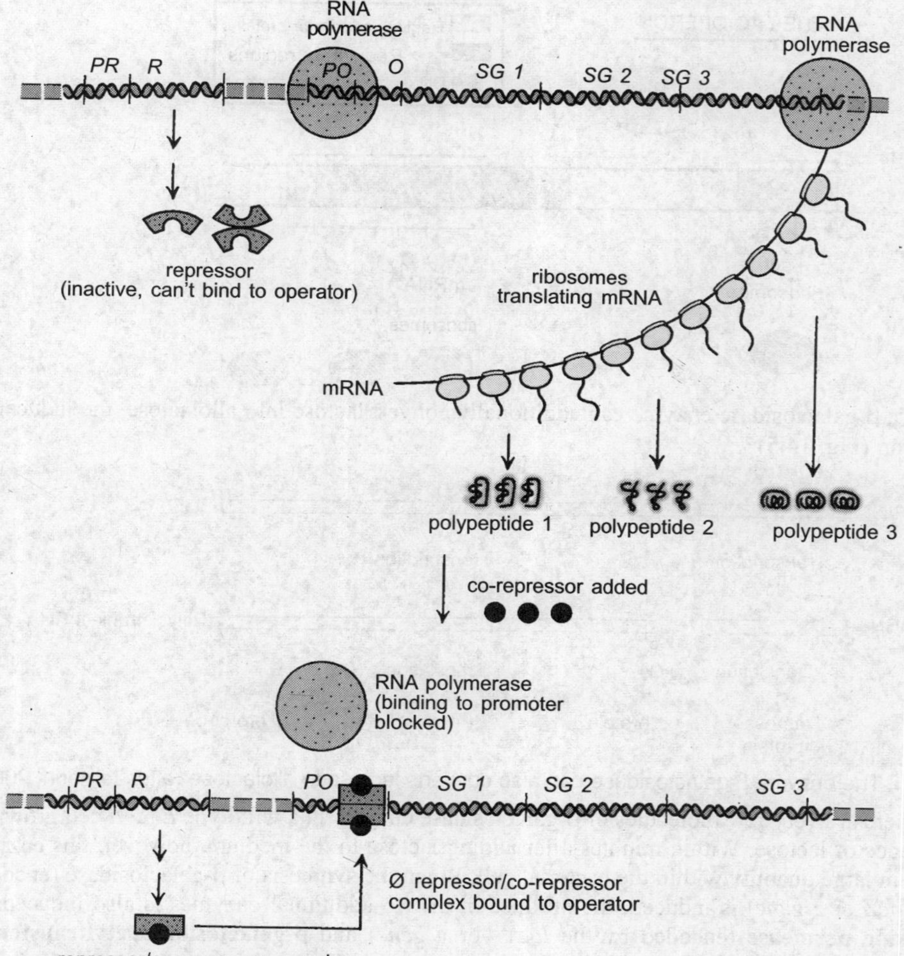

Fig. 14.4. Mode of regulation of gene expression for a repressible operon (*e.g., trp* operon). In this case, the repressor can only bind to the operator in the presence of effector molecule (called a co-repressor, for repressible operons; *e.g.,* tryptophan amino acid in *trp* operon). In its absence, the operator is free, permitting RNA polymerase to bind at the adjoining promoter (*PO*) and to initiate transcription of the structural genes. When co-repressor is added, it forms a complex with the repressor. This repressor – co-repressor complex then binds to the operator (*O*). This, in turn, prevents RNA polymerase from binding at *PO* and transcribing the three structural genes (after Gardner *et al.,* 2002).

Examples of Operons

1. *Lac* Operons (Inducible System)

Lactose (milk sugar, a disaccharide) is a β-galactoside that *E. coli* can use for energy and as a carbon source after it is broken down into glucose and galactose. The enzyme that performs the breakdown is **β–galactosidase:**

$$\underset{\text{(a disaccharide)}}{\text{Lactose}} \xrightarrow{\beta\text{-galactosidase} + H_2O} \text{Galactose} + \text{Glucose}$$

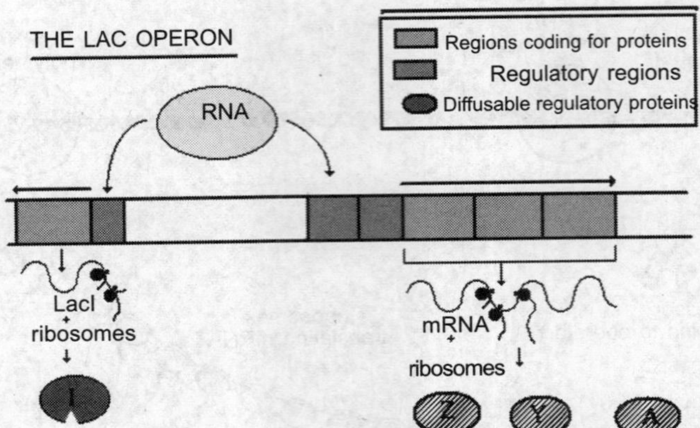

The β-galactosidase enzyme can additionally convert lactose into allolactose, the inducer of *lac* operon (Fig. 14.5)

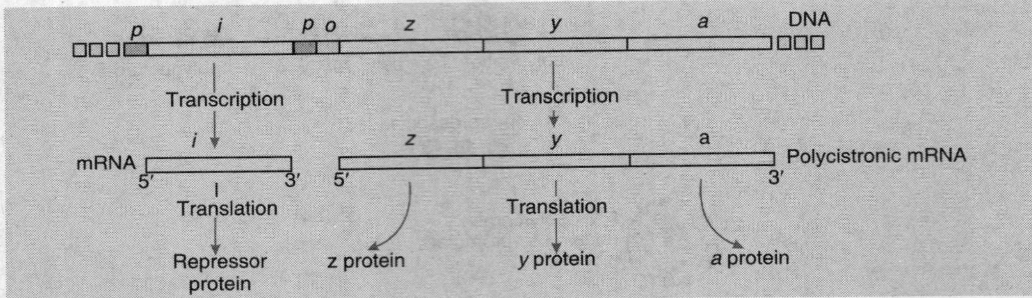

Fig. 14.5. The enzyme β-galactosidase can also converts lactose to allolactose (after Tamarin, 2002).

There are very few molecules of β-galactosidase enzyme in a wild type *E. coli* cell grown in the absence of lactose. Within minutes after adding lactose to the medium, however, this enzyme appears in large quantity within the bacterial cell. When the synthesis of β-galactosidase (encoded by the *lacZ* or *z gene*) is induced, the production of two additional enzymes is also induced, **β-galactoside permease** (encoded by the *lacY* or *y gene*) and **β-galactoside acetyltransferase** (encoded by the *lacA* or *a* gene). The permease (enzyme) is involved in transporting lactose into the cell. The transferase is believed to protect the cell from the build up of toxic products created by β-galactosidase (enzyme) acting on other sugars than lactose, the transferase prevents β-galactosidase from cleaving them (see **Tamarin,** 2002).

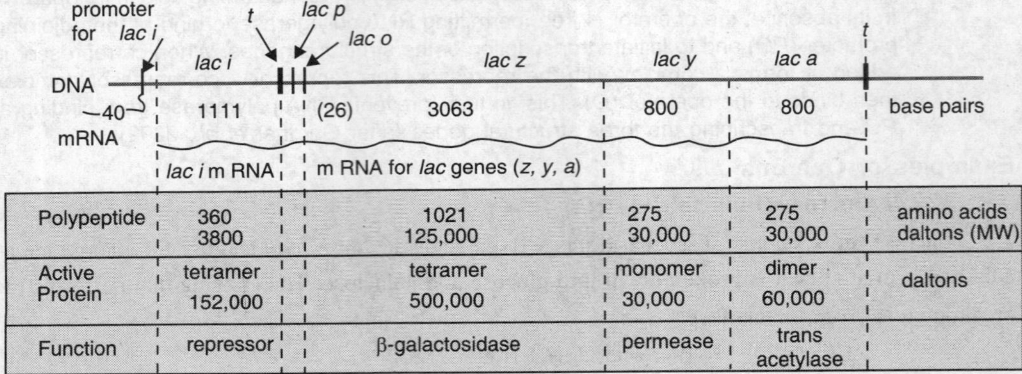

	lac i		lac z	lac y	lac a	
Polypeptide	360 3800		1021 125,000	275 30,000	275 30,000	amino acids daltons (MW)
Active Protein	tetramer 152,000		tetramer 500,000	monomer 30,000	dimer 60,000	daltons
Function	repressor		β-galactosidase	permease	trans acetylase	

Fig. 14.6. The lactose operon of *E. coli* and its regulatory gene (after Lewin, 1990).

The regulator gene. Not only are three *lac* genes (*z, y* and *a*) induced together, but they are adjacent to one another in the *E. coli* chromosome; they are, in fact, transcribed on a single polycistronic messenger RNA (Fig. 14.6 and Fig. 14.7). Induction involves the protein product of another gene, the *regulator* gene, or *i* gene (*lac I*). Although the regulator gene is located adjacent to the three other *lac* genes (*i.e.,* three structural genes), it is totally independent transcriptional entity. The regulator gene specifies a protein, the **repressor,** that interferes with the transcription of the genes involved in lactose metabolism.

The operator. For the repressor protein to exert its influence over transcription, there must be a control element (receptor site) located near the beginning of the β-galactosidase (*lacZ*) gene.

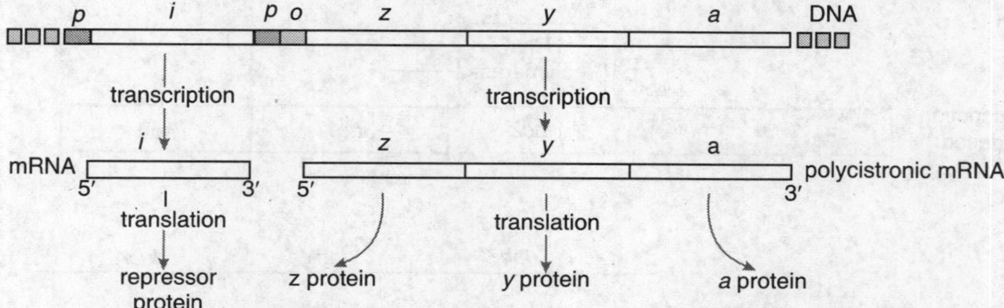

Fig. 14.7. The *lac* operon is transcribed as a multigenic (polycistronic) mRNA. The *z, y* and *a* indicate the *lacZ*, lacY and *lacA* loci. The mRNA transcript is then translated as individual proteins. The regulator gene of *lac* operon is denoted as *i*; the *o* stands for operator and *p* for promoter. Both the operon and the regulator gene have their own promoters (*p*) (after Tamarin, 2002).

This control element is a region referred to as the **operator,** or operator site (Fig. 14.7). the operator site is a sequence of DNA that the product of the regulator gene, the repressor, recognizes. When the repressor is bound to the operator, it interferes with RNA polymerase binding or prevents

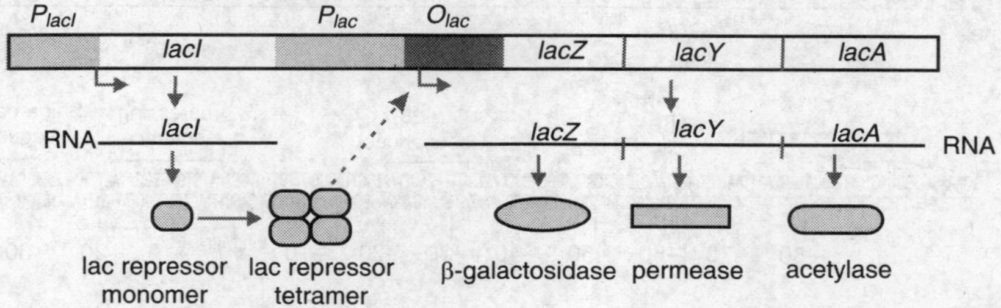

Fig. 14.8. Structure of the lactose operon.

the RNA polymerase from achieving the open complex. In either case, transcription of the operon is prevented (Fig. 14.9). The repressor is released when it combines with an **inducer,** a derivative of lactose called **allolactose** (Fig. 14.8).

Note that the promoter is not only recognized by RNA polymerase but also has elements in the immediate vicinity of the initiation site of transcription. At this stage, operon can be redefined as *a sequence of adjacent genes all under the transcriptional control of the same promoter and operator.*

The nucleotide sequence of the *lac* operator region is shown in Fig. 14.10. The operator in Fig. 14.8 is referred to as the primary operator, *O1*, centered at +11. Two other operator sequence have been found. One, *O2*, is centered at +412. The third operator, *O3,* overlaps the C-terminal end of the *i* gene and is centered at – 82. The structure of the repressor and its interaction with the operator site was worked out recently with X-ray crystallography. The functional repressor is **homotetramer** of the protein product of *i* gene; *i.e,* it is formed from four identical copies of the

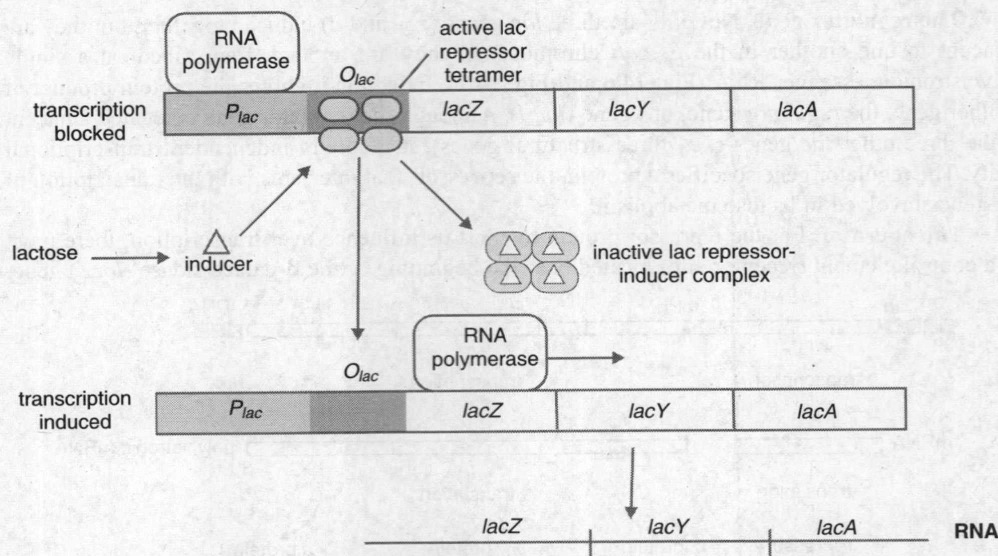

Fig. 14.9. Binding of inducer (allolactose) inactivates the *lac* repressor.

repressor protein. Since each operator site has two-fold symmetry, two repressor monomer proteins bind to each operator site. The monomer is shaped so that it fits into the major groove of the DNA to locate the exact base sequence of the operator; it then binds at the point through electrostatic forces. A tetramer can bind to two of the operator sites at the same time, presumably *O1 and O3* or *O1* and *O2*. In the process, the DNA is formed into a loop (Fig. 14.11).

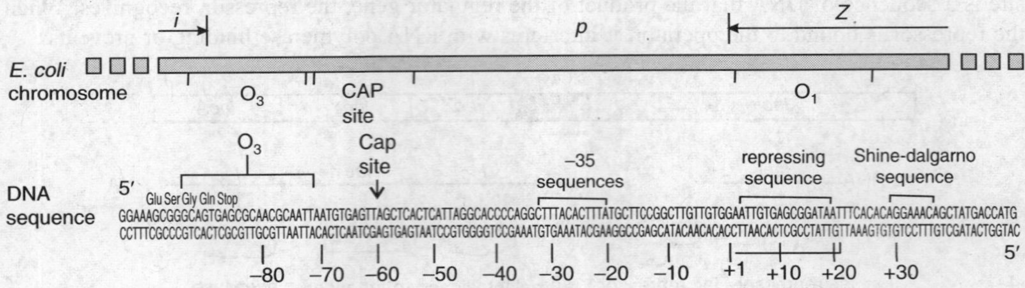

Fig. 14.10. The *lac* operon promoter and operator regions. The base sequence corresponds to the diagram above it. The terminal amino acids of the *i* gene are shown, as well as the initial amino acids of *lacZ* gene. In addition, have been depicted the Shine-Dalgarno sequence of the DNA, the repressor binding region (centered around +10 of the gene), the – 10 and – 35 sequences of the promoter and primary (O_1) and secondary (O_3) operator sites (after Tamarin, 2002).

Induction of the *lac* operon. Before the operon can be "turned on" to produce lactose utilizing enzymes, the repressor will have to be removed from the operator. The repressor is an **allosteric protein**; when it binds with one particular molecule, it changes the shape of the protein, which changes its ability to react with a second particular molecule. Here the first molecule is the inducer allolactose and the second molecule is operator DNA. When allolactose is bound to the repressor, it causes the repressor to change shape and lose its affinity for operator sequences (Fig. 14.11).

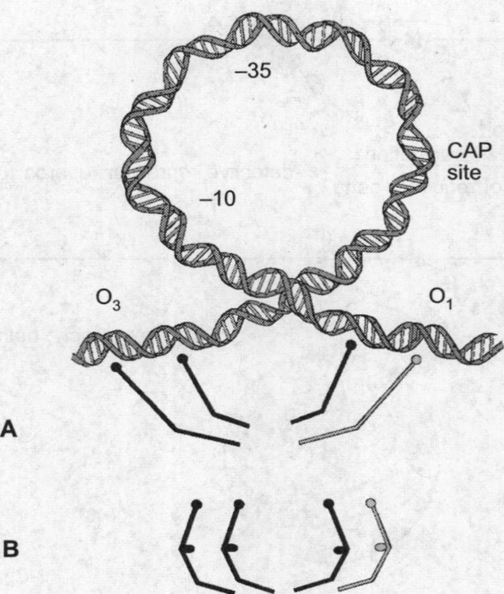

Fig. 14.11. Because the *lac* operator DNA sequences are palindromes (A *palindrome* has the same DNA sequence when one strand is read left to right in a 5′ to 3′ direction and the complementary strand is read right to left in a 5′ to 3′ direction), each half can bind one repressor subunit. A – The tetrameric repressor binds to O_1 and O_3, causing the DNA in between to form a loop. Each of the subunits has been shown in a different shade. The round portion of the subunit in touch with the DNA is the N-terminal end of the repressor subunit; the C–terminal ends form tails that bind the subunits together. Also indicated are the CAP site and the – 10 and – 35 sequences. B – When each of the subunits binds an allolactose molecule (black circles), the shape of the middle portion of the subunit changes, causing the subunit to fall free of the operators (after Tamarin, 2002).

With allolactose (inducer) bound to the repressor, the ability of the repressor to bind to the operator is greatly reduced, by a factor of 10^3. Since no covalent bonds are involved, the repressor simply dissociates from the operator. After the repressor releases from the operator, RNA polymerase can now begin transcription. The three structural genes of *lac* operon are then transcribed and subsequently translated into their respective proteins.

This system of control is very efficient. The presence of the lactose molecule permits transcription of the structural genes of the *lac* operon, which acts to breakdown the lactose. After all the lactose is metabolized, the repressor molecule returns to its original shape and can again bind to the operator. The system is "turned off".

Lac Operon Mutants (Genetic manipulations of *lac* operon)

Verification of *lac* operon system came about through the use of mutants and partial diploids (merozygotes) of the *lac* operon well before DNA sequencing techniques had been developed. The structural (*i.e.,* enzyme specifying) genes of the *lac* operon, *z, y* and *a*, all have known mutant forms in which the particular enzyme does not perform its function. These mutant forms of the enzymes are z^+, y^+ and a^+.

Merozygotes or partial diploids in *E. coli* can be created through **sexduction** because some strains of *E. coli* have the *lac* operon incorporated into an F′ factor. Since F^+ strain can pass the F′ particle into F^- strain, *lac* operon diploids can be formed. By careful manipulation, various combinations of mutations can be looked at in diploid state.

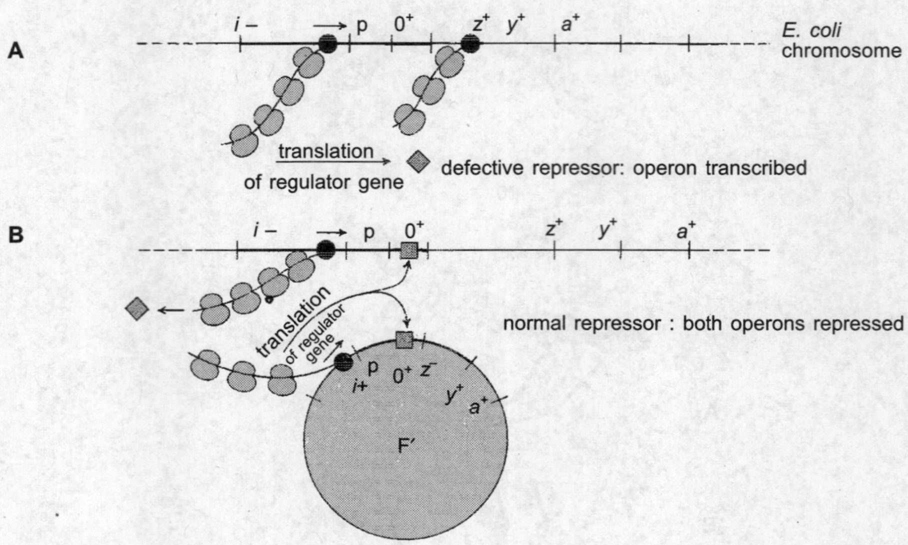

Fig. 14.12. A – A lac operon in *E. coli* with a mutation in the regulator gene (i^-). Transcription and translation of this gene yield a defective repressor; the cell, thus, has a constitutive production of the lac operon. B — The wild-type regulator gene (i^+) is introduced in an F′ factor; there is both a bacterial chromosome and an F′ factor each containing a regulator gene. (The F′ operon carries a mutant z^-allele, allowing geneticists to keep track of the transcriptional control of the chromosomal operon only). In this case, the phenotype is normal (inducible) because enough repressor is produced by the F′ allele (i^+), by transcription and translation, to bind to both operators. RNA polymerase enzyme is shown as solid spheres on the DNA; the wild-type repressor is shown as a square; the mutant repressor, which cannot bind to the operator, is shown as a diamond (after Tamarin, 2002).

Constitutive mutants are mutants in which the three *lac* operon structural genes are transcribed at all times, *i.e.*, they are not turned off even in the absence of lactose. Constitutive production of enzymes can come about in several ways. A defective repressor, produced by a mutant regulator gene, will not turn the system off, nor will a mutant operator that will no longer bind the normal repressor. The regulator constitutive mutants are designated i^- ; the operator constitutive mutants are designated O^c. Both types of mutants produce the same phenotype: constitutive expression of the three *lac* operon genes.

When a new mutant is isolated, it is possible to determine whether it is caused by a regulator or operator mutation. For example, we can determine the exact location of a mutation on the bacterial chromosome by standard mapping techniques or, more recently, by DNA sequencing. Alternatively, the **Jacob** and **Monod** model predicts different modes of actions for the two types of mutations. In merozygotes, a constitutive operator mutation affects only the operon it is physically a part of. Operator mutations are therefore called **cis-dominant**. However, a constitutive *i* – gene mutation, since it works through an altered protein, is recessive to a wild – type regulator gene in the same cell, regardless of which operon (chromosomal or F′ factor) the mutation is on. Constitutive regulator mutations are, therefore, **trans-acting**. (*Note:* If two mutations are on the same piece of DNA, they are in the *cis* configuration. If they are on different pieces of DNA, they are in the *trans* configuration). *Trans*-acting mutations usually work through a protein product that diffuses through the cytoplasm. *Cis*-acting mutants are changes in recognition sequences on the DNA.

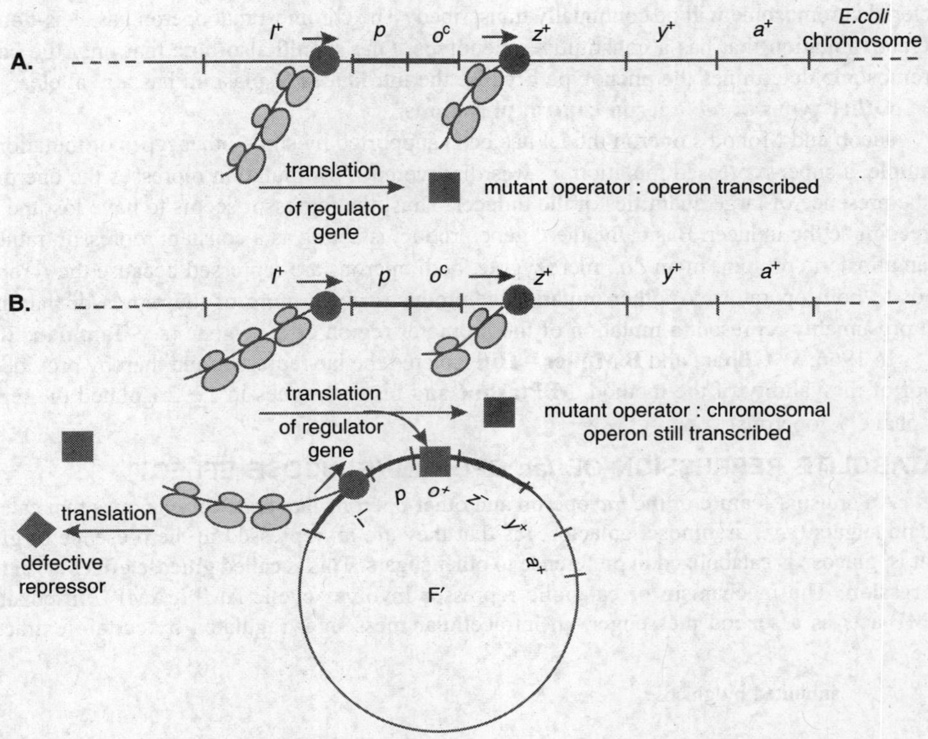

Fig. 14.13. A – A *lac* operon in *E. coli* with a mutation of the operator (O^c). The cell has a constitutive phenotype; the operator cannot bind the wild type repressor protein, and thus, transcription is continuous, even in the absence of lactose. B – The phenotype is unchanged even when a wild-type operator is introduced into the cell in a F′ factor; there is both a bacterial chromosome and a F′ factor; there is both a bacterial chromosome and a F′ factor, each containing an operator. (The F′ operon carries mutant regulator (i^-) and z^-alleles, allowing a geneticist to keep track of the chromosomal operon only). The F′ operator does not change the phenotype of the cell because the wild-type operator exerts no control over the chromosomal operator, which exerts *cis – dominant effect,* another operator on another operon has no effect. RNA polymerase enzyme is shown as solid spheres on the DNA; the wild-type repressor is shown as a square; the mutant repressor, which cannot bind to the operator is shown as diamond (after Tamarin, 2002).

Thus, when the bacterium has a regulator constitutive mutation (i^-); the cell has constitutive production of the operon (Fig. 14.12A). If the wild-type regulator is introduced in an F′ plasmid (Fig. 14.12B), the normal (*i.e.,*inducible) phenotype is restored because the F′ i^+ allele is dominant to the chromosomal mutation – the i^+ gene regulates both the chromosomal and F′ operons. Hence, both operons are inducible. We do not need to be bothered about the other components of the F′ plasmid, because it carries a z^- allele; only the activity of the chromosomal operon will be observed. When, the chromosomal operon carries an operator constitutive mutation; the cell also has constitutive production of the operon (Fig. 14.13). When a wild-type operator is introduced into the cell in an F′ plasmid (Fig. 14.13B), the cell still has the constitutive phenotype because the operator allele on the F′ plasmid does not control the bacterial operon; the *lac* operon on the bacterial chromosome will be continually transcribed. The chromosomal operon has a **cis-dominant** operator mutation that has a constitutive phenotype. One should also note that only the bacterial chromosome determines the phenotype because the introduced F′ plasmid has a z^- allele.

Other types of *lac* operon control mutations.

Jacob and Monod's operon model has been supported by some other types of mutations. For example, a super-repressed mutation, i^s, was discovered. This mutation represses the operon even in the presence of large quantities of the inducer. Thus, the repressor seems to have lost the ability to recognize the inducer. Basically, the *i* gene product is acting as a constant repressor, rather than as an allosteric protein. In an i^s/i^+ merozygote, both operons are repressed because the i^s repressor binds to both operators. Another mutation, i^Q,produces much more of the repressor than normal and presumably represent a mutation of the promoter region of the *i* gene (see **Tamarin**, 2002).

In 1966, **W.Gilbert** and **B.Muller – Hill** isolated the lac repressor and thereby provided final proof of the validity of the method. **M.Ptashne** and his colleagues in 1992 isolated the repressor for phage λ operons.

CATABOLITE REPRESSION OF *lac* OPERON (GLUCOSE EFFECT)

An unusual feature of the *lac* operon and other operons that code for enzymes that catabolize certain sugars (*e.g.,* arabinose, galactose) is that they are all repressed in the presence of glucose. That is, glucose is catabolized in preference to other sugars. This is called **glucose effect** or **catabolic repression**. The mechanism of catabolic repressor involves cyclic AMP(cAMP). In eukaryotes, cAMP acts as a second messenger, an intracellular messenger regulated by certain extracellular

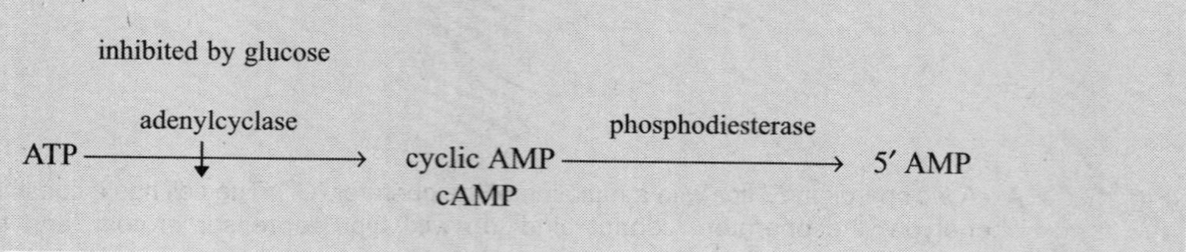

Fig. 14.14. Glucose effect. Glucose uptake lowers the quantity of cAMP (second messenger) in the cell by inhibiting the enzyme adenylcyclase, which converts ATP to cAMP.

hormones. Geneticists were surprised to discover cAMP in *E. coli*, where it works in association with another regulatory protein, the **catabolite activator protein (CAP)**, to control the transcription of certain operons.

When glucose is not present, cAMP combines with CAP, and the CAP-cAMP complex binds to a distal part of the promoter of operons with CAP sites (*e.g.,* the *lac* operon; see Fig. 14.15). This binding clearly increases the affinity of RNA polymerase for the promoter, because without the binding of the CAP-cAMP complex to the promoter, the transcription rate is very low. The uptake of glucose by *E.coli* cells causes the loss of cAMP from the cell, probably by inhibiting adenylcyclase (Fig. 14.14, and thus lowers the CAP-cAMP level. The transcription rate of operons with CAP sites will be reduced (Fig. 14.15). The same reduction of transcription rates has been observed in mutant strains of *E. coli* when this part of the distal end of the promoter is deleted. The binding of CAP-cAMP to the CAP site causes the DNA to bend more than 90 degrees. This bending by itself, may enhance transcription, making the DNA more available to RNA polymerase enzyme.

During process of initiation of transcription, the CAP is in direct contact with RNA polymerase.

This was shown by photocross-linking studies in which CAP was treated with a cross-linking agent that bound the subunit of RNA polymerase when irradiated with UV light. (**Photocrosslinking** is a technique used to determine which moieties (proteins, DNA) are in close proximity during a particular process). For the two proteins to cross-link, they must be in direct contact during the initiation of transcription.

Catabolite repression is an example of positive regulation: Binding of the CAP-cAMP complex at the CAP site enhances the transcription rate of the transcriptional unit. Thus, the *lac* operon is both positively and negatively regulated; the repressor exerts negative control, and the CAP-cAMP complex exerts positive control of transcription.

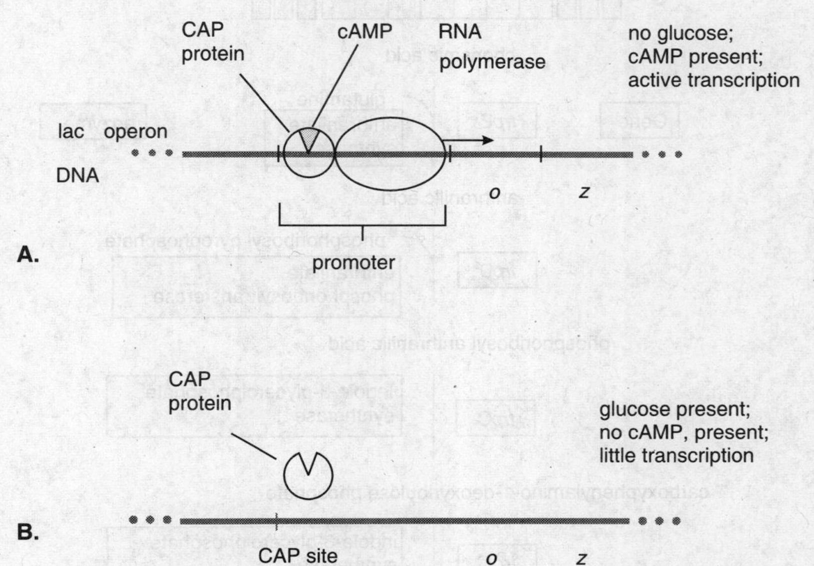

Fig. 14.15. Catabolite repression. When cAMP is present in the cell (no glucose is present), it binds with CAP protein and together they bind to the CAP site in various sugar metabolizing operons, such a *lac* operon shown here. The CAP-cAMP complex increases the transcription of the operon. When glucose is present, it inhibits the formation of cAMP. Thus no CAP-cAMP complex forms, and transcription of the same operon is reduced (after Tamarin, 2002).

2. TRP Operon

The **catabolic operons** or inducible operons are activated when the substrate that is to be catabolized enter the cell. **Anabolic operons** function in the reverse manner. They are turned off (repressed) when their end product accumulates beyond the needs of the cell. Two entirely different mechanisms seem to control the transcription of repressible operons. The *first* mechanism follows the basic scheme of inducible operons and involves the end product of the pathway. The *second* mechanism involves secondary structure in messenger RNA transcribed from an attenuator region of the operon.

(1) Tryptophan synthesis. Tryptophan operon is an example of a repressible system. It contains the five genes that code for the synthesis of the enzymes that build tryptophan amino acid, starting from **chorismic acid** (Fig. 14.16). It has a promoter operator sequence (*p, o*) as well as its own regulator gene (*trpR*).

(2) Operator control. In *trp* operon, the product of *trpR* gene is called **repressor**. It is inactive by itself and it does not recognize the operator sequence of *trp* operon. The repressor only become active when it combines with tryptophan. Thus, when tryptophan builds up, enough is available to bind with and activate the repressor. Tryptophan is, thus, called **corepressor**. The corepressor-repressor complex then recognizes the operator, binds to it, and prevents transcription by RNA polymerase.

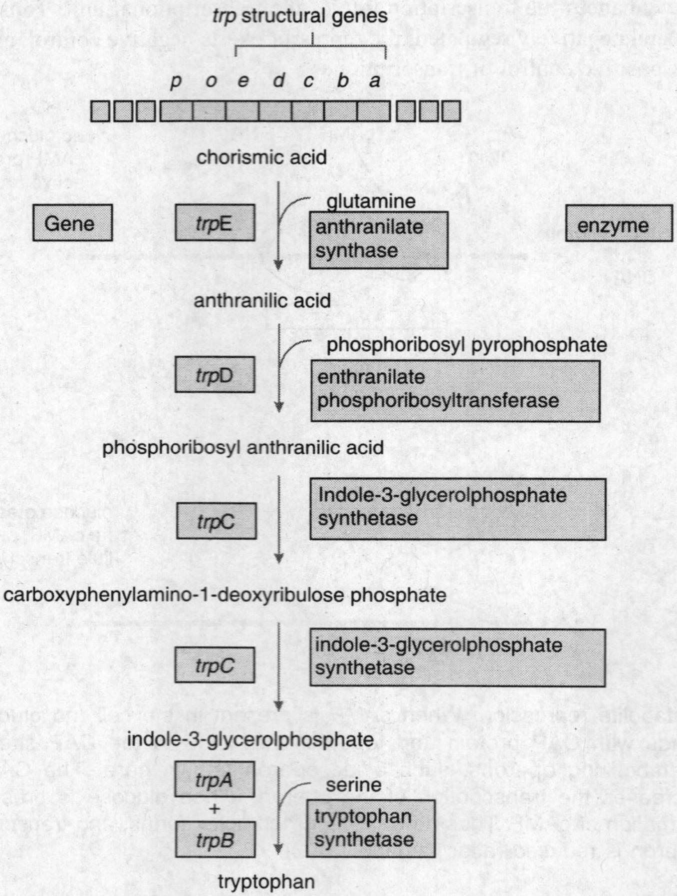

Fig. 14.16. Genes of tryptophan operon (*trp* operon) in *E. coli*. The enzymes they produce control the conversion of chorismic acid to tryptophan. The symbol *o* on the chromosome refers to the *trp* operator, which has its own repressor, the product of the *trpR* gene (after Tamarin, 2002).

After the available tryptophan in the cell is used up, the diffusion process causes tryptophan to leave the repressor, which then detaches from the *trp* operator. Since, the transcription process no longer is blocked, so proceeds normally (*i.e.,* the operon is now derepressed). Transcription continues until enough of the various enzymes are synthesized to again produce an excess of tryptophan. Some tryptophan molecules become available to bind to the repressor and make a functional complex, and the operon is again shut off. This process is repeated to ensure that truptophan is being synthesized as needed (Fig. 14.17). This regulation is modified, however, by the existence of the second mechanism for regulating repressible operons – the attenuation.

(3) Attenuator – controlled trp operon: Details of the second control mechanism of repressible operons have been explained primarily by **C. Yanofsky** and his colleagues (1984), who worked with the tryptophan operon in *E. coli*. This type of operon is controlled by its own **attenuator region** and similar system is also found in operons involved in synthesis of other amino acids such as leucine and histidine. (**Attenuation** means "premature termination").

(*i*) Leader transcript: In the *trp* operon, an attenuator region lies between the operator and the first structural gene (Fig. 14.18). The messenger RNA (mRNA) transcribed from the attenuator region, termed the **leader transcript,** has been sequenced, revealing two surprising and interesting facts.

1. Four subregions of the mRNA have base sequences that are complementary to each other so that three different stem-loop structures can form in the mRNA (Fig. 14.19). Depending on conditions, regions 1-2 and 3-4 can form two stem loop structures, or region 2-3 can form a single stem-loop. When one stem-loop structure is formed, the others are acquired. As we will observe, the particular combination of stem-loop structures determines whether transcription continues.

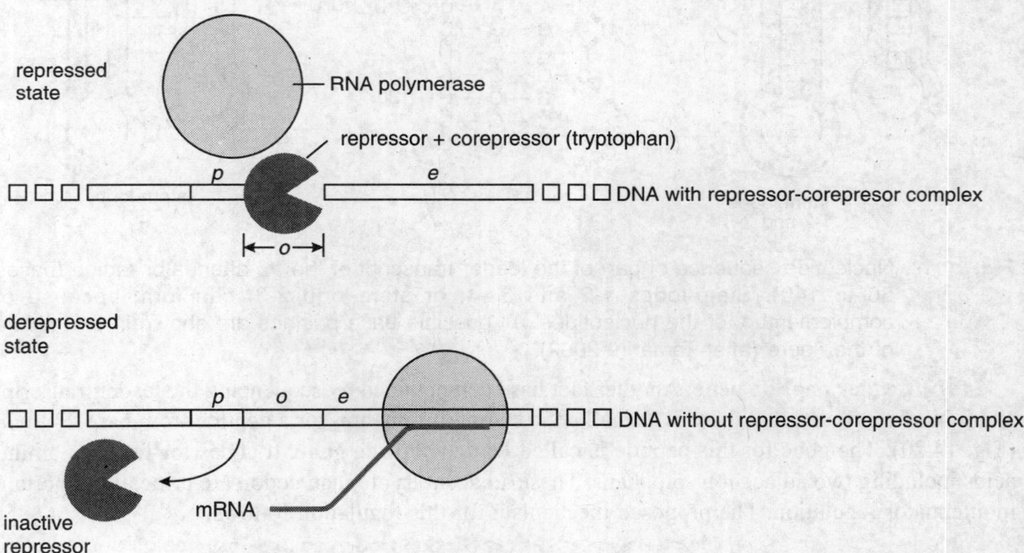

Fig. 14.17. The repressor-corepressor complex binds at the operator and prevents the transcription of the *trp* operon in *E. coli.* Without the co-repressor, the repressor cannot bind, and therefore transcription is not prevented (after Tamarin, 2002).

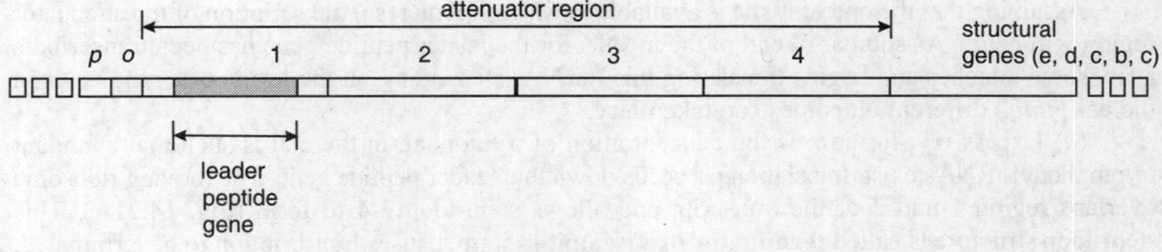

Fig. 14.18. Attenuator region of the *trp* operon, which contains the leader peptide gene. This region is transcribed into the leader transcript (after Tamarin, 2002).

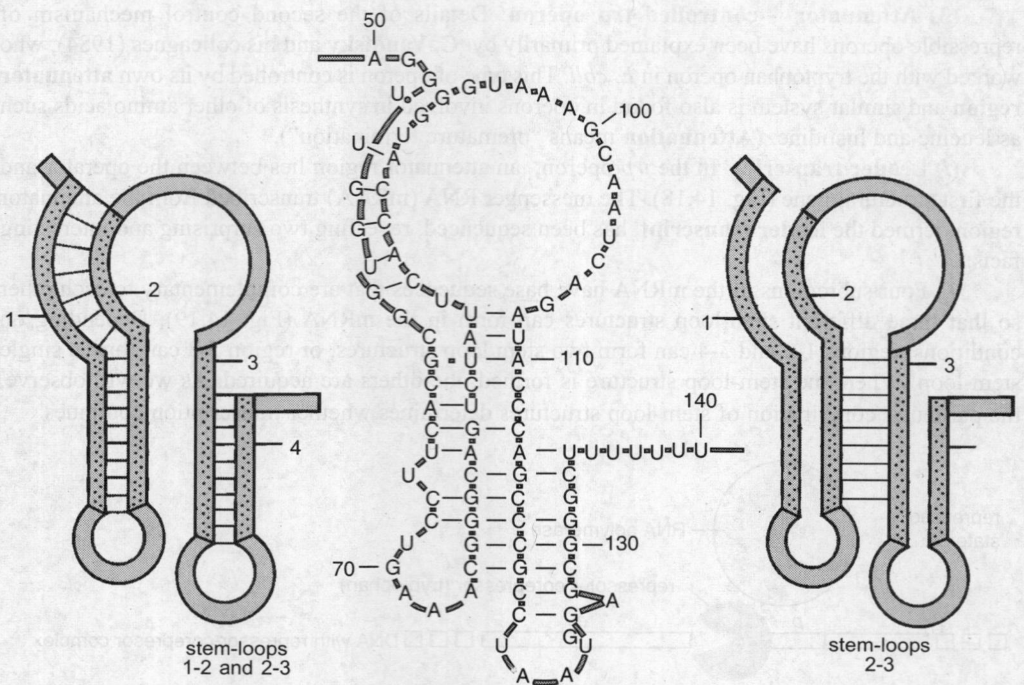

Fig. 14.19. Nucleotide sequence of part of the leader transcript of the *trp* attenuator region (bases 50 to 140). Stem-loops 1-2 and 3 -4, or stem-loop 2-3, can form because of complementary of the nucleotides. All possible base pairings are shown in the middle of the figure (after Tamarin, 2002).

***(ii)* Leader peptide gene**: Another fact has been obtained by sequencing the leader transcript and it has been found that there is a small gene coding infomation for a peptide from base 27 to 68 (Fig. 14.20). The gene for this peptide is called **leader peptide gene**. It codes for fourteen amino acids including two adjacent tryptophans. These adjacent tryptophan codon are critically important in attenuator regulation. The proposed mechanism for this regulation follows.

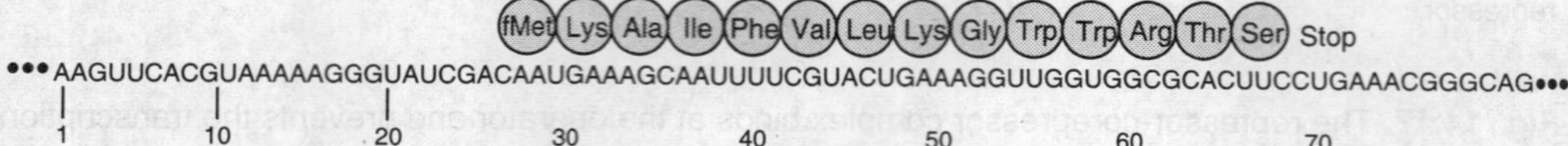

Fig. 14.20. Base sequence of the *trp* leader transcript and the amino acids in these nucleotide code. Note the presence of adjacent tryptophan codons (after Tamarin, 2002).

Assuming that the operator site is available to RNA polymerase, transcription of the attenuator region will begin. As soon as 5′ end of the mRNA for the leader peptide gene has been transcribed, a ribosome attaches and begins translating this mRNA. Depending on the levels of amino acids in the cell, *three* different outcomes can take place.

***(a)* Excess tryptophan**: If the concentration of tryptophan in the cell is such that abundant tryptophenyl-tRNAs exist, translation proceeds down that leader peptide gene. The moving ribosome overlaps region 1 and 2 of the transcript and allows stem-loop 3-4 to form (Fig. 14.21A). This stem-loop structure is called **terminator or attenuator stem**, causes transcription to be terminated. Note that stem-loop 3-4, the terminator stem, followed by a series of uracil (U) – containing bases, is a *rho* independent transcription terminator. Hence, when existing quantities of tryptophan, in the form of tryptophenyl-tRNA, are adequate for translation of the leader peptide 'gene', the transcription is terminated.

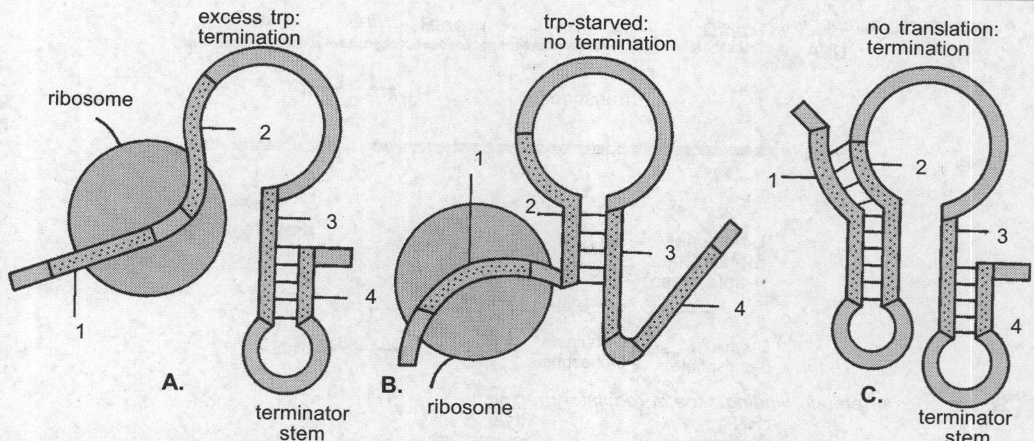

Fig. 14.21. Model for attenuation in the *E. coli trp* operon (**Oxender** *et al.*, 1979). The circle represents the ribosome attempting to translate the leader transcript of Figure 14.19. Under conditions of *excess tryptophan*, the 3-4 stem-loop forms (the terminator stem), terminating transcription. Under conditions of *tryptophan starvation*, the ribosome is stalled, and stem-loop 2-3 forms, allowing continued transcription. Under conditions of *general starvation*, there is no translation, resulting in the formation of stem-loops 1-2 and 3-4, which again results in the termination of transcription (after Tamarin, 2002).

(*b*) Tryptophan starvation: If the quantity of tryptophenyl tRNA is lowered, the ribosome must wait at the first tryptophan codon until it acquires a Trp-tRNA^{Trp} (Fig. 14.21). The stalled ribosome will permit stem-loop 2-3 to form, which precludes the formation of the terminator stem-loop (3-4). In this configuration, transcription is not terminated, so that eventually, the whole operon is transcribed and translated raising the level of tryptophan in the cell. The stem-loop 2-3 structure is referred to as the **preemptor stem**. This is important that the preemptor stem is not a rho-independent transcription terminator and thus, without the rho protein present, will not terminate transcription.

(*c*) General starvation: A final configuration (Fig. 14.21C) is possible. Here, no ribosome interferes with stem formation and presumable stem-loops 1-2 and 3-4 (terminator) form. This configuration also terminates transcription because of the terminator stem. It is believed that this configuration occurs if the ribosome is stalled on the 5′ side of the *trp* codons, which happens when the cell is starved for other amino acids. Presumably, it makes no sense to manufacture tryptophan when other amino acids are in short supply. Hence, the cell can carefully bring up the levels of the various amino acids in the most efficient manner.

The tryptophan operon in bacilli such as *Bacillus subtilis*, is also controlled by attenuation, but secondary structure in the mRNA transport is induced by binding not the ribosome, but a **trp RNA-binding attenuation protein (TRAP) (Antson** *et al.*, 1999).

3. The *ara* Operon

Almost complete mechanism by which the *lac* and *trp* operon of *E. coli* are regulated, are known and supported by an extensive body of experimental data. However, *arabinose (ara)* operon of *E. coli* exhibits much more complex patterns of regulation that are still not completely understood. In the *lac* and *trp* operons, the product of the regulator gene, the repressor, functions in a *negative* manner, turning off transcription of the operon. On the other hand, the catabolic activator protein (CAP) exerts a *positive* control over the *lac* operon by stimulating transcription of the operon. The major regulatory protein of the *ara* operon exhibits both negative and positive regulatory effects on the transcription of the structural genes of the operon, depending on the environmental conditions. Moreover, the regulatory components that control transcription of the *ara* operon include one element that acts from a distance of over 200 nucleotide pairs away from the promoter that it helps to control.

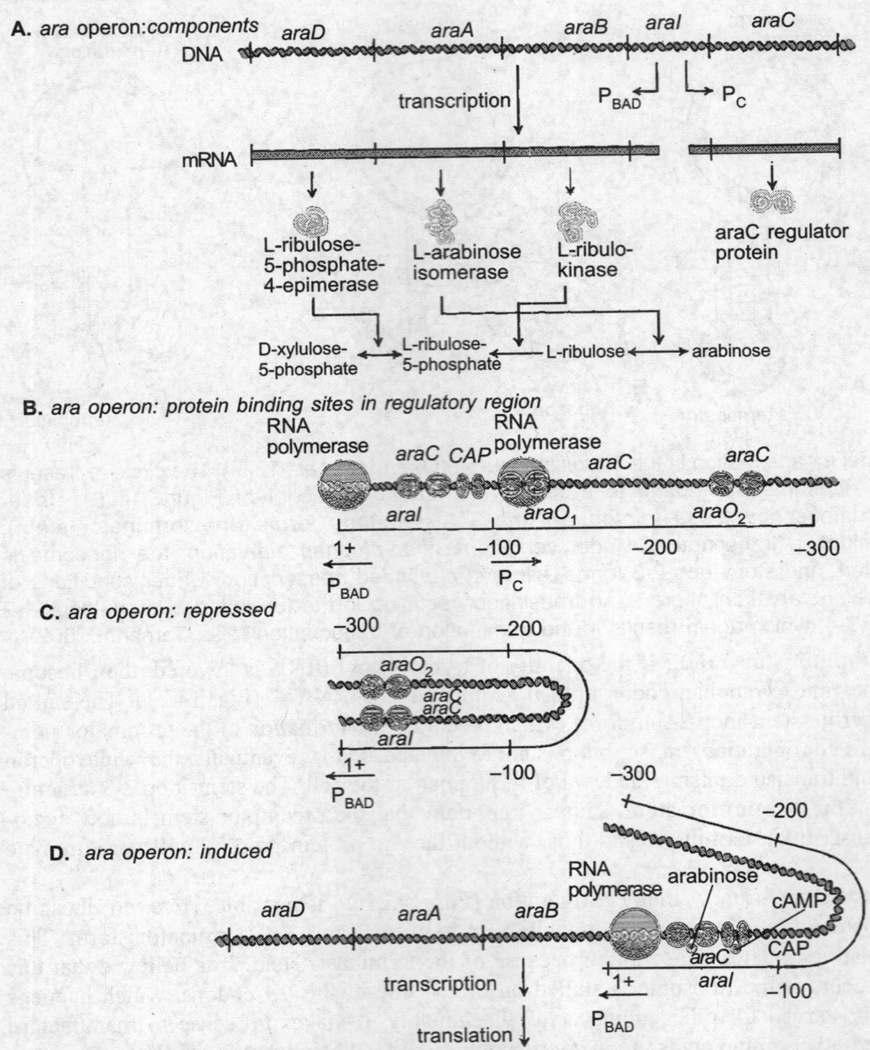

Fig. 14.22. Structure and complex regulation of the arabinose (*ara*) operon of *E. coli*. A – The *araB*, *araA* and *araD* structural genes encode the three enzymes that catalyze the catabolism of arabinose to D-xylulose-5-phosphate. Cotranscription of these three structural genes is initiated at a promoter called P_{BAD}. The *araC* gene encodes a protein that regulates the transcription of the *araBAD* operon. The *araC* protein acts as a repressor of the operon in the absence of arabinose and cAMP; however, it functions as an activator of transcription in the presence of arabinose and cAMP. Transcription of the *araC* gene is initiated at a promoter called Pc. Transcript extension occurs in opposite directions from the P_{BAD} and Pc promoters. The *araI* region contains the binding sites for the *araC* regulator protein and the cAMP – CAP complex. B – Enlargement of the *araI* region showing the positions of the binding sites for *araC* protein, cAMP-CAP and RNA polymerase, and the location of araC protein binding site at $araO_2$. C – Model for the repression of the *ara* operon. *AraC* protein dimers bind at the *ara I* and *araD*$_2$ sites, and then bind to each other to form a DNA loop. RNA polymerase cannot bind at P_{BAD} to initiate transcription when the DNA is in the looped conformation. D – Induction of the *ara* operon takes place when the arabinose – *araC* protein and cAMP – CAP complexes both bind at their *araI* sites. When these complexes are bound at their *araI* sites, the DNA loop structure cannot form, and RNA polymerase is able to bind at P_{BAD} and initiate transcription of the *araB, araA* and *araD* structural genes (after Gardner *et al.*, 2002).

The **arabinose (*ara*) operon** of *E. coli* contains three structural genes (*araB, araA* and *araD*) that encode the three enzymes involved in the catabolism of arabinose (Fig. 14.22A). These three genes are cotranscribed on a single mRNA that is initiated at a promoter called $\mathbf{P_{Bad}}$. Active transport of arabinose into cells is carried out by the products of genes *araE, araF,* and *araG*. These genes are located at sites quite distant from the *araBAD* operon of interest here and will not be further discussed. The major regulatory protein of the *ara* operon (*i.e.*, the *araC protein*) is produced from a transcript that is initiated at a promoter called **Pc**. The Pc promoter is only slightly over 100 nucleotide pairs away from P_{BAD}, but the two promoters initiate transcription in opposite directions (Fig. 14.22 A and B). The *araC* protein acts as a *negative regulator* (a repressor) of transcription of the *araB, araA* and *araD* structural genes from the P_{BAD} promoter in the absence of arabinose and cAMP. It acts as a positive regulator (an activator) of transcription of these genes from the P_{BAD} promoter when arabinose and cAMP are present. Thus, depending on the presence or absence of the effector molecules arabinose and cAMP, the *araC* regulatory gene products may exert either a positive or a negative effect on transcription of the *araB, araA* and *araD* structural genes. Since the *ara* operon is subject to catabolite repression like the *lac* operon and thus to positive control by CAP and cAMP, induction of the *ara* operon depends on the positive regulatory effects of two proteins, the *araC* protein and CAP. The binding sites for these two proteins and for RNA polymerase all appear to be located in a region of the *ara* operon historically called *araI* (*I* for induction), located between the three structural genes of the operon and the regulatory gene (*araC*) (Fig. 14.22A).

Once geneticists working on the problem of regulation of the *ara* operon thought that all the binding sites for the *araC* regulatory protein and cAMP-CAP complex were in the *araI* region. The surprising discovery was that repression of the *ara* operon depended on the binding of *araC* protein at a site called $araO_2$ (*O* for operator, *2* because it was the *second ara operator* identified) located 211 nucleotide-pairs upstream (relative to the direction of the transcription from P_{BAD}) from the *araC* protein binding site in *araI* (Fig. 14.22). (*Note:* operator $araO_1$ – the first *ara* operator to be identified – controls the transcription of the *araC* regulator gene initiated at Pc). The currently accepted model for the repression of the *ara* operon is that the *araC* protein must bind (as a dimer) at both the *araI* site and the $araO_2$ site, and that these proteins then bind to each other to form a DNA loop (Fig. 14.22C).

When the loop structure is formed, it must prevent or interfere with the binding of RNA polymerase at the adjacent promoter (P_{BAD}) of the operon (Fig. 14.22C). In the presence of arabinose and cAMP, the *ara* operon is induced. Moreover, under these conditions, the *araC* protein has been shown to become an activator of transcription of the operon. The arabinose – *araC* protein complex and the cAMP-CAP complex must open the loop by binding at their *araI* sites. This, in turn must permit RNA polymerase to bind at the P_{BAD} site and initiate transcription of the *ara* structural genes (Fig. 14.22D).

Mode of Working of Gene Regulatory Proteins

Gene regulatory proteins must recognize specific nucleotide sequences embedded within its structure. It was originally thought that these proteins might require direct access to the hydrogen bonds between base pairs in the interior of the double helix to distinguish between one DNA sequence and another. It is now clear, however, that the outside of the double helix is studded with DNA sequence information that gene regulatory proteins can recognize without having to open the double helix.

Molecular Recognition

Gene regulatory proteins are found to contain structural motifs that read DNA sequences. Molecular recognition in biology generally relies on an exact fit between the surfaces of two molecules, and the study of gene regulatory proteins has provided some of the clearest example of this principle (see **Alberts** *et al.,* 2002). A gene regulatory protein recognizes a specific DNA sequence because the surface of the protein is extensively complementary to the specific surface features of the double helix in that region. In most cases, the protein makes a large number of contacts with the DNA, involving hydrogen bonds, ionic bonds and hydrophobic interactions. Although individual contact is weak, the 20 or so contacts that are typically formed at the protein – helix interface add together to ensure that the interaction is both highly specific and very strong (Fig. 14.23). In fact, DNA-protein interactions include some of the highest and most specific molecular interactions known in biology. DNA- protein interactions are of the following two types.

1. Helix-turn-Helix Motif

The first DNA-binding protein motif to be recognized was the helix-turn-helix. This was originally identified in bacterial proteins; but later on, this motif has been reported in hundreds of DNA-binding proteins from both eukaryotes and prokaryotes. This protein is constructed from two α-helices connected by a short extended chain of amino acids, which constitutes the "turn" (Fig. 14.24). The two helices are held at a fixed angle, primarily through interactions between the two helices. The C-terminal helix is called the recognition helix because it fits into the major grooves of DNA; its amino acid side chains, which differ from protein to protein, play an important role in recognizing the specific DNA sequence to which the protein binds. Examples of helix-turn-helix proteins include tryptophan repressor, lambda cro, lambda repressor fragment, CAP fragment, etc.

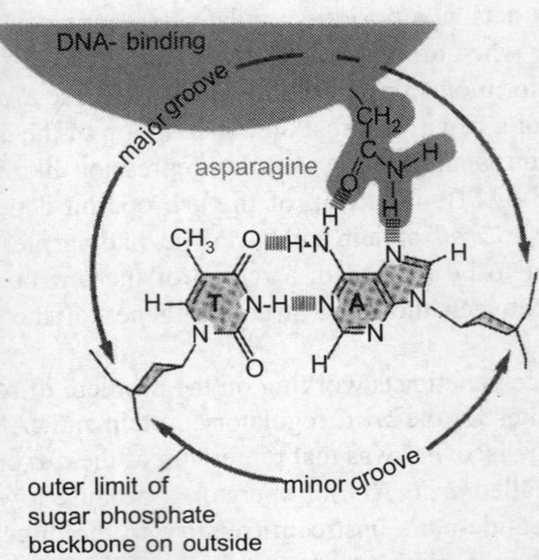

Fig. 14.23. The binding of a gene regulatory protein to the major groove of DNA. Only a single contact is shown. Typically, the protein DNA interface would consist of 10 to 20 such contacts, involving different amino acids, each contributing to the strength of the protein DNA interaction (after Alberts *et al.*,2002).

Homeodomain protein

Soon after the discovery of gene regulatory proteins in bacteria, genetic analyses in the fruitfly *Drosophila* led to the characterisation of an important class of genes, the **homeotic selector genes**, that play a critical part in coordinating fly development. Mutations in these genes cause one body part in the fly to be converted into another, showing that the proteins they encode control critical developmental decisions.

When the nucleotide sequence of several homeotic selector genes of *Drosophila* and higher animals was determined in the early 1980s, each proved to contain an almost identical stretch of 60 amino acids that defines this class of proteins and is termed as the **homeodomain.** When the three – dimensional structure of homeodomain was determined, it was shown to contain a helix-turn-helix motif related to that of the bacterial gene regulatory proteins, providing one of the first indications that the principle of gene regulation established in bacteria are relevant to higher organisms as well. More than 60 homeodomain proteins now have been discovered in *Drosophila* alone, and homeodomain proteins have been identified in virtually all eukaryotic organisms that have been studied, from yeasts to plants to humans.

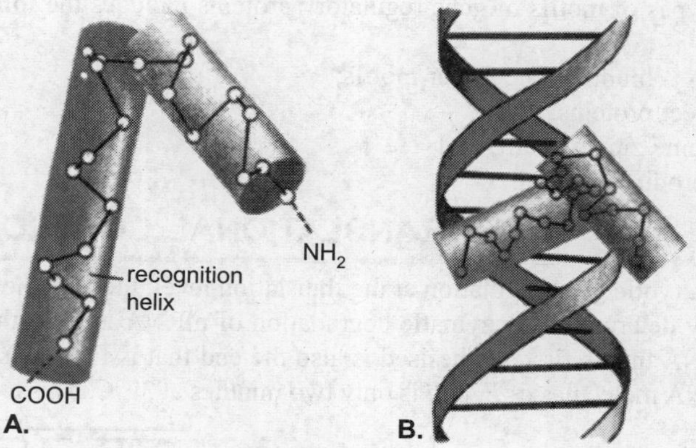

Fig. 14.24. The DNA-binding helix-turn-helix motif. A – Motif with circle which represents the central carbon of an amino acid. The C-terminal α-helix is called the recognition helix because it participates in sequence-specific recognition of DNA. B – The helix of motif fits into the major groove of DNA, where it contacts the edges of the base pairs. The N – terminal α-helix functions primarily as a structural component that helps to position the recognition helix (after Alberts *et al.,* 2002).

The structure of a homeodomain that is bound to its specific DNA sequence, has been shown in Figure 14.25. Where as the helix-turn-helix motif of bacterial gene regulatory proteins is often embedded in different structural contexts, the helix-turn-helix motif of homeodomains is always surrounded by the same structure, suggesting that the motif is always presented in the same way.

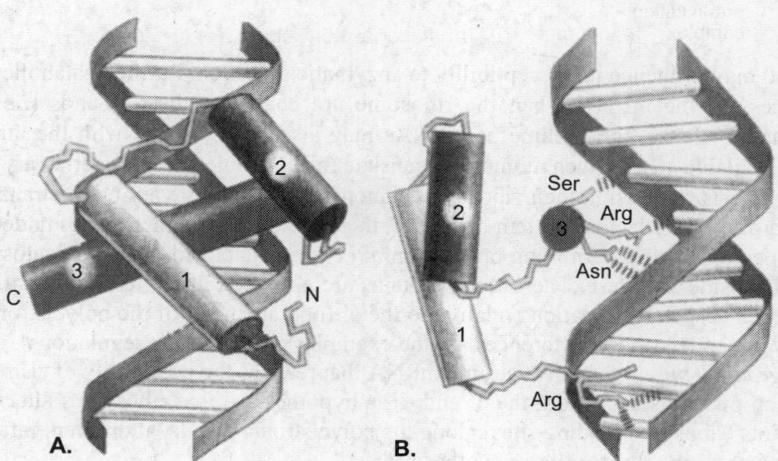

Fig. 14.25. A homeodomain (protein) of yeast bound to its specific DNA sequence. Two different views of the same structure are shown. A – The homeodomain is folded into three α-helices which are packed tightly together by hydrophobic interactions. The part containing helix 2 and 3 closely resembles the helix-turn-helix motif. B – The recognition helix (helix 3) makes important contacts with the major groove of DNA. The asparagine (Asn) of helix 3, for example, contacts an adenine. Nucleotide pairs are also contacted in the minor groove by a flexible arm attached to helix I (after Alberts *et al.,* 2002).

Other types of motifs of gene regulatory proteins includes the following (see **Alberts** *et al.*, 2002).

1. DNA – binding zinc finger motifs;
2. β-sheet proteins;
3. Leucine zipper motif; and
4. Heterodimerization.

14.3. TRANSLATIONAL CONTROL

In prokaryotic gene regulation at the translation level, the lifetime of a mRNA molecule may be genetically determined. Enzymatic degradation of mRNA is from the 5′ to the 3′ end, *i.e.,* the end of the RNA that is first synthesised is also the end that is first degraded. The average lifetime of many mRNA molecules of *E. coli* is only two minutes at 37°C. The specific nucleotide sequences

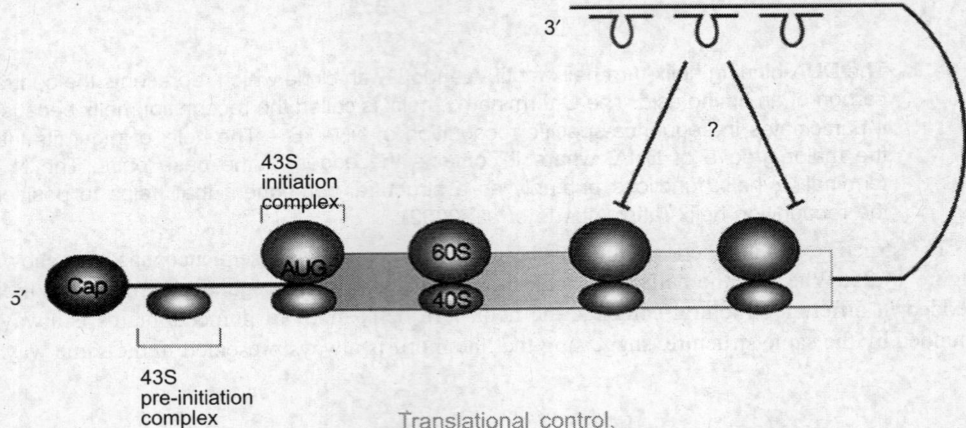

Translational control.

at the 5′ end may influence its susceptibility to enzymatic digestion. Further, catabolic enzymes are denied access to the mRNA when the ribosome are coated at their 5′ ends (*i.e.,* in case of polyribosomes). Hence, the lifetime of mRNAs may also be correlated with the number of free ribosomes available at any given moment to translate mRNA molecules. Bacteria vary their rates of proteins synthesis by varying their ribosomal content rather than by varying the translational rate.

Example: In the lactose system of *E. coli,* there are three structural genes under control of a common operator locus determining production of (1) β – galactosidase, (2) galactoside permease and (3) galactoside acetylase. These three proteins are produced in the respective ratios 1:1/2:1/5, reflecting their respective locations relative to the 5′ (operator) end of the polycistronic mRNA in which they are coded (these differences are the examples of translation regulations). Thus, there is a *polarity gradient* within the polycistronic mRNA that reduces the probability of cistron translation as a function of its distance from the 5′ end. It is hypothesised that ribosomes attach to different starting points (ribosome-binding sites) along the polycistronic mRNA at different rates as reflected by the amounts of the three proteins synthesised.

3. Post-translational control (Feedback Inhibition or End Product Inhibition)

The expression of genes also can be regulated after proteins have been synthesised. This is called **post-translational control of gene action**. Feedback inhibition is a regulatory mechanism which does not affect enzyme synthesis, but rather inhibits enzyme activity (Fig. 14.26). The end product of a biosynthetic pathway may combine loosely (if in high concentration) with the first enzyme in the pathway. This union does not occur at the catalytic site, but it does modify the tertiary structure of an enzyme and, hence, inactivates the catalytic site. This **allosteric transition** of protein molecule blocks its enzymatic activity and prevents overproduction of end products and their intermediate metabolites.

Example: The studies on isoleucine synthesis in *E. coli* (**Umbarger,** 1961) demonstrated that addition of isoleucine (the end product of a five step conversion of threonine) to a culture of the bacteria resulted in immediate blocking of the threonine→isoleucine pathway. In the presence of added isoleucine, the cells preferentially use this exogenous end product (i.e., isoleucine) and their own isoleucine synthesis becomes ceased. Moreover, the production of each of the five enzymes is not interfered with, but action of an enzyme responsible for deamination of threonine to α-ketobutyrate is inhibited by the end product, isoleucine.

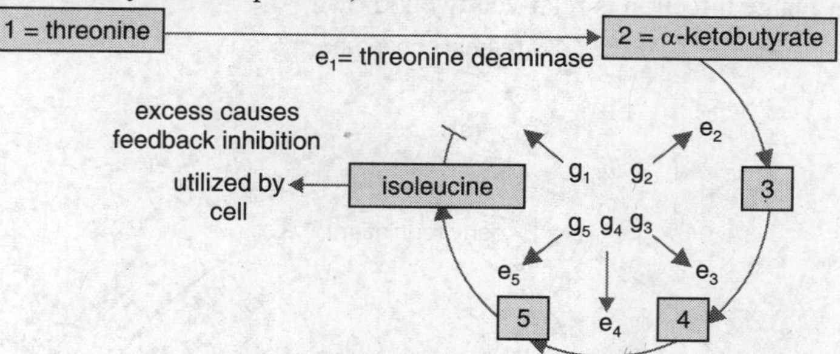

Fig. 14.26. Feedback inhibition in *E. coli*.

14.4. OPERONS OF PHAGE LAMBDA (Regulation of Phage λ Life Cycles)

When a bacteriophage infects a bacterial cell, it has to express its genes in an orderly fashion; some gene products are needed early in infection, and other products are required during later part of infection. **Early genes** usually control phage DNA replication; **late genes** usually determine phage coat proteins and the lysis of bacterial cell. A bacteriophage is considered most efficient if it expresses early genes first and the late genes last in the infection process. Also, temperate phages have the option of entering into lysogeny with the cell (see Box 14.2), here, also, control processes determine which path is taken. One generalisation has been made regarding the phages that their genes are clustered into early and late operons, with separate transcriptional control mechanisms for each.

Box 14.2.

Lysis: Bursting of a cell by the destruction of plasma membrane following an infection by a virus.

Lysogenic: The state of a bacterial cell that has an integrated phage (prophage) in its chromosome.

Lysogenic bacteria: Bacteria harbouring temperate bacteriophages.

Temperate phage: A phage (virus) that invades but may not destroy (lyse) the host (bacterial cell); however, it may continue into the lytic cycle.

Virulent phage: A phage (virus) that destroys the host (bacterial) cell.

Phage λ (= lambda) is one of the best studied bacteriophage. It has a double stranded DNA (chromosome) of about 48,500 base pairs. Since it is a temperate phage, it can exist either vegetatively or as a prophage, integrated into the host chromosome. Life cycle of this phage is suitable for our analysis since its choice of pattern of life cycle is determined by an interesting and complex way. It provides a model system of operon controls. Here, complexity arises from having two conflicting life-cycle choices: one choice favours lytic cycle and other choice prefers nonlytic (*i.e.,* lysogenic) phage of life cycle.

The expression of one of the two life-cycle alternatives, lysogenic or lytic cycles, depends on whether two repressors, *CI* and *cro*, have access to operator site. The **CI repressor** acts to favour lysogeny: it expresses lytic cycle and represses lysogeny. The operator sites, when bound by either *CI* or *cro*, can either enhance or repress transcription. There are certain other control mechanisms which are involved in determining aspects of λ life cycle such as **antitermination** and **multiple promoters** for the same genes (see **Ptashne** and coworkers 1982; **Ptashne**, 1987, 1989).

Structure of Phage λ operons

Phage lambda (λ) shows a complex system of controls of both early and late operons, as well as controls for the decision of lytic infection versus lysogenic integration. The λ genes are grouped into four operons: **left** and **right**, **late** and **repressor**(Fig. 14.27). The *left* and *right operons* contain the genes for DNA replication and recombination and phage integration. The *late operon* contains the genes that determine phage head and tail proteins and lysis of the host cell. The sequence of events following phage infection is relatively well known.

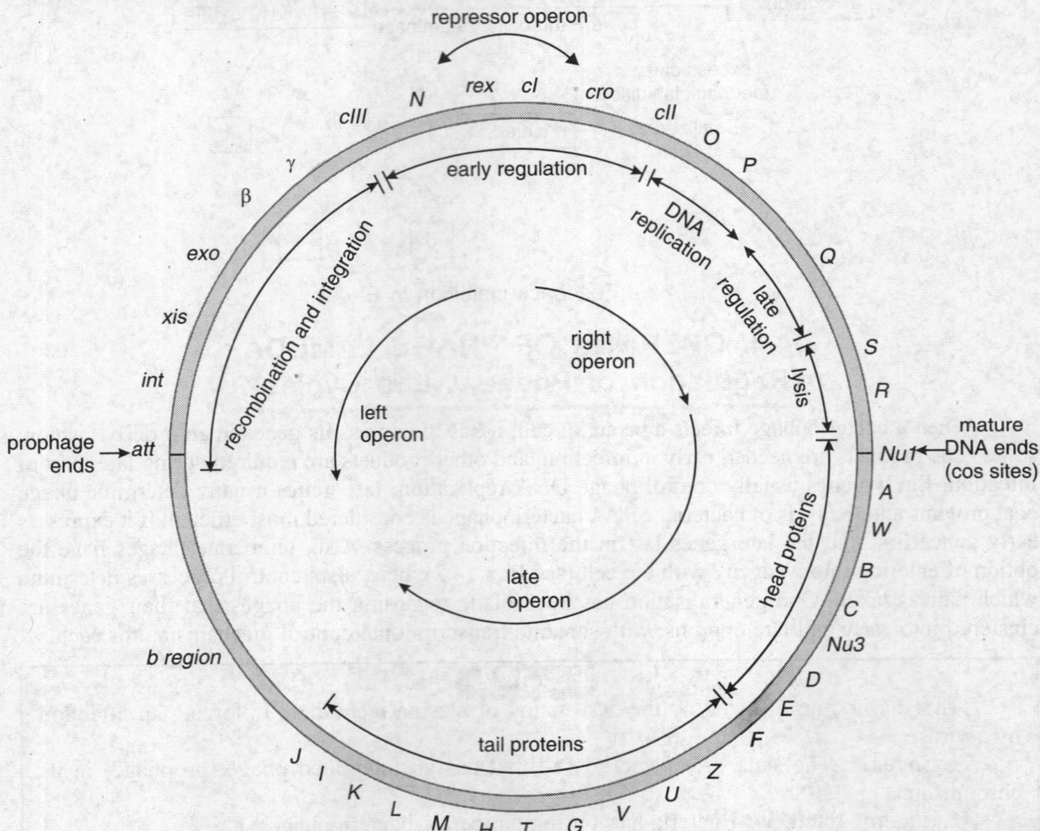

Fig. 14.27. Genetic map of phage λ. There are four operons present: the repressor, left, right and late. The prophage, a linear form integrated in the bacterial chromosome, begins and ends at *att* locus. In mature phage, another linear form of DNA found packed into the phage heads, begins and ends at *Nu1* locus (cos sites) (after Tamarin, 2002).

The genetic map of phage λ (fig. 14.27) is a circle, but the λ chromosome has two linear stages in its life cycle (fig. 14.28). (Note: circular form of DNA is infective cellular form). In one linear form, its DNA is packed within the phage head and in another linear form, DNA of the phage becomes integrated into host chromosome to form a **prophage**. These two linear forms of DNA do not have the same ends (fig. 14.27 and 14.28). The mature DNA which is packed within the phage heads before lysis of the cell, is flanked by *cos* sites. It results from a break in the circular map between the *A* and *R* loci. The prophage is integrated at the *att* site and the circular map is thus broken there at integration.

The homologus integration sites on both λ and the *E.coli* chromosome consists of a 15bp core sequence (called "O" in both), flanked by different sequences on both sides in both the bacterium and the phage (Fig. 14.29). In the phage, the region is referred to as POP′ where P and P′ (P for phage) are two different regions flanking the O core on the phage DNA. In the bacterium, the region is called BOB′ where B and B′ (B for bacterium) are two different regions flanking the O core on the *E.coli* chromosome. Integration, which is a part of the lysogenic life cycle, requires the product of the λ *int* gene, a protein known as *integrase*, and is referred to as **site-specific recombination.** Later excision of the prophage, during introduction, when the phage leaves the host chromosome to enter the lytic cycle, requires both the integrase and the protein product of the neighboring *xis* gene, the **excisionase** (enzyme).

mature DNA
cos A W B
Q S R cos
prophage
gal int xis exo
k I j b
bio
E.coli
E.coli

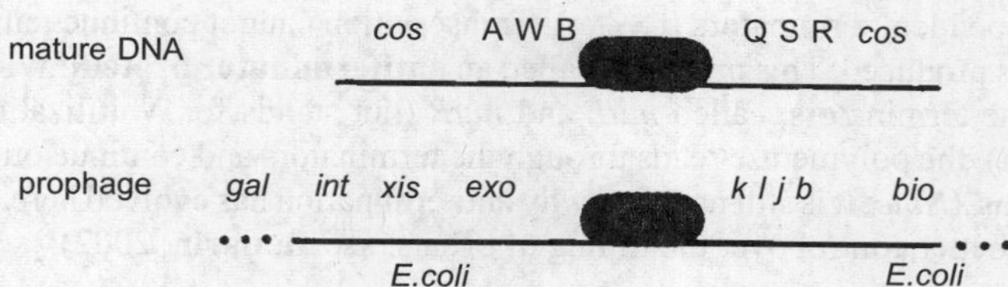

Fig. 14.28. The mature linear DNA (found within phage protein coat) is flanked by *cos* sites. The prophage is flanked by *E. coli* DNA *bio* and *gal* loci (after Tamarin, 2002)

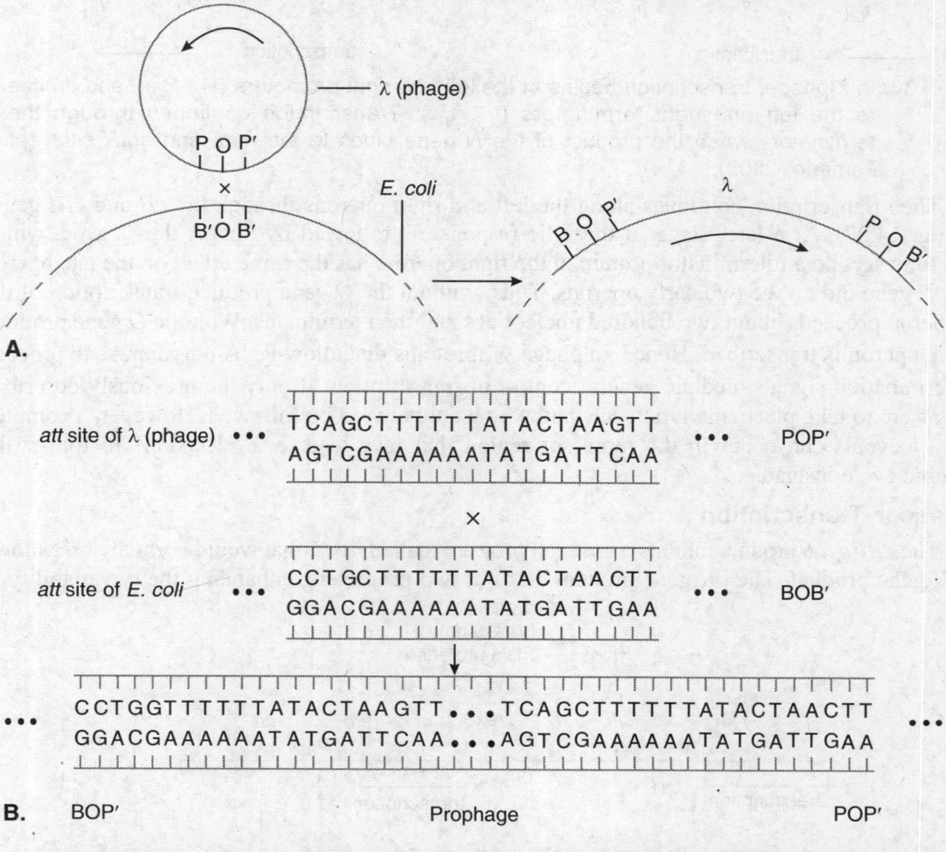

λ (phage)
P O P′
×
B′O B′
E. coli
λ
B O P′
P O B′
A.
att site of λ (phage)
TCAGCTTTTTTATACTAAGTT
AGTCGAAAAAATATGATTCAA
POP′
×
att site of E. coli
CCTGCTTTTTTATACTAACTT
GGACGAAAAAATATGATTGAA
BOB′
CCTGGTTTTTTATACTAAGTT . . . TCAGCTTTTTTATACTAACTT
GGACGAAAAAATATGATTCAA . . . AGTCGAAAAAATATGATTGAA
B.
BOP′
Prophage
POP′

Fig. 14.29. Integration of the I phage into the *E.coli* chromosome requires crossover between the two attach sites, called POP′(Phage) and BOB′ (bacteria). A-General pattern of this site-specific attachment. B-Nucleotide sequences of the various components (after Tamarin, 2002).

Following infection of the *E.coli* cell by a λ phage DNA circularizes, using the complementarity of the *cos* sites. Transcription begins, and within a very short time the phage is producing virus progeny or entering the lysogenic cycle and integrating into the host chromosome.

Early and Late Transcription

When the phage first infects an *E. coli* cell, transcription of the left and right side operons is started at the left (p_L) and right (p_R) promoters, respectively: The *N* (left) and *cro* (right) genes are transcribed (Fig 14.30) and then translated into their respective proteins. Then, transcription stops on both operons at rho-dependent terminators (t_{R1}, t_{L1}). Transcription cannot continue until the protein product of the *N* gene is produced. This protein is called an **antiterminator protein**. When it binds at sites upstream from the terminators, called *nutL* and *nutR* (nut stands for *N* utilization; *L* and *R* stands for left and right), the polymerase reads through the terminators and continue on to transcribe the left and right operons. (*Note*: It is still not clear why antitermination has evolved here, it is believed that it gives the phage better control over the timing of events; see **Tamarin,** 2002).

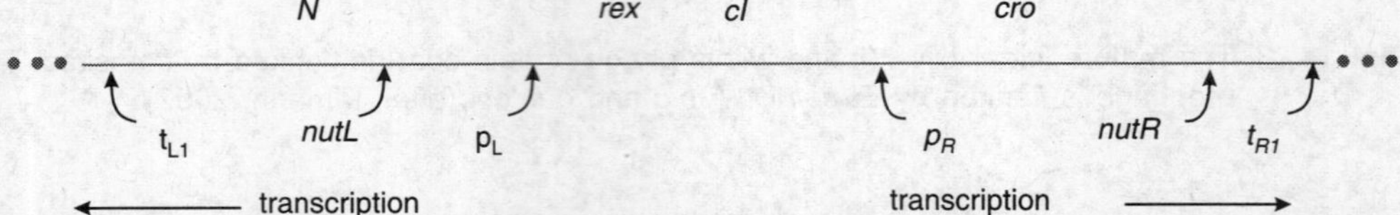

Fig. 14.30. In λ phage, transcription begins at the left and right promoters (P_L, P_R) and proceeds to the left and right terminators (t_{L1}, t_{R1}). Transcription continues through these terminators when the product of the *N* gene binds to the *nutL* and *nutR* sites (after Tamarin, 2002).

Then transcription continues along the left and right operons through the *cII* and *cIII* genes (see Fig. 14.27). At a later stage, if the lytic response is followed by phage, the *Q* gene, which codes for a second antiterminator protein in the right operon, has the same effect on the late operon as the *N* gene did on the two early operons. Thus, without the *Q* gene product, transcription of the late operon proceeds about two hundred nucleotides and then terminates. With the *Q* gene product, the late operon is transcribed. Hence, in phage λ, proteins that allow RNA polymerase to proceed past termination signals mediate general control of transcription. If only the previously described events were to take place, the lytic cycle pathway would always be followed. However, a complex series of events can occur in the repressor region that may lead to a "decision" to follow the lysogenic cycle instead.

Repressor Transcription

The *cIII*-gene product inhibits a host cell protease, called *FtsH*, that would normally breakdown the *cII*-gene product. The *cII*-gene product binds at two promoters, enhancing their availability to

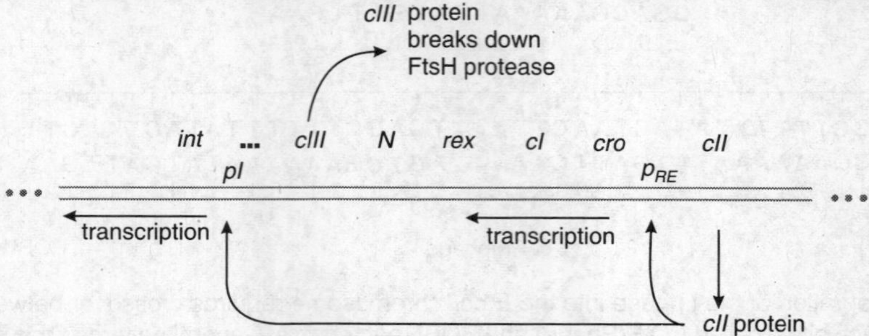

Fig. 14.31. The *cII*-gene product of phage λ binds to the *cI* promoter (p_{RE}) and the *int* promoter (pI), enhancing transcription of those genes. The *cIII* protein breaksdown the Ftsh protease (enzyme) that would normally breakdown the *cII* protein (after Tamarin, 2002).

RNA polymerase, just as the CAP-cAMP product enhances the transcription of the *lac* operon. The *cII* protein binds to the promoters for *cI* transcription and for *int* gene) and *cI* (repressor proteins) are now produced, favouring *lysogeny,* as well as the *cro* gene product, the *antirepressor*, which is a repressor of *cI* and therefore favours the lytic pathway, (*cro* stands for control of repression and other things. These mutants can only undergo lysis without the possibility of lysogeny. Normal λ infections produce turbid plaques, accounted for by lysogenic bacterial growth within the plaques).

Maintenance of Repression

The *cI* gene, with the aid of the *cII*-gene product, is transcribed from a promoter known as p_{RE} (the *RE* stand for repression establishment; Fig. 14.32). Once *cI* is transcribed, it is translated into a protein called the λ *repressor,* which interacts at the left and right operators, O_L and O_R of the left and right operons. When these operators are bound by *cI* protein, transcription of the left and right operons(and therefore also the late operons) ceases. There are several subdivisions of the repression. **First**, lysogeny can be initiated because the *int gene* has been transcribed at the early stage of infection. **Second,** since cII and cIII are no longer synthesised, *cI* transcription from the

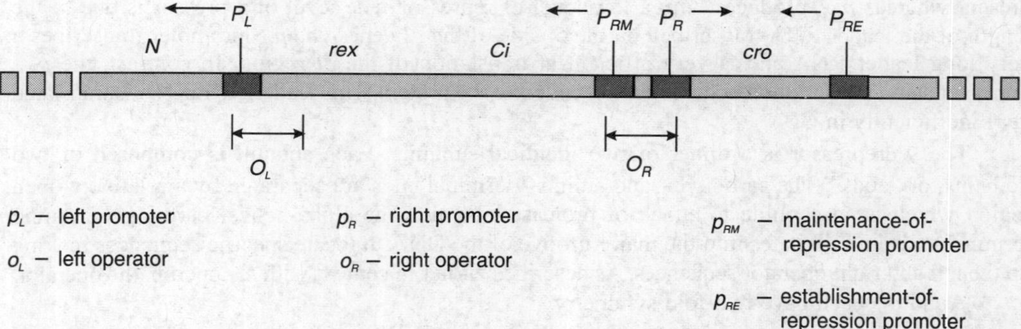

p_L – left promoter
o_L – left operator
p_R – right promoter
o_R – right operator
p_{RM} – maintenance-of-repression promoter
p_{RE} – establishment-of-repression promoter

Fig. 14.32. Early regulation region of phage λ. Two promoters, p_{RE} and p_{RM}, transcribe the *cI* and rex genes. The left operator overlaps the left promoter, and the right operator overlaps both the right promoter and the maintenance of repression promoter (after Tamarin, 2002).

p_{RE} promoter stops. However, p_{RM} (RM stands for repression maintenance), that allows low levels of transcription of the *cI* gene.

In fact, the *cI* gene is found to control its own concentration in the cell. When the right and left operators were sequenced, each was discovered to have three sites of repressor recognition (Fig. 14.33). On the right operator, for example, the right most site (Og1) was found to be most efficient at binding repressor. When repressor was bound at this site, the right operon was repressed, and transcription of *cI* was enhanced (it is similar to enhancement of transcription by the binding of CAP-cAMP at the CAP site in *lac* operon). Excess repressor, when present, however, was also bound by the other two sites within *OR*. The above process results in the repression of the *cI* gene itself. Hence, maintenance levels of cI can be kept within very narrow limit.

A **third** subdivision of repression is the **prevention of superinfection**. That is, bacteria lysogenic for λ phage are protected from further infection by other λ phages because repressor is already present in the cell. Thus, the excess of repressor controls new invading λ phages. (*Note*: for this situation it is normally said that bacterial cells lysogenic for phage λ are immune from infection by additional λ phage). These bacteria are also protected from infection by T_4 phage with *rII* mutants. The *rex*-gene product, the product of the other gene in the repressor operon, controls the protection.

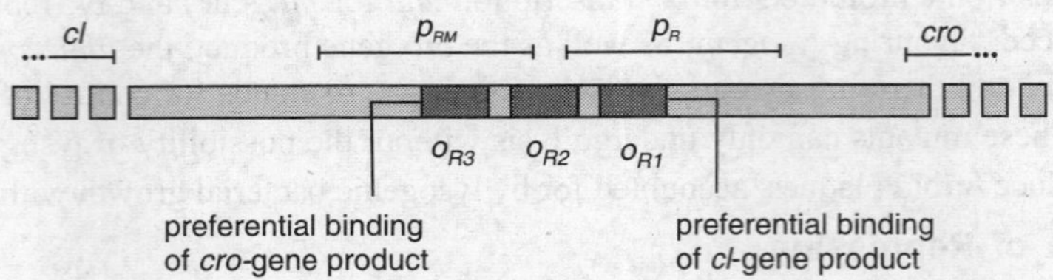

Fig. 14.33. The right operator on the phage λ chromosome overlaps the P_{RM} and P_R promoters. There are three repressor recognition sites within the operator: o_{R1}, o_{R2}, and o_{R3}. Preferential binding by the Cro repressor to o_{R3} and the CI repressor to o_{R1} determines whether transcription occurs to the left or the right (after Tamarin, 2002).

Further the promoters for maintenance and establishment of repression differ markedly in their control of repressor gene expression. When p_{RE} is active, a very high level of repressor is present whereas p_{RM} produces only a low level of repressor. The level of repressor is due to the length of the leader RNA transcribed on the 5′ side of the *cI* gene. The p_{RE} promoter transcribes a very long leader RNA and is very efficient at translation of the *cI* region. In contrast, the p_{RM} promoter starts transcription at the initiation codon of the protein. This leaderless mRNA is translated very inefficiently in *CI*.

The λ repressor is a dimer of two identical subunits. Each subunit is composed of two domains, or "ends". The carboxyl – and amino – terminal ends are separated by a relatively open region which is susceptible to attack of protease enzyme. The alpha-helical parts of the amino terminal ends interdigitates into the major groove of the DNA to locate specific sequences making up the left and right operator sequences. As described earlier in context with *lac* operon, the operator, o_{R1}, o_{R2} and o_{R3} each have two-fold symmetry.

The binding of the λ repressor in o_{R1} enhances the binding of another molecule or repressor into o_{R2}. Together, they enhance p_{RM} transcription, presumably through contact with polymerase. The repressors also block p_R transcription (Fig. 14.33).

Lysogenic versus Lytic Response

Now let us see how does λ phage turn toward the lytic cycle? In this context, the control is exerted by the *cro*-gene product, another repressor molecule that works at the left and right operators in a manner antagonistic to the way the *CI* repressor works. In other words, using the right operator as an example, *cro*-gene product binds preferentially to the left most of the three sites within o_R and represses *cI* but enhances the transcription of *cro* (Fig. 14.33).

The *cro*-gene product can direct the cell toward a lytic response if it occupies the o_R and o_L sites before the λ repressor, or if the λ repressor is removed. From the point of view of phage λ, when would be good time for the *CI* repressor to be removed? From evolutionary point of view, it is expected that a prophage might be at an advantage if it left a host's chromosome and began the lytic cycle when it "sensed" damage to the host. In fact, one of the best ways to induce a prophage to enter the lytic cycle is to direct ultraviolet (UV) light at the host bacterium. UV light causes damage to DNA and induces several repair systems. One, called **SOS repair** makes use of the protein products of the *recA* gene. Among the activities of this enzyme is to cleave the λ repressor in the susceptible region between domains. The cleaved repressor falls free of the DNA, making the operator sites available for the *cro*-gene product. The lytic cycle then follows. Various elements involved in phage λ infection have been tabulated in Table 14.1.

Table 14.1. Elements in phage λ infection (Source: **Tamarin,** 2002).

A. Gene Products	
cI	Repressor protein whose function favours lysogeny
cII	*E*nhances transcription at the p_I and p_{RE} promoters.
cIII	Inhibits the *FtsH* protease
cro	Antirepressor protein that favours lytic cycle
N	Antiterminator acting at *nutR* and *nutL*.
rex	Protects bacterium from infection by T4 *rII* mutants.
int	Integrase for prophage integration
Q	Antiterminator of late operon
FtsH	Bacterial protease that degrades *cII* protein
B. Promoters of	
p_R	Right operon
p_L	Left operon
p_{RE}	Establishment of repression at repressor region.
p_{RM}	Maintenance of repression at repressor region.
p_I	*int* gene.
C. Terminators	
t_{RI}	Terminates after *cro* gene.
t_{LI}	Terminates after *N* gene.
D. Antiterminators	
nutR	In *cro* gene.
nutL	In *N* gene.

Under initial infection, the choice between lysogeny and lytic cycle depends primarily on the *cII* protein, which *gauges* the health activity of the host. After lysogeny is established, it can be reversed by the processes that inactivate the *cI* protein, indicating genetic damage to the bacterium (the SOS response), or an abundance of other hosts in the environment (zygotic induction). In zygotyic induction, the lytic cycle is induced during conjugation, presumably when an Hfr cell sends a copy of the F^- prophage into a F^- -cell. At that point, without repressor present, the prophage can reassess whether to continue lysogeny or enter the lytic cycle.

SOLVED PROBLEMS

Problem 1. From an evolutionary point of view, why do you think *Escherichia coli* (bacterium) evolved a CAP system of positive control of gene expression? Why not just metabolize any and all sugars in the environment as they appear.

Answer: One might think that a bacterium should metabolize any sugar in its environment. However, glucose is the most efficient sugar to metabolize and if it is present, it should be metabolized first. Even under maximal growth, it takes an *E. coli* cell about 20 minutes to grow and divide. If the cells, were to begin to take up and metabolize other sugars, there would be a cost for biosynthesis of new enzymes and the inefficiency of initial steps of the sugar metabolism. In fact, the growth of the cell would slow down under these circumstances and be at an evolutionary disadvantage.

Problem 2. Why would the Rec A protein of *E. coli* cleave the λ repressor?

Answer: Since the cleaving of the λ repressor is a signal to begin the lytic phase of the life cycle of the phage, it seems odd that the lysogenized bacterial cell would be an ally to its own destruction. However, the phenomenon makes much more sense if we realize that the Rec A protein has several other functions critically important to the bacterial cell. The λ phage has evolved the ability to take advantage of the existence of the Rec A protein by evolving a repressor sensitive to it. Evolutionary biologists view this as "coevolution", two interacting organisms evolving to take advantage of or minimize properties of the other. The bacterium, however, might be at disadvantage. Since Rec A has many functions involving interactions with other proteins, it may be highly limited in how it can change.

Problem 3. What are the differences in action of the λ promoters p_{RE} and p_{RM}?

Answer: The promoters p_{RE} and p_{RM} are both promoters of the repressor operon of phage λ. Transcription from these promoters allows production of the *cI* repressor protein, the repressor that favours lysogeny. Initially, the promoter p_{RE} is activated. For it to be a transcription site, it must be activated by the product of the *cII* gene, which lies in the right operon. This promoter, p_{RE}, produces a mRNA with a long leader sequence that is translated very efficiently. Once the repressor binds at the operators of the left and right operons, the *cII* gene is no longer transcribed, and therefore p_{RE} is no longer a site for transcription. However, the repressor gene can still be transcribed from the p_{RM} promoter, which does not need the product of the *cII* gene. This promoter produces a transcript with no leader and thus is translated very inefficiently. At that point, however, only a very small quantity of repressor is needed to maintain lysogeny. Thus, the two promoters are the sites for the initiation of the repressor operons under different circumstances:one early in the infection stage and one after lysogeny is under way.

Questions And Problems

1. How can inducible and repressible enzymes of microorganisms be distinguished?
2. Distinguish between (*a*) repression and (*b*) feedback inhibition caused by the end product of biosynthesis pathway. How do these two regulatory mechanisms complement each other to provide for the efficient regulation of metabolism?
3. In the *lac* operon of *E. coli*, what is the function of each of the following genes or sites: (*a*) regulator, (*b*) operator, (*c*) promoter, (*d*) structural gene *z*, and (*e*) structural gene *y*?
4. What would be the result of inactivation by mutation of the following genes or sites in the *E. coli lac* operon: (*a*) regulator; (*b*) operator; (*c*) promoter; (*d*) structural gene *z* and (*e*) structural gene *y*?
5. Of what biological significance is the phenomenon of catabolite repression?
6. Describe the role of cyclic AMP in transcriptional control in *E. coli.*
7. **J. Beckwith** isolated point mutations that were simulataneously uninducible for the *lac, ara, mal* and *gal* operons, even in the absence of glucose. Provide two different functions that could be missing in these mutations.
8. Describe the interaction of the attenuator and the operator control mechanisms in the *trp* operon of *E. coli* under varying concentrations of tryptophan in the cell. How does attenuator control reacts to shortage of other amino acids?
9. What is the fate of a λ phage entering an *E. coli* cell that contains quantities of λ repressor?
10. What is the fate of a λ phage entering an *E. coli* cell that contains of the *cro*-gene product?
11. Describe the fate of λ phages during the infection process with mutants in the following genes: *cI, cII, cIII, N, cro, att, Q*.
12. What is the fate of λ phages during the infection process with mutants in the following areas: o_{R1}, o_{R3}, p_L, p_{RE}, p_{RM}, p_R, $t_{L1}t_{R1}$, *nutL, nutR*?
13. What are the three different physical forms that the phage λ chromosome can take?
14. How does ultraviolet light (UV) damage induce the phage λ chromosome?

Answers to Problems

1. Both enzyme types are distinguished by studying the synthesis or lack of synthesis of the enzyme in cells grown on chemically defined media. If the enzyme is synthesized only in the presence of a certain metabolite or a particular set of metabolites, it is probably inducible. If it is synthesized in the absence but not in the presence of a particular metabolite or group of metabolites, it is probably repressible.
2. Repression occurs at the level of transcription during enzyme synthesis. The end product, or a derivative of the end product, of a repressible system acts as an effector molecule that usually, if not always, combines with the product of one or more regulator genes to turn off the synthesis of the enzymes in the biosynthesis.
3.

Gene or Regulatory element	Function
1. Regulator gene	Codes for repressor
2. Operator	Binding site repressor
3. Promoter	Binding site of RNA polymerase and CAP –cAMP complex.
4. Structural gene *z*	Encodes β-galactosidase.
5. Structural gene *y*	Encodes β-galactoside permease.

4. (*a*) Constitutive synthesis of the *lac* enzymes;
   (*b*) Constitutive synthesis of the *lac* enzymes;
   (*c*) Uninducibility of the *lac* enzymes;
   (*d*) No β-galactosidase activity;
   (*e*) No β-galactoside permease activity.
5. Catabolite repression has apparently evolved to assure the use of glucose as a carbon source when this carbohydrate is available, rather than less efficient energy sources.
6. Cyclic AMP, combined with CAP protein, attaches to CAP sites enhancing transcription of nonglucose, sugar metabolizing operon in *E. coli*. Glucose inhibits its formation by inhibiting adenylcyclase.
7. One should have an idea about how these operons are controlled. Each operon requires inducer and also the catabolic repression activation system. These mutants could be unable to make cAMP because the adenylcyclase gene is defective. Alternatively, they could be making a defective catabolite activating protein (CAP).
8. The *E. coli trp* operon functions as a normal repressible operon. In addition, attenuation, based on secondary structure and stalling of the ribosome on the leader transcript, can further prevent transcription. Attenuator control can be exerted based on other amino acids if their codons appear in the leader transcript, causing ribosome stalling.

13. The λ chromosome has one circular and two linear forms. The circular form is the infective cellular form. A break at one point (*cos* site) takes place during packaging into the phage head, and a break at another point forms the linear integrative prophage.

MULTIPLE CHOICE QUESTIONS

Choose the correct answer from the four alternatives given.

1. In operon concept, regulator gene functions as
   (*a*) inhibitor (*b*) regulator
   (*c*) repressor (*d*) all of these
2. The operator gene of lac operon is "turned on" when lactose molecule binds to
   (*a*) operator gene (*b*) repressor gene
   (*c*) promotor site (*d*) mRNA
3. The example of inducible operon is
   (*a*) lac-operon
   (*b*) try-operon
   (*c*) arginine operon
   (*d*) both (*a*) and (*b*)
4. How many promoters control the transcription of *E.coli* lac operon?
   (*a*) one (*b*) two
   (*c*) three (*d*) four
5. Which one of the following is not a component of lac-operon model?
   (*a*) promoter (*b*) structural gene
   (*c*) primer gene (*d*) regulator gene
6. In *E.coli*, presence of tryptophan amino acid causes
   (*a*) activation of operon
   (*b*) closure of operon
   (*c*) both of above
   (*d*) none

ANSWERS

1. (*c*) **2.** (*b*) **3.** (*a*) **4.** (*b*) **5.** (*c*) **6.** (*b*).

CHAPTER 15

Regulation of Gene Action in Eukaryotes

Chapter Contents

The different types of cells in the body of a multicellular organism differ in structure and function, but their genes are identical, since all cells are ultimately derived from the zygote by cleavage (mitosis). The problem therefore is: how do cells with identical genetic complements differ so much in structure and function? The answer is that not all genes are active at one time. As development proceeds certain genes become active while others become inactive; *i.e.,* the genes are "switched on" and "switched off" at different times. This process is called **differential gene action**. When genes are active they direct the formation of enzymes which affect certain phenotypic traits.

Further, though a cell has the genes to produce hundreds of enzymes, only the enzymes required at a particular time are produced. Such a control mechanism ensures that the cell is not flooded with unnecessary enzymes. In eukaryotes, genes are active only when their products are required by the cell. The rest of the time they are "switched off". For example, in mouse liver cells, only about 3% of the genes are active (*i.e.,* transcribe RNA), and in brain cells about 9% of the genes are turned on.

In eukaryotes, the following two kinds of controls or regulations of gene expression occur: 1. **Short-term** or **reversible regulation** corresponds to the kind of regulation we studied in bacteria and it represents a cell's response to fluctuations in the environment, especially, it involves changes in activities of concentrations of enzymes as particular substrates and or hormone levels rise and fall. The changes a cell experiences during a cell cycle, particularly the fluctuations in rates of DNA, RNA, and protein synthesis that regularly occur with respect to the time of mitosis can also be placed in this category. 2. **Long-term** or **irreversible regulation** includes the phenomena associated with **determination, differentiation**, or more generally development: it is involved in the numerous steps by which a fertilized egg becomes an organism of, perhaps, trillions of cells with diverse and ultimately quite permanent roles to play in the maintenance of the whole. Short-term regulations also occur in developing and differentiating eukaryotic cells side by side of long-term regulation.

Both of these types of regulations of gene activates in eukaryotes, now, are considered to occur at the following levels involving diverse mechanisms: 1. Regulation at the level of DNA; 2. Regulation at the level of transcription; 3. Regulation at the level of translation; and 4. Regulation at the level of post-translation.

15.1. REGULATION OF GENE ACTION AT THE LEVEL OF GENOME

In eukaryotic cells, it seems that certain classes of genes are transcribed more or less continuously, and only in extreme situations their activities are repressed. For example, genes coding for larger ribosomal RNA (28S or 18S rRNA)) or transfer RNA (tRNA) are present as multiple copies forming **simple multigene family**. These genes are transcribed uniquely by RNA polymerase I for the larger ribosomal RNA or by RNA polymerase III for tRNA and 5S RNA. Although the products of some of these genes, the ribosomes, are used continuously in all cells, it does not confirm that all of these multiple copies are continuously transcribed at maximum rate. Electron micrographs of spread chromatin from nucleoli often show that some of the repetitious rRNA genes are inactive. It is also true that in the nucleated erythrocytes of lower vertebrates such as *Xenopus*, all genes may be turned off (**Maclean** *et al.,* 1972), including those for ribosomal RNA and tRNA. Therefore, it is clear that mechanisms do exist for inactivating sequences even those regarded to be constitutive in normal cells.

Some of the clearest demonstrations that some specific genes are at least available for transcription in different kinds of differentiated cells are provided by *Drosophila* and other organisms (Fig 15.1). For example, the pattern of bands and interbands of **polytene chromosomes** of *Drosophila* does not vary between different larval tissues, yet it is now concluded that the interband regions probably represent **housekeeping genes** which code for essential proteins, that they are expressed in every cell, and that they are retained in a state of permanent decondensation (**Bautz** and **Kabisch**, 1983). Thus, 'housekeeping' genes may be 'left on' for much of the life of the cell when transcription of even the most essential housekeeping genes ceases (*i.e.,* during mitosis).

Further, when we consider the case of the "**cell-specific**" **genes, luxury genes** or **smart genes**, which code for the products only found in specialised tissues, it becomes immediately clear that differential expression is the rule. Whether expression fo gene is measured at the level of the messenger RNA or the protein, genes coding for products such as globin, crystallin, fibroin, ovalbumin, casein ànd immunoglobulin give every indication of complete repression in all but the specialized tissue characterised by their presence.

Thus, at the level of genome (*i.e.,* DNA), the following five modes of regulation are operative:

(*i*) **Situation of total genetic shutdown.** (*a*) During mitotic phase of the cell cycle, chromatin is highly condensed to form chromosomes, and transcriptional activity of all genes is suspended.

(*b*) During meiotic division of germ cells a somewhat similar situation to (a) is evident, although in some rare cases, such as **lampbrush chromosomes** of meiotic diplotene in vertebrates (**Vlad**, 1983), transcription proceeds very actively.

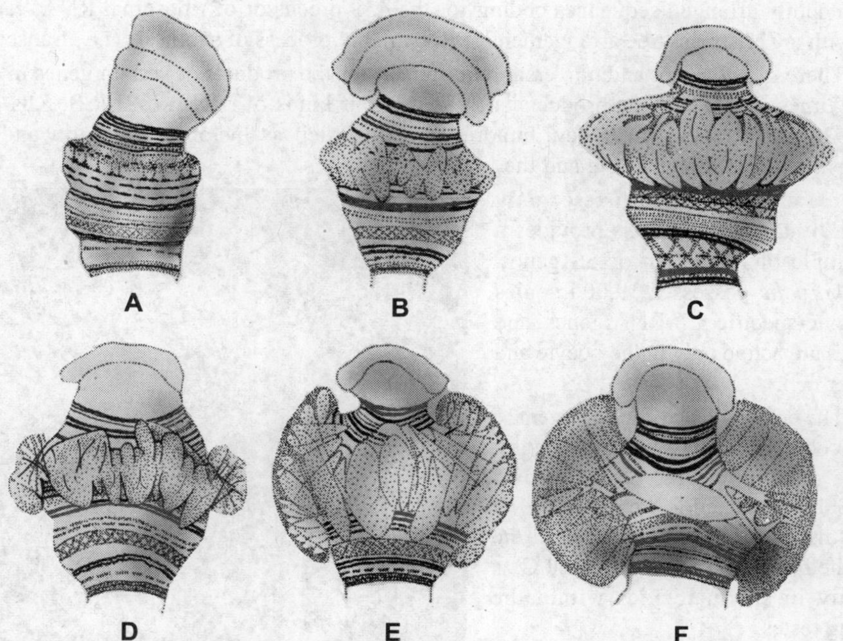

Fig. 15.1. The development of a chromosomal puff in a larval salivary gland cell nucleus of *Chironomus tentans*.

(*c*) The nucleus of mature nucleated erythrocytes of amphibians is transcriptionally inactive. Chromatin in these cells is highly condensed but not organised into discrete chromosomes (**Chegini** *et al.,* 1981). However, transcription can be partially reactivated in these nuclei by transferring them into new cytoplasm or exposing them *in vitro* to altered environmental conditions.

(*d*) In mammalian females, one of the two X chromosomes present in somatic cells undergoes condensation in early embryonic stages to become heterochromatic **sex chromatin** or **Barr body** (Dosage compensation). A variety of experiments indicate that most, though not all, genes of the condensed X chromosome are turned off. In developing oocytes, as opposed to somatic cells, Barr bodies are not present, the activities of both X chromosomes being required for normal oogenesis.

In somatic cells of normal XY males, genes of the single X chromosome remain active and Barr bodies are not found. However, in germ cells, the X chromosome is inactivated prior to spermatogenesis, otherwise, it may prevent sperm maturation and lead to sterility. In one extreme case, that of the creeping vole, *Microtus oregoni*, the X chromosome is eliminated from the germ cells of males by a special process of nondisjunction (see **Farnsworth**, 1988).

(*e*) Sperm cells clearly contain a complete genetic endowment but no transcription occurs until the sperm nucleus is activated within the egg cytoplasm.

(*f*) Complete suspension of transcriptional activity is also known in the following cases: cells of some plant seeds; cells within diapausing *Artemia* gastrulae; cells within inactive organisms such as desiccated *Tardigrada*; nuclei within bacterial and fungal spores; and nuclei within desiccated amoeba cells, as for example in the slime mould *Dictyostelium*.

(*ii*) Evidence for constitutive expression of some genes. (*a*) If the interbands of *Drosophila* polytene chromosomes are correctly interpreted as being loci for "housekeeping" genes, then the evidence is that such chromatin is permanently decondensed and is transcribed at a low but constant rate (**Semehin** *et al.,* 1979).

(*b*) Electron microscopy of spread films of DNA extracted from nucleoli of *X. laevis* oocytes reveals tandemly arranged sequences coding for the 45S precursor of ribosomal RNA, each gene adorned with a Christmas- tree arrangement of RNA in the process of synthesis (*i.e.,* transcription).

(*c*) There is a constant and universal requirement for the products of certain genes in all cells and at all times. These include products such as the four kinds of rRNA-28S, 18S, 5.8s and 5S; tRNA of 20 basic types, and a few hundred proteins such as histones, ubiquitin and lactate dehydrogenase, RNA polymerase, and the like.

(*iii*) Many genes are expressed only in certain tissues. (*a*) *Xenopus* provides a good example of regulation of 5S genes. *Xenopus borealis* possesses 19,000 copies of the oocyte-specific 5S rRNA genes, and these genes are active only in the oocyte and in no other cell.

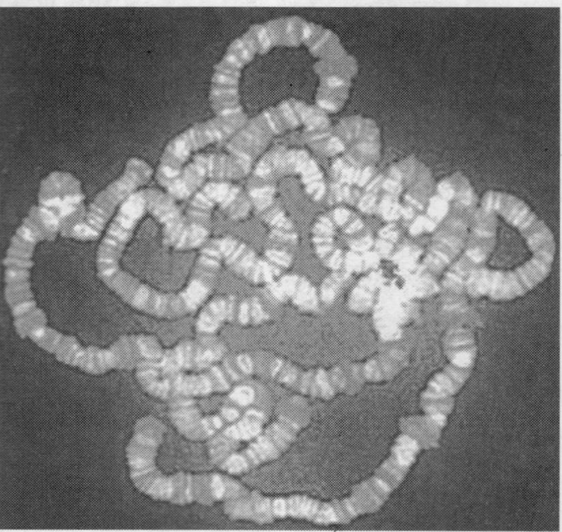

Puffing of polytene chromosomes in *Drosophila.*

(*b*) The enzyme lactate dehydrogenase (LDH) is coded by a small family of genes, each gene determining the structure of a subunit. Subunits A and B are expressed in almost all mammalian cells, but one of the genes in the family, coding for subunit C, is active only in spermatocytes within the developing testes.

(*c*) The **puffing** (*i.e.,* chromatin decondensation) of restricted segments of the polytene chromosomes of *Drosophila, Chironomus* (Fig. 15.1), *etc.*, provides visible evidence of the activity of genes coding for cell-specific products. Certain puffs, known as **heat-shock puffs**, can be induced to appear specifically when salivary glands are exposed to heat shock either *in vivo* or *in vitro*. The correlation between such chromatin decondensation and transcription activity is really proved by autoradiography using tritiated precursors of RNA.

(*iv*) Some DNA is never transcribed in any cell. Analysis of various types of DNA sequences existing in eukaryotic cells reveals that some DNA is comprised of tandemly repeated short sequences that are concentrated in **heterochromatin** such as centromeres of chromosomes and the Y chromosome. Current evidence indicates that much of this DNA is never transcribed in any cell. Some spacer sequences occur between genes, for example, between multiple copies of genes for ribosomal RNA. Such spacer sequences are often taken to be untranscribed, but now it is found that some of the spacer DNA may transcribe nuclear RNA molecules.

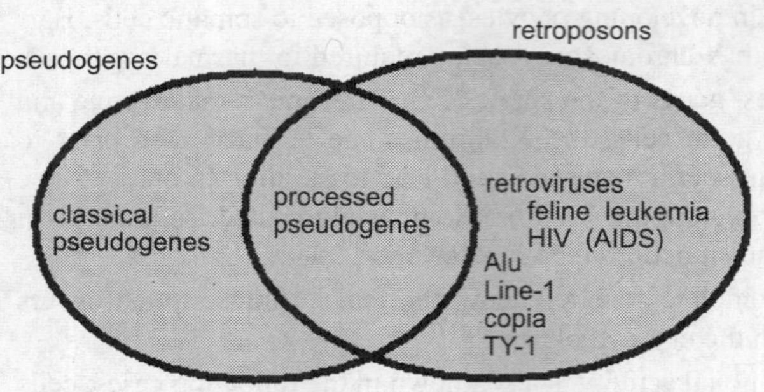

Pseudogenes.

The very large size of the genomes of the higher eukaryotes certainly indicates that much of the DNA is redundant (=repetitious) and probably not utilized as coding or regulatory sequence.

The **pseudogenes** found in many gene families, often presumed to have arisen as cDNA (=complementary DNA) copies of reverse transcribed message. They lack introns and contain many stop signals so that RNA polymerase molecules fail to move very far along them.

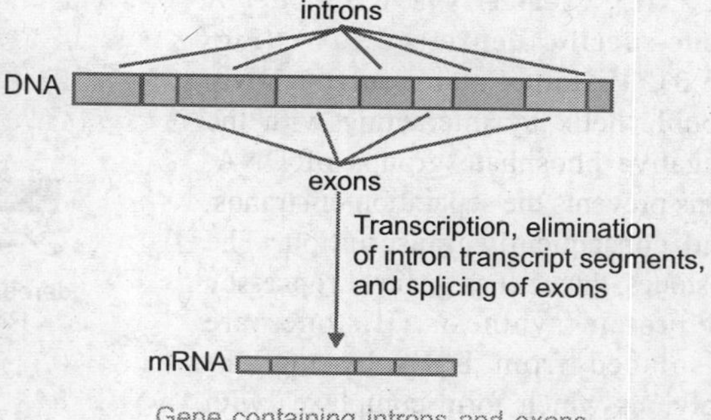

Gene containing introns and exons.

(*v*) **Some DNA is spliced to cause gene rearrangement.** Such a mechanism occurs during expression of immunoglobulin (Ig) genes.

15.2. REGULATION OF GENE ACTION AT THE LEVEL OF TRANSCRIPTION

(*a*) **Chromatin reconstitution experiments.** Chromatin has main components—DNA, histones and non-histones. While it is known since early 1960s that histones may be involved in repressing gene activity, the specific regulation by non-histones was shown only during 1970s. **Gilmour** and **Paul** (1973) performed a chromatin reconstitution experiment to demonstrate the positive role of non-histones in regulation of gene activity (Fig. 15.2). They isolated the chromatin from different tissues separately and then dissociated into DNA, histones and non-histones. This is followed by **chromatin reconstitution** using either the three components derived from the chromatin of the same tissue or by combining the non-histones of one tissue with the DNA and histones of another tissue (Fig. 15.2). From such experiments, it was demonstrated that the mRNA which is synthesized *in vitro* from reconstituted chromatin, mainly depended on the source of non-histone proteins (see **O'Malley** *et al.,* 1977).

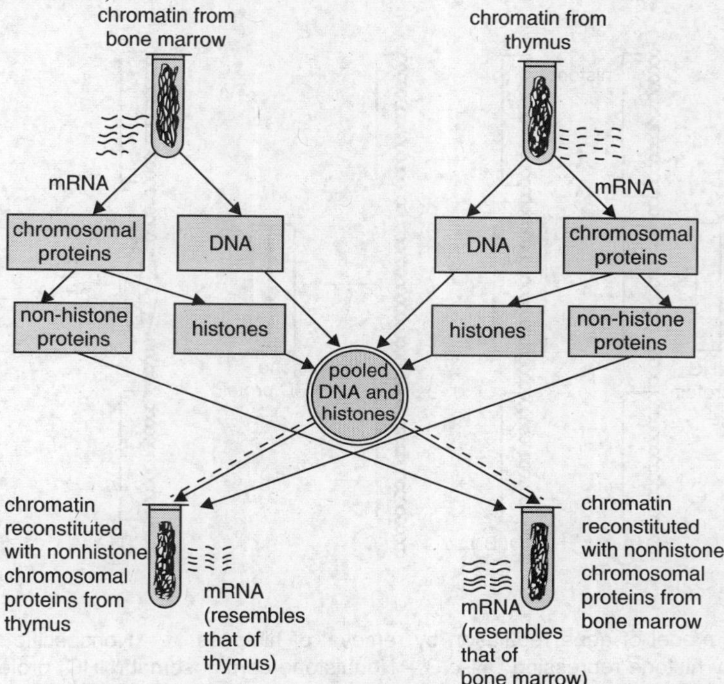

Fig. 15.2. Chromatin reconstitution experiment demonstration in positive role of non-histone proteins in transcription.

(*b*) **Frenster's (1965) model of gene-specific depressor RNA** (Fig. 15.3). Histones stabilise the DNA double helix by interacting with the negative phosphate groups of DNA. This prevents the separation of strands, and consequently transcription. The histones, thus, act as **general repressor** of protein synthesis. Histones are displaced from DNA by nuclear polyions which form complexes with histones. This permits separation of the DNA strands in a random manner. Gene-specific **derepressor RNA** hybridizes with the non-transcribing DNA strand at a particular locus and stabilizes the loop in the open position. This frees the other DNA strand and permits it to transcribe mRNA.

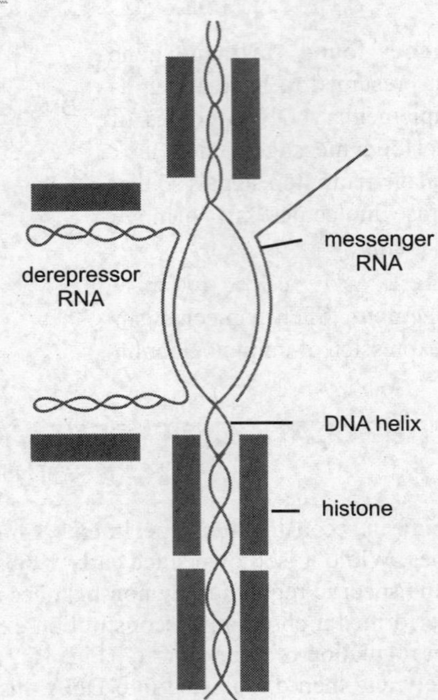

Fig. 15.3. Frenster's model of gene regulation.

(*c*) **Stein's model.** According to **G.S. Stein, J.S. Stein** and **Kleinsmith** (1975), individual NHC (nonhistone chromosomal) proteins recognize certain sites on DNA and bind to it there. They then pulls off the histones repressor from the site, thus, derepressing DNA. Thus, NHC proteins act by derepressing (Fig. 15.4).

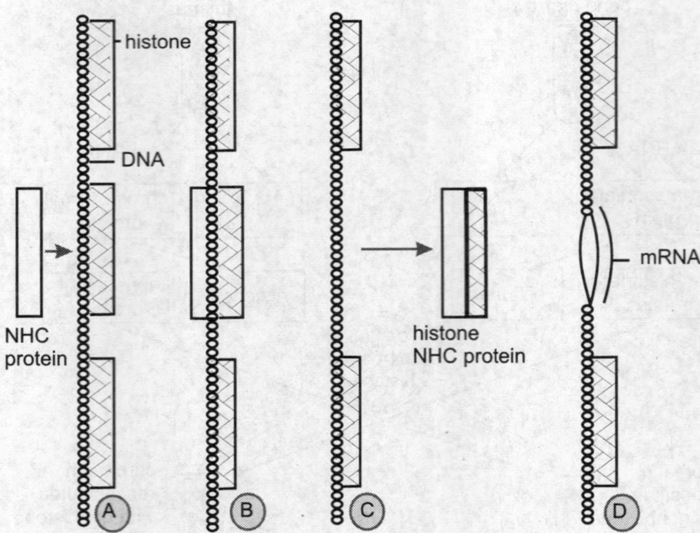

Fig. 15.4. Stein's model of gene regulation by removal of histones. A – Nonspecific masking of DNA by histone repressing DNA; B – Nonhistone chromosomal (NHC) protein binds to specific sites on DNA; C–NHC protein pulls out histone repressor; D – Unmasked DNA is derepressed and transcribes mRNA.

(d) Change in chromatin conformation. Evidently nucleosomes continue to be present on most transcriptionally active DNA sequences, but they are probably reduced in number. Thus, although evidence from some laboratories suggests that the ribosomal genes of the amphibian nucleolus lack nucleosomes, active gene loci in *Drosophila* give a positive reaction to antibodies against H3 and H4, indicating that at least these subunits of the nucleosome persist on such DNA. In fact, in some active genes the nucleosomes are displaced or "phased" in these regions (**Samal** *et. al.,* 1981).

(e) Modification of DNA sequences: DNA methylation. The genomic DNA of higher eukaryotes is modified following replication so that a large proportion of the cystosine (C) residues are present as **5-methylcytosine (5mC)**. However, such methylation has not been detected in the DNA of lower eukaryotes such as yeast and *Dictyostelium*, nor in *Drosophila*. The percentage of methylated C residues in DNA relative to unmethylated C residues is highly variable from less than 1 percent in some insects to over 50 percent in some higher plants and vertebrates. A much greater correlation exists between the methylation or under-methylation of sequence in the vicinity of gene promoters. For example, DNA of sperm is highly methylated, as in the DNA of the oocyte-specific 5S rRNA genes in adult tissues, whereas the sites around the coding regions of genes such as adult globin, ovalbumin, and immunoglobulin are under-methylated in tissues in which they are expressed but are largely methylated in other cells in which they are not expressed.

There is evidence that the cytosine methylation in DNA alters the structure of the double helix in a fundamental way and favours the transistion from B-form to Z-form DNA (**Bele** and **Felsenfed**, 1981). It is possible that B $\leftrightarrow$ Z transition is itself involved in gene regulation and this may be the way in which DNA methylation has its effects on transcription.

(f) Modifications of histones. Histone component of chromatin is subject to three different post-synthetic modifications which have either direct or indirect effect on eukaryotic gene regulation: 1. **Histone methylation** affects only histones H3 and H4 and involves the irreversible methylation of a few lysine residues which alters the hydrophobic nature of the side chain of these histones. 2. **Histone phosphorylation** involves histone H1, phosphorylation affects **serines** and **threonines**, changing them from a state of neutral charge to one of negative charge and is a reversible reaction. The state of phosphorylation of H1 protein varies through the eukaryotic cell cycle, and after H1 phosphorylation, chromatin becomes much more strongly condensed, as it does in mitotic chromosomes. Evidently, activation of the **histone kinase** enzyme that is responsible for H1 phosphorylation may be the first step in the chain of events that leads to eventual chromatin condensation prior to mitotic

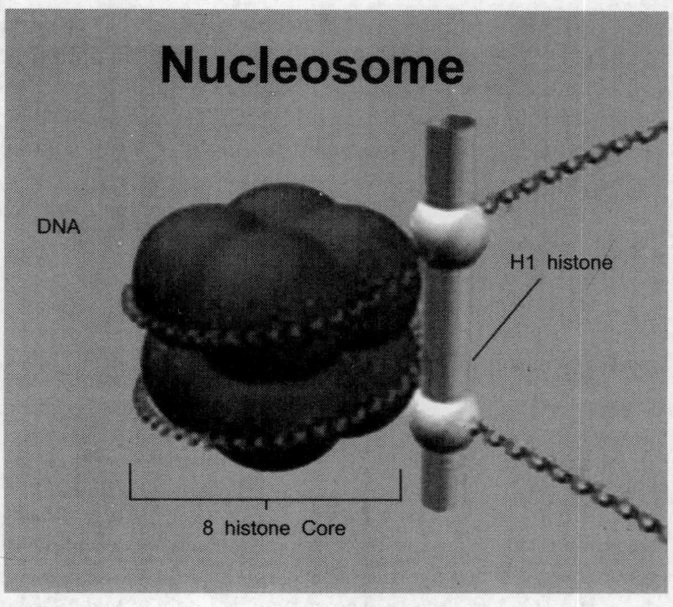

Histone acetylation.

cell division. 3. **Histone acetylation** is of two types. The first is the irreversible acetylation of the amino terminal **serines** of histones H1, H2A and H4. These modifications seem to be associated with histone synthesis. The second is reversible acetylation of **lysine** residues in the amino-terminal regions of histones H2A, H2B, H3 and H4. Acetylation converts the normally basic lysine side chain to a neutral acetyl lysine, and, thus, reduces the net basic charge of the amino-terminal ends of the affected histones. Both H3 and H4 can have up to four lysines in the acetyl

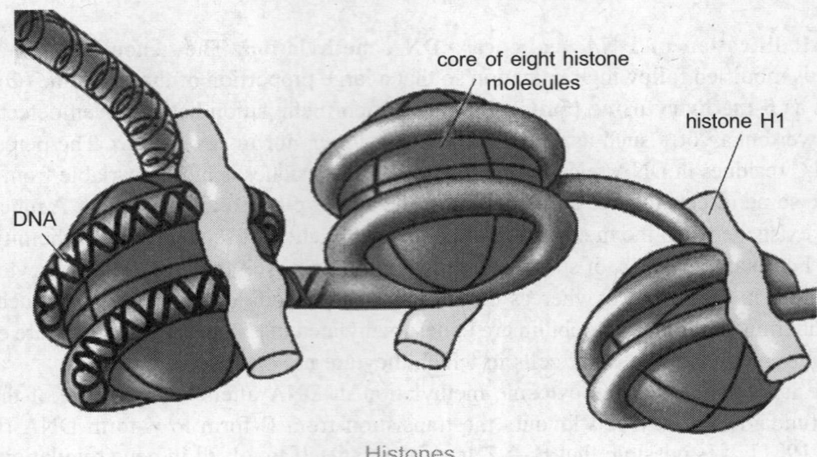

Histones.

form, and there is a strong correlation between this type of histone acetylation, especially **tetracetylation** of all available lysines, and transcriptionally active chromatin. Evidently, the acetylation of the core histone lysines would tend to loosen the nucleosomal structure which brings about the transition from a silent condensed gene to a transcriptionally active and extended one.

(*g*) Transcriptional regulation by protein A24. The **A24** is an unusual hybrid protein, being a complex of histone **H2A** and the non-basic protein **ubiquitin**. The ubiquitin is covalently bound *via* the side chain amino group of lysine 19 of the histone. Some 10 per cent of H2A molecules are in the form of A24 and these specialised histones seem to be confined to interphase chromatin, disappearing as the chromosomes condense. A24 is found highly abundant in the chromatin of active genes.

(*h*) Gene regulatory molecules. Transcription of the eukaryotic genome is believed to be regulated by a variety of specific gene regulatory molecules which are produced by specific regulatory genes or by cytoplasm/cell surface. Examples of such gene regulatory molecules are the following:

(i) RNA polymerases. These enzymes are necessary for transcription and if they are short in supply, they tend to affect it. For example, there is a possible competition for type II polymerase (which is meant for hnRNA and mRNA) by the various promoter sequences that lie upstream of protein coding sequence.

(ii) Endonucleases. These enzymes are likely to affect the transcription, especially *in vitro* cell free system, by introducing nicks into DNA that may serve as initiation sites for some polymerases.

(iii) Topoisomerases, helicases and other DNA helix-destabilizing proteins. Various proteins are known that alter the three-dimensional structure of DNA and render it more available for processing which may affect transcription.

(iv) DNA methylase. The enzyme is likely to make DNA less available for transcription, and factors that antagonise methylation would enhance transcription.

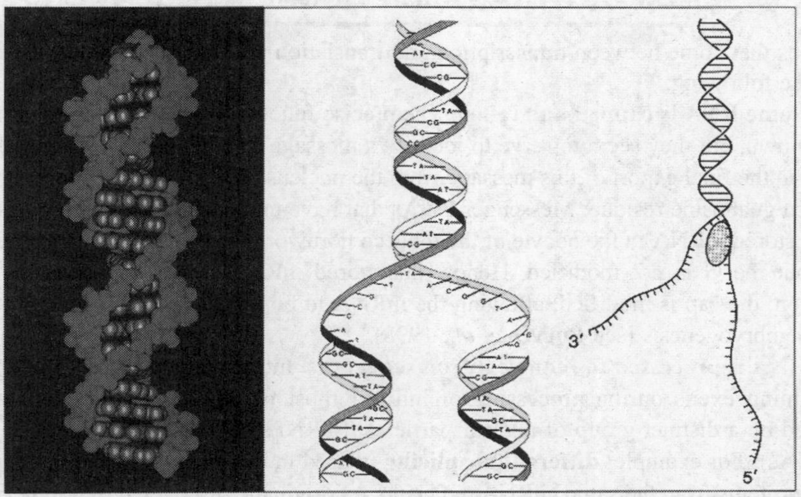

DNA helicases.

(*v*) ***Histone acetylase and deacetylases.*** Such enzymes influence the rate of transcription by modulating acetylation.

(*vi*) ***Factors such as ATP.*** Such molecules may influence the transcription rate by changing the available energy.

(*vii*) ***Ions and small molecules.*** Many ions such as those of **calcium, magnesium** and **manganese** directly affect chromatin conformation, which modulates gene activity.

Britten-Davidson model or gene-battery model of transcription regulation. This model of eukaryotic gene regulation at the level of transcription was proposed by **Britten** and **Davidson** in 1969 and later on elaborated by them in 1973. The gene-battery model (Fig. 15.5) assumes the presence of four classes of sequences: 1. **producer gene** which is comparable to a structural gene of prokaryotic operon; 2. **receptor site** is located adjacent to each producer gene and is comparable to operator gene of prokaryotic operon; 3. **integrator gene** which is comparable to regulator gene and is responsible for synthesis of an **activator RNA** that may or may not give rise to proteins before it activates the receptor site; 4. **sensor site** regulates the activity of integrator gene, which can be transcribed only when the sensor site is activated. The sensor sites are recognised by agents

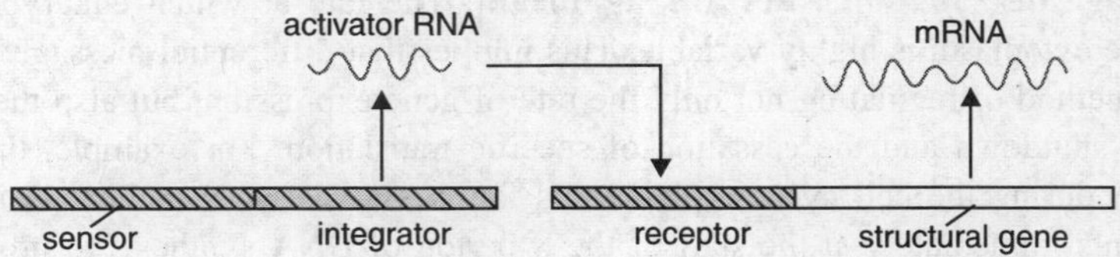

Fig. 15.5. Various components of Britten-Davidson's model for transcription regulation.

which, like hormones and proteins, change the pattern of gene expression. For instance, hormone-protein complex or a transcription factor may bind to a sensor site and cause the transcription of integrator.

In this model, the genes (producer gene and integrator gene) are those sequences which are involved in RNA synthesis, while the receptor and sensor sites help only in recognition without taking part in RNA synthesis. Lastly, in Britten-Davidson's model, a set of structural genes controlled by one sensor site is called the **battery**.

15.3. POST-TRANSCRIPTIONAL REGULATION

Steps that come between transcription and translation are described as **post-transcriptional** and are the following:

1. Some RNA is capped and tailed. The precise functions of capping and tailing of mRNA are not known, but they seem to serve to identify a message or potential message, and tailing may also help in the final export of this message from the nucleus. The mRNA is capped at its 5′ end by a modified guanosine residue. Messenger RNAs that have unmodified caps are not translated. For example, stored mRNA in the oocyte of the tobacco hornworm moth has the guanosine added to its 5′ end, but the G is not modified. Hence, this stored mRNA is not translated. However after fertilization, the cap is modified, allowing the mRNA to be translated to produce proteins needed for early embryogenesis (see **Purves** *et al.,* 1998).

2. RNA is processed to remove intron sequences. Introns removal and **splicing** together of the remaining exons during processing of hnRNA must be absolutely precise. This is in part engineered by a distinct group of nuclear particles (SnRNPs containing U1, U2, U3, U4, U5 and U6 snRNAs). For example, **differential splicing** is used in different lymphocyte cells to produce different proteins from the same hnRNA molecule. As originally discovered by **Early** *et al.,* (1980), the two types of mRNA molecules are produced by part of an intron being omitted from one of the mRNAs but included in the exon splice used to produce the other mRNA. This allows production of two distinct protein both immunoglobulins (Ig), but *one* with a long strand of hydrophobic amino acids at its carboxyl terminus, and the *other* with only a short length of relatively hydrophilic amino acids. The Ig molecule with the long hydrophobic peptide is secreted from the cell. This change in splicing takes place within the life of a single lymphocyte cell and clearly explains the following observation. Immature lymphocytes retain antibody and simply insert the Ig molecules into their plasma membranes, whereas following stimulation with antigen, the same lymphocyte becomes secretory, releasing antibody molecules into circulation.

3. Most RNA is never exported from the nucleus. About 5 per cent of total transcribed RNA never leaves the nucleus. This is explained partly by removal of intron RNA and also by many RNA molecules which break up within the nucleus. The significance of this process is not clear, but some clues about the identification of RNA for export are coming to light. Although not all genes contain introns, most do, and it seems that the presence of some of these introns is essential for RNA export. In other words, introns are used as a means identifying or ticketing the molecules that are to be passed out of the nucleus (see **Maclean** and **Hall**, 1987).

4. Message degradation rates are significant. The rate at which eukaryotic mRNA is degraded in the cytoplasm is highly variable. This implies that differential message breakdown is an important method of regulating not only the rate of gene expression, but also the lag between transcriptional shutdown and the cessation of specific translation. For example, the survival of histone mRNA during the cell cycle explains this fact very clearly. New histone is required in massive amounts immediately at the start of the S period of DNA synthesis to provide the new DNA with nucleosomes. Recently, it is discovered that the restricted availability of histone message is not achieved as a result of transcriptional control alone but by differential breakdown rates for histone message.

Tubulin is a protein that polymerizes to form microtubules, a component of the cytoskeleton. When a large pool of free tubulin is available in the cytoplasm, there is no particular need for the cell to make more of it. Under these conditions, some tubulin molecules bind to tubulin mRNA, and this binding makes such mRNA molecules especially susceptible to breakdown, and less tubulin is made. Other examples illustrate the same mechanism – that the less time an mRNA stays in the cytoplasm, the less it can be translated into protein (see **Purves** *et al.,* 1998).

15.4. TRANSLATIONAL CONTROL

In bacteria, most mRNA molecules are translated about the same number of times with only fairly small variation from gene to gene. In eukaryotes translational regulation occurs in which a mRNA molecule is not translated at all until a signal is received. Translational control may involve the following mechanisms:

1. Extension of lifetime of the mRNA. An important example of translational regulation is that of **informosomes** or **masked mRNA**. Unfertilized eggs are biologically static, but shortly after fertilization many new proteins must be synthesized, for example, the proteins of the mitotic apparatus, the cell membranes, histones for nucleosome formation as well as others. Unfertilized sea urchin eggs store large quantities of mRNA for many months in the form of mRNA-protein particles (=masked mRNA) made during formation of the egg. This mRNA is translationally inactive, but within minutes after fertilization, translation of these molecules begins. Here, the timing of translation is regulated; the mechanism for stabilizing the mRNA, for protecting it against RNases, and for activation are still unknown.

2. Regulation of rate of protein synthesis. This type of regulation also occurs in mature unfertilized eggs. These cells need to maintain themselves but do not have to grow or undergo a change of state. Thus, the rate of protein synthesis in eggs is generally low. This is not due to inadequate supply of mRNA but of a limitation of an as-yet-undefined element, called the **recruitment factor** which apparently interferes with formation of the ribosome-mRNA complex. A good example of translational control is the extension of the lifetime of silk fibroin mRNA in the silkworm *Bombyx mori*. During cocoon formation on the silk gland of the silk worm predominantly synthesises a single type of protein, **silk fibroin**. Since the worm takes several days to construct its cocoon, it is the total amount and not the rate of fibroin synthesis that must be great; the silk worm achieves this by synthesising a fibroin mRNA molecule that is very long lived.

Transcription of the fibroin gene is initiated at a strong promoter by an unknown signal and about 10^4 fibroin mRNA molecules are made in a period of several days (such a synthesis forms an example of transcriptional regulation). A typical eukaryotic mRNA molecule has a lifetime of about three hours before it is degraded. However, the fibroin mRNA survives for several days during which each mRNA is translated repeatedly to yield 10^5 fibroin molecules. Thus, each gene is responsible for the synthesis of 10^9 protein molecules in four days. Altogether the silk gland makes 300μg or 10^{15} molecules of fibroin during this period. If the lifetime of mRNA were extended, either 25 times as many genes would be needed or synthesis of the required fibroin would take about 100 days.

Further, free iron ions (Fe^{2+}) within a mammalian cell are bounded by a storage protein, **ferritin**. When iron is in excess, ferritin synthesis rises dramatically. Yet the amount of ferritin mRNA remains constant. This increase in ferritin synthesis is due to an increased rate of mRNA translation. *When the iron level in the cell is low,* a **translational repressor** protein binds to ferritin mRNA and prevents its translation by blocking its attachment to the ribosome. *When iron level rise,* the excess iron binds to the repressor and alters the three-dimensional structure of the translational repressor, causing it to detach from the mRNA and translation proceeds (**Purves** *et al.,* 1998).

15.5. POST-TRANSLATIONAL MODIFICATION OF PROTEINS TO MAKE THEM ACTIVE ONES

Some proteins are altered after synthesis, usually by partial degradation or trimming, as for example, by the enzymatic removal of the central section of the **proinsulin** molecule to yield the active protein, **insulin**. For their activity, many proteins also depend on being complexed into compound proteins together with other subunits, either the same or different in nature. Such post-translational control mechanisms do play a significant role in determining the activities of differential cells. For example, **haemoglobin** production is highly dependent on the availability of haem to complex with globin protein subunits which may be deficient in cases of iron-dependent anaemia.

15.6. HORMONAL CONTROL OF GENE EXPRESSION

In higher plants and animals, intercellular communication is a very important phenomenon. Signals originating in various glands and /or secretory cells somehow stimulate **target tissue** or **target cells** to undergo dramatic changes in their metabolic patterns. These changes frequently include altered pattern of differentiation that are generally dependent on altered patterns of gene expression. **Peptide hormones** such as insulin, epinephrine, etc., and **steroid hormones** such as estrogen, progesterone, testosterone (in higher animals, *e.g,* mammals) and ecdysone (in insects) control gene expression. In higher animals, hormones are synthesized in various specialized secretory cells (*i.e.,* endocrine cells) and are released into the blood stream. The peptide hormones do not normally enter cells because of their relative large size. Their effects appear to be mediated by receptor proteins located in target-cell membranes and by the intracellular levels of **cyclic AMP (cAMP)** (called **secondary messenger**). The cAMP activates a protein kinase (*e.g.,* A-kinase) which phosphorylates (activates) many specific enzymes. The steroid hormones, on the other hand, are small molecules that readily enter cells through the plasma membrane. Once inside the appropriate target cells, the steroid hormones become tightly bound to **specific receptor proteins** which are

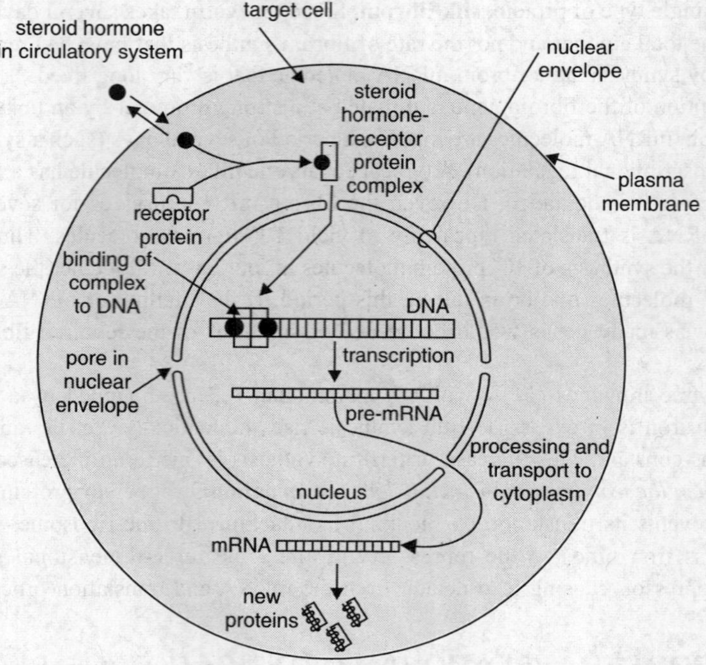

Fig. 15.6. Mode of action of steroid hormones on the gene expression (*i.e.,* transcription) (after Gardner, *et al.,* 1991).

present only in the cytoplasm of target cells. The hormone-receptor protein complexes activate the transcription of specific genes or sets of genes according to following two methods (=hypotheses): 1. The hormone receptor protein complexes interact with specific non-histone chromosomal protein and this interaction stimulates the transcription of the correct genes (**J. Stein, G. Stein** and **L.Kleinsmith**, 1975). 2. The hormone receptor protein complexes activate transcription of target genes by binding to specific DNA sequences present in the *cis*-acting regulatory regions (the enhancers and promoter regions) of the genes (**R.M. Evans**, 1988). In both of these cases, these hormone-receptor protein complexes would function as positive regulators (or "activators") of transcription much like the CAP-cAMP complexes in prokaryotes.

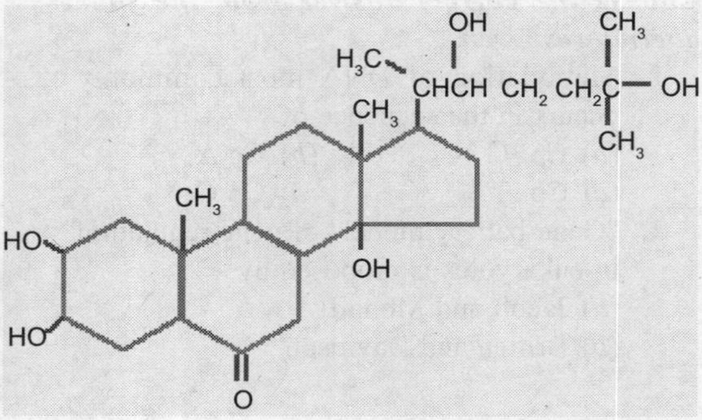

Structure of ecdysone.

During development of dipteran flies (*Drosophila melanogaster* and *Chironomus tentans*) the steroid hormone **ecdysone** is released and triggers moulting. If larvae of these insects are treated with ecdysone at stages of development prior to or between moultings, patterns of chromosome puffing occur that are identical to those occurring during natural moultings. Ecdysone tends to affect the gene expression at the level of transcription.

Questions And Problems

1. Describe the Britten-Davidson's model of regulation of gene activity in eukaryotes.
2. Describe regulation of gene action in eukaryotes at the level of DNA or genetic code.
3. Discuss the relative roles of histones and non-histone proteins in regulation of gene activity in the eukaryotes.
4. In an eukaryotic system, what barriers would an extracellular repressor have to pass through before ultimately binding to DNA?
5. What kinds of mutations would completely eliminate translation of the entire sequence encoding a polypeptide?
6. In what way is the regulation of a gene that is active in a differentiating cell inherently different from regulation of a bacterial gene?
7. Write short notes on the following:
   (i) *lac* operon; (ii) repression;
   (iii) steroid hormones and gene expression; (iv) role of interferon in gene regulation;
   (v) heterochromatin.

Answers To Problems

5. A promoter mutation, which would prevent transcription; or mutation in the start codon, which would prevent initiation.

Multiple Choice Questions

Choose the correct answer from the four alternatives given.

1. Methylation of DNA most commonly occurs in the sequence of
   (*a*) Cp G (*b*) Cp A
   (*c*) Cp T (*d*) Cp C
2. "Gene-battery model" of gene regulation in eukaryotes is proposed by
   (*a*) Jacob and Monod
   (*b*) Britten and Davidson
   (*c*) Beadle and Tatum
   (*d*) Kornberg and Ochoa
3. Fisher and Krebs got Nobel Prize in 1992 for their discovery of
   (*a*) organ and cell transplantation
   (*b*) reversible protein phosphorylation as biological regulatory mechanism
   (*c*) single ion channels
   (*d*) all of the above

Answers

1. (*a*) 2. (*b*) 3. (*b*).

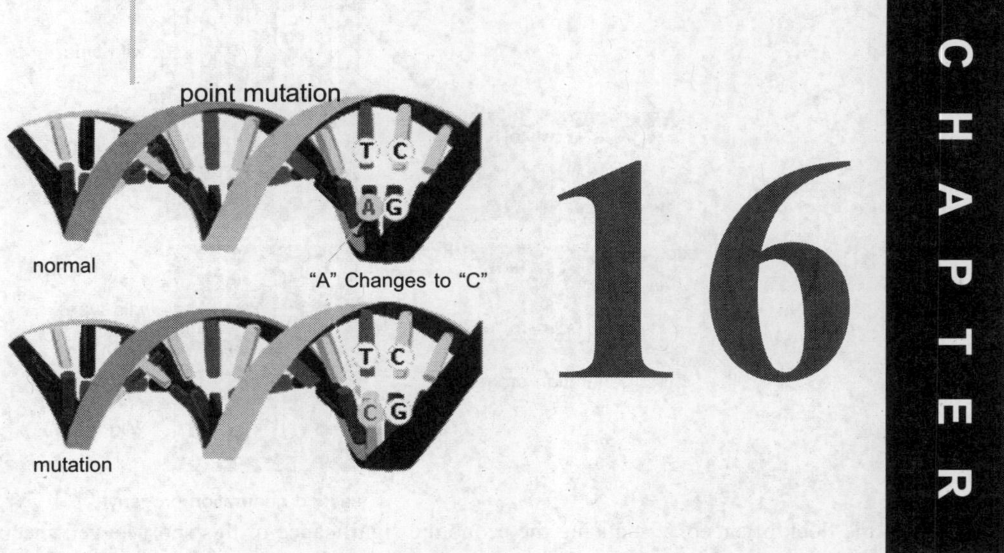

CHAPTER

16

Alteration in the Genetic Material (Point Mutation)

Chapter Contents

16.1. Historical Background
16.2. Occurrence
16.3. Kinds of Mutations
16.4. Mutation Rate
16.5. Method of Detection of Sex-Linked Lethal Mutation
16.6. Practical Application of Mutations
16.7. Significance of Mutation

A gene mutation is abrupt inheritable qualitative or quantitative change in the genetic material of an organism. Since in most organisms, genes are segments of DNA molecule, so a mutation can be regarded as a change in the DNA sequence which is reflected in the change of sequence of corresponding RNA or protein molecules. Such a change may involve only one base/base pair or more than one base pair of DNA. Mutations occur in a **random** manner, *i.e.*, they are not directed according to the requirements of the organism. Most mutations occur **spontaneously** by the environmental effect, however, they can be induced in the laboratory either by radiations, physical factors or chemicals (called **mutagens**). A unicellular organism is more subjected to environmental attacks since it is at the same time a somatic or germ cell. In multicellular organisms, the germ cells are distinct cells, and are relatively protected from the environment. Mutation has a significant role to play in the origin of species or evolution.

16.1. HISTORICAL BACKGROUND

The earliest record of point mutations dates back to 1791, when **Seth Wright** noticed a lamb with exceptionally

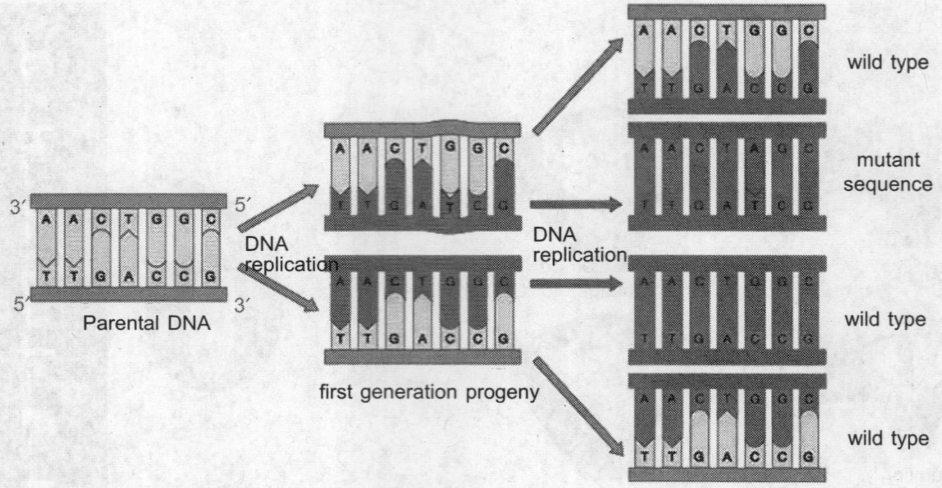

Mutation.

short legs in his flock of sheep. Visualising the economic significance of this short-legged sheep. *i.e.*, short-legged sheep could not cross the low stone fence and damage the crop fields in the vicinity, he produced a flock of sheep, each of which having short legs by employing artificial breeding techniques. The short-legged breed of sheep was known as **Ancon** breed. Later on, the trait of short legs was found to have resulted from a recessive mutation and the short-legged individuals were found to be homozygous recessive.

Hugo de Vries was the first hybridist who used the term "mutation" to describe the heritable phenotypic changes of the evening primrose, *Oenothera lamarckiana*. Many mutations described by de Vries in *O. lamarckiana*, are now known to be due to variation in chromosome number or ploidy and chromosomal aberrations (*viz.* **gross mutations**). The first scientific study of mutation was started in 1910, when **Morgan** started his work on fruitfly, *Drosophila melanogaster* and reported white-eyed male individuals among red-eyed male individuals. The discovery of white-eyed mutants in *Drosophila* was followed by an extensive search of other mutants of *Drosophila* by **Morgan** and his co-workers and other geneticists. Consequently, about 500 mutants of *Drosophila* have been reported by geneticists all over the world. Later on, several cases of mutations have been reported in a variety of microorganisms (*e.g.*, bacteriophages, bacteria (*Escherichia coli*), *Neurospora*, *etc.*, plants (*i.e.*, pea, snapdragon, maize, *etc.*) and animals, (*i.e.*, rodents, fowls, human beings, *etc.*).

16.2. OCCURRENCE

Mutations occur frequently in nature and have been reported in many organisms, *e.g.*, *Drosophila*, mice and other rodents, rats, rabbits, guinea pigs and human beings. In *Drosophila*, mutation causes white and pink eyes, black and yellow body colours, and vestigial wings. In rodents the mutations are responsible for black, white and brown coats. In human beings, the mutations cause variation in hair colour, eye colour, skin pigmentation and several somatic malformations. Various genetical diseases of human beings such as haemophila, colour blindness, phenylketonuria, *etc.*, form other examples of mutation in human beings.

How does a mutation act? Any change in sequence of nucleotides in the DNA will result in the corresponding change in the nucleotide sequence of mRNA. This may result in alignment of different tRNA molecules on mRNA (during protein synthesis). Thus, the amino acid sequence, and, hence, the structure and properties of the enzyme formed will be changed. This defective enzyme or structural protein may adversely affect the trait controlled by the protein. In consequence, a mutant phenotype makes its expression.

16.3. KINDS OF MUTATIONS

There exists a lot of controversy among geneticists about the possible kinds of mutations. Mutations have been classified variously according to different criteria as follows :

1. Classification of Mutation According to Type of Cells

According to their occurrence in somatic and germinal cells, following types of mutations have been classified :

A. Somatic mutations. The mutations occurring in non-reproductive body cells are known as **somatic mutations.** The genetical and evolutionary consequences of somatic mutations are insignificant, since only single cells and their daughter cells are involved. If, however, a somatic mutation occurs early during embryonic life, the mutant cells may constitute a large proportion of body cells and the animal body may be a mosaic of different types of cells. Somatic mutations have often been related with malignant (cancerous) growth. Examples of somatic mutation have been reported in *Oenothera lamarckiana* (by Hugo de Vries) and several other cases including human beings. In human beings, somatic mutation causes several fatal diseases such as *paraoxysmal nocturnal haemoglobinura*, *circumscribed neurofibroma*, *unilateral retinoblastoma* and *heterochromia of the iris.*

B. Gametic mutations. The mutations occurring in gamete cells (*e.g.*, sperms and ova) are called **gametic mutations**. Such mutations are heritable and of great genetical significance. The gametic mutations only form the raw material for the natural selection.

I wide type
ATGACCAGGTC

II base substitution
ATGACTAGGT

III base addition
ATGACACAGGTC

IV base deletion
ATGACAGGTC
missing C

Fig. 16.1. Three types of point mutations. Only the base squence of one DNA strand is shown. Changes are shown in square; the horizontal brackets indicate the affected segment (after Freifelder, 1985).

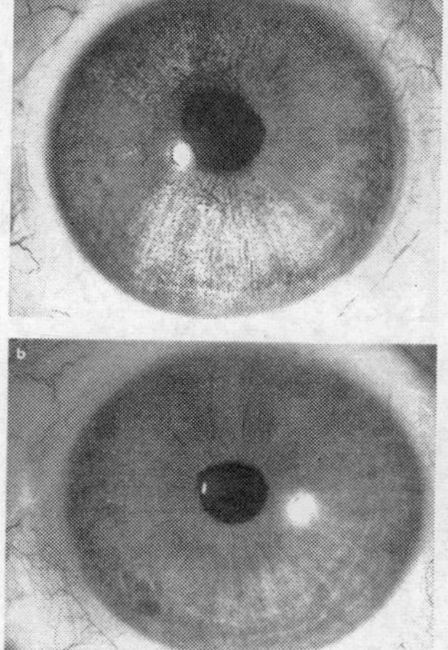
Heterochromia of the iris.

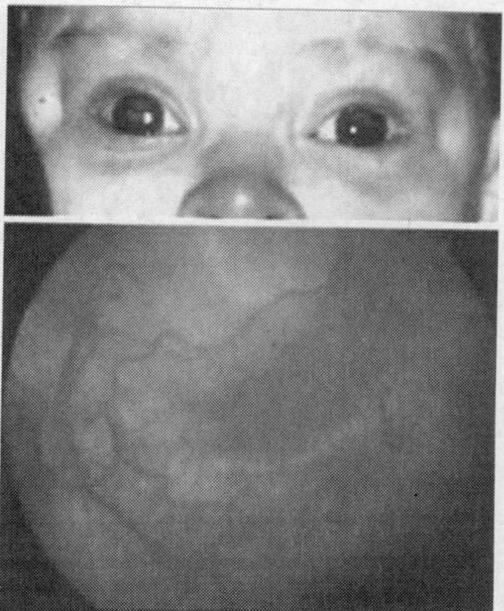
Unilateral retinoblastoma.

SOMATIC MUTATIONS IN HUMANS.

2. Classification of Mutations According to the Size and Quality

According to size, following two types of mutations have been recognised :

A. Point mutation. When heritable alterations occur in a very small segment of DNA molecule, *i.e.*, a single nucleotide or nucleotide pair, then these types of **mutations** are called **"point mutations"**. The point mutations may occur due to following types of subnucleotide change in the DNA and RNA.

1. Deletion mutations. The point mutation which is caused due to loss or deletion of some portion (single nucleotide pair) in a triplet codon of a cistron or gene is called **deletion mutation.** Deletion mutations have been frequently reported in some bacteriophages (Phage T_4).

2. Insertion or addition mutation. The point mutations which occur due to addition of one or more extra nucleotides to a gene or cistron are called **insertion mutations.** The insertion mutations can be artificially induced by certain chemical substances called **mutagens** such as acridine dye and proflavin. A proflavin molecule, it is believed, insert between two successive bases of a DNA strand, thereby stretching the strand lengthwise. At replication, this situation would allow the insertion of an extra nucleotide in the complementary chain at the position occupied by the proflavin molecule.

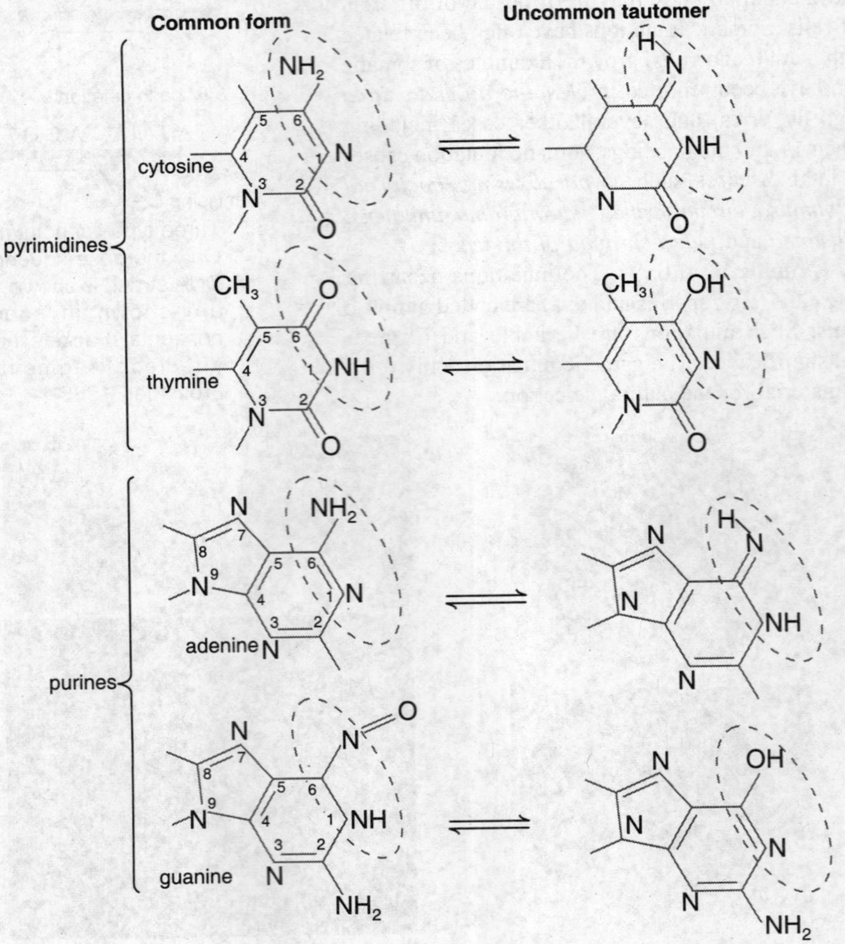

Fig. 16.2. The uncommon form of DNA bases.

The mutations which arise from the insertion or deletion of individual nucleotides and cause the rest of the message downstream of the mutation to be read out of phase, are called **frameshift mutations.** They result in the production of an incorrect, hence, inactive protein, due to which death of the cell may occur.

3. Substitution mutation. A point mutation in which a nucleotide of a triplet is replaced by another nucleotide, is called **substitution mutation.** The substitution mutation affect only a particular triplet codon. Such an altered code word (triplet codon) may designate a different amino acid and may result in the production of a protein with a single amino acid substitution. The substitution mutations alter the phenotype of an organism variously and are of great genetical significance. They may be of following types :

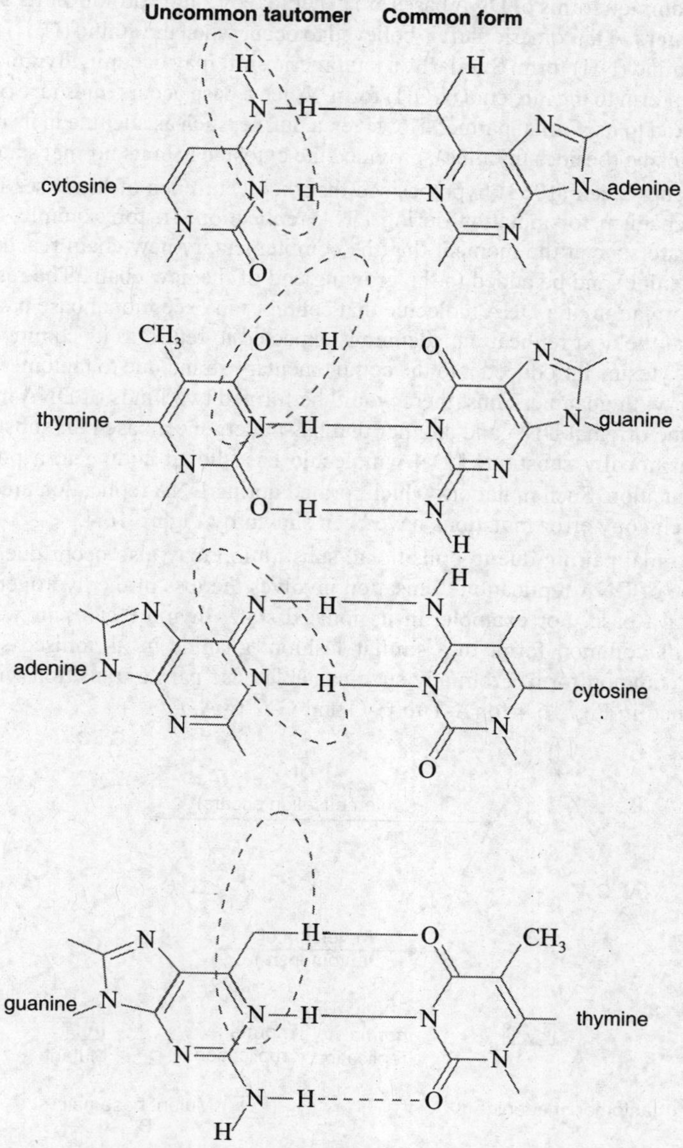

Fig. 16.3. Pairing qualities of rare tautomers of four DNA bases.

(i) **Transition.** When a purine (*e.g.*, adenine) base of a triplet codon of a cistron is substituted by another purine base (*e.g.*, guanine) or a pyrimidine (*e.g.*, thymine) is substituted by another pyrimidine base, (*e.g.*, cytosine) then such kind of substitution is called **transition.** The transitional substitution mutations occur due to tautomerization.

Tautomerization

In a DNA molecule, normally, the purine, adenine (A) is linked to the pyrimidine, thymine (T), by two hydrogen bonds, while the purine guanine (G) is linked to the pyrimidine, cytosine (C) by three hydrogen bonds. Besides the common molecular configurations, each DNA base may have some altered uncommon molecular configuration, as has been shown in the figure 16.2.

Such uncommon forms of DNA bases are generated by single proton shifts and are called **rare states** or **tautomers**. A tautomeric shift is believed to occur when the amino (NH_2) form of adenine is changed to an imino (NH) form. Similarly, a tautomeric shift may occur in thymine changing it form the keto (C = O) form to the rare enol (COH) form. When a base occurs in its rare or tautomeric state, it cannot be linked to its normal partner. However, a purine, such as adenine in its rare state can form a bond with cytosine (besides thymine), provided the cytosine is in its normal state.

Watson and **Crick** (1953) hypothesised that the occurrence of the bases in their rare states provides a mechanism for mutation during DNA replication. If, for example, adenine in an old chain is in its rare state at the moment that the complementary new chain reaches it, cytosine can pair with it (adenine) and be added to the growing end of the new chain. The result of this type of pairing is the formation of a DNA molecule that contains an exceptional base pair. This situation is not stable and at the next replication, adenine is expected to return to its common state and to pair with thymine. Cytosine introduced into the complementary strand due to tautomeric shift in adenine, would then pair with guanine. Thus, there would be formed two kinds of DNA molecules, one that is identical to the original DNA and another that has undergone a base pair substitution of G-C for A-T. This transitionally substituted DNA molecule has altered coding at a point and results in recognisable mutation. Such mutations which formed during DNA replication are called **copy error mutations.** Such copy error mutations have been shown by Figure 16.4.

The abnormal pairing due to transitional substitution may also occur due to **ionisation** of a base at the time of DNA replication. Ionisation involves the loss of the hydrogen from the number one nitrogen of a base. For example, in its ionised state, thymine can pair with guanine, if the guanine is in its common form. In a similar fashion, guanine in its ionised state can pair with thymine in its common form. From any such unstable base pair, a transition will result following the steps outlined in Fig. 16.4 for A-T to G-C and G-C to A-T.

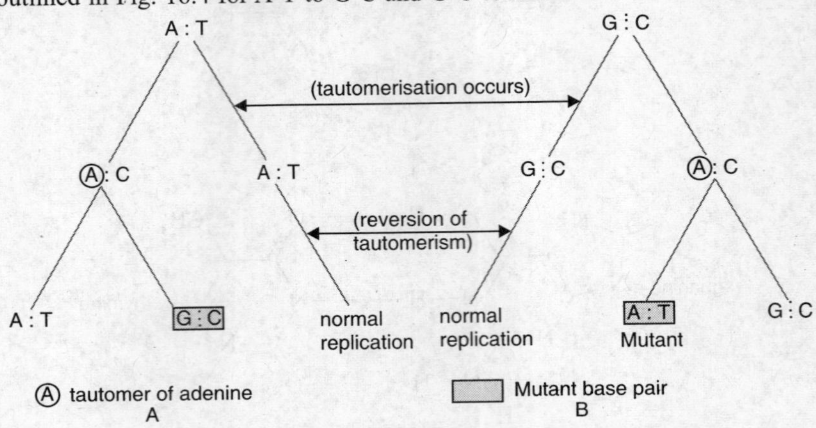

Fig. 16.4. Copy error mutation due to tautomerization of adenine.

Effect of Chemical Mutagens on Nucleotide Sequence

(a) Alteration in Resting Nucleic Acid

1. Deamination. Some chemical substance such as nitrous acid causes transitional mutation due to oxidative deamination of DNA bases. In the process of oxidative deamination, the amino group (NH_2) of a DNA base is replaced by hydroxyl (OH) group by the chemical mutagen. Thus, adenine is deaminated into **hypoxanthine** by nitrous acid as shown in the following figure:

adenine $\xrightarrow{+HNO_2}$ hypoxanthine

By tautomeric shift the hypoxanthine (HX) is converted into more common keto-tautomer which pairs with cytosine. The A:T pair, thus, can be converted to a G : C pair.

Similarly, deamination converts cytosine to uracil, which has pairing properties similar to thymine and in such a case G : C pair would be changed into A : T pair.

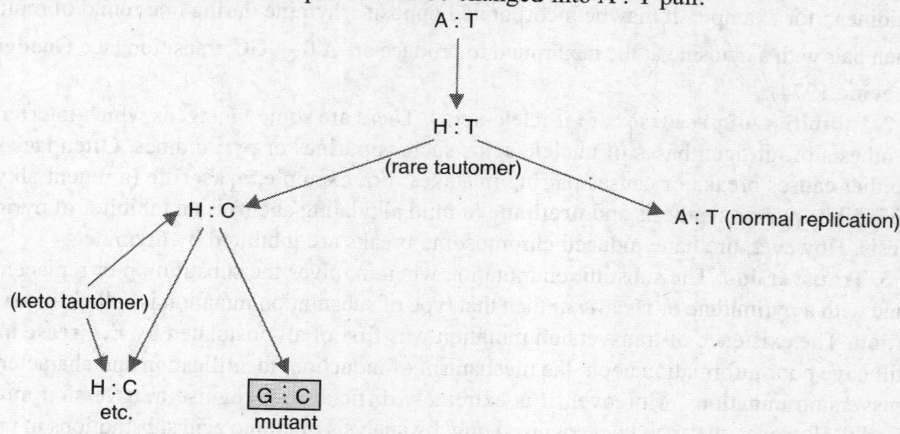

Fig. 16.5. Conversion of A: T pair into G : C pair due to keto-tautomerization of adenine.

2. Hydroxylamine, (HA =NH_2.OH) and hydrazine (HZ =NH_2NH_2). When DNA is treated with hydroxylamine (HA), its cytosine base is the strongest reacting base. Hydroxylamine probably cause hydroxylation of cytosine at amino group giving rise to **hydroxylcytosine,** which then subsequently pair with adenine. Thus, hydroxylamine (HA) induces in DNA a GC $\rightarrow$ AT base pair transition (Fig. 16.6).

The hydrazine affects DNA by breaking of rings of uracil and cytosine giving rise to **pyrazolone** and **3-aminopyrasole**, respectively. The treatment of RNA or DNA with **anhydrous hydrazine** results in the destruction of their pyrimidines.

Fig. 16.6. Conversion of GC pair into AT pair due to conversion of cytosine (C) into hydroxylcytosine (HC).

3. Alkylating agents. Some alkylating agents carry one, two, or more alkyl groups in a reactive form and act as strong mutagens. Examples of some most extensively studied alkylating agents include **diethyl sulphate (DES)**, **dimethylsulphate (DMS), methyl methane sulphonate (MMS), ethyl ethane sulphonate (EES)** and **ethyl methane sulphonate (EMS).** These mutagens produce mutations in the following ways :

(1) They add ethyl or methyl groups to guanine. This makes guanine the base analogue to adenine.

(2) They remove the alkylated guanine. This is known as **depurination.** The loss of the base produces gaps in the DNA chain which may be filled with a wrong base, thus, producing mutation.

(3) The gap may also produce a deletion, causing mutation.

(*b*) Alteration during Replication of Nucleic Acid

1. Base analogues. Certain chemical substances have molecular structure similar to the usual DNA bases that, if they are available, such **analogues** may be incorporated into a replicating DNA strand. For example, **5-bromouracil** (5BU) or its nucleoside **5-bromodeoxyuridine** (5-BUdR) in its usual (keto) form is a structural analogue of thymine (5-methyluracil) and it will substitute for thymine. Thus, an A-T pair becomes and remains A-BU. There is some *in vitro* evidence to indicate the BU immediately adjacent to an adenine in one of DNA strands causes the latter to pair with guanine. But, in its rare (enol) state, 5BU behaves similar to the tautomer of thymine and pairs with guanine. This converts A : T to G : C as shown in Figure 1z6.7.

2-Aminopurine (2-AP) is another base analogue which is a relatively undifferentiated purine that apparently can pair with cytosine and thymine. It is thought that 2-AP acts by "switching" pyrimidines : for example, it may be incorporated opposite thymine during one round of replication and then pair with a cytosine at the next round to produce an $AT \rightarrow GC$ transition (see **Goodenough** and **Levine**,1974).

2. Inhibition of precursors of nucleic acids. There are some mutagens which interfere with the synthesis of nitrogen bases of nucleic acids such as purines or pyrimidines. Often lack of one base either causes breaks or causes pairing mistakes. For example, **azaserine** (a potent alkylating agent) inhibits purine synthesis and **urethane** (a mild alkylating agent) is an inhibitor of pyrimidine synthesis. However, urethane induced chromosome breaks are inhibited by thymine.

3. Transversion. The substitution mutation when involves the substitution or replacement of a purine with a pyrimidine or *vice versa* then that type of substitution mutation is called **transversion mutation.** The existence of transversion mutation was first of all postulated by **E. Freese** in 1959. We still have poor information about the mechanism of induction, identification and characterisation of transversion mutations. Moreover, it is extremely difficult to recognise transversion mutations genetically. However, they can be recognized only by analysis of amino acid substitutions in proteins.

Effects of Physical Conditions on Nucleotide Sequence

High temperature and low pH value are known to affect depurination or loss of purine bases. The removal of a purine from a strand of DNA leaves a gap at that point. At the time of replication, it would be possible for any of the four bases to insert in the complementary newly formed strand. If the inserted nucleotide contained a purine, the complementary strand would contain a transversion.

B. Multiple mutations or gross mutations. When changes involving more than one nucleotide pair, or entire gene, then such mutations are called **gross mutations.** The gross mutations occur due to rearrangements of genes within the genome and may be of the following types :

1. The rearrangement of genes may occur within a gene. Two mutations within the same functional gene can produce different effects depending on gene whether they occur in the *cis* or *trans* position.

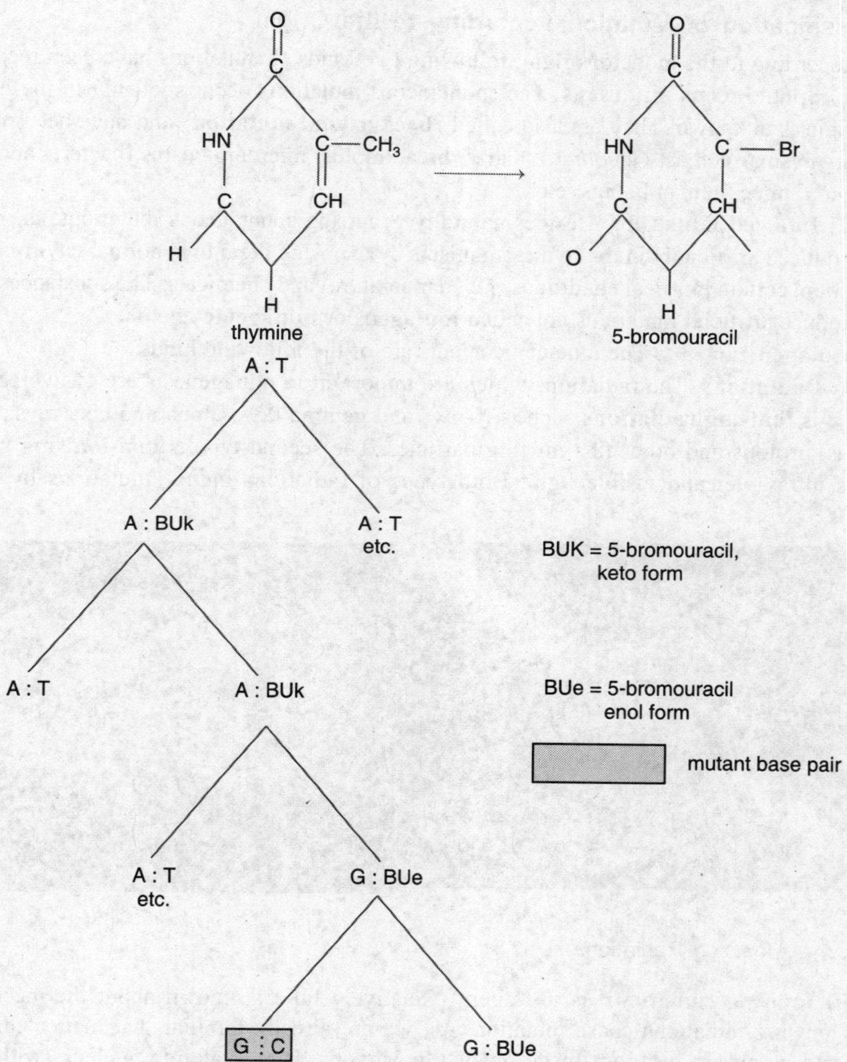

Fig. 16.7. Conversion of A : T base pair into G : C base pair due to keto and enol forms of bromouracil.

2. The rearrangement of gene may occur in number of genes per chromosome. If the number of gene replicas are non-equivalent on the homologous chromosomes, they may cause different types of phenotypic effects over the organisms.

3. Due to movement of a gene locus, new type of phenotypes may be created, especially when the gene is relocated near heterochromatin. The movement of gene loci may take place due to following method :

(i) **Translocation.** Movement of a gene may take place to a non-homologous chromosome and this is known as **translocation.**

(ii) **Inversion.** The movement of a gene within the same chromosome is called **inversion.**

3. Classification of Mutation According to the Origin

According to the mode of origin, following two kinds of mutations have been recognised :

(1) Spontaneous mutations. The spontaneous mutations occur suddenly in the nature and their origin is unknown. They are also called "**background mutation**" and have been reported in many organisms such as, *Oenothera,* maize, bread molds, microorganisms (bacteria and viruses), *Drosophila*, mice, human beings, etc.

(2) Induced mutations. Besides naturally occurring spontaneous mutations, the mutations can be induced artificially in the living organisms by exposing them to abnormal environment such as radiation, certain physical conditions (*i.e.*, temperature) and chemicals. The substances or agents which induce artificial mutations are called **mutagens** or **mutagenic agents.**

Mutagenic agents. The mutagenic agents are of the following kinds:

A. Radiations. The radiations which are important in mutagenesis are of two categories : one type is **ionising radiations** such is X-rays and gamma rays; alpha and beta rays; electrons, neutrons, protons and other fast moving particles. The second type is **non-ionising radiations** such as ultraviolet and visible light. Both types of radiations induce mutations by following methods :

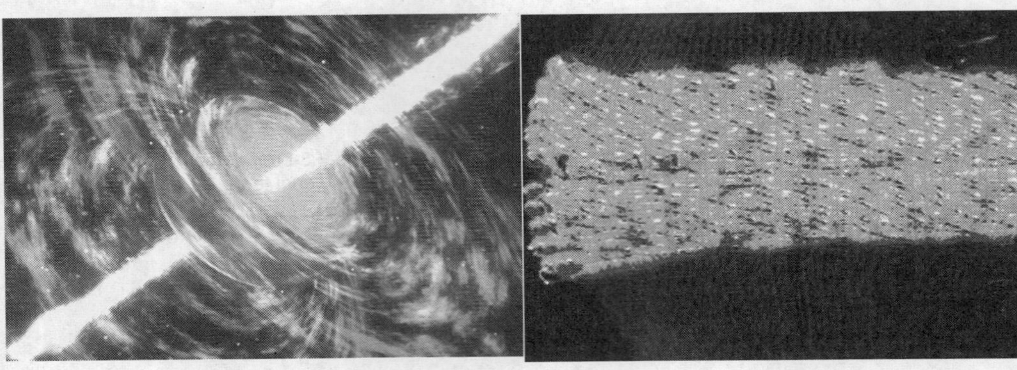

Gamma rays (Ionising Radiations).

Ultraviolet light (Non ionising radiations).

(*i*) Ionising radiations as mutagens. Relatively little is known about the mechanism by which ionising radiations cause mutation. As, we are already familiar that matter composed of atoms and atoms, in turn, are made up of a positively charged atomic nucleus (with neutrons, protons) and a surrounding cons-tellation of negatively charged electrons. The charges of atomic particles remain so balanced that normal atoms are electri-cally neutral. When ionising radiations pass through matter, they dissipate their energy in part through the ejection of electrons from the outer shell of atoms and the loss of these balancing, negati-vely-charged particles (electrons) leaves atoms which are no longer neutral but are positively charged. The positively-charged atom is called **ion.** The ejected electrons move at high speed; knock other electrons free from their respective atoms and when their energy is dissipiated, become attach to other atoms and convert the atoms into negatively charged ions. To achieve their stable configuration (*i.e.*, neutral charge), ions undergo many chemical reactions and during these chemical reactions ionising radiation is thought to cause mutation.

Further, ionizing radiations cause breaks in poly-sugar phosphate backbone of DNA and, thus, causing chromosomal mutations such as break, deletion, addition, inversion and translocation. During breakage of DNA molecule due to ionising radiation, the active role of oxygen is predicted. Because, oxygen is important in the formation of H_2O_2 and H_2O in irradiated water and these products may induce breaks in DNA molecule.

(*ii*) **Non-ionising radiations as mutagens.** The ultraviolet (UV) light is a non-ionizing radiation which may cause mutation. The most effective wavelength of ultraviolet for inducing mutations is about 2,600 A°. This is a wavelength that is best absorbed by DNA and a wavelength at which proteins absorb little energy. When a substance absorbs sufficient energy from the ultraviolet light, some of their electrons are raised to higher energy levels, a state called **excitation.** The excited molecule becomes reactive and mutated and is called **photoproduct.**

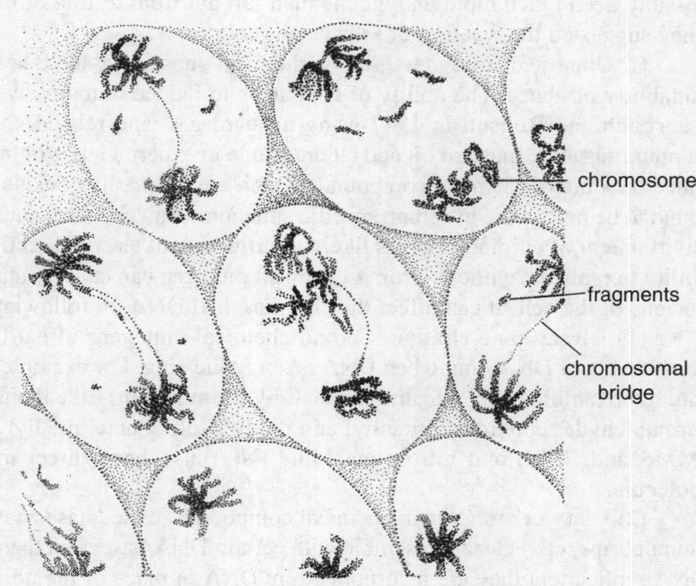

Fig. 16.8. Radiation-induced chromosomal bridges and fragments in cells of X-rayed anthers of *Trillium*.

Dimerization. The ultraviolet radiation produces several effects on DNA, one being the formation of chemical bonds between two adjacent pyrimidine molecules in a polynucleotide and particularly, between adjacent thymine residues as shown in Fig. 16.9. As the two thymine residues associate, or **dimerize** to form a **dimer**, their position in the DNA helix becomes so displaced that they can no longer form hydrogen bonds with the opposing purines and thus regularity of the helix becomes distorted. Thus, dimerization interferes with the proper base pairing of thymine with adenine, and result in thymine's pairing with guanine. This will produce a T-A to C-G transition.

monomers → (thymine, 2800 Å) dimer; dimer → (UV, 2400 Å) monomers

Fig. 16.9. Formation of a dimer of thymine.

B. Temperature as mutagen. The rate of all chemical reactions are influenced by temperature. It is not surprising that temperature can be mutagenic. It is reported that the rate of mutation is increased due to increase in temperature. For example, an increase of 10°C temperature increases the mutation rate two or three-fold. Temperature probably affects both thermal stability of DNA and the rate of reaction of other substances with DNA.

A study of Swedish nudist indicated that the scrotal temperature of human males in ordinary clothing is about 3°C higher than that of nude male. The higher temperature could well increase the mutation rate nearly two-fold, leading the investigators to suggest that the wearing of pants has possibly been much more unhygeinic than fall out from testing of nuclear devices threatens to be. They suggested the wearing of kilts as one solution.

C. Chemical mutagens. Many chemical substances have been responsible to increase the mutability of genes. The ability of chemicals to induce mutation was first of all demonstrated by **Auerbach** and **Robson** in 1947 using mustard gas, and related compounds as the nitrogen and sulphur mustards, mustard oil and chloracetone in experiments with male *Drosophila melanogaster*. Since then many chemical compounds which are ordinarily considered to be non-toxic have been found to be mutagenic in certain specific situations. Any chemical substance that affects the chemical environment of chromosomes is likely to influence, at least indirectly, the stability of DNA and its ability to replicate without error. A chemical mutagen can cause mutation only when it enters in the nucleus of the cell. It can affect the chromosomal DNA by following two ways :

(1) Direct gene change. Certain chemical mutagens affect DNA directly. They affect the constituents of DNA only when DNA is not replicating. For example, nitrous acid converts adenine into hypoxanthine and cytosine to uracil by deamination. Like the nitrous acid, nitrogen mustard, formaldehyde, epoxides, dimethyl and diethyl sulphonate, methyl and ethyl methanesulphonate (MMS and EMS) and nitrosoguanidine (NG) also have direct mutagenic effect on the DNA molecule.

(2) Copy error. Certain chemical compounds, called base analogues (*e.g.*, 5-bromouracil, 2-aminopurine, etc.) closely resemble with certain DNA bases and therefore, act as mutagens. During DNA replication, they are incorporated by DNA in place of the normal DNA bases. Certain other base analogues such as urethane triazine, caffeine (in coffee, tea and soft drinks), phenol and carcinogens, acridines (proflavin, etc.), have mutagenic effects. Certain inorganic substances such as manganese chloride is mutagenic for many organisms, as, they are the compounds which bind calcium and, thus, interfere with the integrity of the chromosome structure.

4. Classification of Mutation According to the Direction

According to their mode of direction, following types of mutations have been recognised :

(A) Forward mutations. In an organism when mutations create a change from wild-type to abnormal phenotype, then that type of mutations are known as **forward mutations.** Most mutations are forward type.

(B) Reverse or back mutations. The forward mutations are often corrected by error correcting mechanism so that an abnormal phenotype changes into wild-type phenotype. They may be of the following types :

***(i)* Single site mutations.** Some reverse mutations change only one nucleotide in the gene and are called **single site mutations.** For example, due to forward mutation the adenine is changed into guanine and backward mutation change guanine into adenine :

$$\text{Adenine} \xrightarrow{\text{forward}} \text{Guanine} \xrightarrow{\text{reverse}} \text{Adenine}$$

***(ii)* Mutation suppressor.** When a mutation occurs at a different site from the site where already primary mutation occurred and that mutated gene reverse the effects of primarily mutated gene, then such (secondary) mutations are called **mutation suppressors.** They may be of following types :

***(a)* Extragenic suppressor.** The extragenic suppressor mutation occurs in a different gene from that of the mutant gene. In *E. coli*, a gene mutation suppressor gene called **rec A** (*rec* for recombination) is known which is necessary for recombination and is found to repair ultraviolet-induced thymine dimers of a gene by a process called **postreplication recombinational repair** (see **Goodenough** and **Levine,** 1974).

***(b)* Intragenic suppressor.** The intragenic suppressor mutation occurs in a different nucleotide within the same gene and shift the reading frame back into register.

(*c*) Photoreactivation. In photoreactivation type reverse mutation, reversal of ultraviolet induced thymine dimers takes place by specific enzymes in the presence of visible light waves. During ultraviolet radiation, a particular enzyme is selectively bound to the bacterial DNA. During photoreactivation the enzyme is activated by visible light and that cleaves the pyrimidine or purine dimers into monomers and restores their original forms.

(*d*) Excision repair or Dark reactivation. In an ultraviolet (UV) induced mutation, the reverse mutation may also occur in the absence of light. According to **Howard Flanders** and **Boyce** (1964), dark reactivation includes following stages: (*i*) An enzyme possibly endonuclease makes a cut in the polynucleotide strand on either side of the dimer which may be formed due to ultraviolet radiation and excises a short, single-strand segment of the DNA. (*ii*) Another enzyme, possibly exonuclease widens the gap produced by the action of the endonuclease. (*iii*) DNA polymerase resynthesises the missing segment, using the remaining opposite strand as a template; and (*iv*) the final gap is closed by some enzymatic rejoining process, (*i.e.*, DNA ligase).

5. Classification of Mutation According to Magnitude of Phenotypic Effect

According to their phenotypic effects following kinds of mutations may occur :

1. Dominant mutations. The mutations which have dominant phenotypic expression are called dominant mutations. For example, in human beings the mutation disease **aniridia** (absence of iris of eyes) occurs due to a dominant mutant gene.

2. Recessive mutations. Most types of mutations are recessive in nature and so they are not expressed phenotypically immediately. The phenotypic effects of mutations of a recessive gene is seen only after one or more generations, when the mutant gene is able to recombine with another similar recessive gene.

3. Isoalleles. Some mutations alter the phenotype of an organism so slightly that they can be detected only by special techniques. Mutant genes that give slightly modified phenotypes are called **isoalleles.** They produce identical phenotypes in homozygous or heterozygous combinations.

4. Lethal mutations. According to their effects on the phenotype, mutations may be classified as lethals, subvitals and supervitals. Lethal mutations result in the death of the cells or organisms in which they occur. **Subvital mutations** reduces the chances of survival of the organism in which they occur. **Supervital mutations,** in contrast, cause the improvement of biological fitness under certain conditions:

6. Classification of Mutation According to Consequent Change in Amino Acid Sequence

1. Missense mutations. They change the meaning of a codon, changing one amino acid into another.

2. Temperature sensitive mutations or T_s mutations. If the substitution produces a protein that is active at one temperature (typically 30°C) and inactive at a higher temperature (usually 40 – 42°C).

3. Nonsense or chain termination mutations. They arise when a codon for an amino acid is mutated into a termination codon (UAG, UAA or UGA), resulting in the production of a shorter protein.

Since, temperature-sensitive and chain termination mutations exhibit the mutant phenotype only under certain conditions, they are called **conditional mutations**; they are the most versatile and useful mutations.

4. Silent mutations. They change a nucleotide but not the amino acid sequence because they affect the *third position* of the codon, which is usually less important in coding. This is a silent mutation because it leaves the protein sequence unchanged.

7. Classification of Mutation According to the Types of Chromosomes

According to the types of chromosomes, the mutations may be of following two kinds:

1. **Autosomal mutations.** This type of mutation occurs in autosomal chromosomes.
2. **Sex chromosomal mutations.** This type of mutation occurs in sex chromosomes.

16.4. MUTATION RATE

The frequency with which genes mutate spontaneously is called **mutation rate.** Most genes are relatively stable and mutation is a rare event. The great majority of genes have mutation rate of 1×10^{-5}, *viz.*, one gamete in 100,000 to one gamete in million would contain a mutation at a given locus. Mutations occur much more frequently in certain regions of the gene than in others. The favoured regions are called **hot spots.** The mutation rate is influenced by various factors which are as follows:

1. Genetic control of mutation rate. There are ample evidences which show that mutation rate is under genetic control, *viz.*, certain genes called **mutator genes** may increase the mutation rate in *Drosophila* (**Demere**, 1937), maize (**Rhoades**, 1938) and *E. coli* (**Goldstein**, 1955). However, certain suppressor genes may decrease the rate of mutation.

In bacteria, as well as in eukaryotes, spontaneous mutations most frequently are caused by transoposons which are segments of DNA that have a tendency to jump around the genome.

2. Viral control of mutation rate. Virus reportedly affect the mutability of host's genes. **Sprague** (1963) experimented with maize suggested that virus may cause mutation. **Baumiliar** (1967) reported that viruses increase the mutation rate in *Drosophila melanogaster*. But, still we do not know how viruses increase the mutability of host genes.

3. Environmental control of mutation rate. There are three major environmental factors that affect mutation rates, *viz*, temperature, certain radiations and chemicals.

16.5. METHOD OF DETECTION OF SEX-LINKED LETHAL MUTATION

H.J. Muller devised an easy method for detecting lethal mutations in the sex chromosomes of *Drosophila*. This is called a CIB method in which a special type of female fly is employed which carries a normal X chromosome and an abnormal X chromosome. The abnormal X chromosome contains an inversion mutation C (which prevents the chromosome to do crossing over with the normal X chromosome, therefore, called **crossover suppressor**), a recessive lethal mutational gene, I and a dominant gene B for bar-eye. In Muller's CIB technique, these CIB female flies are mated with males which are previously treated with some mutagenic agent (such as X-rays) to cause mutation in some of their sperms. The resulting zygotes are of four types and one of these, the CIB male, fail to survive because such embryos contain a recessive lethal which expresses itself when hemizygous. Thus, only one class of male (with CIB X chromosome) male remains to fertilize the F_1 females. Each heterozygous CIB female results from the fertilization of a CIB egg and an irradiated X-bearing sperm and some of these sperms will contain mutated X-chromosomes. Mated F_1 heterozygous CIB females are distributed individually into culture tubes in which each lays fertile eggs and so produces a single F_2 culture. The culture produced by females bearing an induced lethal mutation contain only females; whereas female bearing irradiated X-chromosomes in which no recessive lethal has been induced yield cultures containing some wild-type males. Thus, if in a population of 1000 cultures, 990 contained some males and 10 contained only females, the induced rate of sex-linked, recessive lethal mutations would be 1 per cent.

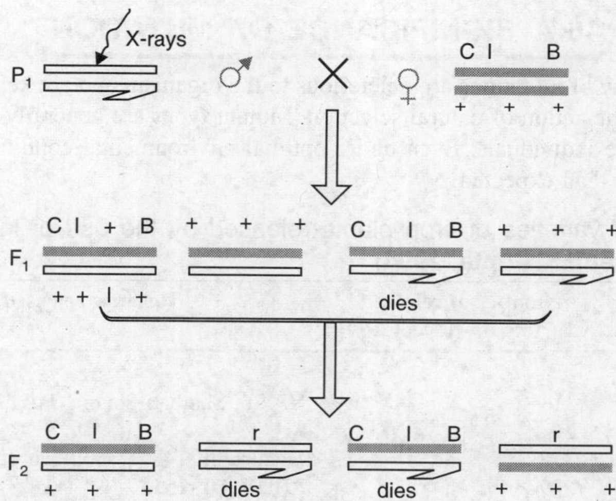

Fig 16.10. Diagram of Muller's CIB technique for detecting sex-linked (X-linked), recessive lethal mutations.

Besides the CIB method, there are many more methods such as **Muller-5 method** (for the detection of sex-linked lethal mutations); **attached X- method** (for the detection of sex-linked visible mutations); **balanced lethal systems** (for the detection of autosomal mutations); and **Stadler's method** and **Singleton's method** (for the detection of specific loci in plants).

16.6. PRACTICAL APPLICATIONS OF MUTATIONS

Mutations are generally deleterious and recessive for the organisms, therefore, majority of them are of no practical value. **A. Gustafsson** has estimated that less than one in 1000 mutants produced may be useful in plant breeding. In India, several useful mutations of various cereals and other crop plants have been developed (see Table 16.1).

M.S. Swaminathan : Father of green revolution of India.

1. Wheat. In bread wheat, many useful mutations have been obtained and utilised in plant breeding, *e.g.*, branched ears, lodging resistance, high protein and lysine content, amber seed colour and awned spikelets. **Dr. M.S. Swaminathan**, one of the most distinguished and legendary figure in the field of cytogenetics and plant breeding in Indian subcontinent, had utilised amber mutation of Mexican wheat variety to develop a new variety of wheat, called **Sharbati Sonora** while working at Indian Agriculture Research Institute (IARI), New Delhi. According to **Dr. N.E. Borlaug** (Nobel Laureate), this variety of wheat paved the way for the Green Revolution in India (see **Gupta**, 1994).

2. Rice. In rice, one of the high yielding varieties **Reimei** was developed through mutations isolated after gamma irradiation. Certain developed mutants of rice are found to contain increased contents of proteins and lysine. In certain other mutant rice, the duration of crop was reduced by as many as 60 days.

3. Barley. In barley, mutations called **erectoides** and **eceriferum** have been induced. These mutants had high yields including several useful characters.

16.7. SIGNIFICANCE OF MUTATION

The vast majority of mutations are deleterious to the organism and are kept at low frequency in the population by the action of natural selection. Mutant types are generally unable to compete equally with wild-type individuals. Even under optimal environmental conditions many mutants appear less frequently than expected.

Table 16.1. List of varieties of crop plants released by the use of induced mutations (Source: **P.K.Gupta,** 1994).

Type of crop	Number of released variety up to year 1991	Varieties released in India
I. Cereals		
1. Bread wheat	113	NP836; Sharbati Sonora, Pusa Lerma; N/I-5643.
2. Durum wheat	25	–
3. Rice	278	Jagannath, IT48, IT 60, Hybrid Mutant 95.
4. Barley	229	RBD-l, DL-253.
II. Legumes	15	Pusa Parvati (French bean).
	136	Hans (pea); Ranjan (lentil); Trombay Vishakhi (pigeon pea); Pant moong 2; TAP -7; MUM-2 (mung bean)
III. Other Crops	223	Pusa Lal Meeruti, S-12 (tomato); Aruna, Sowbhagya (157-B), RC8 (castor bean); MCU-7, Rasmi, Pusa Ageti (cotton); RLM 198, RLM 514 (mustard); New Hybrid Bajra 5 (NH BS); Pusa 46 (pearl millet); Co 997, Co 6608 (sugar cane); MDUI (chili); JRC-7447 (jute).

QUESTIONS AND PROBLEMS

1. What are missense, nonsense, and frameshift mutations and what are the consequences of each ?
2. What do you understand by spontaneous mutations and induced mutations? Discuss variation in mutation rates and frequencies at different loci within an organism.
3. (*a*) What critical evidence is needed to distinguish a gene mutation from a minute deletion, *i.e.*, one too small to be cytologically detectable?
   (*b*) Show how you would determine the rate of mutation from a recessive to a dominant allele in human beings.
   (*c*) Describe what is meant by a tautomeric shift?
   (*d*) Are the mutational consequences of tautomeric shifts base-pair transitions or base pair transversions? Explain, using a specific example.
4. Explain :
   (*i*) Why geneticists find most mutations to be deleterious?
   (*ii*) Why, nevertheless, the mutation process is considered to be the basis of evolutionary progress?
5. Answer each of the following questions as briefly as possible:
   (*a*) Which type of mutation, one induced by a base analog or one induced by proflavin, would you expect to be more deleterious to an organism and why?
   (*b*) What evidence is there that ionisation caused by X-rays need not occur in the gene itself to cause mutation?

(*c*) What are the possible mechanisms by which a gene may change to many different allelic forms?

(*d*) Why are sex-linked lethal mutations easier to detect than autosomal lethals?

6. Three repair mechanisms are known in *E. coli* for the repair of DNA damage (pyrimidine dimer formation) after exposure to ultraviolet light: (1) photoreactivation; (2) excision (dark) repair; and (3) postreplication repair. Compare and contrast these mechanisms, indicating how each achives repair and how the events occurring in each may lead to gene mutations.
7. Why are most mutations in structural genes recessive to their wild-type alleles?
8. Describe different kinds of radiations and chemical mutagens utilised for induction of mutations.
9. Which of the following would be likely to suffer the greatest, and which the least genetic damage from radiation exposure : (*a*) a haploid, (*b*) a diploid, (*c*) a polyploid ?
10. If the mutation rate of a certain gene is directly proportional to the radiation dosage and the mutation rate of *Drosophila* is observed to increase from 3% at 1000 R to 6% at 2000 R. What percentage of mutations would be expected at 3500 R ?
11. The X-linked recessive mutations are more easily studied in appropriate organisms than are autosomal ones. Why ?
12. Discuss the procedure used in the detection of sex-linked lethal mutations by CIB method.
13. Some individuals have a patch of blonde hair in a head of brown hair. What types of mutation would this be ?
14. If a drastic alteration occurred in the structure of one of the genes for 28S rRNA, do you think that the translation of mRNA into protein would cease ? If not, why not ?
15. Explain the difference between a transition and a transversion and give an example of each.
16. What possible explanations can you offer for the reversion of a mutant to the wild-type phenotype?
17. What is the difference between intragenic and intergenic suppression ? Give an illustration of each.
18. How many base pairs would have to be deleted in a mutational event to eliminate a single amino acid from a protein and not change the rest of the protein ?
19. Compare the effects of nitrous acid, hydroxylamine and 5-bromouracil on DNA.
20. The "dotted" gene in maize (Dt) is a "mutator" gene influencing the rate at which the gene for colourless aleurone (*a*) mutates to its dominant allele (A) for coloured aleurone. An average of 7.2 coloured dots (mutations) per kernel was observed when the seed parent was dt/dt, a/a and the pollen parent was Dt/Dt, a/a. An average of 22.2 dots per kernel was observed in the reciprocal cross, How can these results be explained ?
21. Write short notes on the following:

(*a*) Base substitution;	(*b*) Ionising radiation;
(*c*) Balanced lethal;	(*d*) Somatic mutation;
(*e*) Acridine dye;	(*f*) Lethal mutation.

22. What is the difference between substrate and a template transition mutation ?
23. 5-bromouracil, 2-aminopurine, proflavin, ethyl ethane sulphonate and nitrous acid are chemical mutagens. What does each do ?

Answers to Problems

7. Wild-type alleles usually code for complete, functional enzymes or other proteins. One active wild-type allele can often cause enough enzyme to be produced so that normal or nearly normal phenotypes result (dominance). Mutations of normally functioning genes are more likely to destroy the biological activities of proteins. Only in the complete absence of the wild-type gene product would the mutant phenotype be expressed recessiveness.

9. Haploid greatest, polyploid least. Most mutations are recessive, and recessives have only a very low probability of being expressed in polyploids with their multiple sets of chromosomes bearing normal (dominant) alleles. Both dominant and recessive mutations are expressed at once in haploids.
10. 10.5%.
11. Recessive mutations are more easily detected in hemizygous males.
13. Somatic mutation.
14. Translation would not cease since numerous genes for rRNA are present in the genome. The mutation of one of these would probably not interfere with protein synthesis.
16. *Intragenic mutation within the same codon,* either restoring the original amino acid or resulting in the presence of a compatible amino acid; *intragenic mutation within the same cistron,* such as one that restores the normal reading frame; *intergenic direct suppression,* such as alteration in some component directly involved in protein synthesis, for example, tRNA; *intergenic indirect suppression* by an alteration in the cellular milieu.
18. Three, as any other number of deletion (or addition) would cause a frameshift and other amino acid changes.
20. Seed parent contributes two sets of chromosomes to triploid endosperm; one Dt gene gives 7.2 mutations/kernel, two Dt genes increase mutations to 22.2/kernel.
22. In replicating DNA, a transition mutation can occur by tautomerization of a base in the template strand (template transition) or entering the progeny strand (substrate transition).
23. *5-bromouracil* (pyrimidine analogue) and *2-aminopurine* (purine analogue) are incorporated into DNA as thymine and adenine, respectively. However, each undergoes tantomeric shifts more frequently than the normal base. Both cause transitions. *Nitrous acid* also promotes transitions by converting cytosine into uracil, which acts like thymine, and adenine into hypoxanthine, which acts like guanine. *Proflavin* induces insertions and deletions by intercalating and buckling DNA. *Ethyl ethane sulphonate* removes purine rings and thus promotes transitions and transversions.

MULTIPLE CHOICE QUESTIONS

Choose the correct answer from the four alternatives given.

1. Mutation is
   (*a*) change that is inherited
   (*b*) change in parent not inherited
   (*c*) plant growth controlling factor
   (*d*) change which affects the offsprings of F_2 generation
2. Gene mutation is
   (*a*) mutation in the genes of DNA
   (*b*) mutation in phosphodiester linkage
   (*c*) mutation in chromosomes
   (*d*) change in the sequence of nitrogenous bases
3. Mutations which normally happen randomly are considered one of the raw materials for evolution because they
   (*a*) are stable
   (*b*) contribute new variation in organism
   (*c*) cause death of organism
   (*d*) none of these
4. Proflavin and acridine orange induce
   (*a*) transitions
   (*b*) transversions
   (*c*) inversions
   (*d*) frameshift mutations
5. Induction of mutation by X-rays was discovered by
   (*a*) Morgan
   (*b*) Hugu devires
   (*c*) Muller
   (*d*) Luria
6. Who is associated with "Green revolution" in India?
   (*a*) B.P. Pal
   (*b*) M.S. Swaminathan
   (*c*) R.S. Paroda
   (*d*) E.J. Butler

7. Low temperature is mutagenic in
   (*a*) wheat
   (*b*) maize
   (*c*) rice
   (*d*) mustard
8. Frequency of mutation
   (*a*) varies with characters and organisms
   (*b*) can be increased by X-rays
   (*c*) is greatly affected by environmental factors
   (*d*) all of the above
9. Which of the following mutagens can be best used in inducing mutation in microorganisms?
   (*a*) X-rays
   (*b*) β-rays
   (*c*) UV-rays
   (*d*) γ-rays
10. Why are haploids are superior to diploids in study of mutations?
    (*a*) they have shorter life time
    (*b*) smaller number of chromosomes
    (*c*) they allow expression of recessive mutation immediately
    (*d*) obtained in large numbers

ANSWERS

1. (*a*)	2. (*d*)	3. (*b*)	4. (*d*)	5. (*c*)	6. (*b*)
7. (*c*)	8. (*d*)	9. (*c*)	10. (*c*).		

Acridine. A chemical mutagen that intercalates between the buses of a DNA molecule, causing single-base insertions or deletions.

Acentric. A fragment of chromosome without a centromere.

Activation domain. Region of a transcription factor, distinct from the DNA-binding domain, that is responsible for activating transcription.

Activator protein (transcription). Regulatory protein whose binding to DNA leads to an increase in the transcription rate of specific nearby genes.

Activator. A eukaryotic specific transcription factor that binds to an enhancer, often far upstream of a promoter.

Active site. The part of an enzyme where the actual enzymatic function is performed.

Adenine (A). (Gr. *aden*–a gland) A purine (nitrogenous base) occurring in DNA and RNA. It pairs normally with thymine in DNA.

Albinism. Absence of melanin pigment in the iris, skin and hair of an animal; absence of chlorophyll in plants.

Albino. An individual characterized by absence of pigment. Ordinarily applied to animals where skin, hair, and eye are unpigmented and whose skin and eyes are pinkish because of the blood vessels showing through. Also used to describe plants in which a particular pigment (usually the chlorophyll) is absent.

Alkaptonuria. A recessively inherited metabolic disorder in which a defect in the breakdown of tyrosine leads to excretion of homogentisic acid (alkapton) in the urine.

Alkylating agent. An organic compound capable of transferring an alkyl group to other molecules.

Allele. (also **allelomorph; adj., allelic or allelomorphic).** (Gr. allelon- of one another). One member of a group of alternative forms of a gene that can occur at a particular locus on homologus chromosomes.

Allosteric protein. A protein whose shape is changed when it binds a particular molecule. In a new shape, the protein's ability to react to a second molecule is altered.

Allotype. Mutant of a nonvariant part of an immunoglobulin gene that follows the rules of simple Mendelian inheritance.

Allozymes. Forms of an enzyme, controlled by different alleles of the same locus, that differ in electrophoretic mobility.

Alpha satellite. Highly repetitive DNA sequences associated with mammalian centromeres.

Alternative splicing. Various ways of splicing out introns in eukaryotic premessenger RNAs so that one gene produces several different messenger RNA (mRNA) and protein products.

Alu family. A dispersed, intermediately repetitive DNA sequence found in the human genome about 300,000 times. The sequence about 300 bp long. The name Alu comes from the restriction endonuclease that cleaves it.

Ames test. A bacterial test for mutagenicity; also used to screen for potential carcinogen.

Amino acid. A class of organic compounds containing an amino ($-NH_2$) group and a carboxyl ($-COOH$) group, plus a side chain; the basic constructional unit of proteins.

Aminoacidurias. A human disease in which there is an abnormally high excretion in the urine of one or more amino acids.

Aminoacyl-tRNA synthetase. Enzymes that attach amino acids to their proper transfer RNAs (tRNAs).

Amino terminus. The end of a polypeptide chain at which the amino acid bears a free amino group (–NH_2).

Antibiotic – resistant mutant. A cell or organism that carries a mutation conferring resistance to an anitibiotic.

Anibody. A protein produced in response to the presence of some foreign substance in the blood or tissues. It is produced by a B lymphocyte and protects the organism against antigen.

Anticoding strand (also called *antisense strand*). The DNA strand that forms the template for both the transcribed messenger RNA and the coding strand.

Anticodon. A triplet of nucleotides specific to each tRNA, corresponding to a particular amino acid and complementary to the triplet codon to mRNA.

Antigen. (Gr., anti-against+gennam–to produce). A foreign substance, usually protein or protein-polysacchride complex in nature, which elicits the formation of specific antibodies within an organism.

Antimutator mutation. Mutations of DNA polymerase that decrease the overall mutation rate of a cell or of an organism.

Anti-oncogene. A gene that represses malignant growth and whose absence results in malignancy (*e.g.*, retinoblastoma).

Antiparallel strands. Strands, as in DNA that run in opposite directions with respect to their 3′ and 5′ ends.

Antisense RNA. RNA product of *mic* (*m*RNA- *i*nterfering *c*omplementary RNA) genes that regulates another gene by base pairing with, and thus blocking, its messenger RNA.

Anti-sigma factor. A protein that interferes with the action of a sigma factor.

Antiterminator protein. A protein that when bound at its normal attachment sites, lets RNA polymerase read through normal terminator sequences (*e.g.*, the N– and Q– gene products of phage λ).

AP endonucleases. Endonucleases that initiate excision repair at apurinic and apyrimidinic sites on DNA (or An endonuclease that cleaves a DNA strand at any site at which the deoxyribose lacks a base).

Apoptosis. Genetically programmed cell death especially in embryonic development.

Aporepressor. A protein converted into a repressor by binding with a particular molecule.

Archaea. Highly specialized, bacteria like organisms (archaebacteria) that make up the third kingdom of life on earth along with the bacteria (prokaryotes) and the eukaryotes. Identified by **Carl Woese** in 1977 based on ribosomal RNA sequences. Most are thermophilic, halophilic or methanogenic (*i.e.*, found in extreme environments).

A (aminoacyl) site. The site on the ribosome occupied by an aminoacyl -tRNA just prior to peptide bond-formation.

Attenuator region. A control region at the promoter end of repressible amino acid operons that exerts transcriptional control based on the translation of a small leader peptide gene.

Autonomous replicating sequences (ARS). Eukaryotic site of the initiation of DNA replication consisting of an 11-base pair consensus sequence and several other sequences covering 100-200 base pairs.

Autoradiography. A technique in which radioactive molecules make their locations known by exposing photographic plates.

Autoregulation. Regulation of gene expression by the product of gene itself.

Auxotrophs. A mutant microorganism that is unable to synthesize a compound required for its growth but is able to grow if the compound is provided.

Back mutation. The process that causes reversion. A change in a nucleotide pair in a mutant gene that restores the original sequence and hence the original phenotype.

Bacterium. Small, typically one-celled microorganisms characterized by the absence of a formed nucleus. All bacteria form one of the major Kingdom of living organisms.

Bacteriophage. (L., *bacterion*–little rod+Gr., *phagein*–to eat) Virus which infects and may kills bacteria. Also called phage or bacterial virus.

Balbiani rings. The larger polytene chromosomal puffs.

Barr body. A darkly staining body found in the interphase nuclei of female mammals. Consists of the condensed, inactivated X chromosome. It is absent in the nuclei of normal males.

Base. Single-ring (pyrimidine) or double-ring purine, component of a nucleic acid.

Base analog. A chemical so similar to one of the normal bases that it can be incorporated into DNA.

Base pair. A pair of nitrogenous bases, most commonly one purine and one pyrimidine, held together by hydrogen bonds in a double-stranded region of a nucleic acid molecule; commonly abbreviated bp. The normal base pairs in DNA are A-T and G-C.

Base excision repair. The DNA excision repair mechanism that replaces a nucleotide lacking its base with a complete nucleotide. Glycosylases or the environment create the AP (apurinic and apyrimidinic) nucleotides.

Base-flipping. A process whereby enzymes gain access to bases within the DNA double helix by first flipping the bases out of the interior to the outside.

Base-substitution mutation. Incorporation of an incorrect base into a DNA duplex.

β-galactosidase. The enzyme that splits lactose into glucose and galactose; this enzyme is coded by a gene in the *lac operon*.

β-galactoside acetyltransferase. An enzyme that is involved in lactose metabolism and encoded by a gene in *lac* operon.

β-galactoside permease. An enzyme involved in concentrating lactose in the cell; this enzyme is coded by a gene in *lac* operon.

Biochemical pathway. A diagram showing the order in which intermediate molecules are produced in the synthesis or degradation of a metabolite in a cell.

Biochemical genetics. The study of a relationship between genes and enzymes, specifically the role of genes in controlling the steps in biochemical pathways.

Block in a biochemical pathway. Stoppage in a reaction sequence due to a defective or missing enzyme.

Breakage and reunion. The general mode by which recombination occurs. DNA duplexes are broken and reunited in a crosswise fashion according to the Holliday model.

Buoyant density of DNA. A measure of the density or size of DNA determined by the equilibrium point reached by DNA after density gradient centrifugation.

C region (constant region). The region of heavy and light chains of immunoglobulins that does not vary.

CAMP-CRP complex. A regulatory complex consisting of cyclic AMP (cAMP) and the CAP protein; the complex is needed for transcription of certain operons.

Calnexin. A membrane-bound ER protein that forms a protein complex with a newly sythesized glycoprotein and assits in its proper folding.

Calreticulin. A soluble ER protein that forms a protein complex with a newly synthesized glycoprotein and assists in its proper folding.

Cap. A complex structure at the 5′ terminus of most eukaryotic mRNA molecules, having a 5′ - 5′ linkage instead of usual 3′ - 5′ linkage (or a methylated guanosine added to the 5′ end of eukaryotic messenger RNA).

Capsid. The protein shell of a virus.

Capsomere. Protein clusters making up discreate subunits of a viral protein shell (capsid).

Carbon-source mutant. A cell or organism that carries a mutation preventing the use of a particular molecule or class of molecules as a source of carbon.

Carboxyl terminus. The end of a polypeptide chain at which the amino acid has a free carboxyl group (–COOH).

Carcinoma. Tumour arising from epithelial tissue (*e.g.*, glands, breasts, skin, linings of the urogenital and respiratory systems).

Catabolite activator protein (CAP). A protein that, when bound with cyclic AMP can attach to sites on sugar metabolizing operons to enhance transcription of these operons.

Catabolite repression. Repression of certain sugar-metabolizing operons in favour of glucose utilisation when glucose is present in the environment of the cell.

CDNA. See complementary DNA.

Cell cycle. The cycle of cell growth, replication of the genetic material, and nuclear and cytoplasmic division. In eukaryotes, cell cycle is divided into G_1 (gap 1), S (DNA synthesis), G_2 (gap 2) and M (mitosis).

Cell-free system. A mixture of cytoplasmic components from cells, lacking nucleic acids and membranes. Used for in vitro protein synthesis and other purposes.

Central dogma. The concept that genetic information is transferred from the nucleotide sequence in DNA to the nucleotide sequence in an RNA transcript to the amino acid sequence of a polypeptide chain.

Chain elongation. The process of addition of successive amino acids to the growing end of a polypeptide chain.

Chain initiation. The process by which polypeptide synthesis is begun.

Chain termination. The process of ending polypeptide synthesis and releasing the polypeptide from the ribosome; a chain-termination mutation creates a new stop codon, resulting in premature termination of synthesis of the polypeptide chain.

Chargaff's rule. Chargaff's observation that in the base composition of DNA, the quantity of adenine equals the quantity of thymine and the quantity of guanine equals the quantity of cytosine (*i.e.*, equal purine and pyrimidine content).

Charged tRNA. A tRNA molecule to which an amino acid is linked; acylated tRNA.

Chiasmata. X-shaped configurations seen in tetrads during the later stages of prophase I of meiosis. They represent physical crossovers (singular : *Chiasma*).

Chimeras. Individual made up of two or more cell lines in which the cells originated in different zygotes.

Chromatids. The subunits of a chromosome prior to anaphase of meiosis or mitosis. At anaphase of meiosis II or mitosis, when the sister chromatids separate, each chromatid becomes a chromosome.

Chromatin. The aggregate of DNA and histone proteins that makes up a eukaryotic chromosome.

Chromatin assembly factor. Proteins involved in the construction of nucleosomes.

Chromatin remodeling. The change in the structure or positioning of nucleosomes, usually to allow transcription.

Chromatosome. The core nucleosome plus the H1 protein, a unit that includes approximately 168 base pairs of DNA.

Chromomeres. Dark regions of chromatin condensation in eukaryotic chromosomes at meiosis, mitosis, or endomitosis.

Chromosomal painting. A variant of the technique known as fluorescent in situ hybridization. Fluorescent dyes, attached to numerous nucleotide probes, give each human chromosome a different fluorescent signature.

Chromosome. The form of the genetic material in viruses and cells. A circle of DNA in most prokaryotes; a DNA or an RNA molecule in viruses; a linear nucleoprotein complex in eukaryotes.

Chromosome puffs. Diffuse, uncoiled regions in polytene chromosomes where transcription is actively taking place.

Chromosome walking. A technique for studying segments of DNA, larger than can be individually cloned, by using overlapping cloned DNA.

Cis. Meaning "on the near side of", refers to geometric configuration of atoms or mutants on the same chromosome.

Cis configuration. The arrangement of linked genes in a double heterozygote in which both mutations are present in the same chromosome — for example, $a_1\ a_2/ + +$; also called coupling.

Cis dominant. Mutants (*e.g.*, of an operator) that control the functioning of genes on the same piece of DNA.

Cis-trans complementation test. A mating test to determine whether two mutants on opposite chromosomes complement each other; a test for allelism.

Cistron. A DNA sequence specifying a single genetic function as defined by a complementation test; a nucleotide sequence coding for a single polypeptide; synonymous with *gene*.

ClB method. A genetic procedure used to detect X-linked recessive lethal mutations in *Drosophila melanogaster*; so named because one X chromosome in the female parent is marked for with inversion (*C*), a recessive lethal allele (*l*) and the dominant allele for Bar eyes (*B*).

Coding region. The part of a DNA sequence that codes for the amino acids in a protein.

Coding strand. The DNA strand with the same sequence as the transcribed messenger RNA (given U in RNA and T in DNA);

Codon. A sequence of three adjacent nucleotides in an mRNA molecule, specifying either an amino acid or a stop signal in protein synthesis.

Codon preference. The idea that for amino acids with several codons, one or a few are preferred and are used disproportionately. They would correspond with abundant transfer RNAs.

Cohesive end. A single-stranded region at the end of a double-stranded DNA molecule that can adhere to a complementary single-stranded sequence at the other end or in another molecule.

Colchicine. A chemical that prevents formation of the spindle in nuclear division.

Colinearity. The linear correspondence between the order of amino acids in a polypeptide chain and the corresponding sequence of nucleotides in the DNA molecule.

Combinatorial control. Strategy of gene regulation (transcriptional control in eukaryotes) in which a relatively small number of time-and tissue-specific positive and negative regulatory elements are used in various combinations to control the expression of a much larger number of genes.

Complementation. The production of the wild-type phenotype by a cell or an organism that contains two mutant genes. If complementation occurs, the mutants are almost certainly nonallelic.

Complementation group. Cistron (determined by the *cis-trans* complementation test); or a group of mutations that fail to complement one another.

Complementation test. A genetic test to determine whether two mutations are alleles (are present in the same functional gene).

Complementarity. The correspondence of DNA bases in the double helix so that adenine in one strand is opposite thymine in the other strand and cytosine in one strand is opposite guanine in the other. This relationship explains Chargaff's rule.

Conditional mutation. A mutation that results in a mutant phenotype under certain (restrictive) environmental conditions but results in a wildtype phenotype under other (permissive) conditions.

Conformation. Three dimensional shape of a polypeptide or other biological macromolecule.

Consensus sequence. A sequence of the common nucleotides found in many different DNA or RNA samples of homologous regions (*e.g.*, promoters).

Conserved sequence. 1. A base or amino acid sequence that changes very slowly in the course of evolution. 2. A sequence found in many different DNA or RNA samples (*e.g.*, promoters) that is invarient in the sample.

Continuous replication. In DNA, uninterrupted replication in the 5′ to 3′ direction using a 3′ to 5′ template.

Coordinate regulation. Control of synthesis of several proteins by a single regulatory element; in prokaryotes, the proteins are usually translated from a single mRNA molecule.

Core particle. The aggregate of histones and DNA in a nucleosome without the linking DNA.

Co-repressor. A small molecule that binds with an aporepressor to create a functional repressor molecule. 2. The metabolite that when bound to the repressor (of a repressible operon) forms a functional unit that can bind to its operator and block transcription.

Counterselected marker. A mutation used to prevent growth of a donor cell in an Hfr × F^- bacterial mating.

Coupled transcription-translation. In prokaryotes, the translation of an mRNA molecule before its transcription is completed.

CpG islands. Stretches of CG repeats (in which CpG indicates sequential bases on the same strand of DNA, rather than a C-G base pair). These repeats, found in imprinting centers, are important in regulation.

C-value. Amount of DNA in a single (haploid) set of chromosomes.

C-value paradox. Structural and junk DNA create large eukaryotic genomes and large differences in DNA content between eukaryotic species.

Cyclic AMP. A form of AMP (adenosine monophosphate) used frequently as a second messenger in eukaryotic hormone nets and in catabolic repression in prokaryotes.

Cyclin. 1. Family of proteins involved in cell cycle control. 2. One of a group of proteins that participates in controlling the cell cycle. Different types of cyclins interact with the p34 kinase subunit and regulate G_1/S and G_2/M transitions. The proteins are called cyclins because their abundance rises and falls rhythmically in the cell cycle.

Cyclin-dependent kinase (CDK). Family of kinases (phosphorylating enzymes) that, when combined with cyclin, are active in controlling check points in the cell cycle.

Cytosine (C). A nitrogenous pyrimidine base found in DNA and RNA.

Daughter strand. A newly synthesized DNA or chromosome strand.

Degenerate code. A code in which several code words have the same meaning. The genetic code is degenerate because many different codons may specify the same amino acid.

Deletion. Loss of a segment of the genetic material from a chromosome; also called deficiency.

Denatured. Loss of natural configuration (of a molecule) through heat or other treatment. Denatured DNA is single stranded.

Density-gradient centrifugation. A method of separating molecular entities by their differential sedimentation in a centrifugal gradient. The five-carbon sugar present in DNA.

Deoxyribose. The five carbon sugar present in DNA.

Depurination. Removal of purine bases from DNA.

Derepressed. The condition of an operon that is transcribing because repressor control has been lifted.

Dideoxyribose. A deoxyribose sugar that lacks the 3′ hydroxyl group, when incorporated into a polynucleotide chain. It blocks further chain elongation.

Dimerization. The chemical union of two similar molecules.

Diploid. 1. Having each chromosome in two copies per nucleus or cell. 2. A cell or organism with two complete sets of homologous chromosomes.

Directed mutation. A form of mutation that appears to respond to the needs of the cell but may, in fact, be due to hypermutable state of the cell under duress.

Discontinuous replication. In DNA, the replication in short 5′ to 3′ segments, using the 5′ to 3′ strand as a template while going backward away from the replication fork.

D-loop. Configuration found during DNA replication of chloroplast and mitochondrial chromosome where in the origin of replication is different on the two strands. The first structure formed is a displacement loop or D-loop.

DNA. Deoxyribonucleic acid, the macromolecule, usually composed of two polynucleotide chains in a double helix, that is carrier of the genetic information in all cells and many viruses.

DNA gyrase. A topoisomerase that relieves supercoiling in DNA by creating a transient break in the double helix.

DNA ligase. An enzyme that closes nicks or discontinuities in one strand or double-stranded DNA by creating an easter bond between adjacent 3′-OH and 5′-PO_4 ends on the same strand.

DNA methylase. An enzyme that adds methyl groups (–CH_3) to certain bases, particularly cytosine.

DNA polymerase. Any enzyme that catalyzes the synthesis of DNA from deoxynucleoside 5′-triphosphates, using a template strand.

DNA repair. Any of several different processes for restoration of the correct base sequence of a DNA molecule into which incorrect bases have been incorporated whose bases have been chemically modified.

DNA replication. The semiconservative copying of a DNA molecule.

Docking protein. Responsible for attaching (docking) a ribosome to a membrane by interacting with a signal particle attached to a ribosome destined to be membrane bound.

Double helix. The normal structural configuration of DNA consisting of two helices rotating about the same axis.

Duplex DNA. A double-stranded DNA molecule.

Duplication. A chromosome aberration in which a chromosome segment is present more than once in the haploid genome: if the two segments are adjacent, the duplication is a tendem duplication.

Downstream. A convention on DNA related to the position and direction of transcription by RNA polymerase (5′ → 3′). Downstream (in the 3′ direction) is in direction of transcription, whereas upstream (in the 5′ direction) is in the direction from which the polymerase has come.

Downstream promoter element (DPE). A consensus sequence at about + 28 to + 34 of RNA polymerase II promoters that have initiator elements but not TATA boxes.

Editing function. The activity of DNA polymerases that removes incorrectly incorporated nucleotides; also called the proofreading function.

Electrophoresis. A technique used to separate molecules on the basis of their differential rates of movement in response to an applied electric field, typically through a gel.

Elongation complex. The form of RNA polymerase II that actively carries out basal transcription.

Elongation factors (EF-Ts, EF-Tu, EF-G). Proteins necesary for the proper elongation and translocation processes during translation at the ribosome in prokaryotes. Replaced by eEFIα and eEFIβγ in eukaryotes.

Embryoid. A small mass of dividing cells formed from haploid cells in anthers that can give rise to a mature haploid plant.

Embryonic stem cell. Cells in the blastocyst that give rise to the body of the embryo.

Endomitosis. Chromosomal replication without nuclear or cellular division that results in cells with many copies of the same chromosome, such as in the salivary glands of *Drosophila.*

Endonucleases. Enzymes that make nicks internally in the backbone of a polynucleotide. They hydrolyze internal phosphodiester bonds.

Enhancer. An eukaryotic DNA sequence that increases, transcription of a gene by binding specific transcription factors.

Enzyme. A protein that catalyzes a specific biochemical reaction and is not itself altered in the process.

Epigenetic effect. An environmentally induced change in the genetic material that does not cause a change in base pairs.

E (exit) site. Site on the ribosome that depleted transfer RNAs pass through during ejection.

Euchromatin. A region of a chromosome that has normal staining properties and undergoes the normal cycle of condensation; relatively uncoiled in the interphase nucleus (compared with heterochromatin), it apparently contains most of the genes. Presumably the actively transcribing DNA of the chromosomes.

Eukaryote. A cell with a true nucleus (DNA enclosed in a membranous envelope) in which cell division takes place by mitosis or meiosis; an organism composed of eukaryotic cells.

Excision repair. A process whereby cells remove part of a damaged DNA strand and replace it through DNA synthesis, using the undamaged strand as a template.

Exon. 1. The sequence in a gene that are retained in the messenger RNA after the *introns* are removed from the primary transcript. 2. In a gene that has intervening sequences (introns), a region that is actually exported from the nucleus to be expressed or become part of a transfer RNA or ribosomal RNA.

Exon shuffling. The hypothesis put forward by Walter Gilbert that exons code for the functional units of a protein, and that the evolution of new genes proceeds by recombination or the exclusion of exons.

Exonucleases. An enzyme that removes a terminal nucleotide in a polynucleotide chain by cleavage of the terminal phosphodiester bond; nucleotides are removed successively, one by one; usually specific for either DNA or RNA and for either single stranded or double-stranded nucleic acids.

Exportin. Nuclear receptor protein that binds to the nuclear export signal of proteins in the nucleus and then transports the bound protein out through the nuclear pore coplex and into the cytosol.

Feedback inhibition. A post-translational control mechanism in which the end product of a biochemical pathway inhibits the activity of the first enzyme of the same pathway.

Fluctuation test. An experiment by **Luria** and **Delbrick** that compared the variance in number of mutations among small cultures with that among subsamples of a large culture to determine the mechanism of inherited change in bacteria.

Footprinting. A technique to determine the length of nucleic acid in contact with a protein. While in contact, the free DNA is digested. The remaining DNA is then isolated and characterized.

Forward mutation. A change from a wildtype allele to a mutant allele.

Founder effect. Genetic drift observed in a population founded by a small, nonrepresentative sample of a larger population.

F-pili. Sex pili. Hairlike projection on an F^+ or Hfr bacterium involved in anchorage during conjugation.

Frameshift mutation. A mutational event caused by the insertion of deletion of one or more nucleotide pairs in a gene, resulting in a shift in the reading frame of all codons following the mutational site.

Free radical. A highly reactive molecule produced when ionizing radiation interacts with water; free radicals are potent oxidizing agents.

Functional alleles. Mutations that fail to complement each other in a *cis-trans* complementation test.

Gain-of-function mutation. Mutation in which a gene is over-expressed or inappropriately expressed.

Gamete. A mature reproductive cell, having a haploid chromosomal complement, such as a sperm or egg in animals.

Gene. 1. Inherited determinant of the phenotype. 2. The hereditary unit defined experimentally by the complementation test. At the molecular level, a region of DNA containing genetic information, usually transcribed into an RNA molecule that is processed and either functions directly or is translated into a polypeptide chain; a gene can mutate to various forms called alleles.

Gene amplification. Mechanism for creating extra copies of individual genes by selectively replicating specific DNA sequences.

Gene expression. The multistep process by which a gene is regulated and its product synthesized.

Gene family. A group of genes that has arisen by duplication of an ancestral gene. The genes in the family may or may not have diverged.

Gene flow. The movement of genes from one population to another by interbreeding between in the two population.

Gene pool. All of the alleles available among the reproductive members of a population from which gametes can be drawn.

Gene product. A term used for the polypeptide chain translated from an mRNA molecule transcribed from a gene; if the RNA is not translated (for example, ribosomal RNA), the RNA molecule is the product.

Genetic code. 1. The linear sequences of nucleotides that specify the amino acids during the process of translation at the ribosome. 2. The set of 64 triplets of bases (codons) that correspond to the 20 amino acids in protein and the signals for initiation and termination of polypeptide synthesis.

Genotype. 1. The genes that an organism possess. 2. The genetic constitution of an organism or virus, typically with respect to one or a few genes of interest, as distinguished from its appearance, or phenotype.

Goldstein-Hogness box. See TATA box.

Group I introns. Self-splicing introns that require a guanine-containing nucleotide for splicing; the intron is released in a linear form.

Group II introns. Self-splicing introns that do not require an external nucleotide for splicing; the intron is released in a lariat (rope) form.

G-tetraplex. A structure of four guanines that can base pair to form a planar structure that may be involved in novel structure, at the end of eukaryotic chromosome.

Guanine (G). A nitrogenous purine base found in DNA and RNA.

Guide RNA (gRNA). 1. RNA that guides the insertion of uridines (RNA editing) into messenger RNAs in trypanosomes. Found in transcripts from minicircles and maxicircles of DNA in kinetoplasts. 2. The RNA template present in telomerase.

Hammerhead ribozyme. A catalytic RNA, shaped like a hammerhead, capable of splitting other RNA molecules with appropriate complementary sequences.

Helicase. A protein that unwinds DNA, usually at replicating Y-junctions.

Helix-turn-helix motif. Configuration found in DNA-binding proteins consisting of a recognition helix and a stabilizing helix, separated by a short turn.

Heterochromatin. Chromatin that remains tightly coiled (and darkly stained) throughout the cell cycle. It is commonly present adjacent to the centromere and in the telomere, some chromosomes are composed primarily of heterochromatin.

Heterogeneous nuclear mRNA (hnRNA). The original RNA transcripts found in eukaryotic nuclei before post-transcriptional modifications.

Histone. Any of the small arginine and lysine-rich basic protein bounds to DNA in chromatin; the five major histones are designated H1, H2A, H2B, H3 and H4. Each nucleosome core particle contains two molecules each of H2A, H2B, H3 and H4. The H1 histone forms connecting links between nucleosome core particles.

Histone acetyl transferases (HATs). Protein that remodel chromatin by acetylating histones.

Holliday junction. A junction point between two cross-linked DNA double helices. It is an intermediate stage in DNA recombination.

Holoenzyme. The complete enzyme including all subunits. Often used in reference to RNA and DNA polymerases.

Homeo box. A consensus sequence of about 180 base pairs discovered in homeotic genes in *Drosophila.* Also found in other developmentally important genes from yeast to human beings. 2. A DNA sequence motif found in the coding region of many regulatory genes; the amino acid sequence corresponding to the homeo box has a helix-loop-helix structure.

Homoetic (HOX) gene. 1. Any of the group of genes in which a mutation results in the replacement of one body structure by another body structure. 2. Gene that controls the developmental fate of a cell type : mutations of the homeotic gene cause one cell type to follow the developmental pathway of another cell type.

Hot spot. A site in DNA molecule at which the mutation rate is much higher than the rate for most other sites.

Housekeeping gene. A gene that is expressed at the same level in virtually all cells and whose product participates in basic metabolic processes.

Hydrogen bond. A weak noncovalent linkage between two negatively charged atoms in which a hydrogen atom is shared.

Imprinting. A process of DNA modification in gametogenesis that affects gene expression in the zygote; a probable mechanism is the methylation of certain bases in the DNA.

Imprinting center (IC). A region responsible for the control of imprinting. The imprinting mark is almost certainly DNA methylation which is able to turn off gene transcription.

Inborn error of metabolism. A genetically determined biochemical disorder, usually in the form of an enzyme defect that produces a metabolic block.

Induced mutation. A mutation formed under the influence of a chemical mutagen or radiation.

Inducer. A small molecule that inactivates a repressor, usually by binding to it and thereby altering the ability of the repressor to bind to an operator.

Inducible system. A system (a coordinated group of enzymes involved in a catabolic pathway) is inducible if the metabolite it works upon causes transcription of the genes centrolling these enzymes. These systems are primarily prokaryotic operons.

Inducible trancription. Transcription of a gene, or a group of genes, only in the presence of an inducer molecule.

Induction. Activation of an inducible gene; *prophage induction* is the derepression of a prophage that initiates a lytic cycle of phage development.

Initiation. The beginning of protein synthesis.

Initiation codon. The messenger RNA sequence AUG, which specifies methionine, the first amino acid used in the translation process.

Initiation complex. The complex formed for initiation of translation. It consists of the 30S ribosomal subunit, messenger RNA, N- formyl-methionine transfer RNA, and three initiation factors.

Initiator element (inr). A CT-rich area found in RNA polymerase II promoters without TATA boxes.

Initiator proteins. Proteins that recognize the origin of replication on a replicon and take part in primosome construction.

Inosine (I). One of the member of unusual bases found in transfer RNA.

Insertion mutagenesis. Change in gene action due to an insertion event that either changes a gene directly or disrupts control mechanisms.

Insertion sequences (IS). 1. Small, simple transposons. 2. A DNA sequence capable of transposition in a prokaryotic genome; such sequences usually code for their own transposase.

Intercalation. Insertion of a planer molecule between the stacked bases in duplex DNA.

Intercalary heterochromatin. Heterochromatin other than centromeric heterochromatin, dispersed throughout eukaryotic chromosomes.

Intergenic suppression. A mutation at a second locus that apparently restores the wild-type phenotype to a mutant at a first locus.

Internal ribosomal entry site. Sequence in eukaryotic messenger RNA, that allows ribosomes to initiate translation at a point other than the 5′ cap.

Interrupted-mating technique. 1. A mapping technique in which bacterial conjugation is disrupted after specified time intervals. 2. In an Hfr × F cross, a technique by which donor and recipient cells are broken apart at specific times, allowing only a particular length of DNA to be transferred.

Intervening sequence (intron). DNA sequences within a gene that are transcribed but removed prior to translation. Introns are found primarily in eukaryotic cells.

Intra-allelic complementation. The restoration of activity in an enzyme made of subunits in a heterozygote of two mutants that, when homozygous, are not active. Caused by the interaction of the subunits in the protein.

Intragenic suppression. A second change within a mutant gene that results in an apparent restoration of the original phenotype.

Inverted repeat sequence. A nucleotide sequence read in opposite orientations on the same double helix, often found at the ends of transposable elements.

Ionizing radiation. Electromagnetic or particular radiation that produces ion pairs when dissipative its energy in matter.

In vitro. Biological or chemical work done in the test tube (literally, "in glass") rather than the living systems.

Isozymes. Different electrophoretic forms of the same enzyme. Unlike alloenzymes, isozymes are due to differing subunit configurations rather than allelic differences.

Kilobase pairs (kb). Unit of length of a duplex DNA molecule; equal to 1000 base pairs.

Lac operon. 1. The inducible operon, including three loci involved in the uptake and breakdown of lactose. 2. Group of adjoining bacterial genes that code for enzymes involved in lactose metabolism and whose transcription is selectively inhibited by the lac repressor.

Lactose permease. A enzyme responsible for transport of lactose from the environment into the bacteria.

Lagging strand. 1. Strand of DNA being replicated discontinuously 2. The DNA strand whose complement is synthesized in short fragments that are ultimately joined together.

Lampbrush chromosomes. Chromosomes of amphibian oocytes having loops that are suggestive of a lampbrush.

Leader. The length of messenger RNA from the 5′ end to the initiation codon.

Leader peptide gene. A small gene within the attenuator control region of a repressible amino acid operon. Translation of the gene tests the concenteration of amino acids in the cell.

Leader transcript. The messenger RNA transcribed by the attenuator region of a repressible amino acid operon. The transcript is capable of several alternative stem-loop structures, depending on the translation of a short leader peptide gene.

Leading strand. Strand of DNA being replicated continuously.

Leucine zipper. Configuration of a DNA binding protein in which leucine residues on two helices interdigitate, in zipper fashion to stabilize the protein.

Linker DNA. A small segment of DNA that contains a restriction site. It can be added to blunt-ended DNA to give that DNA a particular restriction site for cloning.

Locus. The site or position of a particular gene on a chromosome.

Long interspersed elements (LINEs). Sequences of DNA, up to seven thousand base pairs in length, interspersed in eukaryotic chromosomes in many copies.

Loss-of-function mutation. A mutation that eliminates gene function; also called *null mutation.*

Major groove. In B-form DNA, the larger of two continuous indentations running along the outside of the double helix.

Masked mRNA. Messenger RNA that cannot be translated until specific regulatory substances are available; present in eukaryotic cells, particularly eggs; storage mRNA.

Messenger RNA (mRNA). 1. A complementary copy of a gene that is translated into a polypeptide at the ribosome. 2. An RNA molecule transcribed from a DNA sequence and translated into the amino acid sequence of a polypeptide.

Metabolic pathway. A set of chemical reactions that take place in a definite order to convert a particular starting molecule into one or more specific products.

mic RNA (=mRNA interfering complementary RNA) or anti-sense RNA. By the help of mic RNA cells exercise control of gene expression at the level of translation. Thus, mic RNA can be used to stop translation of mRNA in cells at will. For example in nematode *Caenorhabditis elegans,* expression of gene *lin4* (regulating larval development) is controlled by another gene *lin14* coding for two transcripts (*viz.,* mic RNAs having 22 and 61 nucleotides, respectively).

Microsatellite DNA. Repeats of very short sequences of DNA, such as CACACACA, dispersed throughout the eukaryotic genome. The loci can be studied by polymerase chain reaction (PCR) amplification.

Minimal medium. A culture medium for microorganisms that contains the minimal necessities for growth of the wild-type; its composition is precisely known. Minimal medium contrasts with *complex medium* or *broth*, which is an extract of biological material (vegetables, milk, meat, that contains a large number of compounds and the precise composition of which is unknown.

Minor groove. In B-form DNA, the smaller of two continuous indentations running along the outside of the double helix.

Mismatch repair. Removal of one nucleotide from a pair that cannot properly, hydrogen-bond, followed by replacement with a nucleotide that can hydrogen-bond.

Missense mutation. Mutations that change a codon for one amino acid into a codon for a different amino acid.

Mitochondrion. The DNA-containing eukaryotic cellular organelle in which Krebs cycle and electron transport reactions take place.

Mitosis. The process of nuclear division in which the replicated chromosomes divide and the daughter nuclei have the same chromosome number and genetic composition as the parent nucleus.

Mixed families. Groups of four codons sharing their first two bases and coding for more than one amino acid.

Molecular chaperone. A protein that aids in the folding of a second protein. The chaperone prevents proteins from forming structures that would be inactive.

Molecular evolutionary clock. A measurement of evolutionary time in nucleotide substitutions per year.

Molecular imprinting. The phenomenon in which there is differential expression of a gene depending on whether it was maternally or paternally inherited.

Molecular mimicry. The situation in which one type of molecule resembles another type in order to function. For example, the prokaryotic ribosomal release factors, RF1 and RF2, mimic the structure of a transfer RNA.

Monocistronic. Usually referrring to a messenger RNA that carries the information for only one gene (cistron).

Mutability. The ability to change.

Mutagen. An agent that is capable of increasing the rate of mutation.

Mutant. Any heritable biological entity that differs from wild type, such as mutant DNA molecule, mutant allele, mutant gene, mutant chromosome, mutant cell, mutant organism, or mutant heritable phenotype; also a cell or organism in which a mutant allele is expressed.

Mutant screen. A type of genetic experiment in which the geneticists seek to isolate multiple new mutations that affect a particular trait.

Mutation. A heritable alteration in a gene or chromosome; also, the process by which such an alteratioin happens.

Mutation rate. The proportion of mutations per cell division in bacteria or single-celled organisms or the proportion of mutations per gamate in higher organisms.

Mutator mutations. Mutations of DNA polymerase that increase the overall mutation rate of a cell or of an organism.

Muton. A term Benzer coined for the smallest mutable site within a cistron.

Negative regulation. Regulation of gene expression in which mRNA is not synthesized until a repressor is removed from the DNA of the gene.

N-end rule. The life span of a protein is determined by its amino-terminal (N-terminal) amino acid.

Nick. A single-strand break in a DNA molecule.

Nitrous acid. HNO_2, a chemical mutagen.

Nonhistone proteins. The proteins remaining in chromatin after the histones are removed.

Nonsense codon. One of the messenger RNA sequences (UAA, UAG, UGA) that signals the termination of translation.

Nonsense mutation. A mutation that changes a codon specifying an amino acid into a stop codon, resulting in premature polypeptide chain termination; also called a chain termination codon.

Nuclease. An enzyme that breaks phosphodiester bonds in nucliec acid molecules.

Nuclease-hypersensitive site. A region of a eukaryotic chromosome that is specifically vulnerable to nuclease attack because it is not wrapped as nucleosomes.

Nucleic acid. A polymer composed of repeating units of phosphate-linked five-carbon sugars to which nitrogenous bases are attached.

Nucleolar organizer. The chromosomal region around which the nucleolus forms; site of tandem repeats of the major ribosomal RNA gene.

Nucleolus (pl. nucleoli) The globular, nuclear organelle formed at the nucleolar organizer. Site of ribosome construction.

Nucleoprotein. The substance of eukaryotic chromosomes consisting of proteins and nucleic acids.

Nucleoside. A sugar-base compound that is a nucleotide precursor. Nucleotides are nucleoside phosphates.

Nucleosomes. 1. Arrangements of DNA and histones forming regular spherical structures in eukaryotic chromatin. 2. The basic repeating subunits of chromatin each consisting of a *core particle* composed of two molecules each of four different histones around which a length of DNA containing about 145 nucleotide pairs is wound, joined to an adjacent core particle by about 55 nucleotide pairs of *linker DNA* associated with a fifth type of histone.

Nucleotide. Subunits that polymerize into nucleic acids (DNA or RNA). Each nucleotide consists of a nitrogenous base, a sugar, and one or more phosphate groups.

Nucleotide excision repair. The DNA excision repair mechanism responsible for repairing thymine dimers and other lesions. Enzymes excise a short segment of one of the DNA strands and then repair and ligate the DNA.

Nutritional mutation. A mutation in a metabolic pathway that creates a need for a substance to be present in growth medium or that eliminates the ability to utilise a substance present in the growth medium.

Okazaki fragments. Any of the short segment strand of a newly replicated DNA produced during a discontinuous replication of the lagging strand; also called a precursor fragment.

Oligonuclotide primer. A short, single-stranded nucleic acid synthesized for use in DNA sequencing or as a primer in the polymerase chain reaction.

Oncogene. 1. A gene that can initiate tumor formation 2. Gene capable of transforming a cell. Oncogenes are found in the active state in retroviruses and transformed cells and in the inactive state in nontransformed cells, in which they are called proto-oncogenes.

One-gene-one enzyme hypothesis. Hypothesis of Beadle-Tatum that states that one gene controls the production of one enzyme. Later modified to the concept that one cistron controls the production of one polypeptide.

Open reading frames (ORFs). 1. Sequence of codons between the initiation and termination codons in a gene 2. In the coding strand of DNA or in mRNA, a region containing a series of codons uninterrupted by stop codons and therefore capable of coding for a polypeptide chain.

Operator. A regulatory region in DNA that interacts with a specific repressor protein in controlling the transcription of adjacent structural genes. When the operator is complexed with the repressor, transcription is prevented.

Operon. A collection of adjacent structural genes regulated by an operator and a repressor.

Origin-recognition complex (ORC). A complex of six proteins that bind to the eukaryotic autonomously replicating sequences (ARS). Needed for the initiation of DNA replication in concert with other proteins.

Palindrome. A sequence of words, phrases or nucleotides that reads the same regardless of the direction from which one starts; the sites of recognitions of type II restriction endonucleases.

Parent strand. In DNA replication, the strand that served as the template in a newly formed duplex.

Pedigree. 1. A diagram representing the familial relationships among relatives. 2. A representation of the ancestry of the individual or family. 3. A family tree.

Peptide bond. A covalent bond between the amino group ($-NH_2$) of one amino acid and the carboxyl group ($-COOH$) of another.

Peptidyl transferase. The enzymatic center responsible for peptide bond formation during translation at the ribosome.

Permissive temperature. A temperature at which temperature sensitive mutants are normal.

Permissive condition. An environmental condition in which the phenotype of a conditional mutation is not expressed; contrasts with the nonpermissive or restrictive condition.

Phenocopy. A phenotype that is not genetically controlled but looks like a genetically controlled phenotype.

Phenotype. 1. The observable attributes of an organism. 2. The observable properties of a cell or an organism, which result from the interaction of the genotype and the environment.

Phenylalanine hydroxylase. The enzyme, efficient in phenylketonuria, that converts phenylalanine into tyrosine.

Phenyl ketonuria (PKU). A hereditary human condition resulting from the inability to convert phenylalanine into tyrosine; causes severe mental retardation unless treated in infancy and childhood via a low-phenylalanine diet.

Phosphodiester bond. A diester bond linking two nucleotides together (between phosphoric acid and sugars) to form the nucleotide polymers DNA and RNA.

Photoreactivation. The process whereby dimerized pyrimidines (usually thymines) in DNA are restored by an enzyme (deoxyribodipyrimidine photolyase) that requires light energy.

Pilli (fimbraie). Hairlike projections on the surface of bacteria; Latin for "hair".

Plasmid. 1. A autonomous, self-replicating genetic particle of bacteria usually of double stranded DNA. 2. An extrachromosomal genetic element that replicates independently of the host chromosome. Some plasmids, such as the F factor, may become integrated into the host chromosome.

Plastid. A chloroplast prior of the development of chlorophyll.

Pleiotropic effect. Any phenotypic effect that is a secondary manifestation of a mutant gene.

Pleiotropy. The phenomenon whereby a single mutant affects several apparently unrelated aspects of the phenotype.

Point mutation. Small mutation that consists of a replacement, addition, or deletion of one or a few bases.

Polarity. Meaning "directionality" and referring either to an effect seen in only one direction from a point of origin or to the fact that linear entities (such as single strand of DNA) have ends that differ from each other (*i.e.*, 5′ – to 3′ orientation of a strand of nucleic acid).

Poly-A tail. A sequence of adenosine nucleotides added to the 3′ end of many eukaryotic messenger RNAs.

Polycistronic mRNA. Prokaryotic messenger RNAs that contain several genes within the same messenger RNA transcript.

Polymerase cycling. The process by which a DNA polymerase III enzyme completes an Okazaki fragment, release it and begins synthesis of the next Okazaki fragment.

Polymerized. Formed into a complex compound by linking together smaller elements.

Polymorphic gene. A gene for which there is more than one relatively common allele in a population.

Polynucleotide chain. A polymer of covalently linked nucleotides.

Polynucleotide phosphorylase. An enzyme that can polymerize diphosphate nucleotides without the need for a primer. The function of this enzyme *in vitro* is probably in its reverse role as an RNA exonuclease.

Polypeptide chain. A polymer of amino acids linked together by peptide bonds.

Polysome. The configuration of several ribosomes simultaneously translating the same messenger RNA. Shortened form of a term *polyribosome*.

Polytene chromosome. 1. Large chromosome, seen, for example, in *Drosophila* salivary glands, consisting of many chromatids formed by rounds of endomitosis. Synapsis of homologous chromosomes occurs during the process. 2. A giant chromosome consisting of many identical strands laterally opposed and in register, exhibiting a characteristic pattern of transverse banding.

Positive regulation. Mechanism of gene regulation in which an element must be bound to DNA in an active form to allow transcription. Positive regulation contrasts with negative regulation, in which a regulatory element must be removed from DNA.

Posttranscriptional modifications. Changes in eukaryotic messenger RNA made after transcription has been completed. These changes include additions of caps and tails and removal of introns.

Precursor fragment. See Okazaki fragment.

Preemptor stem. A configuration of leader transcript messenger RNA that does not terminate transcription in the attenuator-controlled amino acid operons.

Pre-initiation complex (PIC). The form of the RNA polymerase II enzyme with general transcription factors bound equivalent to the *E.coli* holoenzyme. Phosphorylation of the enzyme then allows transcription to begin.

Pribnow box. Consensus sequence of TATAAT in prokaryotic prometers centred at the position –10.

Primary transcript. An RNA copy of a gene; in eukaryotes, the transcripts must be processed to form a translatable mRNA.

Primase. An enzyme that creates a messenger RNA primer for Okazaki fragment initiation.

Primer. 1. In nucleic acids, a short RNA or single-stranded DNA segment that functions as a growing point in polymerization. 2. In DNA replication, a length of double stranded DNA that continue, as a single stranded template in the 3′ to 5′ direction.

Primosome. A complex of two proteins, a primase and helicase, that initiates RNA primers on the lagging DNA strand during DNA replication.

Prion. Infectious agent responsible for several neurological diseases (scrapie, kuru, Creutzfeld-Jakob syndrome, mad-cow disease). It is a protein that lacks DNA or RNA.

Probe. In recombinant DNA work, a radioactive nucleic acid complementary to a region being searched for in a restriction digest or genomic library.

Processivity. The ability of an enzyme to repetitively continue its catalytic function without dissociating from its substrate.

Product molecule. The end result of a biochemical reaction or a metabolic pathway.

Prokaryote. An organism that lacks a nucleus.

Promoter. A DNA sequence at which RNA polymerase binds and initiates transcription.

Prophage. A temperate phage replicating with the host that can later initiate the lytic cycle.

Protein. Macromolecule that consists of one or more polypeptides folded into a conformation specified by the linear sequence of amino acids. Proteins play important roles as enzymes, structural proteins, motility proteins and regulatory proteins.

Protein splicing. Removal of amino acid sequence called inteins from a polypeptide chain accompanied by splicing together of the remaining polypeptide segments, called exteins.

Proteosome. A barrel shaped cellular organelle for protein breakdown involving the ubiquitin pathway.

Pseudoalleles. Genes that are functionally but not structurally allelic. Within gene families, pseudoalleles are alleles that are not expressed.

P (peptidyl) site. The site on the ribosome occupied by the peptidyl-tRNA just before peptide bond formation.

P transposable element. A *Drosophila* transposable element used for the induction of mutations, germ-line transformations and other types of genetic engineering.

Purines. Nitrogenous bases of which guanine and adenine are found in DNA and RNA.

Pyrimidines. Nitrogenous bases of which thymine is found in DNA, uracil in RNA and cytosine in both.

Pyrimidine dimer. Two adjacent pyrimidine bases, typically a pair of thymines, in the same polynucleotide strand, between which chemical bonds have formed; the most common lesion formed in DNA by exposure to ultraviolet light.

Quaternary structure. The association of polypeptide subunits to form the final structure of a protein.

Reading frame. The phase in which successive triplets of nucleotides in mRNA form codons; depending on the reading frame, a particular nucleotide in an mRNA could be in the first, second or third position of a codon. The reading frame actually used is defined by the UAG codon that is selected for chain initiation.

Read-through. Transcription or translation beyond the normal termination signals in DNA or RNA, respectively.

Recessive. An allele (phenotype) that does not express itself in the heterozygous condition.

Recombination. The nonparental arrangement of alleles in progeny that can result from either independent assortment or crossing over.

Recon. A term Benzer coined for the smallest recombinant unit within a cistron.

Recruitment. The process in which a transcriptional activator protein interacts with one or more components of the transcription complex and attracts it to the promoter.

Redundancy. The feature of the genetic code in which an amino acid corresponds to more than one codon; also called degeneracy.

Regulatory gene. A gene primarily involved in control of the production of another gene's product.

Reinitiation. The initiation of translation by a ribosome that has just completed translation of a region of the messenger RNA upstream of the current point initiation.

Relaxed mutant. A mutant that does not exhibit the stringent response under amino acid starvation.

Release factors (RF1 and RF2). Proteins in prokaryotes responsible for termination of translatioin and release of the newly synthesized polypeptide when a nonsense codon appears in the A site of the ribosome.

Repetitive DNA. DNA made up of copies of the same nucleotide sequence.

Replication fork. In a replicating DNA molecule, the region in which nucleotides are added to growing strands.

Replication-coupling assembly factor. A protein complex in first flies that assembles new nucleosomes.

Replication origin. The base sequence at which DNA synthesis begins.

Replicons. A replicating genetic unit including a length of DNA and its site for the initiation of replication.

Replisome. The DNA-replicating structure at the Y-junction, consisting of two DNA polymerase III enzymes and a primosome (primase and DNA helicase).

Repressible system. A coordinated group of enzymes involved in a synthetic pathway (anabolic) is repressible if excess quantities of the end product of the pathway lead to the termination of transcription of the genes for the enzymes. These systems are primarily prokaryotic operons.

Repressor 1. A protein that binds specifically to a regulatory sequences adjacent to a gene and blocks transcription of the gene. 2. The protein product of a regulator gene that acts to control transcription of inducible and repressible operons.

Restrictive temperature. A temperature at which temperature sensitive mutants display the mutant phenotype.

Retrotransposons. Transposable genetic elements found in eukaryotic DNA that move through the reverse transcription of an RNA intermediate.

Retinoblastoma. 1. A childhood cancer of retinoblast cells caused by the inactivation of an anti-oncogene. 2. An inherited disease caused by a mutant in the tumor-suppressor gene located in chromosome band *13q14*. Inheritance of one copy of the mutation results in mutiple malignancies in retinal cells of the eyes in which mutation becomes homozygous — for example, through a new mutation or mitotic recombination.

Reverse genetics. Procedure in which mutations are deliberately produced in cloned genes and introduced back into cells or the germ line of an organism.

Reverse mutation. A mutation that undoes the effects of a preceding mutation.

Rho-dependent terminator. A DNA sequence signalling the termination of transcription; termination requires the presence of the rho-protein.

Rho-independent terminator. A DNA sequence signalling the termination of transcription; the rho-protein is not required for termination.

Rho-protein. A protein involved in the termination of transcription.

Ribosome recycling factor (RRF). A protein needed to prepare ribosomal subunits that have just finished translating a messenger RNA for another cycle of translation.

Ribosomes. Organelles at which translation takes place. They are made up of two subunits consisting of RNA and proteins.

Ribosome-binding site. The base sequence in a prokaryotic mRNA molecule to which a ribosome can find to initiate protein synthesis; also called the Shine-Dalgarno sequence.

Ribosome tRNA-binding sites. The tRNA binding sites on the ribosome to which tRNA molecules are bound. The aminoacyl site receives the incoming charged tRNA, the peptidyl site holds the tRNA with the nascent polypeptide chain, and the exit site holds the outgoing uncharged tRNA.

Ribosomal RNA (rRNA). RNA components of the subunits of the ribosomes.

Ribosomal translocation. Movement of the ribosome along a molecule of messenger RNA in translation.

Ribozyme. 1. An RNA molecule able to catalyze one or more biochemical reactions. 2. Catalytic or autocatalytic RNA.

RNA. Ribonucleic acid, a nucleic acid in which the sugar constituent is ribose. Typically, RNA is single-stranded and contains the four base-adenine, cytosine, guanine and uracil.

RNA editing. The insertion of uridines into messenger RNAs after transcription is completed; controlled by guide RNA. May also involve insertion of cytidines in some organisms or possible deletion of bases.

RNA interference. Ability of short RNA molecules (siRNA, or microRNAs) to inhibit gene expression by triggering mRNA degradation, inhibiting mRNA translation, or inhibiting transcription of the gene coding for a particular mRNA.

RNA phages. Phages whose genetic material is RNA. They are the simplest phages known.

RNA polymerase. 1. An enzyme that makes RNA by copying the base sequence of DNA strand. 2. The enzyme that polymerize RNA by using DNA as a template. (Also known as *transcriptase* or *RNA transcriptase*).

RNA polymerase I. Type of eukaryotic RNA polymerase present in nucleolus that synthesizes an RNA precursor for three of four types of rRNA (*e.g.*, 28S rRNA, 18S and 5.8S rRNA).

RNA polymerase II. Type of eukaryotic RNA polymerase rpesent in the nucleoplasm that synthesizes pre-mRNA and most of the snRNA (small nuclear RNA).

RNA polymerase III. Type of eukaryotic RNA polymerase present in the nucleoplasm that synthesizes pre-mRNA and 5S rRNA.

RNA primer. Short RNA fragment, synthesized by DNA primase, that serves as an initiation site for DNA synthesis.

RNA processing. Conversion of an initial of RNA transcript into a final RNA product by the removal, addition, and or chemical modification of nucleotide sequences.

RNA replicase. A polymerase enzyme that catalyzes the self-replication of single-stranded RNA.

RNA splicing. 1. Excision of introns and joining of exons. 2. Excision of introns from a primary RNA transcript to generate the mature, functional form of the RNA molecule.

Rolling-circle replication. A model of DNA replication that accounts for a circular DNA molecule producing linear daughter double helices.

Satellite DNA. Highly repetitive eukaryotic DNA primarily located around centromeres. Satellite DNA usually has a different buoyant density than the rest of the cell's DNA. It also occurs in mitochondrial or chloroplast DNA.

Scaffold. Protein-containing eukaryotic chromosomal structure that remains when DNA and histones have been removed. Believed to be responsible in part for the compaction of chromatin.

Scanning hypothesis. Proposed mechanism by which the eukaryotic ribosome recognizes the initiation region of a messenger RNA after binding the 5′ capped end of it. The ribosome scans the messenger RNA for the initiation codon.

Selected marker. A genetic mutation that allows growth in selective medium.

Selection. In evolution, intrinsic differences in the ability of genotypes to survive and reproduce; in plant and animal breeding; the choosing of organism with certain phenotypes to be parents of the next generation; in mutation studies, a procedure designed in such a way that only a desired type of cell can survive, as in selection for resistance to an antibiotic.

Selective medium. A culture medium enriched with a particular strain of organisms.

Selfish DNA. DNA sequences that do not contribute to the fitness of an organism but are maintained in the genome through their ability to replicate and transpose.

Semiconservative replication. The usual mode of DNA replication, in which each strand of a duplex molecule serves as a template for the synthesis of a new complementary strand, and the daughter molecules are composed of one old (parental) and one newly synthesized strand.

Sexduction. A process whereby a bacterium gains access to and incorporates foreign DNA brought in by a modified F factor during conjugation.

Shine-Dalgarno hypothesis. A proposal that prokaryotic messenger RNA is aligned at the ribosome by complementarity between the messenger RNA, upstream from the Initiation codon and the 3′ end of the 16S ribosomal RNA.

Short interspersed elements (SINEs). Sequences of DNA interspersed in eukaryotic chromosomes in many copies. *Alu*, a three-hundred base-pair seqence, is found about half a million times in human DNA.

Shunting. Process in which the first initiation codon on a messenger RNA is bypassed for an initiation codon further down the messenger. The process is probably guided by secondary structure in the messenger.

SiRNA. Class of double stranded RNAs about 23 nucleotides in length that silence gene expression; act by either promoting the degradation of mRNAs with precisely complementary sequences or by inhibiting the transcription of gene centraining precisely complementary sequences.

Sickle-cell anemia. A severe anemia in human beings inherited as an autosomal recessive and caused by an amino acid replacement in the β-globin chain; heterozygotes tend to be more resistant to *falciparum malaria* than are normal homozygotes.

Sigma factors. The protein that gives promoter-recognition specificity to the RNA polymerase core enzyme of bacteria.

Signal hypothesis. The major mechanism whereby a protein that must insert into or across a membrane are synthesized by a membrane-bound ribosome. The first thirteen to thirty-six amino acids synthesized termed as *signal peptide*, are recognised by a *signal recognition particle*, that draws the ribosome to the membrane surface. The signal peptide may be removed later from the protein.

Signal transduction pathway. A pathway in which the action of kinase enzymes that free transcription factors, or some other action, translates and environmental signal into some form of gene action.

Silent mutation. A mutation that has no phenotypic effect.

Single-strand binding proteins. Proteins that attach to single-stranded DNA, usually near the replicating Y-junction, to stabilize the single strands.

Single-stranded DNA. A DNA molecule that consists of a single polynucleotide chain.

Small nuclear ribonucleoproteins (snRNPs). Small nuclear particles that contain short RNA molecules and several proteins. They are involved in intron excision and splicing and in other aspects of RNA processing.

Small nucleolar ribonucleoprotein particles (snoRNPs). Particles composed of RNA and protein found in the nucleolus that modify ribosomal RNAs, particularly by converting some uridines to pseudouridines and methylating some ribose sugars.

SnoRNA. Group of small nucleolar RNAs that bind to complementary regions of a pre-rRNA and target specific sites for methylation or cleavage.

SnRNA. A small nuclear RNA molecule that binds to specific proteins to form a snRNP, which in turn assembles with other snRNPs to form a spliceosome.

Somatic cell. Any cell of a multicellular organism than the gametes and the germ cells from which gametes develop.

Somatic mutation. A mutation arising in a somatic cell.

SOS box. The region of the promoters of various genes that the LexA repressor recognizes. Release of repression results in the induction of the SOS response.

SOS response. Repair systems (RecA, Uvr) induced by the presence of single-stranded DNA that usually occurs from postreplicative gaps caused by various types of DNA damage. The RecA protein, stimulated by single-stranded DNA is involved in the inactivation of the LexA repressor, thereby inducing the response.

Splice acceptor. The 5′ end of an exon.

Splice donor. The 3′ end of an exon.

Spliceosome. Protein-RNA complex that removes introns in eukaryotic nuclear RNAs.

Spontaneous mutation. A mutation that happens in the absence of any known mutagenic agent.

Start codon. An mRNA codon, usually AUG, at which polypeptide synthesis begins.

Stem-loop structure. A lollipop-shaped structure formed when a single-stranded nucleic acid molecule loops back on itself to form a complementary double helix (stem), topped by a loop.

Stepladder gel. A DNA-sequencing gel. The numerous bands in each lane give the appearance of a stepladder.

Stop codon. One of three mRNA codons – UAG, UAA and UGA – at which polypeptide synthesis stops.

Stringent factor. A protein that catalyzes the formation of an unusual nucleotide (guanosine tetraphosphate) during the stringent response under amino acid starvation.

Stringent response. A translational control mechanism in prokaryotes that represses transfer RNA and ribosomal RNA synthesis during amino acid starvation.

Substrate molecule. A substance acted on by an enzyme.

Super coiling. Negative or positive coiling of double-stranded DNA that differs from the relaxed state.

Suppressor gene. A gene that, when mutated apparently restores the wild-type phenotype to a mutant of another locus.

Svedberg unit. A unit of sedimentation during centrifugation. Abbreviation is S, as in 50S.

Synexpression group. A group of eukaryotic genes that are involved in the same function or pathway and are induced together.

Synthetic medium. A chemically defined substrate that microorganisms are grown upon.

TACTAAC box. A consensus sequence surrounding the lariat branch point of eukaryotic messenger RNA introns.

TATA-bind protein. A protein (part of TFIID) that binds to the TATA motif in the promoter region of a gene in eukaryotes.

TATA box. An invariant DNA sequence at about-25 in the promoter region of eukaryotic genes; analogous to the Pribnow box in prokaryotes.

T DNA. Transposable element found in *Agrobacterium tumefaciens* which produces crown gall tumors in a wide variety of dicotyledonous plants.

Tautomeric shift. Reversible shifts of proton positions in a molecule. The bases in nucleic acids shift between the keto and enol forms or between the amino and imino forms.

TBP-associated factors (TAFs). Proteins that bind with the TATA-binding protein to form TFIID. They aid in the selectivity of TFIID.

T-cell receptors. Surface proteins of T cells that allow the T cells to recognize host cells that have been infected.

Telocentric chromosome. A chromosome whose centromere lies at one of its ends.

Telomerase. An enzyme that adds specific nucleotides to the tips of chromosomes to form the telomeres.

Telomeres. 1. The ends of linear chromosomes. 2. The tips of chromosomes, containing a DNA sequence required for stability of the chromosome ends.

Temperate phage. A phage that can enter into lysogeny with its host.

Temperature-sensitive mutant. An organism with an allele that is normal at a permissive temperature but mutant at a restrictive temperature.

Template. 1. A pattern serving as a mechanical guide. In DNA replication, each strand of the duplex act as a template for the synthesis of a new double helix. 2. A strand of nucleic acid whose base sequence is copied in a polymerization reaction to produce either a complementary DNA or an RNA strand.

Terminal redundancy. In bacteriophage T4, a short duplication found at opposite ends of the linear molecule; the duplicated region differs from one phage to the next.

Terminator sequence. A sequence in DNA that signals the termination of transcription to RNA polymerase.

Terminator stem. A configuration of the leader transcript that signals transcriptional termination in attenuator-controlled amino acid operons.

Tertiary structure. The further folding of a protein, bringing α helices and β sheets into three-dimensional arrangements.

Theta (θ) replication. Bidirectional replication of a circular DNA molecule, starting from a single origin of replication.

Theta structure. An intermediate structure formed during the replication of a circular DNA molecule.

30-nm fibre. The level of compaction of eukaryotic chromatin, resulting from coiling of the extended, nucleosome-bound DNA fiber.

3′-end. The end of a DNA and RNA strand that terminates in a sugar and so has a free hydroxyl group on the number-3′ carbon.

Thymine (T). A nitrogeneous pyrimidine base found in DNA.

Time of entry. In an Hfr × F^- bacterial mating the earliest time that a particular gene in the Hfr parent is transferred to the F^- recipient.

Ti plasmid. A plasmid that is present in *Agrobacterium tumefaciens* and is used in genetic engineering in plants.

T-loop. A loop that forms at the end of mammalian telomeres by the interdigitation of the 3′ free end into the DNA double helix.

Topoisomerase. An enzyme that can relieve (or create) supercoiling in DNA by creating transitory breaks in one (type I) or both (type II) strands of the helical backbone.

Topoisomers. Forms of DNA with the same sequence but differing in their linkage number (coiling).

Totipotent. The state of a cell that can give rise to any and all adult cell type, as compared with a differentiated cell whose fate is determined.

Trailer. The length of messenger RNA from the nonsense codon to the 3′ end or in polycistronic messenger RNAs, from a nonsense codon to the next-gene's leader.

Trait. Any aspect of the appearance, behaviour, development, biochemistry, or other feature of an organism.

Transcript. An RNA strand that is produced from, and is complementary is base sequence to, a DNA template strand.

Transcription. The process by which the information contained in a template strand of DNA is copied into a single-stranded RNA molecule of complementary base sequence.

Transcription complex. An aggregate of RNA polymerase (consisting of its own subunits) along with other polypeptide subunits that makes transcription possible.

Transcription factors. Eukaryotic proteins that aid RNA polymerase in recognising promoters.

Trancriptional activator protein. Positive control element that stimuletes transcription by binding with particular sites in DNA.

Transfer operon (tra). Sequence of loci that impart the male (F-pillus producing) phenotype on a bacterium. The male cell can transfer the F plasmid to an F^- cell.

Transfer RNA (tRNA). 1. Small RNA molecules that carry amino acids to the ribosome for polymerisation 2. A small RNA molecule that translates a codon into an amino acid in protein synthesis; it has a three-base sequence, called the *anticodon,* complementary to a specific codon in mRNA, and a site to which a specific amino acid is bound.

Transformation. A process whereby prokaryotes take up DNA from environment and incorporate it into their genomes or the conversion of a eukaryotic cell into a cancerous one.

Transition mutation. A mutation in which a purine-pyrimidine base pair replaces a base pair in the same purine-pyrimidine relationship.

Translation. The process of protein synthesis on a ribosome where in the nucleotide sequence in messeneer RNA determines the primary structure of the protein.

Translocase (EF-G). Elongation factor in prokaryotes necessary for proper translocation at the ribosome during the translation process. Replaced by eEF_2 in eukaryotes.

Translocation. A part of the translation process in which the messenger RNA is shifted one codon in relation to the ribosome.

Translocation channel (translocon). A protein-lined pore or channel in a membrane through which nascent proteins are transported during translation.

Transposable genetic element. 1. A DNA sequence capable of moving (transposing) from one location to another in a genome. 2. A region of the genome, flanked by inverted repeats, a copy of which can be inserted at another place; also called *transposon* or a *jumping gene*.

Transposon tagging. Insertion of a transposable element that contains a genetic marker into a gene of interest.

Trinucleotide repeat. A tandemly repeated sequence of three nucleotides; genetic instability in trinucleotide repeats is the cause of a number of human hereditary diseases.

Triplet code. A code in which each codon consists of three bases.

tRNA. *See* transfer RNA.

trp RNA-binding attenuation protein (TRAP). Protein that can bind to the attenuation region of the messenger RNA of the tryptophan operon in *Bacillus subtilis,* causing a terminator stem to form and halting further transcription.

Tumor-suppressor gene. 1. A gene whose absence predisposes to malignancy; also called an *anti-oncogene.* 2. Genes that normally prevent unlimited cellular growth, when both copies of the gene are mutated, cellular transformation follows. Examples are the p53 gene and the genes for retinoblastoma and Wilm's tumor.

Ubiquitin. A peptide of twenty-six amino acid residues that enzymes attach to proteins that the proteasome will degrade.

Uninemic chromosome. A chromosome consisting of one DNA double helix.

Unique DNA. A length of DNA with no repetitive nucleotide sequences.

Unusual bases. Other bases, in addition to adenine, cytosine, guanine and uracil, found primarily in transfer RNAs.

Upstream. A convention on DNA related to the position and direction of transcription by RNA polymerase (5′→3′). Downstream (or 3′ to) is in the direction of transcription whereas upstream (5′ to) is in the direction from which the polymerase has come.

Upstream element. A sequence of about twenty AT-rich bases centered at –50 in promoters of prokaryotic genes, that are expressed strongly.

Uracil (U). A nitrogeneous base found in RNA.

Watson-Crick base pairing. Base pairing in DNA or RNA in which A pairs with T (or U in RNA) and G pairs with C.

Wild type. 1. The phenotype of a particular organism when it is first seen in nature. 2. The most common phenotype of genotype in a natural population also.

Wobble. Referring to the reduced constraint over the third base of an anticodon as compared with the other bases, thus allowing additional complementary base pairings.

X chromosome. A chromosome that plays a role in sex determination. It is present in two copies in the homogametic sex and in one copy in the heterogametic sex.

Xeroderma pigmentosum. In humans, an inherited defect in the repair at ultraviolet-light damage to DNA, associated with extreme sensitivity to sunlight and multiple skin cancers.

Z DNA. A left-handed form of DNA found under physiological conditions in short GC segments that are methylated. It may be involved in regulating gene expression in eukaryotes.

Zygote. A product of the fusion of a female gamete and a male gamete in sexual reproduction; a fertilized egg.

Selected Reading

Alberts, B., A. Johnson, J. Lewis, M. Raff, K. Roberts and P. Walter, 2002	*Molecular Biology of the Cell*, Garland Science, New York.
Becker, W.M., L.J. Kleinsmith and J. Hardin, 2006	*The World of the Cell,* Dorling Kindersley (India) Pvt. Ltd., New Delhi.
Brown, T.A., 1998	*Genetics : A molecular approach*, Nelson Thornes Ltd., Cheltenham.
Brown, T.A. 2002	*Genomes,* John Wiley & Sons, Inc., Publication, New York.
Burns, G.W., 1969, 1972	*The Science of Genetics,* The Macmillan Company, New York.
Cooper, G.M. and R.E. Hausman, 2007	*The Cell : A molecular approach,* ASM Press, Washington, D.C.
DeRobertis, E.D.P., W. Nowinski and F.A. Saez, 1970	*Cell Biology*, W.B. Saunders Company, Philadelphia.
DeRobertis, E.D.P., F.A. Saez and and E.M.F. DeRobertis, 1975	*Cell Biology*, W.B. Saunders Company, Philadelphia.
Dupraw, E.J., 1969	*Cell and Molecular Biology,* Academic Press, New York.
Gardner, E.J., M.J. Simmons and D.P. Snustad, 2001	*Principles of Genetics*, John Wiley & Sons, Inc., New York.
Gilbert, S.F., 2000	*Development Biology*, Sinauer Associates Inc., Publishers, Sunderland, Massachusetts.
Goodenough, V. and R.P. Levine, 1974	*Genetics*, Holt, Reinhard and Wiston, Inc., New York.
Gupta, P.K., 2004	*Cell and Molecular Biology*, Rastogi Publications, Meerut
Karp, Gerald, 2002	*Cell and Molecular Biology*, John Wiley & Sons, Inc., New York.
Lodish, H., A. Berk, P. Matsudaira, C.A. Kaiser, M. Krieger, M.P. Scott, S.L. Zipursky and J. Darnell, 2004	*Molecular Cell Biology,* W.H. Freeman and Company, New York
Mathews, C.K., K.E.V., Holde and K.V. Ahern, 2000	*Biochemistry*, Pearson Education (Singapore) Pte., Ltd.,

Powar, C.B., 2007 — *Cell Biology,* Himalaya Publishing House, Mumbai.

Purves, W.K., G.H. Orians, H.C. Heller and D. Sadava, 1998 — *Life : The Science of Biology*, W.H. Freeman and Company, Utah, U.S.A.

Raven, P.H., C.B. Johnson, J.B. Losos and S.B. Singer, 2005 — *Biology*, Tata-McGraw-Hill Publishing Company Limited, New Delhi, India.

Strachan, T. and A.P. Read, 1996 — *Human Molecular Genetics,* BIOS Scientific Publishers Limited, Oxford, U.K.

Strickberger, M.W., 1968 — *Genetics*, Macmillan Co., New York.

Tamarin, R.H., 1996 — *Principles of Genetics*, Wm. C. Brown Publishers, Boston.

Tamarin, R.H., 2002 — *Principles of Genetics*, Tata-McGraw Hill Publishing Company Limited, New Delhi, India.

Turner, P.C., A.G. McLennan, A.D. Bates and M.R.H. White, 2000 — *Instant Notes : Molecular Biology*, BIOS Scientific Publishers Ltd., Oxford, U.K.

Verma, P.S. and V.K. Agarwal, 2008 — *Cell Biology*, *Genetics*, *Molecular Biology* and *Evolution,* S. Chand & Co., Ltd., New Delhi.

Verma P.S. and V.K. Agarwal, 2009 — *Genetics*, S. Chand & Co. Ltd., New Delhi.

Watson, J.D., 1970 — *Molecular Biology of the Gene*, W.A. Benjamin, New York.

A

B

C

D

G

T